MAMMALIAN METABOLISM
OF
PLANT XENOBIOTICS

RONALD R. SCHELINE

Department of Pharmacology,
University of Bergen,
Norway

1978

ACADEMIC PRESS
LONDON   NEW YORK   SAN FRANCISCO
A Subsidiary of Harcourt Brace Jovanovich, Publishers

ACADEMIC PRESS INC. (LONDON) LTD.
24/28 Oval Road
London NW1

United States Edition published by
ACADEMIC PRESS INC.
111 Fifth Avenue
New York, New York 10003

Library of Congress Catalog Card Number: 78–54538
ISBN: 0–12–623350–0

Printed in Great Britain by
J. W. Arrowsmith Ltd., Bristol

MAMMALIAN METABOLISM

OF

PLANT XENOBIOTICS

PREFACE

An understanding of the metabolic fate of foreign compounds is of importance in many fields of scientific endeavour. This subject, which may conveniently be termed xenobiochemistry, has therefore expanded rapidly during the past few decades as the use of xenobiotics has become increasingly widespread. A natural result of this development has been the growth of the scientific literature devoted to xenobiochemistry and related subjects. This has been particularly marked during the past five to ten years with the result that several primary and review journals as well as numerous books, the latter including both comprehensive treatises and review series, are now available. Much of this information deals with specific topics rather than, for example, with the metabolism of specific compounds. Thus, the investigator who requires a convenient source of information on the pathways of metabolism of a particular xenobiotic often finds that no recent compilation of these data is available. I have personally felt this need on many occasions and this factor has been a major reason for my decision to summarize current knowledge on the metabolism of at least some of the multitude of xenobiotics. This intention has been strengthened by the demonstrated value of Professor R. T. Williams' classic book "Detoxication Mechanisms" which, although published nearly 20 years ago, remains a continually useful source of information on the metabolism of specific xenobiotics. This, I believe, furnishes a clear indication of the need for this type of book, however, the magnitude of the subject today precludes a single author from making a comprehensive survey or from such a survey being completed in anything less than a series of volumes. My approach has therefore been to treat one segment of this large field, the limits being dictated by my interests in the metabolism of xenobiotics of plant origin.

Although the above choice establishes the general framework of this book, it is obvious that different approaches might have been taken with regard to many of the details. I wish to stress that the underlying aim has been to present a compilation of metabolic data which, while not necessarily exhaustive in coverage, is nonetheless sufficiently detailed to contain the data likely to be needed by those interested in the metabolic pathways of plant xenobiotics. This decision has obviously placed constraints on other options which as a consequence could not be developed without unduly increasing the length of the book. Therefore, ancillary material dealing with botanical sources, pharmacological and toxicological properties and other uses of the various compounds is largely omitted. It is my sincere hope that the deficiencies incurred as a result of this approach will be more than compensated for by the more detailed coverage of the metabolic data that has been possible as a result. It is also felt that because information on these ancillary matters is fairly readily available in the literature, its inclusion in the present volume would offer little gain.

I have attempted to maintain a reasonably consistent style throughout the book, however, the contents of the various chapters have obviously had some influence on the manner in which each chapter is developed. Therefore, while Chapter 1 is meant to give a brief account of the general features of the most common pathways of metabolism seen with plant xenobiotics, the relative amount of general information in the other chapters varies in accordance with the degree to which they cover specific compounds or larger groups of compounds. However, the unifying theme is the metabolism of specific plant xenobiotics. The term plant xenobiotics is not precise and it therefore seems worthwhile to define it in accordance with its use in this book. In brief, it can be roughly equated with the phrase secondary plant compounds, these include those substances which are apparently not essential to the functioning of the living plant cell. The text will therefore not deal with the mammalian metabolism of the simple sugars and organic acids which form part of the glycolytic and tricarboxylic acid cycles, the basic nucleotides, the 20-odd amino acids of proteins and the common lipids. Furthermore, it should be noted that attention has been confined to xenobiotics from higher plants, the lower classes including algae and fungi have been excluded.

I am grateful to the University of Bergen for granting me a sabbatical leave during which the literature review was largely concluded and the writing of the book begun. Some of the work during this period was carried out in London and I wish to thank the Norwegian Research Council for Science and the Humanities for a travel grant to defray my expenses. Also, Professor D. Robinson, Department of Biochemistry, Queen Elizabeth College, University of London kindly provided me with office facilities during my stay in London. I am grateful to Professor D. V. Parke, Department of Biochemistry, University of Surrey for his helpful comments and suggestions and to Inger Johanne Andresen and her staff at the University of Bergen medical library for their unfailing assistance in providing me with much of the necessary literature. I thank Eli Tepstad and Astrid Hetle for their help in compiling my collection of literature references and in proofreading. Further valuable help, mainly with many of the earlier German references, was given by Linda P. S. Francis. Finally, I thank Solfrid Dybvik for her excellent help in typing the manuscript.

June, 1978   RONALD R. SCHELINE

CONTENTS

1

METABOLIC REACTIONS OF PLANT XENOBIOTICS

The realization that xenobiotic compounds may undergo metabolic change in the body is not new, dating instead from the first half of the nineteenth century. The tremendous strides in this field which have taken place during the past two or three decades sometimes obscure the fact that the study of the metabolism of xenobiotics has deep roots in the past. In fact, it developed concurrently with the subjects of organic chemistry and biological chemistry and these three areas of investigation often intermingled and nourished each other in those formative years. The most significant advances which were made in the early studies of xenobiotic metabolism dealt with various conjugation reactions and it is noteworthy that many of the most important of these were discovered in the 1800s. These include the reactions of hippuric acid formation, glucuronide conjugation, sulphate conjugation, mercapturic acid formation and methylation. This historical viewpoint is pertinent to the subject of this book, the mammalian metabolism of plant xenobiotics, because the first of these reactions to be discovered concerned the metabolism of a plant compound. This is the conjugation of benzoic acid, a compound found in many berries and numerous balsamic substances, with the amino acid glycine to form hippuric acid, a reaction which was reported in the early 1840s. Of course, the subsequent development of the study of xenobiotic metabolism did not proceed along lines dictated by the source, natural or synthetic, of the particular foreign compound. Instead, metabolism studies of xenobiotics from both sources have been interwoven in the development of this large interdisciplinary subject.

We now know that most xenobiotic compounds are metabolized partly or entirely to one or usually several metabolic products. However, this need not invariably be the case and it is important to remember that certain structural features may confer a high degree of metabolic stability on a compound. This property is brought about either by the high polarity of the compound or by its volatility. The first situation is illustrated by several types of plant compounds including dicarboxylic acids (e.g. tartaric acid) and compounds containing quaternary ammonium groups (e.g. trigonelline or the curare alkaloids). Some lower aliphatic alcohols represent natural compounds with high volatility and which therefore are excreted unchanged to a relatively large extent. However, in the great majority of cases the xenobiotic will undergo chemical change, usually as a result of enzyme catalysed transformations but sometimes also due to spontaneous reactions. A very large number of transformations are now known and, with regard to the reactions taking place in the tissues, it is usual to classify them in four groups: oxidations, reductions, hydrolyses and conjugations (syntheses). The general trend of these metabolic processes is the conversion of liophilic compounds to more polar hydrophilic derivatives and Williams (1959) proposed that it was convenient to regard this as a biphasic phenomenon. Thus, phase I reactions include oxidations, reductions and hydrolyses whereas the phase II reactions include the various conjugations. The phase I reactions commonly introduce hydroxyl or carboxyl groups into the molecule and these derivatives or their intermediates may show either increased or decreased biological activity. The products of these reactions then serve as substrates for the phase II reactions, the latter generally resulting in metabolites with less biological activity which are very often readily excreted from the body.

The aim of this chapter is to present a short introduction to the pathways of metabolism seen with plant xenobiotics, firstly the reactions carried out in the tissues and then those effected by the microflora of the gastro-intestinal tract. The subject of the tissue reactions may be approached from several points of view. For example, classification can be based on whether or not the reaction site is hepatic or extra-hepatic or, perhaps, whether or not the transformation is carried out by microsomal or non-microsomal enzymes. However, the following summary of tissues metabolism is based on the four reaction types noted above as it is felt that this more chemically-orientated approach is more in keeping with the underlying theme found in the remainder of the book.

I. Reactions of the Tissues

A. OXIDATIONS

Oxidations furnish the largest and most varied group of phase I metabolic reactions. They have an exceedingly important function in the metabolism of a wide range of xenobiotics, partly because of the low substrate specificity seen with many oxidative reactions. Oxidations similarly play a key role in the metabolism of a multitude of plant xenobiotics and nearly all of the known types of oxidations also occur with these naturally occurring compounds. Oxidations are carried out in numerous tissues of the body and the enzymes responsible are located in various cellular fractions. These enzymes include oxidases and dehydrogenases located in the soluble and mitochondrial fractions of tissue preparations, however it is now abundantly clear that the most important are the mono-oxygenase systems which are present in many tissues, the most important and best known being the liver. These mono-oxygenases are associated with the hepatic endoplasmic reticulum which, due to its lipoidal character, preferentially allows for the metabolism of lipoidal compounds. Homogenization of liver tissue leads to the fragmentation of the membranes of the endoplasmic reticulum and their transformation into small vesicles known as microsomes. These artifacts, which are isolated by differential centrifugation, contain the mono-oxygenases and some other enzyme systems as well (e.g. hydrolytic and conjugative enzymes). As the microsomes are commonly employed in *in vitro* studies of xenobiotic metabolism, it is usual to speak of microsomal reactions or, in the present context, microsomal oxidations.

Microsomal oxidations of xenobiotics have been intensely studied during the past two decades with the result that the mono-oxygenase systems involved are now reasonably well understood. The general reaction involved in these oxidations is as follows:

$$RH + NADPH + O_2 + H^+ \rightarrow ROH + NADP^+ + H_2O$$

Oxidation therefore requires the reduced form of nicotinamide adenine dinucleotide phosphate (NADPH) and molecular oxygen. One atom of the latter is incorporated into the xenobiotic, the other being reduced and appearing as water. Because of this aspect of the mechanism, the mono-oxygenases have also been termed mixed function oxidases. Substrate oxidation is dependent on the availability of NADPH which, in turn, maintains the necessary flow of reducing equivalents in the chain of coupled redox reactions which is involved in the microsomal oxidations. NADPH-cytochrome c reductase is the enzyme responsible for the transfer of electrons in this sequence. However, the central role in the oxidizing system is taken by the CO-binding haemoprotein known as cytochrome P-450. It binds with substrate in both its oxidized and reduced forms and in the latter case functions as a carrier in the activation of oxygen. This complex then dissociates to give oxidized substrate, water and cytochrome P-450 in its oxidized (Fe^{3+}) form. Further details on the mechanism of microsomal oxidations are found in the reviews by Estabrook (1971) and by Hutson (1970, 1972, 1975, 1977). The reviews by Hutson are comprehensive surveys of the mechanisms of biotransformation and therefore provide detailed information on many of the topics summarized in this chapter.

1. Aliphatic Hydroxylation

a. ω-Oxidation. Among the wide variety of aliphatic hydroxylations known to be carried out by the microsomal mono-oxygenase system, that occurring at the terminal methyl group is a very common type. This reaction, commonly termed ω-oxidation, gives rise to primary alcohols which, as noted below, may be further oxidized by other systems to the corresponding aldehydes and acids. The metabolism of n-heptane (1) illustrates ω-oxidation.

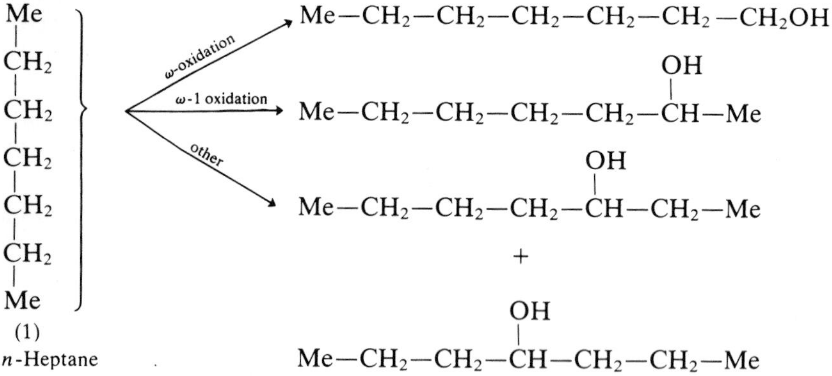

b. ω-1 Oxidation. Another common type of aliphatic hydroxylation is ω-1 oxidation, whereby the penultimate C atom is hydroxylated. This pathway is often favoured over ω-oxidation and, in the case of n-heptane shown above, considerably more of the secondary alcohol than of the primary alcohol is formed by the microsomal system. Further oxidation in this case leads to the ketone rather than the aldehyde and acid.

c. ω-n-Oxidation. The reactions shown above with n-heptane indicate that hydroxylation can occur at other aliphatic sites as well. This phenomenon of multiple hydroxylation of an aliphatic moiety is also observed with several cannabinoids. The n-pentyl side chain of cannabidiol (2) is transformed at all five possible sites to give 1"-, 2"-, 3"-, 4"- and 5"-hydroxycannabidiol.

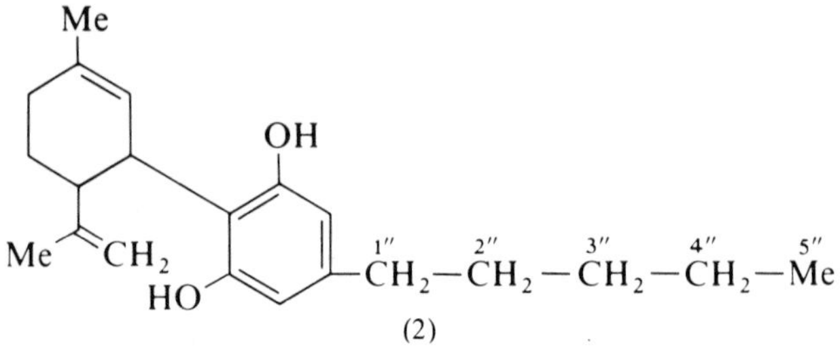

(2)

Cannabidiol

d. Benzylic hydroxylation. Benzylic hydroxylation may also be termed α-oxidation. The oxidation of a methyl substituent in aromatic compounds illustrates its simplest form, as exemplified by the oxidation of *p*-cymene (3) to an alcoholic intermediate (4) which then undergoes further oxidation to cumic acid (5). Another form of benzylic hydroxylation occurs with methylated heterocyclic compounds. The hydroxylated intermediate (7) is undoubtedly formed in the conversion of 4-methylthiazole (6) to thiazole-4-carboxylic acid (8). Yet another type of benzylic hydroxylation is that which takes place at the benzylic position in longer side chains. The hydroxylation of cannabidiol (2) to 1″-hydroxy-cannabidiol noted above illustrates this as does the β-hydroxylation of some phenethylamines including *p*-tyramine (9).

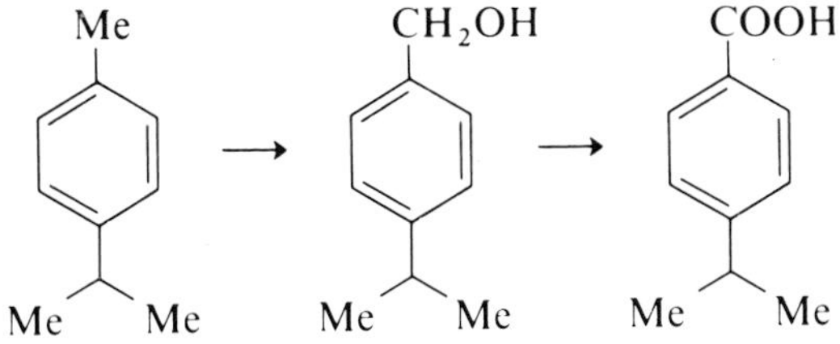

Me CH$_2$OH COOH

Me Me Me Me Me Me

(3) (4) (5)
p-Cymene Cumic acid

Me CH$_2$OH COOH

N N N
S S S

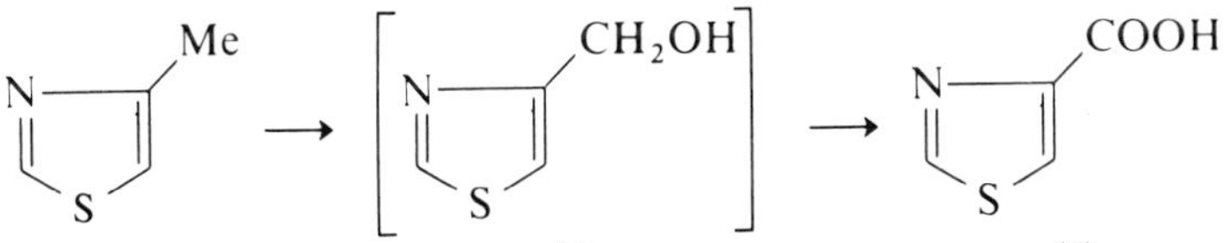

(6) (7) (8)
4-Methylthiazole Thiazole-4-carboxylic acid

HO—⟨ ⟩—CH$_2$—CH$_2$—NH$_2$ ⟶ HO—⟨ ⟩—CH—COOH
 OH

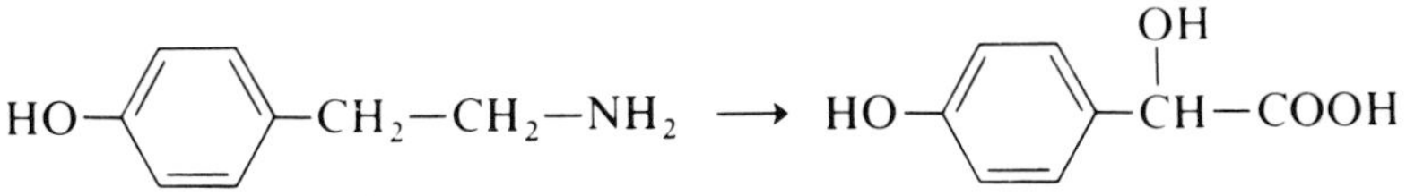

(9)
p-Tyramine

e. Allylic hydroxylation. Allylic hydroxylation is an important type of aliphatic oxidation which is observed with many plant xenobiotics. It occurs with several allylbenzene derivatives, as illustrated by the conversion of estragole (10) to its 1′-hydroxy derivative (11). Other compounds which are metabolized via this pathway include geranic acid, citronellol and pyrethrin I. It should also be noted that, in addition to these compounds which undergo allylic hydroxylation of aliphatic moieties, some alicyclic and heterocyclic compounds are metabolized in this manner. Both *d*-limonene and Δ^1-tetrahydrocannabinol contain an allylic structure as part of their alicyclic ring systems. Allylic hydroxylation occurs readily with these compounds as shown with *d*-limonene (12). Hydroxylation in this case occurs at an aliphatic rather than an alicyclic position. However, other examples show oxidation in the ring and these are included in Part A.2 of this section.

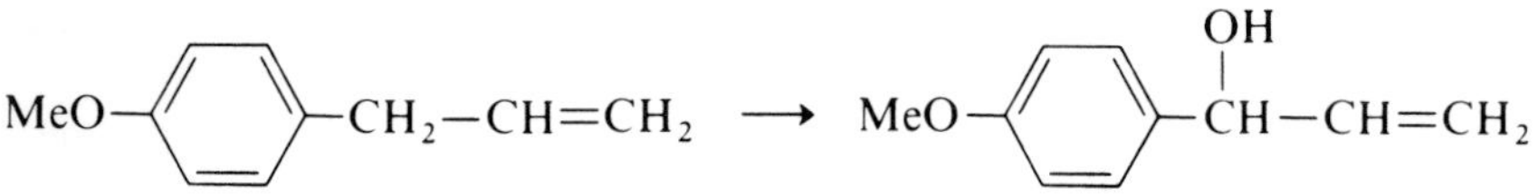

(10)

Estragole

(11)

1′-Hydroxyestragole

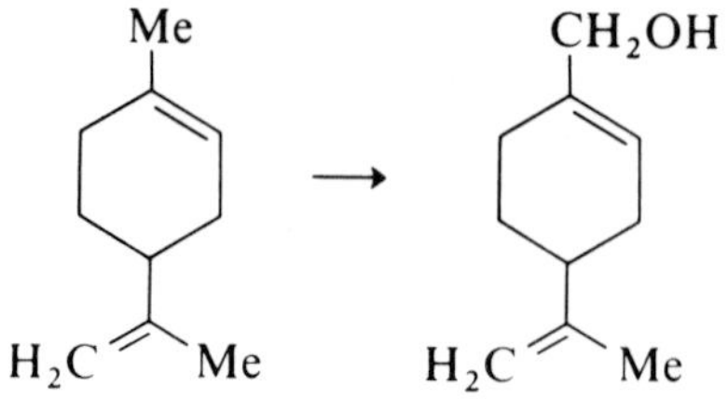

(12)

d-Limonene

f. Tertiary alcohol formation. The final type of aliphatic hydroxylation to be noted is tertiary alcohol formation, as seen with *p*-cymene (3).

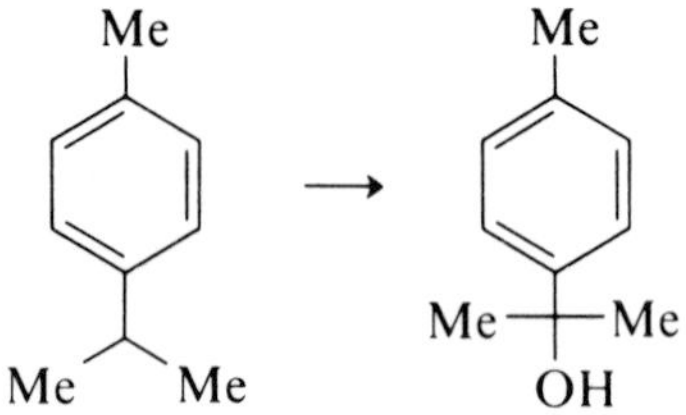

(3)

p-Cymene

2. Alicyclic Hydroxylation

Hydroxylation is a characteristic reaction with many cyclohexane derivatives and related compounds. It is common that oxidation occurs at multiple sites with the result that a mixture of isomeric alcohols is formed.

In some cases the vicinal diols are produced. Alicyclic ring hydroxylation is involved in the metabolism of (+)-camphor (see also Part A.3 of this section) and also gives rise to the two alcohols formed from the fully saturated derivative bornane (camphane) (13). The illustration shows that axial rather than equatorial hydroxylation occurs in this case. This results in the formation of the *endo* alcohols.

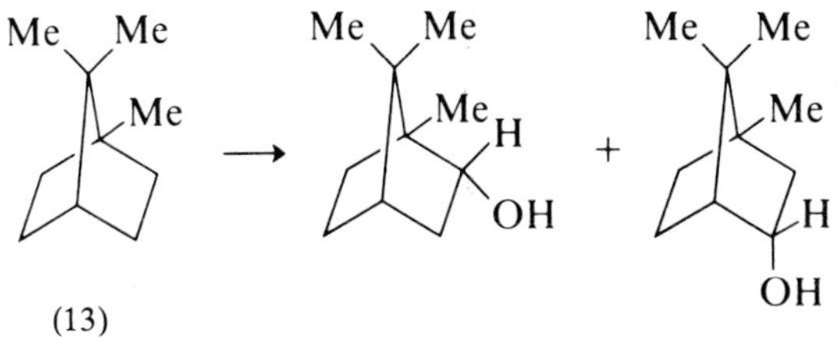

(13)

Bornane   Borneol   Epiborneol

It was mentioned above in Part A.1e that allylic hydroxylation occurs with some compounds in which the allylic moiety is part of a cyclic structure. In some cases the oxidative attack involves an aliphatic group attached to the ring. This was shown to be the case with *d*-limonene (12). A similar example is Δ^6-tetrahydrocannabinol (14), however this compound is interesting because the oxidation occurs both on the aliphatic group (at C7) and in the ring (at C5). Hydroxylation at the latter position produces both the α- and β-hydroxy derivatives. Analogous hydroxylation reactions occur with Δ^1-tetrahydrocannabinol (27).

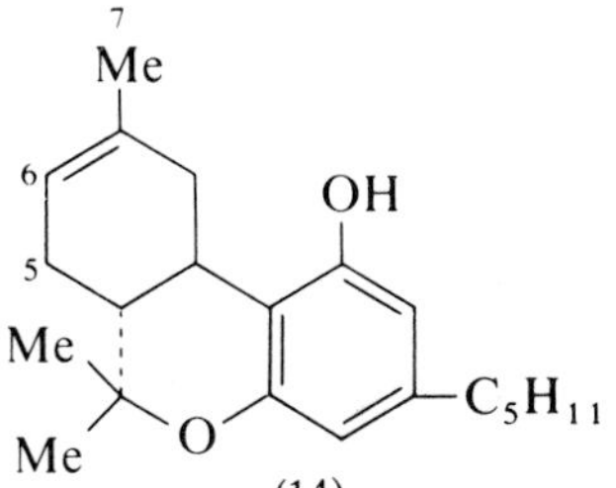

(14)

Δ^6-Tetrahydrocannabinol

Another example of allylic hydroxylation at a cyclic position is seen with β-ionone. The quinuclidine structure in quinine and its stereoisomer quinidine (15) contains an allylic moiety and recent evidence indicates that these compounds are metabolized partly via allylic hydroxylation to 3-hydroxy derivatives.

(15)
Quinine and Quinidine

3. Other Hydroxylations

Included in this section are hydroxylations which occur at C atoms located α- to a ring nitrogen or α- or β- to a carbonyl group. This classification is somewhat arbitrary as the first type is closely related to the oxidative dealkylations (Part A.12) which proceed via oxidation of carbon α- to N, O or S, thereby producing unstable hydroxyalkyl intermediates. Also, the β-oxidation of carboxylic acids covered below in Part A.9 of this section involves intermediates which are hydroxylated in the β-position.

Several examples of hydroxylation α- to an heterocyclic nitrogen are found in Chapter 9. These include the oxidation of some simple compounds such as 2,3-dimethyl pyrazine (16) which is metabolized partly to its 5-hydroxy derivative (17). A more well-known example involves the

(16)   (17)
2, 3-Dimethylpyrazine

metabolism of nicotine (18) to cotinine (20) which probably proceeds via the 5′-hydroxy intermediate (19). Examples of more complex alkaloids which are hydroxylated α- to the nitrogen include securinine and the

(18)   (19)   (20)
Nicotine   Cotinine

quinoline derivatives cinchonine and quinine. In the former case a fully saturated ring is hydroxylated whereas aromatic rings are involved with the quinoline derivatives.

Hydroxylation α- to a carbonyl group is seen in the metabolism of (+)-camphor (21). This ketone was mentioned above in Part A.2 in connection with alicyclic hydroxylation. The latter oxidation results in the formation of 5-hydroxycamphor, however 3-hydroxycamphor is also a metabolite of (+)-camphor. Coumarin (22) is hydroxylated in both the 3- and 4- positions, thus illustrating hydroxylation α- and β- to a carbonyl

(21)

Camphor

(22)

Coumarin

group. α-Hydroxylation of a complex cyclic structure is seen with rotenone while the aromatic ketone zingerone is hydroxylated in the methylene group β- to the carbonyl function.

4. Epoxidation

There has been growing realization during the past decade that epoxide formation is an important and widespread metabolic reaction. Epoxides are formed from aliphatic groups containing a double bond (i.e. alkenes) and from aromatic compounds. In the latter case epoxidation of a formal double bond in the aromatic structure results in the formation of an arene oxide. It is mainly in this area of arene oxide formation that attention has been focused as it has been demonstrated that this phenomenon is the key to the understanding of most aromatic hydroxylation reactions, that it is an important step in the formation of many glutathione conjugates and that it is involved in the production of toxic and carcinogenic metabolites of many substances including polycyclic hydrocarbons. The first two of these subjects are treated briefly in this and a subsequent part of Section I of this chapter, however comprehensive recent reviews of the literature, especially that dealing with arene oxides, are available (see Daly, 1971 and Jerina and Daly, 1974, 1977).

   a. Epoxidation of alkenes. Several plant xenobiotics containing double bonds are known to be metabolized by epoxidation. This reaction can take place with several types of double bonds including those present in olefinic side chains, those formed by a methylene group attached to a ring system or those present in cyclic structures. Examples of the first type are seen with *n*-octene, styrene and several alkenebenzene derivatives, as illustrated by that occurring with safrole (23). It seems reasonable to assume

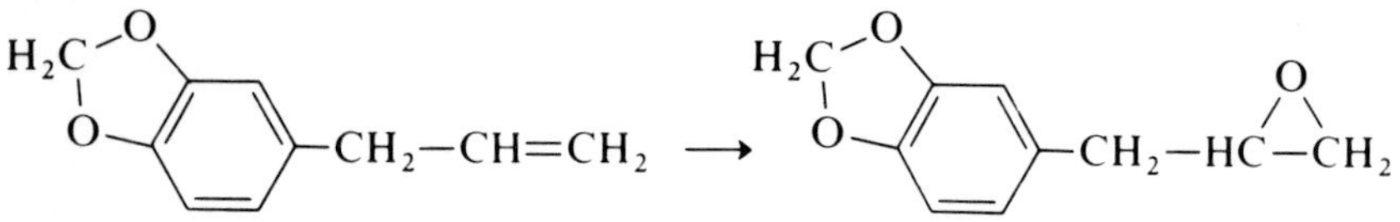

(23)

Safrole

that epoxide intermediates are involved in the conversion of the isopropenyl group in *d*-limonene (12) to several glycol derivatives.

The second type involves the epoxidation of a methylenic double bond and is undoubtedly the initial reaction in the conversion of camphene (24) to camphene glycol (25).

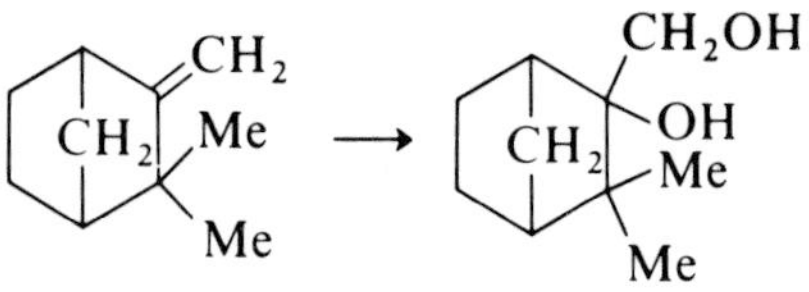

(24)   (25)

Camphene   Camphene glycol

The third type of alkene group undergoing epoxidation is found as part of a cyclic structure. An epoxide intermediate (26) has been proposed to be the precursor of an alcoholic metabolite of *d*-limonene (12). A noteworthy similar example is the epoxidation of Δ^1-tetrahydrocannabinol (27) which is converted to a 1, 2-epoxide derivative (28).

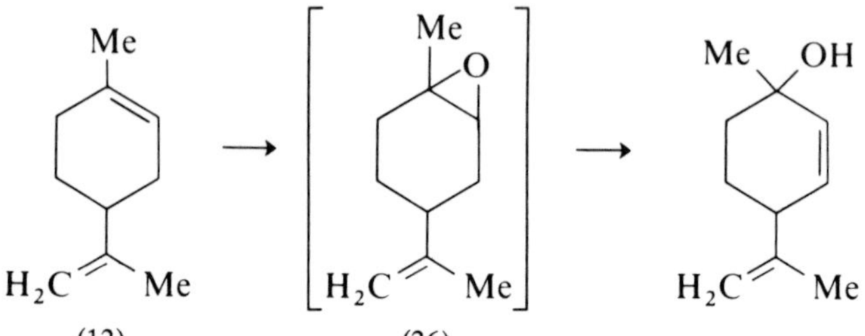

(12)   (26)

d-Limonene

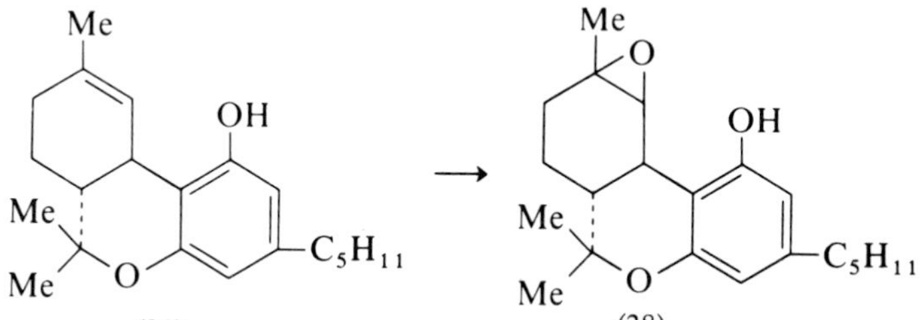

(27)
Δ'-Tetrahydrocannabinol

(28)

b. Arene oxide formation and its significance in aromatic hydroxyla-tion. The importance of arene oxide formation in the oxidation of aromatic compounds was underlined by the discovery of the NIH shift. The latter phenomenon involves the migration and retention of aromatic ring sub-stituents during the metabolic conversion of these compounds to phenols. These findings indicated that aromatic hydroxylation, rather than being a result of an insertion reaction, is based upon a mechanism involving addi-tion. This addition, which may be considered as an aryl epoxidation, results in the formation of an arene oxide which may then be chemically or metabolically transformed along several different pathways. The relevant routes in the present context are illustrated below in a general fashion. This

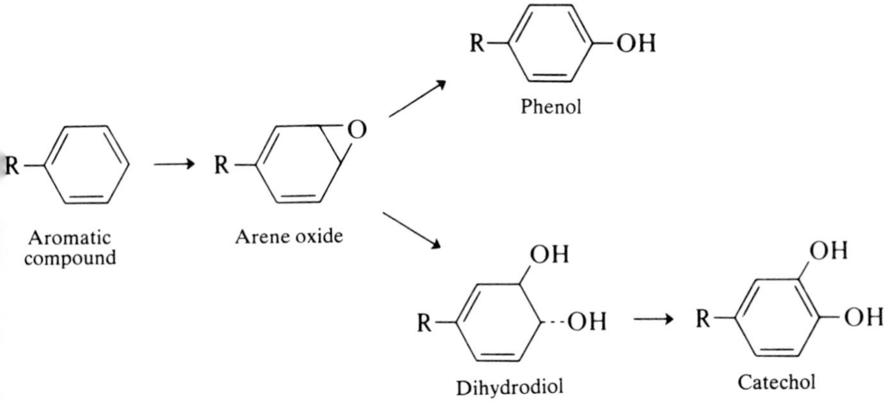

Aromatic compound

Arene oxide

Phenol

Dihydrodiol

Catechol

shows that catechol formation may be a consequence of arene oxide formation rather than sequential hydroxylations. In this case the arene oxide is hydrated (see Part C.3 below on epoxide hydratase) to give the *trans*-dihydrodiol which can then be dehydrogenated to give a catechol. Small amounts of catechols are commonly detected as metabolites of compounds containing suitable aromatic groups. The phenol formed from isomerization of the arene oxide is the *para* isomer, however formation of the isomeric arene oxide will lead to production of some *ortho* derivative.

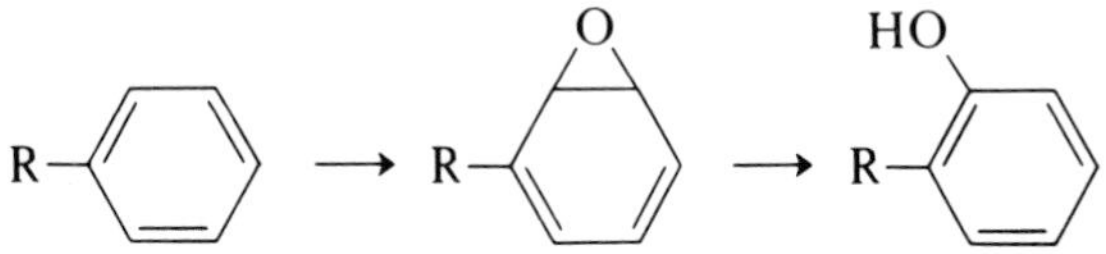

While variations in the production of these phenolic metabolites are dependent upon steric and electronic factors as well as the particular mono-oxygenase systems involved, it is known that electron-donating substituents tend to enhance the rate of oxidation and that bulky groups can diminish oxidation at adjacent sites. The mechanism resulting in aromatic hydroxylation thus shows similarity to electrophilic substitution reactions and many compounds, having activated rings, are converted predominantly to metabolites hydroxylated in the *para*-position. In spite of the clear and abundant evidence implicating arene oxide intermediates in the conversion of aromatic compounds to phenols, it must also be noted that oxidation at certain positions of some substrates is know to occur via the aforementioned mechanism of direct insertion. Thus, the formation of *meta*-hydroxylated products appears to result from this direct type of aromatic hydroxylation.

5. *Demethylenation*

Demethylenation is a reaction carried out by the microsomal enzymes which results in the scission of the methylenedioxy group, probably via an unstable hydroxy intermediate, to give a catechol. The extent to which this

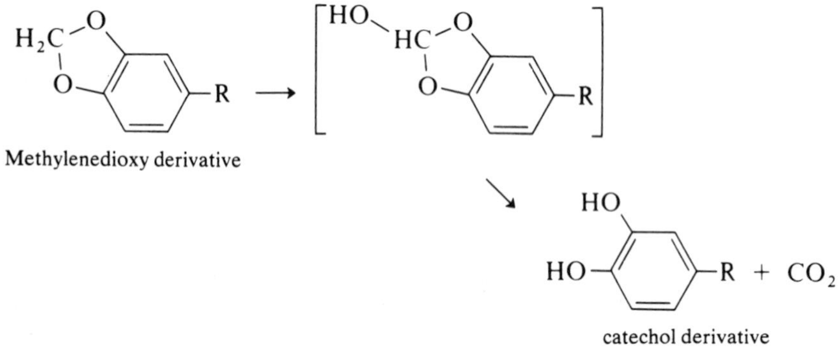

Methylenedioxy derivative

catechol derivative

reaction takes place varies considerably with different substrates. Piperonal or its acid derivative, piperonylic acid, are converted to catechols to only a minor extent whereas this reaction is extensive with safrole.

6. *Alcohol Oxidation*

The metabolic oxidation of alcohols is carried out by enzymes designated oxidoreductases. In contrast to the oxidative reactions described above which are catalysed by NADPH-dependent mono-oxygenases located in the endoplasmic reticulum, alcohol oxidation is to a large extent carried out by soluble liver enzymes. The best known of these is liver alcohol dehydrogenase (also termed NAD oxidoreductase). This enzyme has been studied mainly in connection with its role in ethanol metabolism, however it is also responsible for the oxidation of many xenobiotic alcohols of various types. The general reaction shown by primary alcohols is:

$$R-CH_2OH + NAD^+ \rightleftarrows R-\overset{\displaystyle O}{\underset{\displaystyle H}{C}} + NADH + H^+$$

Alcohol   Aldehyde

It is noteworthy that this reaction is reversible, allowing aldehydes (see Part B.2) and ketones (see Part B.3) to be reduced. Since the reaction is pH-dependent, *in vitro* systems show a shift in the equilibrium to the carbinol form as the pH is lowered to neutrality. However, the formation of the aldehydes is usually favoured *in vivo* because these products can be further oxidized to acids.

As noted above, a wide variety of alcohol substrates are oxidized by alcohol dehydrogenase. Maximum activity is found among primary aliphatic alcohols, the activity peak occurring with 1-butanol. Branched chain primary aliphatic alcohols are also oxidized. Another important group of substrates of alcohol dehydrogenase includes benzyl alcohol and its derivatives. As discussed in Chapter 3 (Section I.C) several plant alcohols of this type including saligenin, vanillyl alcohol and benzyl alcohol itself are extensively oxidized in animals. Secondary alcohols may also be oxidized by alcohol dehydrogenase to form ketones. However, dehydrogenation of these compounds proceeds much slower than with primary alcohols and many substrates including secondary aromatic alcohols are inactive.

Many soluble oxidoreductases besides alcohol dehydrogenase are present in mammals, however they often have narrow substrate specificities and are certainly more important in the metabolism of normal rather than xenobiotic compounds. It is also noteworthy that oxidoreductases localized in the microsomal rather than the soluble fraction of liver cells have been reported. These seem to be important in the oxidation of some cyclic and secondary aromatic alcohols. Details of the mammalian oxidation of alcohols by the various enzyme systems and especially of the properties, substrate specificity and stereochemistry of liver alcohol dehydrogenase are available in the review by McMahon (1971).

7. Aldehyde Oxidation

Aldehydes may enter the body as such or they may be formed metabolically, usually by the oxidation of primary alcohols as noted above or as a result of oxidative deamination (see Part A.15 of this section). Regardless of their origin they are metabolically very reactive, mainly undergoing oxidation to the corresponding carboxylic acid derivative and sometimes further metabolism prior to their excretion from the body.

Three enzymes or groups of enzymes have hitherto been implicated in the mammalian oxidation of aldehydes. These are the aldehyde dehydrogenases, aldehyde oxidase and xanthine oxidase. The relative importance of these three types in the *in vivo* oxidation of xenobiotic aldehydes has not been clarified, however it seems that the dehydrogenases may be the most important. It is known that liver aldehyde dehydrogenase oxidizes many aliphatic and aromatic aldehydes. The reaction involved is illustrated by the conversion of benzaldehyde (29) to benzoic acid (30).

$$\text{(29)} \quad + \text{ NAD}^+ + \text{H}_2\text{O} \longrightarrow \text{(30)} + \text{ NADH} + \text{H}$$

(29)
Benzaldehyde

(30)
Benzoic acid

The other enzymes, aldehyde oxidase and xanthine oxidase, are molybdo-flavoproteins which are similarly located in the soluble fraction of liver homogenates and, in the case of xanthine oxidase, in cows' milk. They have many similar properties including a substrate specificity for aldehydes and certain N-heterocyclic compounds. The review of McMahon (1971) briefly discusses the role of these three enzyme systems in the oxidation of aldehydes.

8. Dehydrogenation

The oxidation of both alcohols and aldehydes is partly dependent on dehydrogenation reactions. These reactions are treated above and the present discussion will therefore be limited to some lesser known examples of this metabolic conversion. One of these involves the dehydrogenation of the —CHOH—CHOH-moiety of *trans*-dihydrodiols to catechols. As noted above in Part A.4.b, the conversion of arene oxides to dihydrodiols is one feature of the metabolism of these important epoxide intermediates of many aryl groups. The initial steps are carried out by microsomal enzymes, however the dehydrogenation reaction leading to the enol which

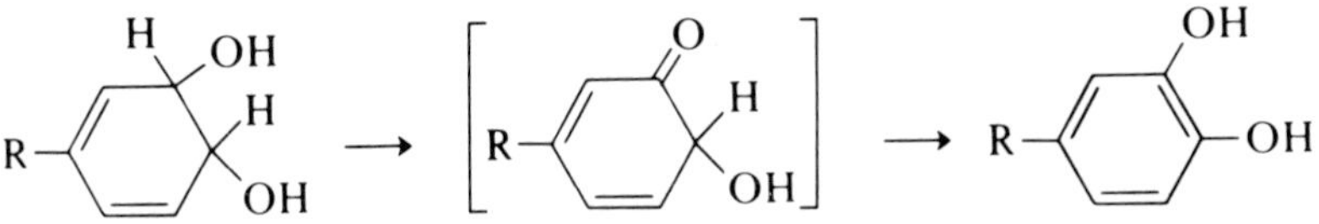

Dihydrodiol derivative   Catechol derivative

then rearranges to the catechol is catalysed by a soluble liver enzyme system.

Other dehydrogenation reactions occur with the $-CH_2-CH_2-$moiety in straight chain or cyclic structures. This is seen with several phenylpropionic acid derivatives which are dehydrogenated to the corresponding cinnamic acids. This reaction can perhaps more properly be considered the initial reaction in the process of β-oxidation which is illustrated below in Part A.9 of this section. Another dehydrogenation of this type is seen with harmalol (31) which is metabolized partly to harmol (32). An interesting dehydro-

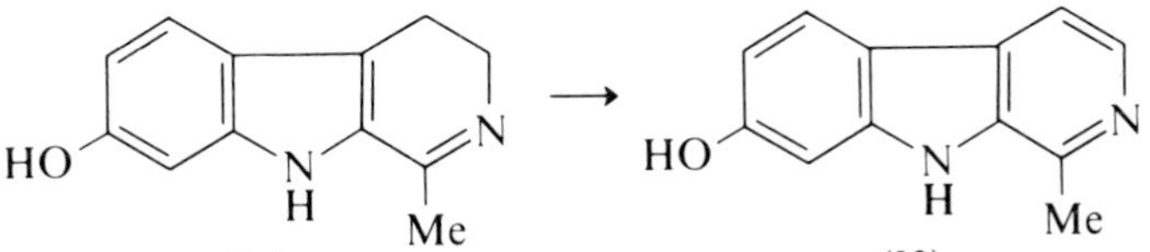

(31)   (32)
Harmalol   Harmol

genation reaction is that shown with several pyrrolizidine alkaloids. The conversion of heliotrine (33) to a pyrrole derivative, dehydroheliotrine (34), by liver microsomal enzymes is believed to be the key reaction responsible for the toxicity shown by this and other pyrrolizidine alkaloids (see Chapter 9, Section VII).

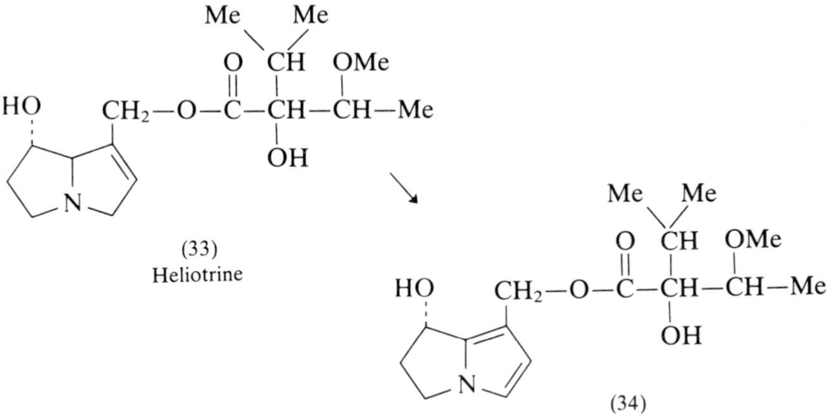

(33)
Heliotrine

(34)
Dehydroheliotrine

9. β-Oxidation

β-Oxidation is a well-known metabolic reaction by virtue of its importance in fatty acid oxidation. The basic features in this sequence have been known since the pioneering studies of Knoop and of Embden and Dakin at the beginning of this century. The details of β-oxidation including especially the involvement of acyl-*S*-Co A intermediates are readily available in textbooks of biological chemistry and need not be outlined here. β-Oxidation of plant xenobiotics is most commonly encountered in the metabolism of cinnamic acids and their various precursors including dihydrocinnamic acids and related aromatic C_6—C_3 aldehydes and alcohols. With cinnamic acid (35) itself this pathway leads to the formation of benzoic acid. The

(35)
Cinnamic acid   Benzoic acid

various intermediates, often as minor urinary metabolites, can sometimes be detected and the overall pathway starting with dihydrocinnamic acid (36) can be illustrated as follows:

(artifact)

10. *Aromatization*

Although several of the dehydrogenation reactions described above in Part A.8 produce aromatic metabolites (e.g. the conversion of dihydrodiols to catechols), the term aromatization is customarily employed to denote the formation of an aromatic acid from certain cyclohexane carboxylic acids. Two plant acids which are aromatized in this fashion are quinic acid and shikimic acid, however it is noteworthy that the sequence of reactions leading to benzoic acid involves enzymes found in both the gut microflora and the tissues. This subject is discussed more fully in Chapter 5 (Section I.B) and while the data presently available are insufficient to allow a final conclusion to be drawn, it is nonetheless clear that tissue enzymes can play an important role in the actual aromatization step. Thus, bacterial enzymes are responsible for the conversion of the hydroxylated cyclohexane derivatives to cyclohexanecarboxylic acid (37) whereas the latter compound can be metabolized by the liver to benzoic acid which is finally excreted as hippuric acid (38) following conjugation with glycine. Aromatization has been shown to be catalysed by a liver mitochondrial enzyme and the sequence of tissue reactions is as follows:

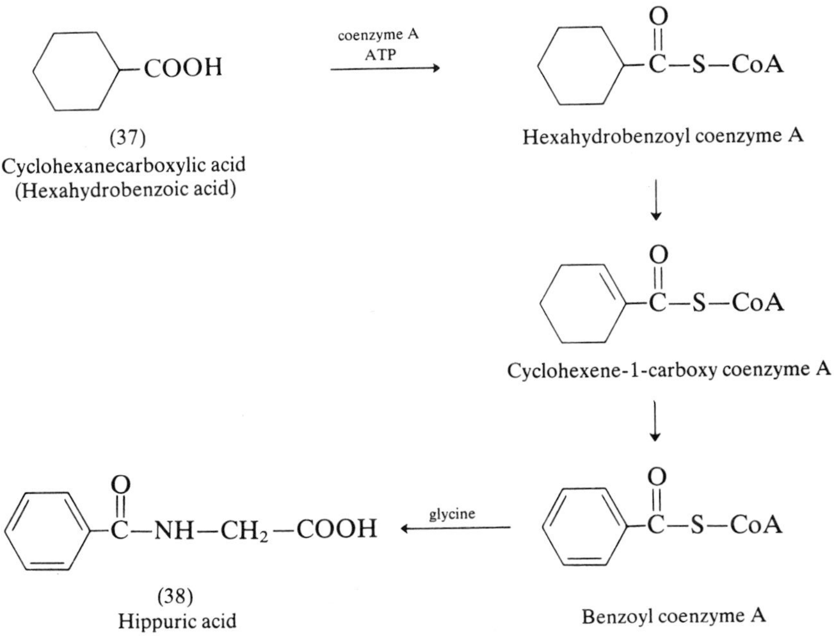

(37)
Cyclohexanecarboxylic acid
(Hexahydrobenzoic acid)

coenzyme A
ATP

Hexahydrobenzoyl coenzyme A

Cyclohexene-1-carboxy coenzyme A

(38)
Hippuric acid

glycine

Benzoyl coenzyme A

However, Parke (1968) noted that the aromatizing enzyme system, which is active in guinea pig and rabbit mitochondria, is absent in similar preparations from the mouse, cat, dog and man even though man is known to extensively aromatize quinic acid *in vivo*.

11. *Phenol oxidation*

Polyhydric phenols may be oxidized to quinones. The naturally occurring antioxidant nordihydroguaiaretic acid (39) is converted to its *o*-quinone derivative (40) in rats.

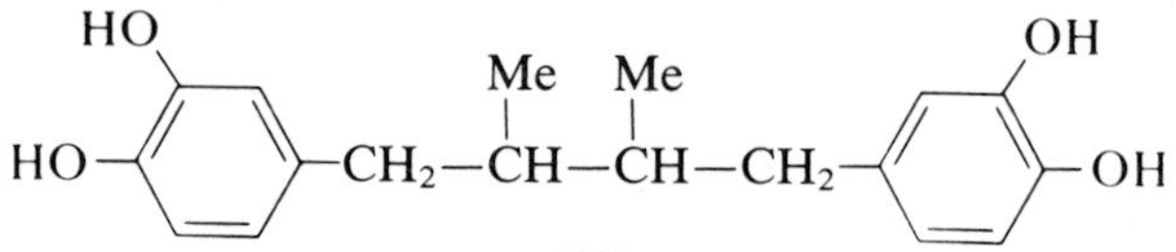

(39)
Nordihydroguaiaretic acid

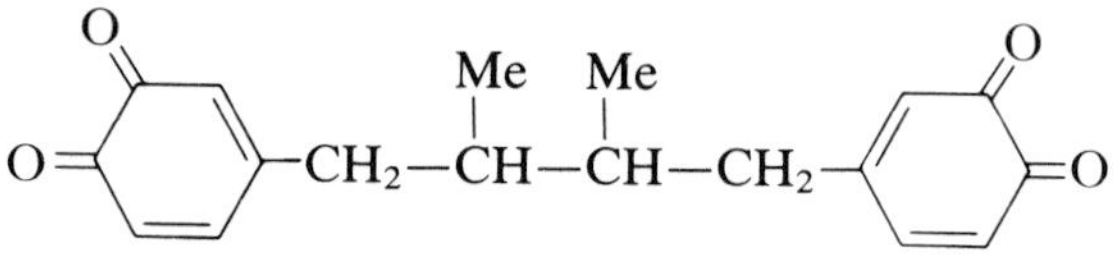

(40)

12. *Dealkylation*

The oxidative removal of alkyl groups attached to nitrogen, oxygen or sulphur atoms is a common metabolic reaction of foreign compounds. Although the methyl group, its higher homologues and some larger substituents may be metabolized in this manner, the many plant xenobiotics containing *N*-, *O*- or *S*-alkyl groups are usually methyl derivatives. Therefore, dealkylation in this context may conveniently be limited to oxidative demethylation. This reaction is carried out by enzyme systems located in the microsomal fraction of various tissues, mainly liver, and the general reaction involved can be represented as follows:

$$R-X-Me \rightarrow [R-X-CH_2OH] \rightarrow R-XH + HCHO$$

where $X = N$, O or S

The formaldehyde produced in this reaction is further oxidized and excreted from the body as respiratory CO_2. The mechanism makes it clear that in the case of *O*-methyl compounds the oxidative attack is at the carbon atom and that the oxygen atom is retained rather than the entire methoxyl moiety being lost. Numerous examples of oxidative demethylation are found in subsequent chapters and it is felt that no advantage is gained here by listing these. Instead, representative examples of the three types of demethylations are chosen to illustrate these oxidations. Further details on *N*-, *O*- and *S*- dealkylation of xenobiotics by the microsomal systems are available in the review by Gram (1971).

a. N-Demethylation. Microsomal *N*-demethylation is a relatively non-specific reaction which takes place with several classes of compounds containing the *N*-Me group, including both straight chain and cyclic secondary and tertiary amines. Ephedrine (41) is demethylated to nor-ephedrine (42) in many animal species.

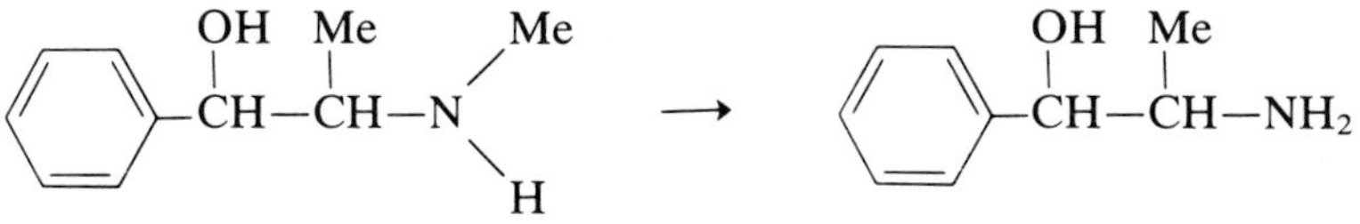

(41)
Ephedrine

(42)
Norephedrine

A well-known example of *N*-demethylation of a cyclic tertiary amine is that of morphine which is converted to normorphine. Conflicting reports have appeared on the ability of the *N*-demethylating systems of liver microsomes to metabolize methylated xanthine alkaloids (theophylline, theobromine and caffeine), however this activity has been detected in some investigations (see Chapter 9, Section IX).

b. O-Demethylation. The *O*-demethylation of aromatic methyl ethers is a common reaction of plant xenobiotics which is noteworthy because it may be carried out both in the liver by the microsomal mono-oxygenases and in the intestine by the enzymes of the microflora. The latter phenomenon is discussed below in this chapter (Section II.B.3) where it is

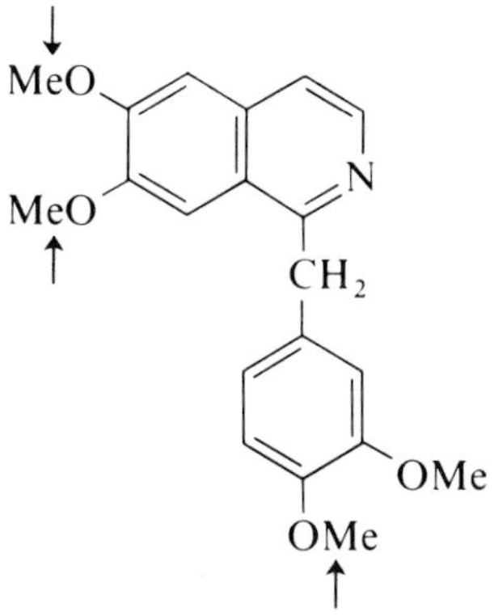

(43)
Papaverine

seen that bacterial O-demethylation generally takes place with a more restricted range of substrates. Some evidence suggests that this bacterial metabolism occurs with compounds of simpler chemical structure and that more complex ethers such as methoxylated derivatives of alkaloids are resistant. This is the case with papaverine (43) which, however, undergoes microsomal demethylation. Thus, methoxyl groups at three positions (see arrows) are cleaved to give the corresponding monodemethylated metabolites. Codeine furnishes a well-known example of the oxidative O-demethylation of a single methyl ether group. This common structural feature is found in the alkenebenzene derivatives estragole (10) and anethole which are very extensively metabolized to the corresponding phenols in rats.

$$MeO-\langle\text{C}_6\text{H}_4\rangle-CH_2-CH=CH_2 \longrightarrow HO-\langle\text{C}_6\text{H}_4\rangle-CH_2-CH=CH_2$$

(10)   Chavicol

Estragole

In addition to the abundant literature which shows that aromatic methyl ethers are cleaved by microsomal mono-oxygenases, a few examples are known in which these systems O-demethylate aliphatic ethers (e.g. corynantheidine and the pyrrolizidine alkaloid heliotrine).

   c. *S-Demethylation.* The microsomal demethylation of thioethers is the least studied of these three types of dealkylations and there is evidence indicating that the S-demethylating systems differ from those involved in the metabolism of N- and O-methyl compounds. Nevertheless, the metabolism of methanethiol by rat liver microsomes requires the presence of NADPH and O_2 and gives H_2S and formaldehyde as metabolites. This system also demethylates S-methylcysteine.

13. *N-Oxidation*

The metabolic oxidation of nitrogen in xenobiotics may be conveniently divided so that it encompasses, on the one hand, the N-hydroxylation of primary and secondary amines to hydroxylamines and, on the other, the oxidation of tertiary amines to amine oxides. Much interest in the former area results from the fact that some amines, especially arylamines, are converted to more toxic N-hydroxy intermediates (see Weisburger and Weisburger, 1971, 1973). In fact, N-hydroxylation is largely a reaction of arylamines and the knowledge that plant amines are generally aliphatic rather than aromatic indicates that this metabolic pathway will not be encountered very often with these xenobiotics. However, some aliphatic amines (e.g. ephedrine and norephedrine) are metabolized by liver micro-some preparations to N-hydroxy derivatives. Beckett (1977) recently reviewed some aspects of the mechanisms involved in the formation of these metabolites, mainly from aliphatic amines.

The formation of N-oxides is a more commonly encountered reaction involving the metabolism of plant xenobiotics containing nitrogen. This reaction occurs with tertiary aliphatic or cyclic amines and also with aromatic nitrogen heterocyclic compounds. The first type of N-oxidation occurs with numerous alkaloids and is illustrated by the oxidation of arecoline (44). The second type, involving the nitrogen in an aromatic heterocyclic ring, is observed with cotinine (20).

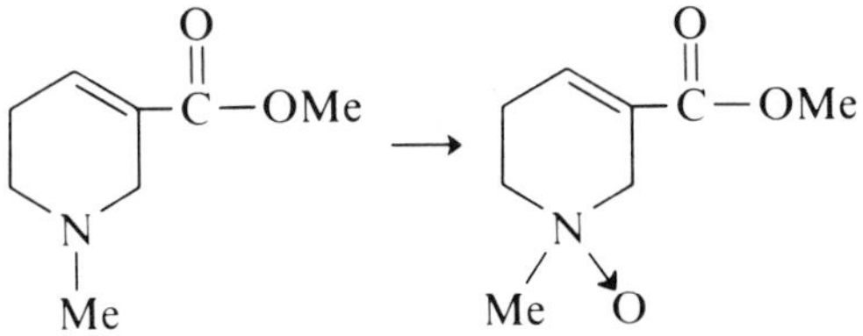

(44)
Arecoline

Arecoline-N-oxide

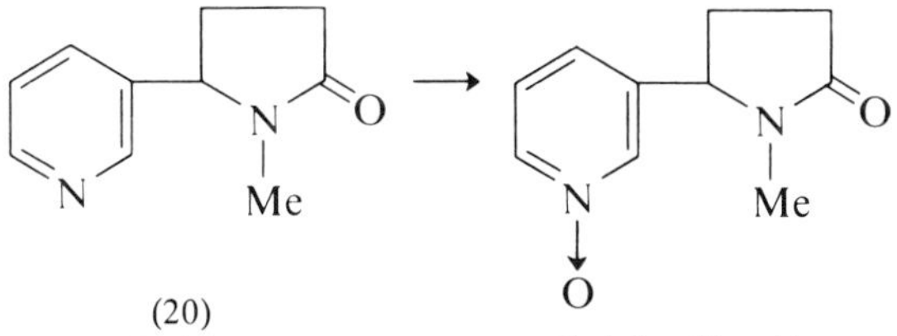

(20)
Cotinine

Cotinine-N-oxide

N-Oxidation is carried out mainly by the enzymes of the hepatic endo-plasmic reticulum and requires NADPH and O_2. However, results from inhibitor studies indicate that these systems differ from those commonly involved in mono-oxygenase reactions. Similar to that observed with N-hydroxylation, the formation of N-oxides may also produce metabolites of enhanced toxicity. This is not, however, a general rule and alkaloidal N-oxidation often leads to less toxic products. Several articles in the proceedings of a symposium (Bridges et al., 1971) deal with N-oxide formation and this field was reviewed by Bickel (1969) and by Weisburger and Weisburger (1971).

14. S-Oxidation

Xenobiotics containing a thioether moiety either in straight chain or heterocyclic structures may undergo S-oxidation. This metabolic pathway which leads to the formation of the sulphoxide and sulphone derivatives may be represented as follows:

$$
\underset{\text{Sulphide}}{\text{R--S--R}'} \rightarrow \underset{\text{Sulphoxide}}{\overset{\displaystyle \overset{O}{\uparrow}}{\text{R--S--R}'}} \rightarrow \underset{\text{Sulphone}}{\overset{\displaystyle \overset{O}{\uparrow}}{\underset{\displaystyle \underset{O}{\downarrow}}{\text{R--S--R}'}}}
$$

Little information on this pathway with plant xenobiotics is available but it is probably involved in the metabolism of diethyl disulphide and its reduction product, ethanethiol. The latter compound is S-methylated and this unsymmetrical sulphide is then metabolized partly to ethyl methyl sulphone.

15. *Oxidative Deamination*

The oxidative deamination of amines is carried out by both microsomal and mitochondrial enzymes. The latter include the various monoamine oxidases and diamine oxidases which are involved in the metabolism of many endogenous substrates in addition to xenobiotics. The plasma furnishes a further source of amine oxidase activity. Thus, catecholamines and tryptamines are substrates of monoamine oxidase which also catalyses the oxidation of other aralkylamines and of alkylamines. An important exception to this is seen with α-substituted amines (e.g. ephedrine) which are deaminated by microsomal enzymes. Similarly, diamine oxidases oxidize histamine as well as several aliphatic diamines. The general reaction in the oxidative deamination of amines results in the formation of the corresponding aldehyde and ammonia as follows:

$$
\text{R--CH}_2\text{--NH}_2 \rightarrow \text{R--CH=NH} \rightarrow \text{R--C} \underset{\displaystyle \text{H}}{\overset{\displaystyle O}{\big\langle}} + \text{NH}_3
$$

Substrate specificity was one of the many subjects dealing with the monoamine oxidases reviewed by Zeller (1971). Excellent substrates are usually found among derivatives of methylamine containing a single alkyl, aryl or aralkyl substituent. The aromatic groups may be either carbocyclic or heterocyclic. It is noteworthy that derivatives which are monomethylated (secondary amines) and dimethylated (tertiary amines) also serve as substrates, however the reaction rates are usually greatly decreased as substitution increases. The role of the various amine oxidases in the metabolism of particular plant amines is discussed in Chapter VIII, Section I.

B. REDUCTIONS

The metabolic reduction of xenobiotics by tissue enzymes is much less common than oxidation. This is especially the case with plant xenobiotics as the two most characteristic reactions, nitro and azo reduction, are virtually never encountered among these compounds. However, reductions of some aldehydes, ketones and sulphur compounds from plants may sometimes occur and it is also possible that the reduction of double bonds may be carried out by tissue enzymes. It should be noted, however, that double bond reduction and, indeed, many other types of reductions are characteristic reactions of the gastrointestinal microflora.

1. Double Bond Reduction

The subject of double bond reduction by tissue enzymes is poorly understood. Several terpenoid compounds containing unsaturated bonds have been reported or suggested to be metabolized to reduced derivatives. Carvone and pulegone have been cited in this regard, however the experimental evidence for their double bond reduction is tenuous. α-Phellandrene (45) appears to be metabolized partly to phellandric acid (46)

	(45)	(46)
	α-Phellandrene	Phellandric acid

which contains only a single ring double bond, however this result was obtained using sheep and the participation of the rumen microflora must therefore be suspected. A reduced acidic metabolite (48) was reported to be formed in dogs given the acyclic terpenoid citral (47), however the site of this reaction is similarly unclear.

	(47)	(48)
	Citral	

Cinnamic acid derivatives are a common class of plant compounds which are known to be metabolized partly by double bond reduction to the corresponding dihydro derivatives. It is not clear to what extent, if any, tissue reductases are involved in this reaction, however several factors indicate that it is not extensive. Firstly, the opposite reaction, dehydrogenation, which is described above in Part A.8 takes place in the tissues. Secondly, the reductive reaction is a well-known reaction of the gut microflora which has been reported with many cinnamic acid derivatives. Thirdly, unpublished results obtained in experiments with germ-free rats given caffeic acid (Scheline and Midtvedt, 1970) indicated that dihydrocaffeic acid, normally found as a urinary metabolite of caffeic acid in conventional rats, was not excreted. However, Masri *et al.* (1962) found that the interconversion of caffeic and dihydrocaffeic acids and also the reduction of ferulic and isoferulic acids occurred readily when these cinnamic acid derivatives were incubated aerobically with rat liver slices. Also, Ranganathan and Ramasarma (1974) reported that mitochondrial enzymes from rat liver are able to carry out the reduction of *p*-coumaric acid to phloretic acid. This reaction was greatly favoured under anaerobic incubation conditions and its significance *in vivo* is not known.

2. Aldehyde Reduction

The reduction of xenobiotic aldehydes in the tissues may be catalysed by various oxidoreductases including liver alcohol dehydrogenase. As noted above in Part A.6, the latter enzyme brings about the oxidation of alcohols to aldehydes and the reduction of these carbonyl compounds to carbinols. Under *in vivo* conditions, however, the equilibrium favours the aldehyde form and the role of this and probably other alcohol dehydrogenases in the body is to function as dehydrogenases rather than reductases.

McMahon (1971) described several enzyme systems which appear to function as true reductases and which may be involved in the reduction of xenobiotic aldehydes. These have been isolated from liver, kidney cortex and intestinal mucosa. In some cases both aliphatic and aromatic aldehydes are reduced but other systems appear to utilize only aromatic substrates. Vanillin (49) furnishes an example of an aromatic aldehyde which is reduced in the tissues. Shiobara (1977) found that the presence of a

(49)

Vanillin   Vanillyl alcohol

carboxyl group in the *ortho* position in aromatic aldehydes greatly pro-
moted the reduction of the aldehyde to the corresponding alcohol. In
contrast, the *meta* and *para* isomers were preferentially oxidized.

3. Ketone Reduction

Ketones, unlike aldehydes, are not susceptible to further oxidation and it is
therefore not surprising that the reductive pathway is more commonly seen
with this type of carbonyl compound. McMahon (1971) listed several
reductase systems obtained from liver, kidney or erythrocytes which are
capable of catalysing the reduction of various ketones. One of these
systems, the aromatic aldehyde-ketone reductase found in rabbit liver and
kidney cortex, reduces acetophenone derivatives and some alicyclic
ketones but not aliphatic ketones. Another system from dog erythrocytes
or human liver catalyses the reduction of α,β-unsaturated ketones. It is
noteworthy that the former system reduced acetophenone mainly to $(-)$-
methylphenylcarbinol as this stereoisomer is the metabolite which is
excreted in the urine of rabbits given the ketone. Hydroxylated derivatives
of acetophenone appear not to undergo reduction to carbinols, however
this reaction is observed with zingerone (50), a higher homologue.

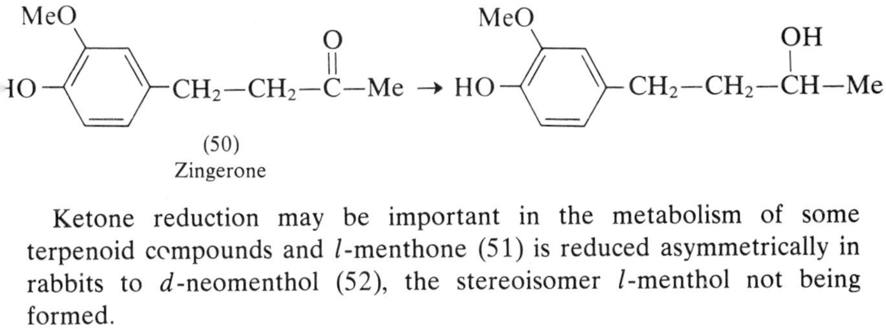

(50)
Zingerone

Ketone reduction may be important in the metabolism of some
terpenoid compounds and *l*-menthone (51) is reduced asymmetrically in
rabbits to *d*-neomenthol (52), the stereoisomer *l*-menthol not being
formed.

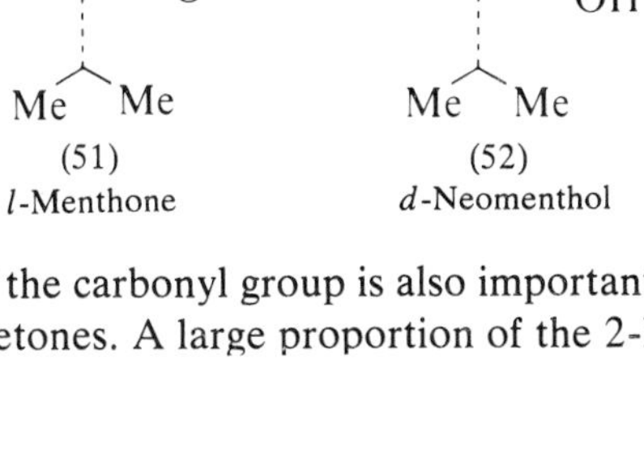

(51)
l-Menthone

(52)
d-Neomenthol

The reduction of the carbonyl group is also important in the metabolism
of aliphatic plant ketones. A large proportion of the 2-heptanone given to

rabbits is excreted in the urine as the glucuronide conjugate of the corresponding carbinol. However, the enzyme systems involved in aliphatic
ketone reduction have not been elucidated.

4. *Sulphur Compound Reduction*

The metabolic reduction of disulphides in the body, although little studied,
appears to be a general reaction of these compounds. Thus, diethyl
disulphide (53) is converted partly to ethanethiol (54).

$$Me-CH_2-S-S-CH_2-Me \rightarrow Me-CH_2-SH$$

(53)   (54)

Diethyl disulphide   Ethanethiol

Little is known about the reduction of other sulphur compounds in the
body. Dimethyl sulphoxide is reduced to dimethyl sulphide in the cat
(Distefano and Borgstedt, 1964), however it is not known if this is a
general reaction of sulphoxide derivatives. The site of this reduction was
not determined, however the data are more easily understood if one
assumes tissue metabolism. On the other hand, some sulphoxides may be
reduced by the gut microflora.

C. HYDROLYSES

This group of reactions shares with the reductions described above a more
limited role in the metabolism of plant xenobiotics than that shown by the
oxidative reactions. When it does occur, however, hydrolysis may be a
metabolic reaction which proceeds rapidly and extensively, as with many
esters. It is also noteworthy that hydrolytic reactions may be carried out
both by enzymes of the tissues and of the gut microflora.

1. *Ester Hydrolysis*

The most widely studied of the hydrolyses effected by tissue enzymes is
that of ester hydrolysis. Investigations dealing with the various enzymes
capable of hydrolysing esters have given rise to a large and complex
subject which, not unexpectedly, has often dealt with the metabolism of
normal substrates (e.g. acetylcholine). Conclusions about the esterases
involved in the hydrolysis of particular xenobiotics are often difficult to
make, however it is well known that esterases vary considerably from tissue
to tissue and also that considerable species differences exist. An often
cited example of this is seen with atropine which is poorly hydrolysed in
several species including the mouse, rat and man whereas extensive
hydrolysis occurs in the guinea pig and often in rabbits. Atropine

hydrolysis in the latter species is interesting because of the genetically determined serum esterase which hydrolyses this alkaloid in some rabbit populations. This reaction is dicussed in more detail in Chapter 9, Section III which also covers the hydrolysis of a chemically related compound, cocaine. Cocaine-hydrolysing activity is also pronounced in rabbit serum, however it appears not to be identical to that responsible for atropine hydrolysis. Interestingly, cocaine (55) contains two ester linkages and most

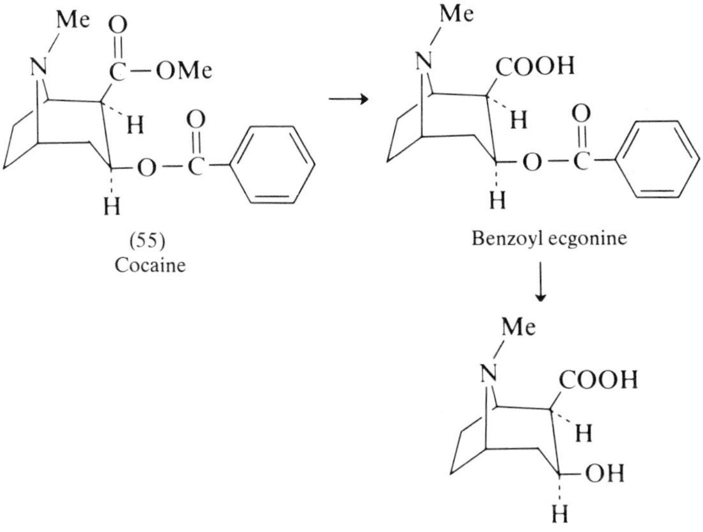

(55)
Cocaine

Benzoyl ecgonine

Ecgonine

evidence suggests that the metabolic sequence usually involves the initial hydrolysis of the methyl ester group followed by loss of the benzoyl moiety. However, the alternative sequence has also been reported.

While serum esterases may be involved in xenobiotic hydrolysis, those at other sites are also of importance. This is seen with reserpine, an alkaloidal diester which is hydrolysed by esterases located in the intestinal mucosa and in the liver. Although liver esterases associated with the soluble fraction are well known, the reserpine-hydrolysing activity is found in the microsomal fraction.

Many plant esters are relatively simple aliphatic or aromatic esters and their metabolic hydrolysis may occur very rapidly. This was noted by Fahelbum and James (1977) with methyl cinnamate which was not detectable in the peripheral blood of rats or rabbits given the ester orally. Only traces of the ester were found in the portal blood of rats. Thus, no qualitative or significant quantitative difference was found between the metabolism of the ester and the parent acid. This investigation also pointed towards multiple sites of ester hydrolysis as pronounced activity was shown by rat liver homogenates or duodenal scrapings whereas little hydrolysis was noted with blood or serum.

The classification and types of esterases present in human tissues were reviewed by LaDu and Snady (1971). Hutson (1970) summarized information on esterases and their types.

2. Amide Hydrolysis

Metabolic hydrolysis occurs with both aliphatic and aromatic amides, however at a rate which is generally much less than that seen with the corresponding esters. Interestingly, the hydrolysis of amides is mediated by liver microsomal carboxylesterases. Plant xenobiotic amides are not common and relatively little metabolic data is available on this group of compounds. Colchicine contains a Me—CO—NH-moiety which has been reported to undergo amide hydrolysis in man. This reaction also occurs with some simpler amides of various amines linked to the γ-carboxyl group of glutamic acid. Amide hydrolysis is involved in the metabolism of nicotine which leads via cotinine and ring-opened intermediates to several pyridine derivatives.

3. Epoxide Hydrolysis

Plant xenobiotics containing the epoxide moiety are not common, however this group may be formed metabolically from numerous compounds containing aryl or olefinic groups. In the former case the arene oxide formed may isomerize to the corresponding phenol as described above in Part A.4.b. Other possible fates for the epoxides include their covalent binding to tissue components or their reaction with glutathione (see Part D.6), however a competing pathway involves their hydration by epoxide hydratase (also referred to as epoxide hydrase or epoxide hydrolase) to vicinal diols. The following general reactions illustrate this reaction for arene oxides and aliphatic epoxides:

Arene oxide   Dihydrodiol

Aliphatic epoxide   Diol

Epoxide hydratase likewise converts alicyclic epoxides to 1,2-diols. The cyclic diol products invariably have the *trans*-configuration. Mammalian epoxide hydratases are associated with the microsomal fraction and are found mainly in the liver, however recent evidence (Oesch *et al.*, 1977) indicates that some activity is present in nearly all tissues in rats. This subject was extensively reviewed by Oesch (1973).

Epoxide hydratases are involved in the metabolism of several epoxide metabolites of plant xenobiotics including styrene oxide which is converted to phenyl glycol. Several alkenebenzene derivatives (e.g. estragole, safrole and elemicin) are converted to metabolites containing the 1,2-diol moiety in place of the original double bond. Epoxide intermediates have been demonstrated and it is therefore evident that diol formation is mediated by the epoxide hydratases.

4. *Glycoside Hydrolysis*

β-Glucuronidase and numerous other glycosidases are widespread in mammalian tissues. They may have functions in the metabolism of mucosubstances, however xenobiotic glycosides or metabolically formed glucuronides of xenobiotics ordinarily do not undergo significant metabolism by these enzymes. Much of this lack of activity may be related to the intracellular distribution of the enzymes. These are located in the mitochondrial and microsomal fractions and the generally highly polar glycosides would not be expected to readily penetrate into these sites. However, important exceptions are sometimes encountered and it is known that many of the sugar residues of cardiac glycosides are hydrolysed by tissue enzymes (see Chapter 6, Section III). It is noteworthy that the terminal glucose unit of the natural digitalis glycosides is removed by bacterial rather than tissue enzymes, the latter being responsible for cleavage of the remaining rather unusual sugars attached to the triterpenoid nucleus.

The gastrointestinal tract is a much more important site of hydrolysis with the majority of plant glycosides. This aspect of xenobiotic metabolism by the gut microflora is discussed in Section II.A.1 in this chapter.

D. Conjugations

Conjugations are synthetic reactions (sometimes called phase II reactions) which generally result in metabolites of reduced biological activity. These pathways differ from those summarized above in that a functional group of the xenobiotic is combined with an endogenous substrate, often derived from carbohydrate or protein sources, to form the conjugate. The endogenous conjugating moiety does not ordinarily react directly with the xenobiotic. It is instead transferred from various coenzymes, the most common being the uridine diphosphate coenzymes involved in glucoside and glucuronide formation and the adenosine coenzymes involved in sulphate conjugation and methylations. Alternatively, the xenobiotic itself may sometimes be activated, as observed with carboxylic acids which are converted to coenzyme A derivatives before being conjugated with glycine and other amino acids.

1. *Glucuronic Acid Conjugation*

The most common and perhaps the most generally important of the synthetic reactions is that of conjugation with glucuronic acid. This derives from the fact that glucuronic acid may be transferred to a wide variety of xenobiotics containing various types of functional groups, from the relative ease with which it can be made available from carbohydrate sources and from the ability for relatively large amounts of aglycone to be conjugated. The main features of the steps in the formation of glucuronides are shown below:

$$\text{Glucose-1-phosphate} + \text{UTP} \xrightarrow{\text{pyrophosphorylase}} \text{UDPG} + \text{pyrophosphate}$$

$$\text{UDPG} + 2\,\text{NAD}^{+} \xrightarrow{\text{UDPG-dehydrogenase}} \text{UDPGA} + 2\,\text{NADH} + 2\,\text{H}^{+}$$

$$\text{UDPGA} + \text{RXH} \xrightarrow{\text{UDP-glucuronyltransferase}} \text{RX-glucuronic acid} + \text{UDP}$$

where

$$\text{UTP} = \text{uridine triphosphate}$$

$$\text{UDPG} = \text{uridine diphosphate glucose}$$

$$\text{UDPGA} = \text{uridine diphosphate glucuronic acid}$$

$$\text{RXH} = \text{xenobiotic (X = O, COO, NH or S)}$$

Detailed information on these metabolic steps and especially on the glucuronyltransferases which catalyse the final reaction is available in reviews by Dutton (1966, 1971) and Dutton *et al.* (1977). It is, however, useful to note that UDP-glucuronyltransferase is associated predominantly with the endoplasmic reticulum, with the highest conjugating activities being noted in the liver. Nevertheless, recent findings make it increasingly clear that many other sites may also be involved. The fact that activity is high in the intestine may be significant when xenobiotics which contain conjugatable moieties are administered orally.

It was noted above that glucuronides may be formed from many types of xenobiotics. This fact becomes apparent in subsequent chapters which describe the glucuronidation of numerous plant xenobiotics, either with reactive groups already existing or with those produced as a result of metabolic change. No advantage is gained by compiling lists of specific xenobiotics which undergo conjugation with glucuronic acid, however a summary of the types of groups which form conjugates is useful.

a. O-Glucuronides. Glucuronides linked through oxygen are formed from phenols, alcohols and carboxylic acids and may be divided into several types. The ether type is formed from phenols and from primary, secondary and tertiary alcohols. Less common types of *O*-glucuronides formed from hydroxy compounds include enol and hydroxylamino types. The enol type occurs with the —CH=COH-moiety, an example being the glucuronide formed from the coumarin metabolite, 4-hydroxycoumarin. *N*-Hydroxy metabolites are sometimes formed from nitrogen compounds and these may undergo conjugation. A further type of *O*-glucuronide is the ester type which occurs with aromatic and primary, secondary and tertiary aliphatic carboxylic acids. The aromatic acids may be either carbocyclic or heterocyclic and the aliphatic acids may also include aralkyl types. Many xenobiotics contain more than one of these functional groups and this can result in the formation of more than one glucuronide derivative. Thus, phenolic benzoic acids are converted to both the ether and ester glucuronides and 4-hydroxybenzoic acid has additionally been reported to form a diglucuronide.

b. N-Glucuronides. Several different types of *N*-glucuronides are known (see Smith and Williams, 1966), however some including those formed with sulphonamide or carbamate groups have little relevance with regard to plant xenobiotic conjugation. Other *N*-glucuronides are formed with aromatic amines and with nitrogen heterocyclic compounds. It is noteworthy that some *N*-glucuronides, including those derived from aromatic amines, may be formed spontaneously from the aglycone and free glucuronic acid.

c. S-Glucuronides. Some thiol compounds from *S*-glucuronides.

d. C-Glucuronides. Few examples of the formation of *C*-glucuronides are known, however Δ^6-tetrahydrocannabinol has been reported to be converted by a rabbit UDP-glucuronyl transferase to such a product (see Chapter 7, Section II.A).

2. Glucoside Conjugation

Dutton *et al.* (1977) summarized the growing literature which indicates that xenobiotics may be conjugated with glucose as well as with glucuronic acid. Reactions involving UDP-glucose rather than UDP-glucuronic acid were previously thought to be limited to lower animals (e.g. insects). This viewpoint is no longer valid, however it is clear that the glucosides are quantitatively much less important metabolites than the corresponding glucuronides. As noted in Chapter 7, Section IV.G, several isoflavones are converted *in vitro* to monoglucoside conjugates.

3. Sulphate Conjugation

Sulphoconjugates are found which have the sulphate group linked to O, N or S, however the first type is the most common. This type is a sulphate ester, often termed ethereal sulphate, with the general formula $ROSO_3^-$. It is therefore a half-ester of sulphuric acid, often being isolated as the potassium salt. The most common type of sulphate ester formed from xenobiotics is that of aryl sulphates which are formed from phenolic compounds. These substrates are generally carbocyclic, however heterocyclic phenols also give rise to ethereal sulphates. Additionally, alkyl sulphate esters are known to be formed from several short chain primary alcohols and from some polyols. A third type sometimes encountered in metabolic studies of xenobiotics is that of the sulphamates which are formed from aromatic amines.

Conjugation with sulphate involves the formation of an activated form of sulphate which is formed in a series of reactions involving ATP as follows:

$$SO_4^{2-} + ATP \underset{}{\overset{\text{enzyme 1}}{\rightleftharpoons}} APS + \text{pyrophosphate}$$

$$APS + ATP \xrightarrow{\text{enzyme 2}} PAPS + ADP$$

$$PAPS + RXH \xrightarrow{\text{enzyme 3}} RXSO_3^- + PAP$$

where

      ATP = adenosine triphosphate

      APS = adenosine-5'-phosphosulphate

   PAPS = 3'-phosphoadenosine-5'-phosphosulphate

    RXH = xenobiotic (X = O or N)

enzyme 1 = ATP sulphate adenlytransferase (ATP sulphurylase)

enzyme 2 = ATP adenylyl sulphate-3'-phosphotransferase (APS-kinase)

enzyme 3 = various sulphotransferases

The enzymes involved in the formation of PAPS are located in the soluble fraction of the cell. This is also the case with the sulphotransferases involved in the formation of sulphoconjugates of small molecules including xenobiotics. It is possible for more than one sulphate group to be transferred if the substrate contains more than one functional group. This is seen with 2,6-dimethoxyphenol which, following hydroxylation, is excreted partly as 2,6-dimethoxyquinol disulphate.

An important feature of sulphate conjugation is the fact that it is rate-limited by the endogenous level of inorganic sulphate. Thus, the fraction of the dose of a xenobiotic excreted as its sulphate conjugate will decrease as the dose increases. Concurrent administration of several substrates of the sulphotransferases will lead to a reduction in the sulphate ester formation with any one of these because of competition for the available sulphate. Administration of sulphate precursors (e.g. cysteine) promotes sulphoconjugation, thereby reducing or abolishing this competition as well as increasing the ratio of the sulphate:glucuronide conjugates excreted. Further information on the sulphate conjugation reactions is available in the reviews by Roy (1971) and Dodgson (1977).

4. *Methylation*

Methylation is another common metabolic reaction using, like sulphoconjugation, an adenosine coenzyme. In this case the methyl group is transferred from S-adenosylmethionine which is formed as follows:

$$\text{Methionine} + \text{ATP} \rightarrow S\text{-adenosylmethionine} + \text{pyrophosphate} + PO_4^{3-}$$

The final step is carried out by a variety of transferase enzymes which convert various amines, phenols or thiols to the corresponding methylated derivatives. These methylation reactions share the common characteristic seen with most conjugation reactions of involving both endogenous and xenobiotic compounds.

$$S\text{-adenosylmethionine} + \text{RXH} \xrightarrow{\text{methyl transferase}} \text{RXMe} + S\text{-adenosylhomocysteine}$$

where

$$\text{RXH} = \text{xenobiotic } (X = O, N\text{- or } S)$$

a. N-Methylation. Several N-methyltransferases have been reported including phenylethanolamine N-methyltransferase which is involved in the methylation of adrenalin and related compounds, imidazole N-methyltransferase which methylates histamine and a non-specific N-methyltransferase which methylates a wide variety of primary, secondary and tertiary amines. Nornicotine (56) is a secondary amine which is N-methylated to nicotine (18), the latter compound capable of undergoing further methylation to the corresponding quaternary derivative (57). The non-specific N-methyltransferase is located in the soluble supernatant

(56)   (18)   (57)
Nornicotine   Nicotine   Isomethylnicotinium ion

fraction of tissue preparations, the rabbit lung being an especially active source. A summary of the properties of several N-methyltransferases is found in the review by Axelrod (1971).

   b. O-Methylation. Among the O-methyltransferases described, phenol-O-methyltransferase and especially catechol-O-methyltransferase (COMT) are noteworthy in relation to xenobiotic metabolism. The former enzyme is located in the microsomal fraction of many mammalian tissues including liver and lung, however the methylation of monohydric xenobiotic phenols does not appear to be a commonly encountered metabolic pathway. The 3-O-methylation of morphine to give codeine may occur to a small extent and a methyl ether of Δ^6-tetrahydrocannabinol was reported to be formed in rats.

   Contrariwise, the O-methylation of catechol derivatives is often seen, especially with the wide variety of plant xenobiotics which contain this structural feature. COMT is the transferase involved and is present mainly in the soluble fraction of many tissues including the liver.

   The methylation of catechols leads to the formation of mono- rather than dimethyl ethers and the orientation of the reaction is an important feature. With 3,4-dihydroxyphenyl compounds, methylation occurs predominantly in the *meta* position although some *para*-methylation also occurs. The extent of the latter reaction varies according to the side chain present and can be appreciable in some cases (e.g. 3,4-dihydroxyacetophenone). Interestingly, the ability of O-methyl ethers to be demethylated (see Part A.12.b of this section) allows for the interconversion of the isomeric O-methylated catechols as a result of remethylation of the dihydroxy intermediate. The tendency for p-demethylation to predominate over m-demethylation will also promote the excretion of 3-methoxy-4-hydroxyphenyl metabolites rather than their iso derivatives. With pyrogallol derivatives the middle of the three hydroxyl groups is methylated. Axelrod (1971) reviewed the O-methyltransferases.

   c. S-Methylation. Thiol derivatives may be methylated, this reaction probably being one of the steps in the metabolism of ethanethiol.

5. Amide Synthesis

Amide synthesis is a metabolic reaction by which some carboxylic acids are linked by an amide bond to the α-amino group of various amino acids. The conjugation of carbocyclic and heterocyclic aromatic acids is commonly seen, the best known example being the formation of hippuric acid from benzoic acid and glycine. The discovery of this conjugate dates from the earliest days of biological chemistry and its name derives from its isolation from horse urine (Greek *hippos* = horse) rather than from its chemical properties. However, the suffix -uric has been traditionally used to denote the glycine conjugates of many acids. Amide synthesis occurs also with various aralkyl acids. These include phenylacetic acid and its derivatives and many β-substituted acids. The latter group is of interest as several plant xenobiotics are $C_6—C_3$ acids (e.g. cinnamic acid derivatives). An additional type is encountered with the carboxylic acid metabolites of some terpenoid compounds. Thus, glycine conjugates are formed from phellandric acid, a metabolite of α-phellandrine and from perillic acid, a metabolite of *d*-limonene.

a. *Glycine conjugation.* The most common of the amide syntheses occurs with glycine. This and other amide syntheses differ from most other conjugations in that the xenobiotic rather than the transferring agent is activated. Cumic acid (5) is a major metabolite of *p*-cymene which can then undergo conjugation with glycine to form cuminuric acid (58) as follows:

$$\underset{\substack{\text{Me}\quad\text{Me}}}{\underset{|}{\overset{\displaystyle\overset{O}{\|}}{C}-S-CoA}} \quad +H_2N-CH_2-COOH \rightarrow \underset{\substack{\text{Me}\quad\text{Me}}}{\underset{|}{\overset{\displaystyle\overset{O}{\|}}{C}-NH-CH_2-COOH}}$$

Glycine

(58)
Cuminuric acid

These reactions take place in the mitochondria and the first two steps are probably carried out by the same activating system which is responsible for the formation of coenzyme A derivatives of fatty acids. Glycine conjugation is carried out in the liver and kidney, the latter site often making an appreciable contribution to the overall formation. Indeed, the kidneys appear to be the sole site of glycine conjugation in dogs. Further details of the conjugation of acids with glycine are available in the reviews by Williams (1959, pp. 348–353) and by Weber (1971).

   b. Conjugation with other amino acids. Although the mammalian conjugation of carboxylic acids with amino acids most commonly involves glycine, several other compounds are also utilized. These include glutamine and taurine and, in rare cases, some dipeptides. The nature of the amino acid utilized depends on the animal species and considerable attention has recently been devoted to this subject. The data available on non-primates and primates were reviewed by Hirom *et al.* (1977) and Smith and Caldwell (1977), respectively.

   Conjugation with glutamine occurs with phenylacetic acid and its derivatives and is generally restricted to man and other primate species. It is not, however, seen in the lower primates (e.g. prosimians) although small amounts of glutamine conjugates have been reported in the cat and ferret. Phenylacetic acid derivatives are not common plant xenobiotics but they sometimes arise metabolically. This is seen with some flavonoids (e.g. flavonols) which are degraded to C_6-C_2 phenolic acids, however glutamine conjugates of these metabolites have not been reported. On the other hand, mescaline is known to be metabolized partly to 3,4-dihydroxy-5-methoxyphenylacetic acid which is excreted as its glutamine conjugate in man. Indole-3-acetic acid is another arylacetic acid conjugated with glutamine in man, apes and higher monkeys.

   Taurine conjugation is similarly a reaction of arylacetic acids, however it has a more widespread occurrence in that it occurs in both anthropoid and prosimian primates and also in carnivores (e.g. cat, dog, ferret). Taurine conjugation is often an important alternative route of comjugation in carnivores.

6. *Glutathione Conjugation*

It has long been recognized that some xenobiotics are excreted as their N-acetylcysteine derivatives, these conjugates being known as mercapturic acids. It was later shown that the source of the cysteine in these conjugates is the tripeptide glutathione (glutamylcysteinylglycine). The overall metabolic pathway involves a series of steps as follows:

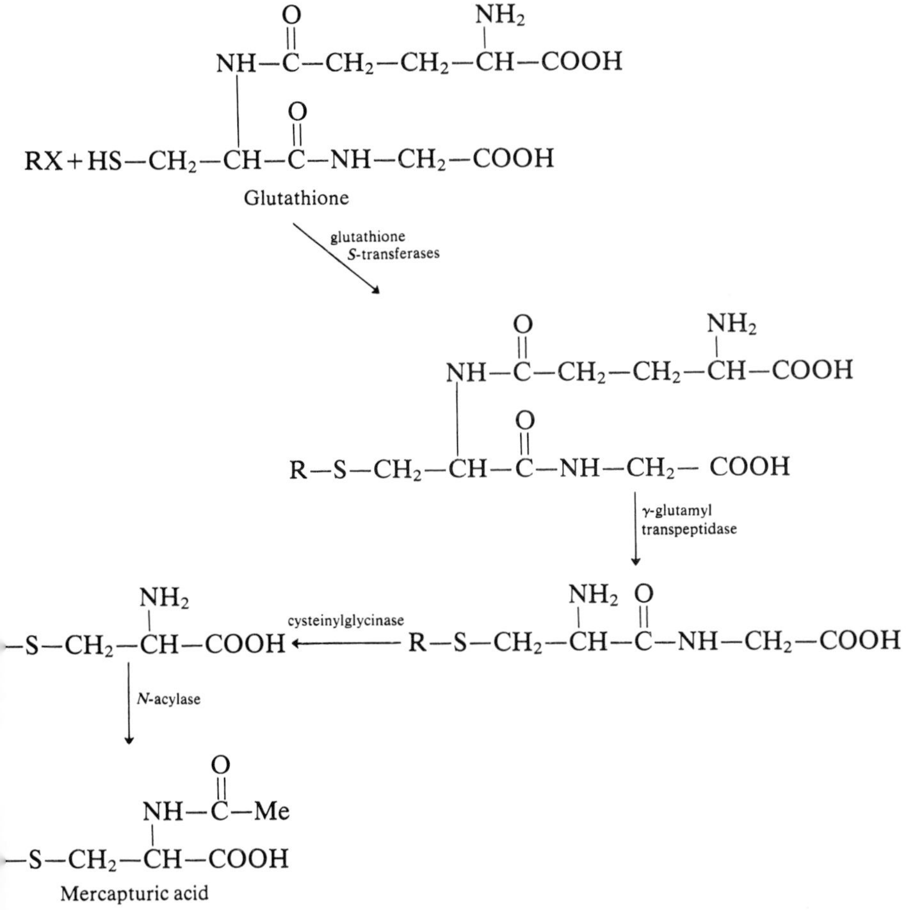

$$
\begin{array}{c}
\underset{\text{O}}{\overset{\text{NH}_2}{}} \\
\text{NH--C--CH}_2\text{--CH}_2\text{--CH--COOH} \\
\mid \quad \overset{\text{O}}{\underset{\|}{}} \\
\text{RX + HS--CH}_2\text{--CH--C--NH--CH}_2\text{--COOH}
\end{array}
$$

Glutathione

glutathione
S-transferases

$$
\begin{array}{c}
\overset{\text{O}}{\underset{\|}{}} \quad \overset{\text{NH}_2}{\mid} \\
\text{NH--C--CH}_2\text{--CH}_2\text{--CH--COOH} \\
\mid \quad \overset{\text{O}}{\underset{\|}{}} \\
\text{R--S--CH}_2\text{--CH--C--NH--CH}_2\text{-- COOH}
\end{array}
$$

γ-glutamyl
transpeptidase

$$
\overset{\text{NH}_2}{\mid} \quad \overset{\text{O}}{\underset{\|}{}} \\
\text{R--S--CH}_2\text{--CH--C--NH--CH}_2\text{--COOH}
$$

cysteinylglycinase

$$
\overset{\text{NH}_2}{\mid} \\
\text{--S--CH}_2\text{--CH--COOH}
$$

N-acylase

$$
\overset{\text{O}}{\underset{\|}{}} \\
\text{NH--C--Me} \\
\mid \\
\text{--S--CH}_2\text{--CH--COOH}
$$

Mercapturic acid

Reaction of a large number of compounds containing an electrophilic carbon atom with the nucleophile glutathione is usually catalysed by various glutathione S-transferases, however this conjugation can also proceed non-enzymically in some cases. An example of the latter situation is seen with the alkaloid arecoline. These glutathione conjugates are not excreted in the urine, but may be often found as biliary metabolites as is also the case with the lower peptide derivatives. The products excreted in the urine are the mercapturic acids and, occasionally, the cysteine derivatives.

The glutathione-S-transferases are soluble, cytoplasmic enzymes found in several tissues including kidney and especially liver. The knowledge that these enzymes catalyse the reaction of glutathione with a wide variety of substrates led to the classification of the enzymes on the basis of the chemical type of substrate involved. Thus, the nomenclature included alkyl, aralkyl, aryl, alkene and epoxide transferases. However, recent findings obtained with purified transferases show that these have overlapping specificities, a fact which has led to the introduction of a new nomenclature in which the transferases from rat liver are designated by capital Latin letters (see Jakoby et al., 1976) and those from human liver by lower case Greek letters (see Habig et al., 1976). A large number of the substrates of the glutathione-S-transferases so far studied include polycyclic aromatic hydrocarbons, compounds containing halogen or nitrogen groups or compounds which differ in other ways from the naturally occurring xenobiotics. This important metabolic pathway may therefore be of more limited occurrence with compounds of the latter type, however certain groups of plant xenobiotics are noteworthy as they have either a demonstrated or potential ability to form glutathione conjugates and to be excreted as mercapturic acids. α,β-Unsaturated compounds furnish one such group and cinnamaldehyde, several unsaturated aliphatic aldehydes and parasorbic acid are examples of plant xenobiotics which are substrates for the enzyme-catalysed conjugation with glutathione and which may be metabolized partly to mercapturic acids (Chasseaud, 1976). Other plant xenobiotics contain olefinic moieties which are converted to epoxides. The latter type of compound is known to undergo glutathione conjugation and this pathway may therefore be significant in the metabolism of this group of plant compounds. It is a minor metabolic route with styrene but seems not to occur with a higher homologue, p-methoxyallylbenzene (estragole). Several recent reports indicate that conjugation with glutathione is also a feature of the metabolism of some thiocyanate and isothiocyanate derivatives. In the latter case a mercapturic acid is the ultimate urinary metabolite, however reaction of glutathione with the thiocyanates results in loss of the carbon and nitrogen atoms as inorganic cyanide and incorporation of the remainder of the RSCN compound into an asymmetric disulphide derivative. The details of these reactions of glutathione with organic thiocyanates and isothiocyanates are given in Chapter 10, Section VI.

A wealth of new information on the conjugation of glutathione with xenobiotics is now available as a result of the increasing awareness of the importance of this pathway not only in the metabolism of a wide variety of xenobiotics but also because of the important role of glutathione in eliminating reactive and toxic intermediates formed metabolically. These findings have been summarized in a number of useful reviews including those by Boyland (1971), Chasseaud (1973, 1976) and Grover (1977). Arias and Jakoby (1976) edited a book derived from a conference devoted to the metabolism and function of glutathione.

7. Acetylation

As with the amino acid conjugations described above in Part D.5, acetylation results in the formation of metabolites containing the amide bond. However, in the latter case the acidic moiety is the endogenous component and this reaction is therefore one which occurs with amino compounds. While formylation and succinylation reactions have been reported in rare cases, acetylation is the usual reaction. This pathway utilizes acetyl coenzyme A as follows:

$$R-NH_2 + Me-\overset{\overset{\displaystyle O}{\|}}{C}-S-CoA \xrightarrow{\text{N-acetyltransferase}} R-NH-\overset{\overset{\displaystyle O}{\|}}{C}-Me + HS-CoA$$

While the liver is a major site of N-acetylation, this activity is observed in many other tissues as well. Studies on arylamine acetylation show that considerable variations in enzyme activity exist between various tissues and also among different animal species. For example, high, moderate and low activities may be found in rabbits, man and dogs, respectively. Genetic and developmental factors are also known to influence the rate and pattern of xenobiotic acetylation. Further information on the N-acetylation enzymes is available in the review of Weber (1971).

Acetylation occurs with several types of primary amino compounds, however the majority of investigations have dealt with groups which have little relevance to the metabolism of plant xenobiotics (e.g. aromatic amines, sulphonamides and hydrazine derivatives). Nevertheless, plant compounds are found among the additional substrate groups including some amino acids and derivatives of aliphatic and alicyclic amines. An important example of amino acid acetylation is that seen with the cysteine derivatives of glutathione conjugates which are converted to mercapturic acid (N-acetylcysteine) derivatives. This reaction is deficient in guinea pigs and results in a low excretion of mercapturic acid in this species. Among compounds containing an aliphatic amino group, the aralkyl amines tyramine and mescaline are metabolized partly by N-acetylation. Isojuripidine is a complex triterpenoid compound which contains an amino group attached to an alicyclic ring. The *N*-acetyl derivative is the major metabolite found in the plasma and liver of isojuripidine-treated animals.

8. *Thiocyanate Formation*

This reaction is noteworthy because it occurs with a single substrate, inorganic cyanide ion. The reaction is as follows:

$$\text{CN}^- \; + \; \text{S}_2\text{O}_3^{2-} \; \xrightarrow{\text{sulphur transferase}} \; \text{SCN}^- \; + \; \text{SO}_3^{2-}$$

| Cyanide | Thiosulphate | Thiocyanate | Sulphite |

The enzyme involved, sulphur transferase, is also called rhodanese and is found in most animal tissues, however the liver is a particularly good source. The sulphur required in this reaction is derived from endogenously formed thiosulphate. Neither cysteine nor glutathione serve as sulphur donors. Organic cyanides do not themselves form thiocyanate derivatives, however they may be metabolized to cyanide which can then be converted to thiocyanate. An excellent example of this is seen with the cyanogenetic glycosides which liberate HCN upon metabolic degradation. Glucosinolates are another class of plant xenobiotic which may give rise to HCN. In this case the parent compounds are converted to several types of products including organic cyanides which, via oxidative reactions, may liberate HCN. Further details on HCN-production from these two types of compounds are found in Chapter 8, Section II.

II. Reactions of the Gastrointestinal Microflora

The data summarized above in Section I make it clear that the animal organism possesses a wide variety of tissue enzymes which carry out the metabolism of plant xenobiotics. However, during the past 10–20 years there has been a growing realization that some of the metabolic reactions seen with these and other xenobiotics may also be effected by the micro-organisms normally present in the gastrointestinal tract and, perhaps of greater interest, that certain reactions occur which are unique to the gut microflora. About 30 different types of these reactions have so far been identified and, of these, roughly a half are known to take place with plant xenobiotics. While the examples given below therefore illustrate only some of these bacterial reactions, they nevertheless belong to the main groups of reactions including hydrolyses, removal of groups and other degradations and reductions. Comprehensive treatment of the gut microfloral metabolism of xenobiotics in general is available in the reviews by Scheline (1973,

1978) which also include information on the gastrointestinal microflora itself.

There are several reasons for the increased interest in xenobiotic metabolism by the gut microflora. These include the fact that increases in biological activity may be brought about and that these may have pharmacological or toxicological significance. Furthermore, these gastrointestinal reactions may be of importance in the enterohepatic circulation of xenobiotics, in the metabolic differences observed between various species, strains or individuals and in the metabolic adaptation to compounds administered repeatedly. These subjects may be exemplified by findings obtained with plant xenobiotics.

Owing to the fact that metabolism by the gut microflora involves a pronounced tendency for hydrolytic, degradative and reductive reactions to take place, it is perhaps not surprising that metabolites of increased biological activity will often be formed. In fact, some of the most striking examples of this phenomenon are found with plant xenobiotics, especially glycosides. A classic example of pharmacological activation is that seen with the anthraquinone cathartics. As discussed in Chapter 4, Section III.B the sugar residues serve to reduce absorption in the upper intestine and transport the glycoside to the lower bowel. Bacterial glycosidases in this region will then hydrolyse the compounds to cathartic principles. Glycoside hydrolysis also furnishes an excellent example of a type of bacterial reaction which has toxicological significance. Thus, cyanogenetic glycosides are hydrolysed to products which release HCN (see Chapter 8, Section II) and the glucoside cycasin is converted to a hepatotoxic and carcinogenic aglycone, as described subsequently in this chapter.

It is now well established that intestinal microorganisms often play a key role in in the enterohepatic circulation of compounds. This is particularly the case with glucuronide conjugates which are formed in the body from a large number of xenobiotics and their metabolites. These conjugates will often be excreted in the bile (see Smith, 1973; Klaassen, 1975) and will then be exposed eventually to the microflora of the lower gut which harbours numerous microorganisms capable of producing β-glucuronidase (Hawksworth et al., 1971). Hydrolysis of the glucuronide will, by virtue of the large changes in physico-chemical properties which generally result, allow for absorption of the aglycone from the gut and/or its further bacterial metabolism. It may also influence the relative proportions of metabolites excreted by the urinary and faecal routes. A closely related point of possible importance is the knowledge that the production of bacterial metabolites from xenobiotics which are rapidly and extensively absorbed after oral administration depends upon this mechanism of enterohepatic circulation. This phenomenon has been shown to occur in rats with the aldehyde vanillin, some of which is excreted in the urine as simple phenols arising from the intestinal reduction or decarboxylation of biliary conjugates of vanillin and its primary oxidized and reduced metabolites (Strand and Scheline, 1975). This mechanism also explains the urinary excretion in rats of several O-demethylated metabolites of the ginger principle zingerone (Monge *et al.*, 1976).

Variations in the metabolism of xenobiotics by the gastrointestinal microorganism may sometimes be responsible for species variations in metabolism. A well-known example of this is seen with quinic acid which undergoes extensive aromatization in man and old world monkeys but little or no aromatization in new world monkeys and many lower animal species (Adamson *et al.*, 1970). Similar results have been reported with some flavonoid compounds. As noted in Chapter 7, Section IV.C, the flavanone naringin is excreted as its aglycone naringenin and the glucuronide of the latter when fed to humans but, when given to rats, is much more extensively degraded and excreted partly as phenolic acids which are characteristic bacterial metabolites of many flavonoids. Similarly, the flavanone glycoside hesperidin is degraded somewhat differently in humans than in rats or rabbits. In other cases, metabolic differences have been reported in different groups of a single animal species. Scheline (1968a) found that coumarin was metabolized partly to melilotic acid in rats, a reaction carried out by the gut microflora and involving hydrogenation of the heterocyclic ring followed by ring scission. However, it was noted that similar earlier studies looked for but did not find this metabolite.

It is clear that large, even absolute, differences in the gut microfloral metabolism of xenobiotics may occur between conventional, specific pathogen-free and germ-free animals. Also, age and dietary factors may influence the bacterial metabolism. Changes in the gut microflora are largest during early life and Fischer and Weissinger (1972), Kent *et al.* (1972) and Fischer *et al.* (1973) demonstrated that intestinal β-glucuronidase activity is very low in newborn rats and humans but that this capacity increases with age. Measurements in rats showed a concomitant increase in the anaerobic microflora. Several specific examples of the influence of diet were noted in the two reviews cited in the beginning of this section.

The metabolic adaptation of the gut microflora to a xenobiotic given repeatedly is an interesting phenomenon because it may result in the development of new pathways of metabolism. Thus, a form of induction in a qualitative rather than a quantitative sense occurs. The most outstanding example of this phenomenon is that which occurs with cyclamate, the synthetic sweetening agent (see Renwick, 1977). Results obtained hitherto do not suggest that this form of metabolic development is very common, nonetheless other examples are known including some involving plant xenobiotics. Griffiths and Smith (1972a) noted that administration of the flavonoid robinin (kaempferol-3-robinoside-7-rhamnoside) resulted in an increased excretion of *p*-hydroxyphenylacetic acid, the product of bacterial degradation, when administered to rats previously given a diet containing robinin or its aglycone kaempferol.

As noted above, the metabolism of plant xenobiotics by the gut microflora, while not covering all of the reaction types now known, nevertheless includes numerous examples of the main reaction groups. The following summary illustrates the various reactions in these groups.

A. HYDROLYSES

1. *Glycoside Hydrolysis*

One of the most significant and well-documented reactions carried out by the microflora of the gastrointestinal tract is that of glycoside hydrolysis. When dealing with xenobiotic compounds in general, the importance of this reaction is often associated with the fact that numerous compounds of various types are converted in the body to glucuronide conjugates. These conjugates will thereafter often be subjected to hepatic excretion into the bile and subsequently come into contact with the gut bacteria. Not unexpectedly, this phenomenon is observed with many plant xenobiotics and their metabolites. The metabolism of the aromatic aldehyde vanillin in rats furnishes an interesting example. It is well known that vanillin is extensively oxidized and that vanillic acid, either free or in conjugated forms, is its main metabolic product. Some reduction of the aldehyde to vanillyl alcohol also occurs and Strand and Scheline (1975) found that small amounts of several simple phenols, known to be formed via reduction of the aldehyde or alcohol or decarboxylation of the carboxylic acid by the intestinal microorganisms, were excreted in the urine. They demonstrated that the formation of these urinary phenols was dependent upon the excretion in the bile of the glucuronides of vanillin and its alcoholic and acidic metabolites. These conjugates, after coming into contact with the microflora, were hydrolysed and the liberated aglycones then served as substrates for the reactions of reduction, decarboxylation and O-demethylation which gave rise to the simple phenols ultimately absorbed and excreted in the urine. Other examples of this interplay between the glucuronidation of plant xenobiotics, the biliary excretion of these conjugates and the subsequent metabolism by the gut microflora are included in the subsequent chapters and it is evident that this possibility must be given adquate recognition. Of course, a more direct method for exposing glucuronides to the effects of bacterial metabolism involves the ingestion of naturally occurring plant glucuronides, however combination of plant phenolics with this hexuronic acid is much less common than that seen with the plant sugars noted below.

The great majority of glycosides contain common plant sugars, either singly or as di- or trisaccharides. The most common sugar involved is D-glucose, however D-galactose, D-xylose, L-rhamnose and L-arabinose are also encountered, either singly or as component sugars of most of the di- and trisaccharides. It is noteworthy that the metabolism of many glycosides has been included in subsequent chapters together with the metabolic data on the corresponding aglycones and related compounds. This approach seems reasonable in view of the similarity often seen in the metabolism of aglycones and their glycosides. This is typically observed with flavonoid compounds, an excellent example being furnished with the flavonol quercetin and rutin, its 3-rutinoside derivative. The *in vitro* degradation of rutin to *m*-hydroxyphenylpropionic acid, a prominent metabolite of quercetin as well, by the rat intestinal bacteria was reported by Booth and Williams (1963). Scheline (1968b) found that the aglycone was also detectable in such incubates. Similar studies by Griffiths and Smith (1972a,b) showed the bacterial hydrolysis of several other flavonoid glycosides including the flavonols myricitrin (myricetin-3-rhamnoside) and robinin (kaempferol-3-robinobioside-7-rhamnoside), the flavone apiin (apigenin-7-apioglucoside), the flavanone naringin (naringenin-7-rhamnoglucoside), the anthocyanin pelargonin (pelargonidin-3,5-diglucoside) and the dihydrochalcone phloridzin (phloretin-2'-glucoside).

Metabolism by the gut microflora is a central feature with other types of glycosides including the anthraquinone, cardiac and cyanogenetic glycosides. In these cases the subject is included in later sections of the text and these examples will therefore not be covered here. The metabolism of another glycoside, cycasin, is not treated subsequently and this can conveniently be included at this point as it furnishes an excellent example of the possible significance of this bacterial reaction.

(59)
Cycasin

(60)
Methylazoxymethanol

Cycasin (methylazoxymethanol-β-D-glucoside) (59) is the toxic constituent of cycad meal which is obtained from the nuts of plant of the *Cycas* genus. The product is used as a food in some areas of the world and thereby poses a potential toxicological problem since cycasin is known to have hepatotoxic and carcinogenic properties in rats (see Yang and Mickelsen, 1969). Kobayashi and Matsumoto (1965) reported that the intraperitoneal injection of cycasin did not lead to toxicity and that the glucoside was excreted almost quantitatively in the urine. Oral administration resulted in toxic effects and a much lower urinary excretion of unchanged compound. Since injection of the aglycone, methylazoxymethanol (60) also proved to be toxic, it was clear that this property is associated with the latter compound rather than with the glucoside itself. Laqueur (1964) and Spatz et al. (1966) studied cycasin toxicity and excretion in conventional and germ-free rats. They found that the latter group lacked the β-glucosidase needed to hydrolyse cycasin and that these animals excreted it unchanged without showing signs of toxicity. These findings clearly implicated the gut microflora and Spatz et al. (1967) reported that the toxic properties of cycasin reappeared when germ-free rats were monocontaminated with various strains of bacteria possessing β-glucosidase activity. Furthermore, the degree of toxicity observed and the amounts of unchanged cycacin excreted in the urine were consistent with the *in vitro* levels of β-glucosidase activity measured in the various bacteria. Thus, *Streptococcus faecalis* contained high enzyme levels and its use as a monocontaminant resulted in relatively low excretion of unchanged cycasin and severe liver injury.

The above examples make it clear that glycoside hydrolysis by the gut microflora is an important feature in the biological disposition of this class of compounds. The microorganisms responsible for this reaction are consequently a subject of interest. Hawksworth et al. (1971) investigated this question in detail using many strains of six different groups of intestinal bacteria (*Bacteroides*, bifidobacteria, clostridia, enterobacteria, enterococci and lactobacilli) and several substrates having α- or β-glucoside, α- or β-galactoside or β-glucuronide linkages. Enterococci produce large amounts of β-glucosidase and β-galactosidase, but little β-glucuronidase, whereas enterobacteria (mainly *Escherichia coli*) produce much β-glucuronidase and β-galactosidase, but little β-glucosidase. Nevertheless, when the total numbers of the various intestinal bacteria were taken into account, it was concluded that β-glucuronides are hydrolysed in the rat mainly by *Bacteroides*, bifidobacteria and lactobacilli and β-glucosides mainly by the first two groups. The subject of the hydrolysis of various types of glycosides by intestinal bacteria is discussed in the monograph by Drasar and Hill (1974).

A further interesting example of glycoside hydrolysis is seen in the metabolism of the xanthone derivative mangiferin (61). When given orally to rabbits it is converted to euxanthic acid (62), the 7-glucuronide of

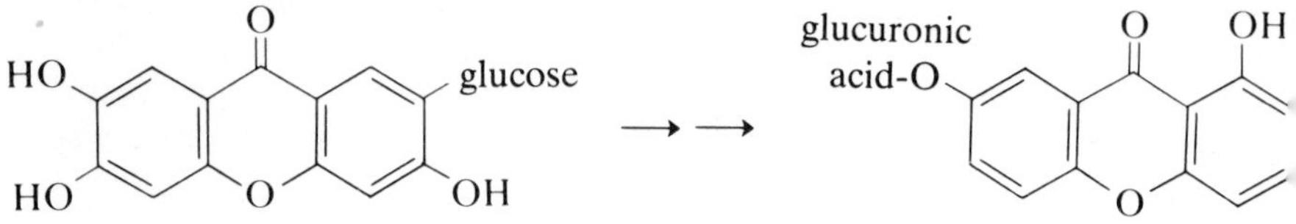

(61)
Mangiferin

(62)
Euxanthic acid

euxanthone. This metabolic sequence involves the cleavage of a *C*-gluco-side residue and although the site of this reaction has not been ascertained, it seems reasonable to believe that the gut microflora is responsible.

2. *Ester Hydrolysis*

Subsequent chapters include several examples of esters which are hydro-lysed by the gut microflora. These include compounds of varying but generally fairly high degrees of structural complexity and often with substituent groups which tend to retard their absorption from the intestine. Chlorogenic acid (pKa = 2·7) is a relatively simple compound of this type which will exist nearly exclusively in an ionized form in the intestine. Among the more complex types are several higher terpenoid derivatives including asiaticoside which loses its esterified sugar moieties as a result of bacterial metabolism. The loss of the terminal sugar residue in some natural cardiac glycosides was noted above. Some of these compounds

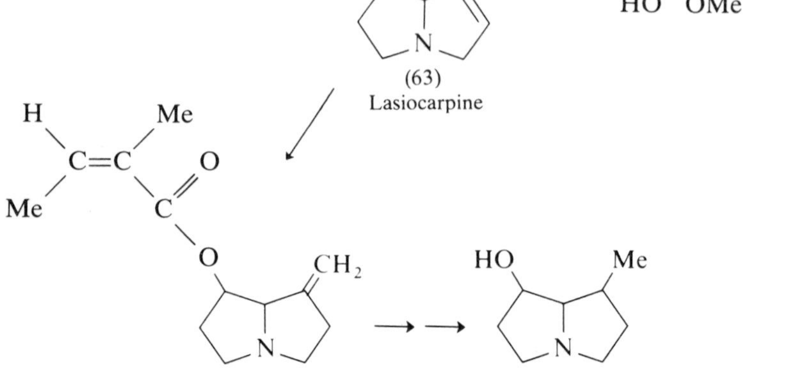

(63)
Lasiocarpine

(64)

including lanatoside A and C also contain acetylated digitoxose moieties which may also be hydrolysed by the gut microflora. An interesting example of bacterial hydrolysis of esters is seen with the pyrrolizidine alkaloid lasiocarpine (63). This compound contains two ester linkages, one of which undergoes reductive fission and the other ester hydrolysis to give the hydroxylated methylpyrrolizidine (64) as a result of metabolism by rumen microorganisms.

B. REACTIONS INVOLVING REMOVAL OF GROUPS AND OTHER DEGRADATIONS

Although not all types of the degradative reactions known to be carried out by the gastrointestinal microorganisms with xenobiotics in general have been shown to occur with plant compounds, the most striking examples of reactions of this group are nevertheless those which take place with these naturally occurring compounds. It is within this area that we find pathways of metabolism which show the greatest differences from those taking place in the tissues.

1. *Dehydroxylation*

Although both *C*- and *N*-dehydroxylation by the gut microflora have been demonstrated, only the former type is presently known to occur with plant xenobiotics. This reaction has been extensively studied with catecholic derivatives, especially catechol acids. The metabolism of the latter is discussed in detail in Chapter 5, Section I.C and only the highlights of their dehydroxylation need be given here. Discovery of the dehydroxylation reaction dates from the 1950s when it was shown that homoprotocatechuic acid (65) is partly metabolized in rats and rabbits to *m*-hydroxyphenylacetic acid (66). It was later found that while *p*-hydroxylation is the major dehydroxylation pathway, a small amount of *m*-dehydroxylation to give *p*-hydroxyphenylacetic acid (67) also occurs. In addition to these

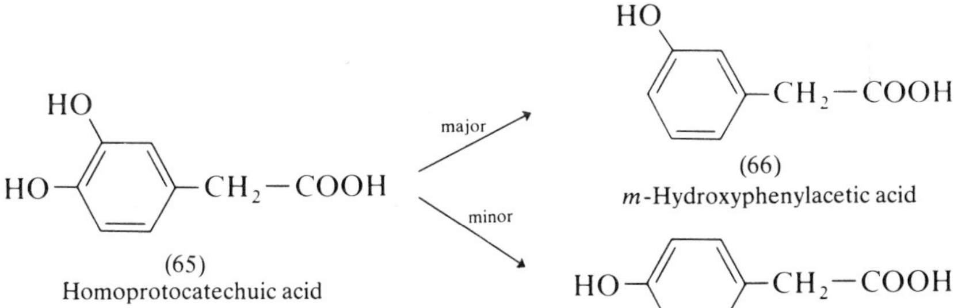

(65)
Homoprotocatechuic acid

major

minor

(66)
m-Hydroxyphenylacetic acid

(67)
p-Hydroxyphenylacetic acid

findings with this C_6—C_2 catecholic acid, numerous studies with homologous acids have furnished similar results. Dehydroxylation is not extensive with the C_6—C_1 acid, however the formation of *m*-hydroxyphenyl derivatives of C_6—C_3 acids (both caffeic and dihydrocaffeic acid) and even the C_6—C_5 acid (Scheline, 1970) has been well documented.

Although the dehydroxylation of catechol acids has been the most commonly studied reaction of this type, other examples are known which involve other types of compounds. Some may have a catechol moiety, as with the (+)-catechin metabolite δ-(3,4-dihydroxyphenyl-γ-valerolactone which is found in the urine together with the corresponding 3-hydroxyphenyl derivative. In other cases the sole phenolic hydroxyl group may be removed as found with some tyrosine metabolites (Curtius *et al.*, 1976), however the significance of the bacterial dehydroxylation of monohydric plant phenolics has not been assessed. The aromatization of quinic and shikimic acids, which is discussed below, is an interesting reaction sequence which involves the removal of several alicyclic hydroxyl groups.

2. Decarboxylation

Decarboxylation of plant xenobiotics by the gut microflora is a reaction which takes place with some phenolic acids. Typical examples of this degradation are seen with 4-hydroxybenzoic acid (68) or with ferulic acid (69). These reactions indicate that simple phenols may be formed from

(68)
4-Hydroxybenzoic acid   Phenol

(69)
Ferulic acid   4-Vinylguaiacol

both benzoic and cinnamic acid derivatives. Decarboxylation occurs also with appropriate C_6—C_2 (phenylacetic) acids, however C_6—C_3 acids of the phenylpropionic acid-type are resistant to this reaction. An essential structural requirement is the presence of a free *p*-hydroxyl group. Thus, the closely related 4-methoxy derivatives will not be decarboxylated unless first *O*-demethylated. The decarboxylation reaction proceeds best when a sole hydroxyl group is present and it may be inhibited to various degrees by further substituent groups. Strong or even complete inhibition may occur when 2-hydroxy or 3,5-dimethoxy groups are also present.

3. Dealkylation

O-Demethylation of plant xenobiotics and their metabolites is a commonly occurring bacterial reaction. It is observed with numerous aromatic methyl ethers including those of some simple phenols, alcohols, aldehydes, ketones and acids as well as with various types of flavonoids. Hydroxy-methoxyphenyl derivatives seem particularly susceptible to O-dealkylation and the bacterial formation of catechol derivatives from a variety of chemical classes is described in subsequent chapters. Single methoxyl groups are quite resistant to demethylation which takes place much more readily with di- and trimethoxy derivatives, especially when the substituents are located vicinally. The extensive metabolism of sinapic acid (70) to 3,5-dihydroxyphenylpropionic acid (71) in rats provides an excellent illustration of this type of dealkylation which, in this case, is accompanied by p-dehydroxylation and double bond reduction.

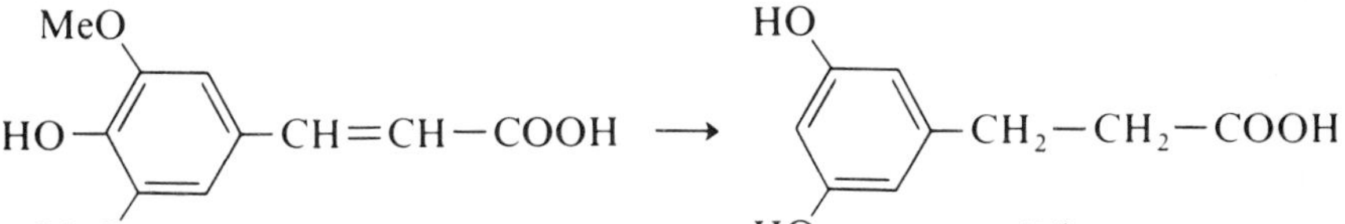

$$\text{MeO} \quad \text{HO} \quad \text{CH}=\text{CH}-\text{COOH} \longrightarrow \quad \text{HO} \quad \text{CH}_2-\text{CH}_2-\text{COOH}$$

(70)
Sinapic acid

(71)
3,5-Dihydroxyphenylpropionic acid

It is also known that some structurally more complex methoxylated plant xenobiotics are dealkylated by the gut microflora. This is observed with several flavonoid compounds including hesperidin (72) which is degraded to m-hydroxyphenylpropionic acid (73), this metabolite being derived from

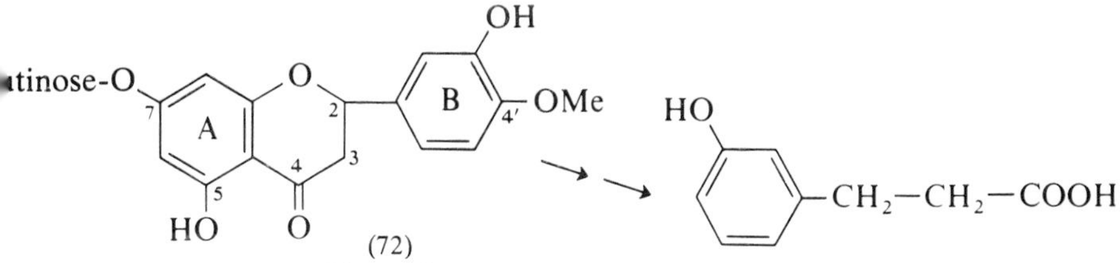

(72)
Hesperidin

(73)
m-Hydroxyphenylpropionic acid

the B-ring and the carbon atoms at positions 2, 3 and 4. It is likely that the demethylation reaction occurs at an early stage in this metabolic sequence since it is known that a free 4'-hydroxy group is required for fission of the heterocyclic ring (see Part B.4 below). O-Demethylation of some isoflavonoids by intestinal bacteria has also been reported. This is illustrated by the conversion of formononetin (74) to daidzein (75) which, interestingly, involves the cleavage of an isolated methoxyl group.

HO— ... —OMe → HO— ... —OH

(74)   (75)
Formononetin   Daidzein

The methoxyl group is one of the most common substituents found in plant xenobiotics and it would therefore be of interest to assess the wider significance of O-demethylation by the gut flora. Little pertinent information is available, however Smith and Griffiths (1974) found that the methoxylated alkaloids papaverine, quinine and reserpine did not undergo dealkylation when incubated with the rat caecal microflora.

4. Heterocyclic Ring Fission

The degradation of numerous flavonoids by the gut microflora furnishes the prime example of heterocyclic ring fission. This transformation is noteworthy because it involves a series of reactions which generate fairly simple products (usually phenolic acids) from relatively complex precursors. Flavonoid degradation consequently furnishes the most extreme example of the difference existing between tissue and gut microflora metabolism of xenobiotics. The metabolism of flavonoids is discussed in detail in Chapter 7, Section IV which shows that bacterial involvement is of fundamental importance with many classes including flavones, flavonols, flavanones, catechins and dihydrochalcones. Other classes, including anthocyanins and isoflavonoids, may also be subject to varying degrees of bacterial metabolism. Compounds possessing free 5- and 7-hydroxyl groups (A-ring) and a 4'-hydroxyl group (B-ring) are those which undergo ring fission and the absence of these groups or their replacement by methoxyl groups reduces or abolishes the reaction.

Among the numerous examples of flavonoid metabolism hitherto reported, that noted in the foregoing section on O-demethylation illustrated the overall reaction whereby hesperidin is converted to m-hydroxy-phenylpropionic acid. The individual steps in this sequence are not well understood, however results obtained with related flavonols indicate that a major pathway, although not necessarily the sole one, probably involves the catabolism of the A-ring to CO_2. Another well-known example of flavonoid degradation is seen with (+)-catechin (76). m-Hydroxy-

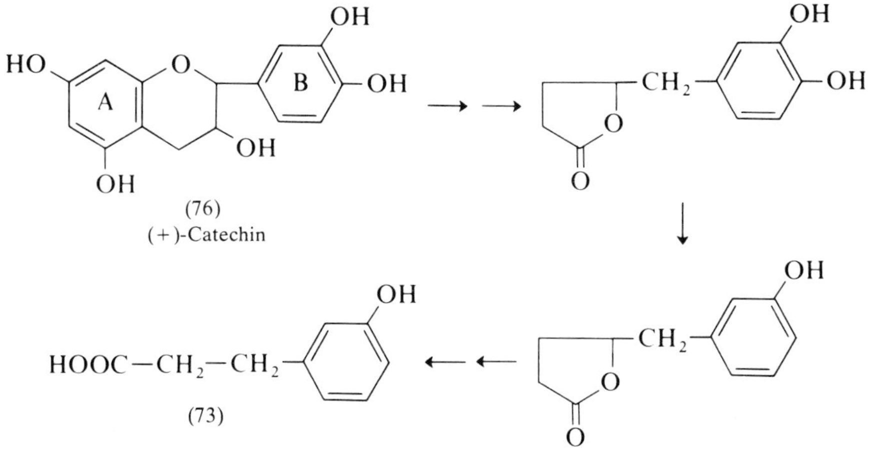

(76)
(+)-Catechin

(73)

phenylpropionic acid (73) is also in this case a major bacterial metabolite, however the details of the reaction sequence are somewhat clearer and several intermediates have been identified. Degradation of the A-ring in (+)-catechin is also known to involve the production of CO_2.

Another example of heterocyclic ring fission is that seen with coumarin and some derivatives including umbelliferone (77). Studies of this transformation indicate that the key step is actually reduction of the heterocyclic

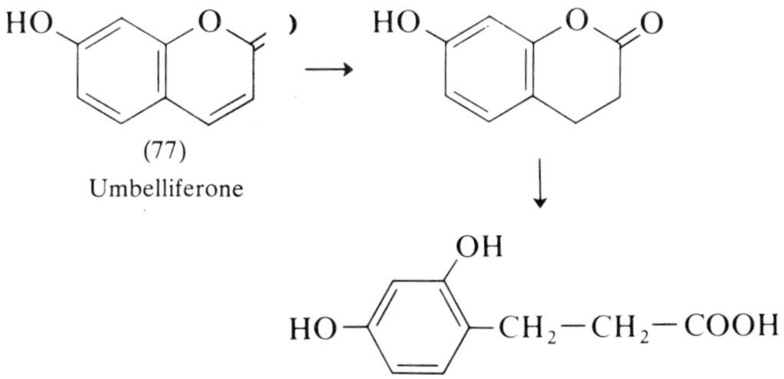

(77)
Umbelliferone

ring to the corresponding 3,4-dihydro derivative and that the actual scission then probably takes place spontaneously.

C. REDUCTIONS

The reduction of xenobiotics by the gastrointestinal microflora occurs with a wide variety of compounds. Among the numerous types of reductions demonstrated, those involving double bonds, alcohols, aldehydes, ketones,

N-oxides, sulphoxides as well as the reactions of aromatizations and ester fission have been shown to be involved in the reduction of plant xenobiotics and their metabolites.

1. *Double Bond Reduction*

The best documented example of double bond reduction by the gut flora deals with the hydroxycinnamic acids, e.g. *o*-coumaric, *p*-coumaric, caffeic, ferulic and sinapic acids. Results from both animal studies and *in vitro* investigations using intestinal bacteria have shown that these unsaturated C_6-C_3 acids undergo reduction to the corresponding phenylpropionic acids. This subject is covered in detail in Chapter 5, Section I.C and the bacterial reduction of cinnamic acid derivatives can at this point be illustrated by the example of caffeic acid (78).

HO
IO⟨⟩—CH=CH—COOH ⟶ HO—⟨⟩—CH$_2$—CH$_2$—COOH
HO

(78)
Caffeic acid   Dihydrocaffeic acid

Another double bond reduction carried out by intestinal bacteria is that observed with the 1-methylene group in some pyrrolizidine derivatives. As described in Chapter 9, Section VII, some pyrrolizidine alkaloids are metabolized in the sheep rumen to *l*-goreensine (79) which is further reduced to the corresponding 1-methyl derivative (80). Similarly, the vinyl group of 4-vinylcatechol (81), a minor metabolite of caffeic acid, undergoes

(79)
l-Goreensine   (80)

HO
HO—⟨⟩—CH=CH$_2$ ⟶ HO—⟨⟩—CH$_2$—Me
HO

(81)
4-Vinylcatechol   4-Ethylcatechol

reduction. The cleavage of the heterocyclic ring in coumarin and some of its derivatives was noted above to involve an initial double bond reduction.

Examples of the reduction of unsaturated monocyclic terpenes have also been described. These are mentioned above in the section on tissue reactions, however the site of reduction of these compounds has not been ascertained and it is quite possible that the gut microflora may also be involved in these reactions.

2. Alcohol Reduction

Rat intestinal microorganisms are capable of reducing certain benzyl alcohol derivatives to the corresponding hydrocarbons (Scheline, 1972a). This conversion is illustrated by the formation of 4-methylguaiacol (82) from vanillyl alcohol (81). While the structural requirements of this reduction

MeO   MeO

HO—⟨ring⟩—CH_2OH → HO—⟨ring⟩—Me

(81)   (82)

Vanillyl alcohol   4-Methylguaiacol

have not been studied in detail, it appears that the presence of a free p-hydroxyl group is essential. It should be noted that this group may sometimes be formed metabolically. Whether or not this type of reduction occurs with other types of alcohols remains to be seen, however homovanillyl alcohol which is a phenylethanol derivative and the next higher homologue of vanillyl alcohol was not reduced to 4-ethylguaiacol when similarly incubated anaerobically with the rat intestinal microflora (Scheline, 1972b). As described in Chapter 3, Section I.C, the administration of vanillyl alcohol to rats results in the urinary excretion of some 4-methylguaiacol and its O-demethylated derivative, 4-methylcatechol, thus showing that this bacterial reduction has significance in vivo.

3. Aldehyde Reduction

Similar to that noted above with some benzyl alcohol derivatives, several benzaldehyde derivatives were shown to be reduced by the gut microflora (Scheline, 1972a). The reduction of salicylaldehyde (83) illustrates this reaction. This example additionally illustrates that a free p-hydroxyl group is not essential for the reduction as is the case with the related benzyl alcohols. In fact, aldehyde reduction was observed in compounds containing one or more hydroxyl and/or methoxyl groups at various ring positions. Reduction of the unsubstituted parent compound benzaldehyde and some

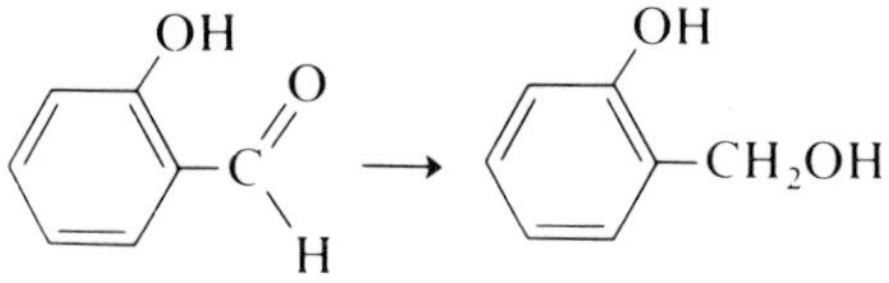

(83)
Salicylaldehyde   Salicyl alcohol

of its *p*-alkyl derivatives also occurs under these conditions (Scheline, 1972b).

4. *Ketone Reduction*

The reduction of plant ketones by the gut microflora has been only superficially studied and the possible importance of this reaction is difficult to assess. Scheline (1973) reported some preliminary findings which suggested that neither acetovanillone (84) nor zingerone (50) were reduced

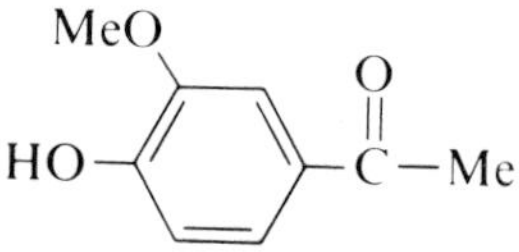

(84)
Acetovanillone

to the corresponding carbinols when incubated anaerobically with mixed cultures of rat caecal microorganisms. However, a subsequent detailed investigation of the metabolism of zingerone in rats demonstrated that its reduction to the corresponding carbinol as well as to smaller amounts of the *O*-demethylated carbinol can be carried out *in vitro* by the gut microflora (Monge *et al.*, 1976). Additional data nevertheless indicated that this intestinal reduction of zingerone furnishes only a minor amount of the total carbinol produced when the ketone is administered to rats.

In addition to the bacterial reduction of a ketone to a carbinol, complete reduction to a hydrocarbon is also possible. This transformation is observed with the isoflavonoid daidzein (75) which is reduced to equol (85) when incubated with microorganisms from sheep rumen fluid or rat caecal contents (see Chapter 7, Section IV.G).

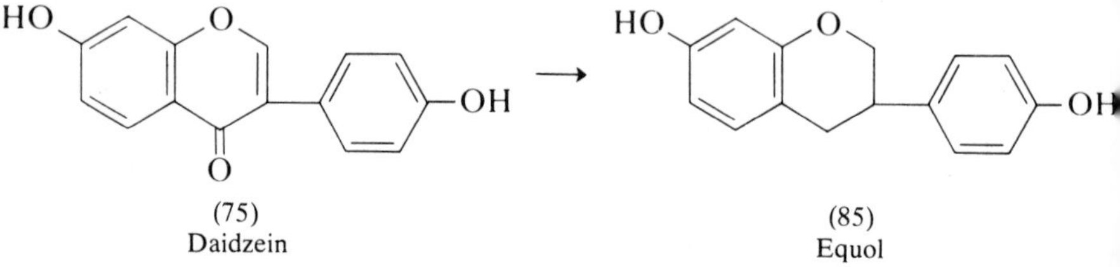

(75)
Daidzein

(85)
Equol

5. N-Oxide Reduction

Some alkaloids occur naturally as N-oxide derivatives. This situation is perhaps best exemplified by the pyrrolizidine alkaloids which include many such oxidized products. Reduction of the N-oxide to the parent alkaloid is possible, as illustrated by the conversion of heliotrine-N-oxide (86) to heliotrine (33) when incubated anaerobically with microorganisms from sheep rumen fluid.

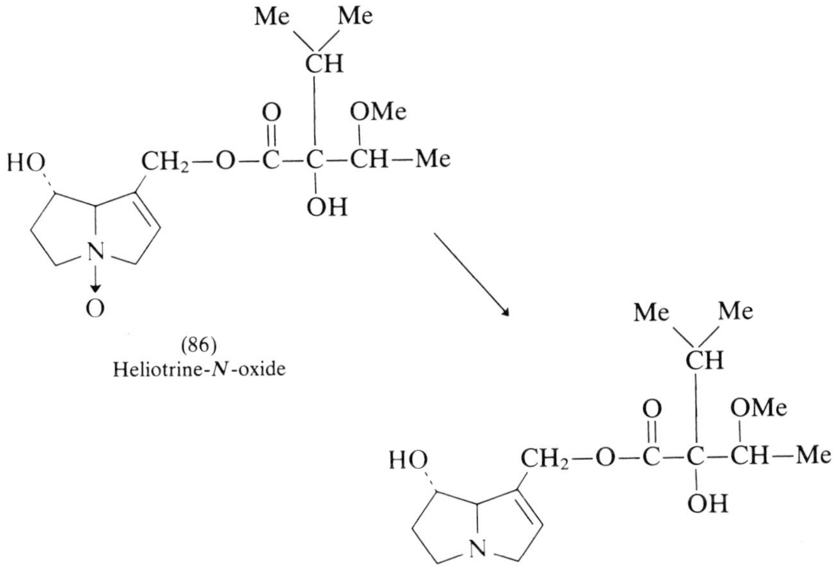

(86)
Heliotrine-N-oxide

(33)
Heliotrine

Numerous nitrogen-containing xenobiotics including some plant compounds are known to be converted in the animal body to N-oxides (see Jenner, 1971). This is true of nicotine and, as described in Chapter 9, Section II.B, its N-oxide derivative is reduced to nicotine by intestinal microorganisms.

6. Sulphide and Sulphoxide Reduction

The occurrence of these reactions has been little studied, however it is known that both S-methylcysteine (88) and its sulphoxide (87) are reduced to volatile sulphur compounds by the rumen microflora. As described in Chapter 10, Section V, it is probable that the sulphoxide is first converted to the sulphide after which methanethiol (89), dimethyl sulphide (90) and dimethyl disulphide (91) are formed.

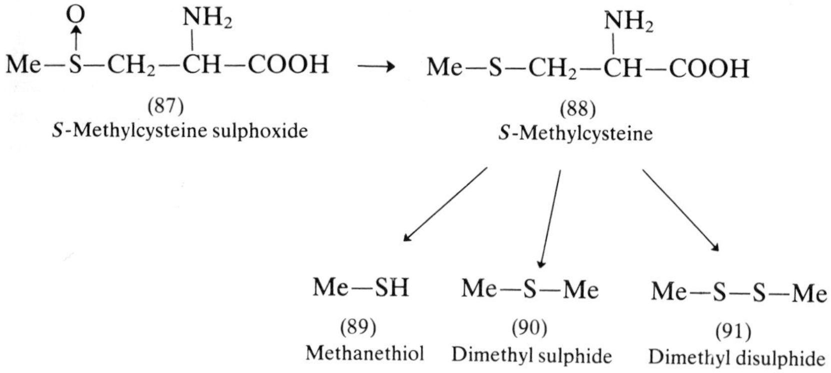

$$\underset{\substack{(87)\\ \textit{S}\text{-Methylcysteine sulphoxide}}}{\overset{\overset{\displaystyle O}{\uparrow}}{Me-S}-CH_2-\overset{\overset{\displaystyle NH_2}{|}}{CH}-COOH} \longrightarrow \underset{\substack{(88)\\ \textit{S}\text{-Methylcysteine}}}{Me-S-CH_2-\overset{\overset{\displaystyle NH_2}{|}}{CH}-COOH}$$

$$\underset{\substack{(89)\\ \text{Methanethiol}}}{Me-SH} \qquad \underset{\substack{(90)\\ \text{Dimethyl sulphide}}}{Me-S-Me} \qquad \underset{\substack{(91)\\ \text{Dimethyl disulphide}}}{Me-S-S-Me}$$

7. Aromatization

A few cyclic plant acids including quinic acid (92) undergo conversion to aromatic products when fed to animals. Inclusion of this reaction in the present section is somewhat arbitrary since the initial metabolic steps are not fully understood. It is here that metabolism by the gut microflora

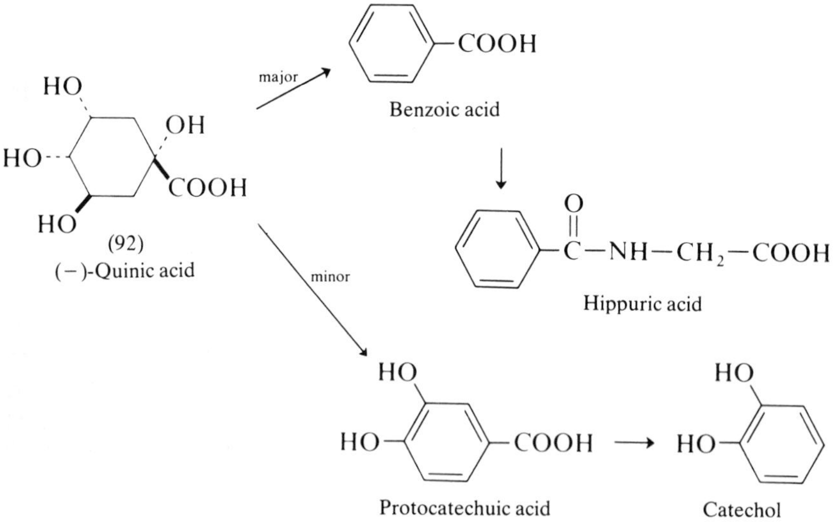

(92)
(−)-Quinic acid

Benzoic acid

Hippuric acid

Protocatechuic acid   Catechol

occurs and the final conjugative step is clearly a tissue reaction. The extent to which the bacteria are involved in the sequence leading to benzoic acid is not entirely clear but, as discussed in Chapter 5, Section I.B, some evidence has been obtained indicating that aromatic compounds (e.g. benzoic acid and catechol) are formed bacterially. However, studies with the closely related shikimic acid (93) showed that the bacteria are responsible for the replacement of the hydroxyl groups with hydrogen, commonly defined as a reductive step, and the reduction of the 1,2-double bond to give cyclohexane carboxylic acid (37). The subsequent aromatization to benzoic acid was shown to be carried out by tissue enzymes.

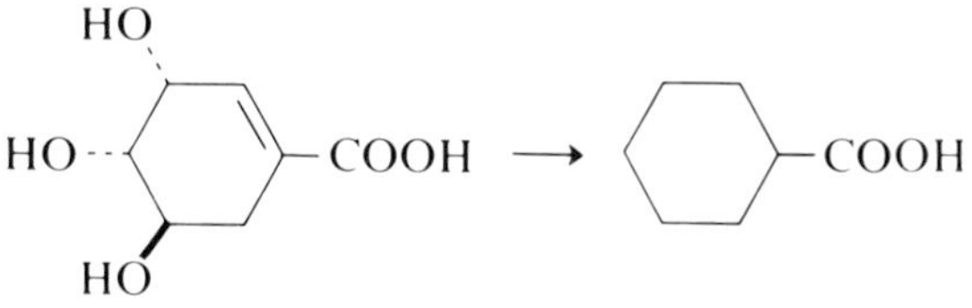

(93)	(37)
(−)-Shikimic acid	Cyclohexanecarboxylic acid

8. *Ester Group Fission*

Examples of this reaction are seen with the pyrrolizidine alkaloids heliotrine and lasiocarpine which are converted by sheep rumen microorganisms to 1-methylene derivatives of pyrrolizidine. This degradation of lasiocarpine is illustrated above in Part A.2. of this section which deals with the bacterial hydrolysis of plant esters.

References

Adamson, R. H., Bridges, J. W., Evans, M. E. and Williams, R. T. (1970). *Biochem. J.* **116**, 437–443.

Arias, I. M. and Jakoby, W. B. (Eds) (1976). "Glutathione: Metabolism and Function". Raven Press, New York.

Axelrod, J. (1971). *In* "Handbook of Experimental Pharmacology. Concepts in Biochemical Pharmacology" (B. B. Brodie and J. R. Gillette, Eds), Vol. XXVIII/2, pp. 609–619. Springer-Verlag, Berlin, Heidelberg, New York.

Beckett, A. H. (1977). *In* "Drug Metabolism—from Microbe to Man" (D. V. Parke and R. L. Smith, Eds), pp. 33–42. Taylor and Francis, London.

Bickel, M. H. (1969). *Pharmac. Rev.* **21**, 325–355.

Booth, A. N. and Williams, R. T. (1963). *Biochem. J.* **88**, 66P–67P.

Boyland, E. (1971). *In* "Handbook of Experimental Pharmacology. Concepts in Biochemical Pharmacology" (B. B. Brodie and J. R. Gillette, Eds), Vol. XXVIII/2, pp. 584–608. Springer-Verlag, Berlin, Heidelberg, New York.

Bridges, J. W., Gorrod, J. W. and Parke, D. V. (Eds) (1971). "Proceedings of the Symposium on Biological Oxidation of Nitrogen in Organic Molecules". *Xenobiotica* **1**, 313–571.

Chasseaud, L. F. (1973). *Drug Metab. Rev.* **2**, 185–220.

Chasseaud, L. F. (1976). *In* "Glutathione: Metabolism and Function" (I. M. Arias and W. B. Jakoby, Eds), pp. 77–114. Raven Press, New York.

Curtius, H. C., Mettler, M. and Ettlinger, L. (1976). *J. Chromat.* **126**, 569–580.

Daly, J. (1971). *In* "Handbook of Experimental Pharmacology. Concepts in Biochemical Pharmacology" (B. B. Brodie and J. R. Gillette, Eds), Vol. XXVIII/2, pp. 285–311. Springer-Verlag, Berlin, Heidelberg, New York.

Distefano, V. and Borgstedt, H. H. (1964). *Science* N.Y. **144**, 1137–1138.

Dodgson, K. S. (1977). *In* "Drug Metabolism—from Microbe to Man" (D. V. Parke and R. L. Smith, Eds), pp. 91–104. Taylor and Francis, London.

Drasar, B. S. and Hill, M. J. (1974). "Human Intestinal Flora" pp. 54–71. Academic Press, London, New York, San Francisco.

Dutton, G. J. (1966). *In* "Glucuronic Acid, Free and Combined" (G. J. Dutton, Ed.), pp. 185–299. Academic Press, New York and London.

Dutton, G. J. (1971). *In* "Handbook of Experimental Pharmacology. Concepts in Biochemical Pharmacology" (B. B. Brodie and J. R. Gillette, Eds), Vol. XXVIII/2, pp. 378–400. Springer-Verlag, Berlin, Heidelberg, New York.

Dutton, G. J., Wishart, G. J., Leakey, J. E. A. and Goheer, M. A. (1977). *In* "Drug Metabolism—from Microbe to Man" (D. V. Parke and R. L. Smith, Eds), pp. 71–90. Taylor and Francis, London.

Estabrook, R. W. (1971). *In* "Handbook of Experimental Pharmacology. Concepts in Biochemical Pharmacology" (B. B. Brodie and J. R. Gillette, Eds), Vol. XXVIII/2, pp. 264–284. Springer-Verlag, Berlin, Heidelberg, New York.

Fahelbum, I. M. S. and James, S. P. (1977). *Toxicology* **7**, 123–132.

Fischer, L. J. and Weissinger, J. L. (1972). *Xenobiotica* **2**, 399–412.

Fischer, L. J., Kent, T. H. and Weissinger, J. L. (1973). *J. Pharmac. exp. Ther.* **185**, 163–170.

Gram, T. E. (1971). *In* "Handbook of Experimental Pharmacology. Concepts in Biochemical Pharmacology" (B. B. Brodie and J. R. Gillette, Eds), Vol. XXVIII/2, pp. 334–348. Springer-Verlag, Berlin, Heidelberg, New York.

Griffiths, L. A. and Smith, G. E. (1972a). *Biochem. J.* **128**, 901–911.

Griffiths, L. A. and Smith, G. E. (1972b). *Biochem. J.* **130**, 141–151.

Grover, P. L. (1977). *In* "Drug Metabolism—from Microbe to Man" (D. V. Parke and R. L. Smith, Eds), pp. 105–122. Taylor and Francis, London.

Habig, W. H., Kamisaka, K., Ketley, J. N., Pabst, M. J., Arias, I. M. and Jakoby, W. B. (1976). *In* "Glutathione: Metabolism and Function" (I. M. Arias and W. B. Jakoby, Eds), pp. 225–231. Raven Press, New York.

Hawksworth, G., Drasar, B. S. and Hill, M. J. (1971). *J. med. Microbiol.* **4**, 451–459.

Hirom, P. C., Idle, J. R. and Millburn, P. (1977). *In* "Drug Metabolism—from Microbe to Man" (D. V. Parke and R. L. Smith, Eds), pp. 299–329. Taylor and Francis, London.

Hutson, D. H. (1970). *In* "Foreign Compound Metabolism in Mammals" (D. E. Hathway, Ed), Vol. 1, pp. 314–395. Chemical Society, London.

Hutson, D. H. (1972). *In* "Foreign Compound Metabolism in Mammals" (D. E. Hathway, Ed.), Vol. 2, pp. 328–397. Chemical Society, London.

Hutson, D. H. (1975). *In* "Foreign Compound Metabolism in Mammals" (D. E. Hathway, Ed.), Vol. 3, pp. 449–549. Chemical Society, London.

Hutson, D. H. (1977). *In* "Foreign Compound Metabolism in Mammals" (D. E. Hathway, Ed.), Vol. 4, pp. 259–346. Chemical Society, London.

Jakoby, W. B., Habig, W. H., Keen, J. H., Ketley, J. N. and Pabst, M. J. (1976). *In* "Glutathione: Metabolism and Function" (I. M. Arias and W. B. Jakoby, Eds), pp. 189–202. Raven Press, New York.

Jenner, P. (1971). *Xenobiotica* **1**, 399–418.

Jerina, D. M. and Daly, J. W. (1974). *Science* N.Y. *185*, 573–582.

Jerina, D. M. and Daly, J. W. (1977). *In* "Drug Metabolism—From Microbe to Man" (D. V. Parke and R. L. Smith, Eds), pp. 13–32. Taylor and Francis, London.

Kent, T. H., Fischer, L. J. and Marr, R. (1972). *Proc. Soc. exp. Biol. Med.* *140*, 590–594.

Klaassen, C. D. (1975). *In* "Critical Reviews in Toxicology" (L. Golberg, Ed.), Vol. 4, pp. 1–29. CRC Press, Cleveland.

Kobayashi, A. and Matsumoto, H. (1965). *Archs Biochem. Biophys.* *110*, 373–380.

La Du, B. N. and Snady, H. (1971). *In* "Handbook of Experimental Pharmacology. Concepts in Biochemical Pharmacology" (B. B. Brodie and J. R. Gillette, Eds), Vol. XXVIII/2, pp. 477–499. Springer-Verlag, Berlin, Heidelberg, New York.

Laqueur, G. L. (1964). *Fedn Proc. Fedn Am. Socs exp. Biol.* *23*, 1386–1388.

McMahon, R. E. (1971). *In* "Handbook of Experimental Pharmacology. Concepts in Biochemical Pharmacology" (B. B. Brodie and J. R. Gillette, Eds), Vol. XXVIII/2, pp. 500–517. Springer-Verlag, Berlin, Heidelberg, New York.

Masri, M. S., Booth, A. N. and DeEds, F. (1962), *Biochim. biophys. Acta* *65*, 495–505.

Monge, P., Scheline, R. and Solheim, E. (1976). *Xenobiotica* *6*, 411–423.

Oesch, F. (1973). *Xenobiotica* *3*, 305–340.

Oesch, F., Glatt, H. and Schmassmann, H. (1977). *Biochem. Pharmac.* *26*, 603–607.

Parke, D. V. (1968). "The Biochemistry of Foreign Compounds", p. 60. Pergamon Press, Oxford, London, Edinburgh, New York, Toronto, Sydney, Paris, Braunschweig.

Ranganathan, S. and Ramasarma, T. (1974). *Biochem. J.* *140*, 517–522.

Renwick, A. G. (1977). *In* "Drug Metabolism—from Microbe to Man" (D. V. Parke and R. L. Smith, eds), pp. 169–189. Taylor and Francis, London.

Roy, A. B. (1971). *In* "Handbook of Experimental Pharmacology. Concepts in Biochemical Pharmacology" (B. B. Brodie and J. R. Gillette, Eds), Vol. XXVIII/2, pp. 536–563, Springer-Verlag, Berlin, Heidelberg, New York.

Scheline, R. R. (1968a). *Acta pharmac. tox.* *26*, 325–331.

Scheline, R. R. (1968b). *Acta pharmac. tox.* *26*, 332–342.

Scheline, R. R. (1970). *Biochim. biophys. Acta* *222*, 228–230.

Scheline, R. R. (1972a). *Xenobiotica* *2*, 227–236.

Scheline, R. R. (1972b). Unpublished results.

Scheline, R. R. (1973). *Pharmac. Rev.* *25*, 451–523.

Scheline, R. R. (1978). *In* "Extrahepatic Metabolism of Drugs and Other Foreign Compounds" (T. E. Gram, Ed.). Spectrum Publications. Jamaica, N.Y.

Scheline, R. R. and Midtvedt, T. (1970). *Experientia* *26*, 1068.

Shiobara, Y. (1977). *Xenobiotica* *7*, 457–468.

Smith, G. E. and Griffiths, L. A. (1974). *Xenobiotica* *4*, 477–487.

Smith, R. L. (1973). "The Excretory Function of the Bile". Chapman and Hall, London.

Smith, R. L. and Caldwell, J. (1977). *In* "Drug Metabolism—from Microbe to Man" (D. V. Parke and R. L. Smith, Eds), pp. 331–356. Taylor and Francis, London.

Smith, R. L. and Williams, R. T. (1966). *In* "Glucuronic Acid, Free and Combined" (G. J. Dutton, Ed.), pp. 457–491. Academic Press, New York and London.

Spatz, M., McDaniel, E. G. and Laqueur, G. L. (1966). *Proc. Soc. exp. Biol. Med.* **121**, 417–422.

Spatz, M., Smith, D. W. E., McDaniel, E. G. and Laqueur, G. L. (1967). *Proc. Soc. exp. Biol. Med.* **124**, 691–697.

Strand, L. P. and Scheline, R. R. (1975). *Xenobiotica* **5**, 49–63.

Weber, W. W. (1971). *In* "Handbook of Experimental Pharmacology. Concepts in Biochemical Pharmacology" (B. B. Brodie and J. R. Gillette, Eds), Vol. XXVIII/2, pp. 564–583. Springer-Verlag, Berlin, Heidelberg, New York.

Weisburger, J. H. and Weisburger, E. K. (1971). *In* "Handbook of Experimental Pharmacology. Concepts in Biochemical Pharmacology" (B. B. Brodie and J. R. Gillette, Eds), Vol. XXVIII/2, pp. 312–333. Springer-Verlag, Berlin, Heidelberg, New York.

Weisburger, J. H. and Weisburger, E. K. (1973). *Pharmac. Rev.* **25**, 1–66.

Williams, R. T. (1959). "Detoxication Mechanisms". Chapman and Hall, London.

Yang, M. G. and Mickelsen, O. (1969). *In* "Toxic Constituents of Plant Foodstuffs" (I. E. Liener, Ed.), pp. 159–167. Academic Press, New York and London.

Zeller, E. A. (1971). *In* "Handbook of Experimental Pharmacology. Concepts in Biochemical Pharmacology" (B. B. Brodie and J. R. Gillette, Eds), Vol. XXVIII/2, pp. 518–535. Springer-Verlag, Berlin, Heidelberg, New York.

2

METABOLISM OF HYDROCARBONS

I. Aliphatic Hydrocarbons

The aliphatic hydrocarbons most commonly found in plants are long-chain n-alkanes with the general formula $Me-(CH_2)_n-Me$. The lowest natural member of this group is n-heptane which is especially abundant in the terpentines obtained from various species of *Pinus*. More commonly, these n-alkanes are compounds with chain lengths ranging from C_{25} to C_{35} and it is noteworthy that the odd numbered members are much more abundant than are the even numbered. This fact is apparently related to their pathway of synthesis which involves decarboxylation of fatty acids with even numbers of carbon atoms. Branched hydrocarbons are also found, usually in small quantities. Branching commonly occurs near the end of the chain (methyl group in the ω-1 or ω-2 positions). Of course, many branched-chain hydrocarbons are acyclic terpenoids which are based on the isoprene molecule and which are therefore $(C_5)_n$-compounds. Although there is no reason to believe that the pathways and mechanisms involved in the mammalian metabolism of these terpenoid compounds differ significantly from those involved in the metabolism of other compounds containing similar chemical groups, examples of terpenoid metabolism are grouped separately in this book. A further type of aliphatic hydrocarbon from plants includes compounds containing unsaturated groups. Olefinic groups are most common, however a few hydrocarbons containing acetylenic groups are also known.

Considerable interest has been devoted recently to the mammalian metabolism of several n-alkanes, with the result that the reaction products and the mechanisms by which they are formed are well understood in some cases. The report by McCarthy (1964) appears to have been influential in directing attention to the possibility of n-alkane metabolism. This study showed that both **n-hexadecane** ($Me-(CH_2)_{14}-Me$) and n-octadecane ($Me-(CH_2)_{16}-Me$) were converted by ω-oxidation in rats and goats to fatty acids of the same chain lengths. These products could then undergo further reactions including alterations in chain length or incorporation into lipid fractions, especially liver phospholipids. The metabolism of

n-octadecane in rats was confirmed by Popović (1970). Two hours after administration of the $[1-{}^{14}C]$-labelled compound (200 mg/kg, p.o. or i.v.), most of the radioactivity was found in the liver, especially as fatty acids of lecithin. Respiratory ${}^{14}CO_2$ was also detected during this period. The localization of the enzymes metabolizing n-hexadecane and the mechanisms involved have also been investigated. Mitchell and Hübscher (1968) reported that it was metabolized via cetyl alcohol (n-hexadecanol) to palmitic acid (n-hexadecanoic acid) by the mucosa of guinea pig small intestine. Activity was located mainly in the microsomal fraction. These same reaction products were shown by Kusunose et al. (1969) to be formed by the microsomal fraction from mouse liver. Similar fractions from mouse lung or kidney showed little or no activity, respectively. This investigation indicated that the ω-hydroxylation of n-hexadecane is catalysed by a mono-oxygenase system requiring NADPH and oxygen.

Several studies dealing with the enzymic metabolism of lower alkanes, especially **n-heptane**, have also been reported. Das et al. (1968) studied the hydroxylation of n-heptane by rat liver microsomes and concluded that this reaction and those of laurate hydroxylation, amino pyrine demethylation and testosterone hydroxylation proceed by way of a common microsomal hydroxylating system involving NADPH-cytochrome c reductase and cytochrome P-450. This microsomal system was further studied by Frommer et al. (1972) who found that all four isomeric alcohols were formed. In these in vitro experiments, ω-1 hydroxylation was the main reaction and the relative amounts of 2-, 1-, 3- and 4-heptanol formed were approximately $27:5:4:1$. This pattern of multiple product formation was also found with several lower alkanes including some having branched chains (Frommer et al., 1970). Recently, Di Vincenzo et al. (1976) reported that ω-1 oxidation occurs with n-hexane in guinea pigs. Following a dose of 250 mg/kg (i.p.), the serum contained both 5-hydroxy-2-hexanone and 2,5-hexanedione. It seems reasonable to assume that this pathway, involving symmetrical ω-1 oxidations, may also be operative with the higher n-alkanes. In the experiments with n-heptane (Frommer et al., 1972), the results obtained with preparations from rats treated with inducing agents or with inhibitors of the mono-oxygenase system indicated that at least three activities are involved in its microsomal hydroxylation. Similar results were obtained with n-hexane (Frommer et al., 1974). Lu et al. (1970) reported on the properties of a solubilized form of the liver microsomal system which hydroxylated a series of n-alkanes from C_6 to C_{16}. Further information on the microsomal hydroxylation of alkanes of low chain length has been presented by Ichihara et al. (1969). Preparations from livers of mice, rats, rabbits and cattle oxidized hexane, octane and decane and the data indicated that the last compound was metabolized via decanol to decanoic acid and decamethyleneglycol.

The above results indicate that numerous normal and branched alkanes ranging from the lower homologues to the C_{16} and C_{18} compounds undergo oxidation, initially to alcohols. It seems likely that this general reaction should also be involved in the metabolism of the higher homologues of plant alkanes.

Very little is known of the metabolism of higher plant olefins and acetylenes. Neubauer (1901) reported that rabbits given **n-1-octene** (1-octylene) $(Me-(CH_2)_5-CH=CH_2)$ excrete material conjugated with glucuronic acid. This finding indicates that oxidation of the alkene has taken place. Recently, the metabolism of two octene isomers, n-1-octene and n-4-octene, was studied using rat liver microsomes and NADPH in the presence of oxygen (Maynert et al., 1970). These compounds were oxidized to epoxides which were then hydrolysed by epoxide hydratase to n-octane-1,2-diol and n-octane-4,5-diol, respectively. A similar metabolic sequence was found by Watabe and Yamada (1975) with the synthetic homologue, n-1-hexadecene $(Me-(CH_2)_{13}-CH=CH_2)$.

II. Monoterpenoid Hydrocarbons

A. ACYCLIC TERPENE HYDROCARBONS

This is a fairly limited group of C_{10} olefinic compounds, the best known being **myrcene** (1). Unfortunately, metabolic data are lacking for this group although a closely related and partially reduced derivative, 2,6-dimethyl-2,6-octadiene (dihydromyrcene) (2), was studied by Kuhn et al. (1936). When rabbits were given 25 g of compound (2) (2×5 g daily, p.o.), 1·5 g of the optically inactive "Hildebrandt acid" (3) was isolated from the urine as the only detected metabolite. It seems evident that the pathway leading to this dicarboxylic acid initiates with allylic hydroxylation at one of

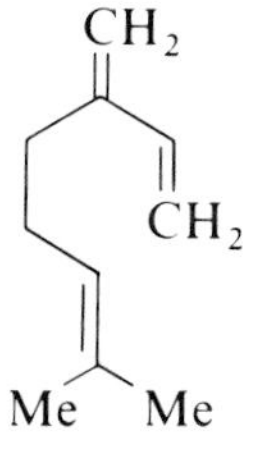

(1)
Myrcene

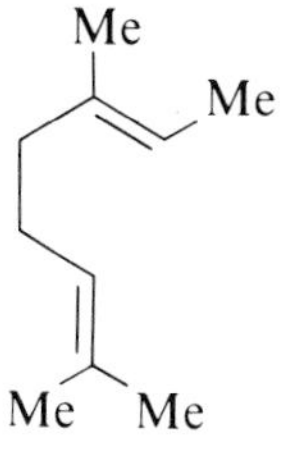

(2)

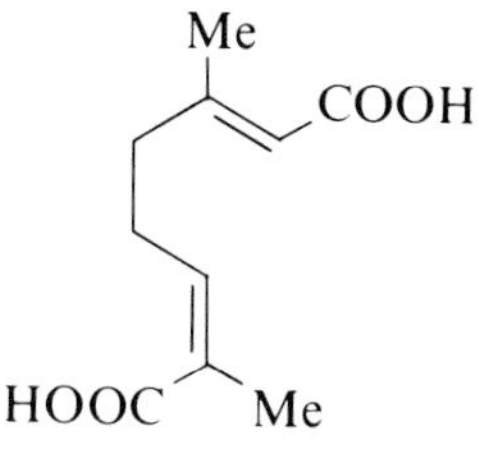

(3)

the two terminal allylic moieties. It also seems likely that much of the 90–95% of the dose unaccounted for may consist of intermediate products, especially alcohols which will be excreted mainly as glucuronide conjugates rather than in the free form as is the case with the dicarboxylic acid metabolite.

B. Monocyclic Terpene Hydrocarbons

Only three compounds from this group have been studied. These are p-menthene (4), α-phellandrene (p-mentha-1,5-diene) (5) and limonene (p-mentha-1,8-diene) (6). Most of the reports on these compounds are of older date with the result that only fragmentary knowledge is available. However, the metabolism of the last compound has recently been extensively investigated in a number of mammalian species.

(4)	(5)	(6)
p-Menthene	α–Phellandrene	Limonene

Neubauer (1901) administered **p-menthene** (4) (0·8 g/kg, p.o.) to rabbits and reported the urinary excretion of material conjugated with glucuronic acid. In similar experiments Hämäläinen (1912) found no evidence for reduction of the double bond but showed that hydroxylation and glucuronidation occurred. Oxidation was reported to take place in the position *ortho* to the methyl group, giving either p-menthen-2-ol or p-menthen-6-ol.

α-Phellandrene (5) differs from p-menthene in the number and position of the double bonds, a difference which appears to have considerable bearing on its metabolic fate. Fromm and Hildebrandt (1901) reported that rabbits given α-phellandrene excreted in the urine a glucuronide material which, on acid hydrolysis, gave the aromatic compound p-cymene (7) and a phenolic compound. Later studies employed sheep and Harvey (1942) found that acid-hydrolysed urines contained p-cymene, carvotanacetone (8), phellandric acid (9) and possibly the corresponding aldehyde, phellandral. Harvey felt that the phenolic metabolite reported by Fromm and Hildebrandt may have been due to an impurity in the sample of α-phellandrene used and, in fact, the two samples employed by Harvey were only 80 and 95% pure. Wright (1945) isolated the glycine conjugate of phellandric acid, termed phellanduric acid (10), from the urine of sheep dosed with α-phellandrine. The urine also contained material conjugated with glucuronic acid which, in agreement with the earlier study by Harvey, was converted to p-cymene and carvotanacetone upon acid hydrolysis. These results indicate that α-phellandrene is metabolized along at least two pathways, one of which involves both reduction of a ring double bond and oxidation of the methyl group to give phellandric acid (9) and its glycine conjugate, phellanduric acid (10). The other pathway involves ring hydroxylation to give a polyol which is excreted as a glucuronide conjugate. This conjugate, upon treatment with hot acid, is hydrolysed and then dehydrated. Loss of two or three molecules of water produces carvotanacetone and p-cymene, respectively. Based on an analogy with the known behaviour of other p-menthane triols, Wright postulated that the polyol is p-menthane-2,3,6-triol (11).

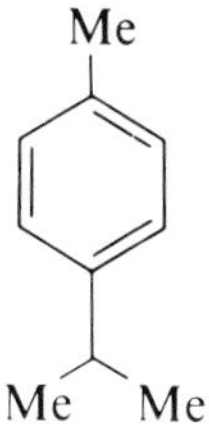

(7)

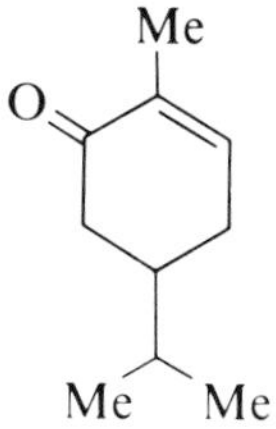

(8)

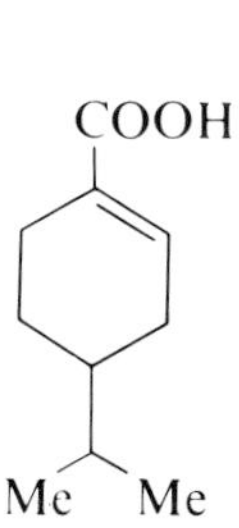

(9)

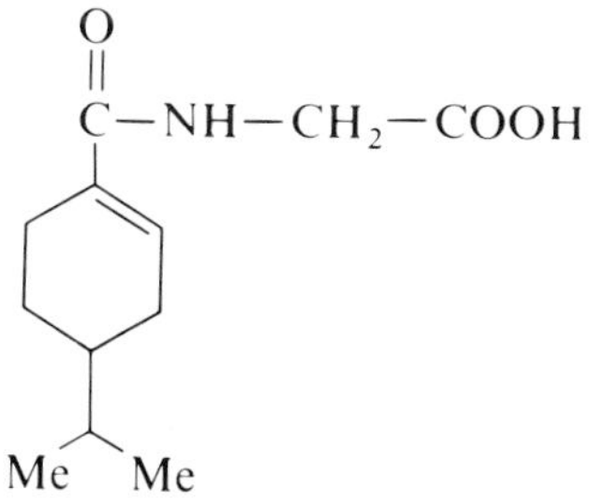

(10)

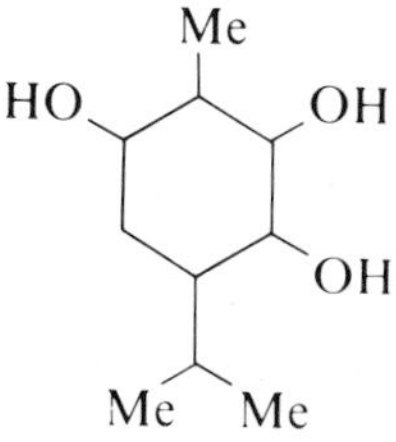

(11)

The metabolism of **limonene** (6) was the subject of an early investigation by Hildebrandt (1902) who reported that rabbits oxidized the methyl group at C7 to a carboxyl group and hydroxylated the molecule at some other position to form a metabolite which was excreted as a glucuronic acid conjugate. No further reports on limonene metabolism appeared for a period of more than 60 years. Then Wade *et al.* (1966) discovered that human urine contained, in conjugated form, a monoterpenoid compound named uroterpenol. It was shown to be p-menth-1-ene-8,9-diol (see Fig. 2.1, Structure (21)) and the evidence obtained indicated that it originated from dietary limonene. While some uroterpenol was excreted in the free form, most appeared to be conjugated with glucuronic acid, an assumption which was subsequently confirmed (Dean et al., 1967). Smith et al. (1969) reported large increases of uroterpenol excretion over normal values following ingestion by man of limonene or limonene-containing products.

Recently, several extensive investigations of the metabolism of d-limonene have been carried out. Some of these employed the [9-^{14}C]-labelled compound and studied its absorption, distribution and excretion in rats (Igimi et al., 1974), its metabolism in rabbits (Kodama et al., 1974) and its metabolism in six species including man (Kodama et al., 1976). The metabolic fate of d-limonene in rats has also been studied by Regan and Bjeldanes (1976). These exemplary studies, employing a range of modern analytical techniques, clearly show that multiple pathways of metabolism are involved. Oxidation at C7 to form the carboxylic acid reported by Hildebrandt is regularly seen and the formation of metabolites with hydroxyl groups at one or more of five different aliphatic and alicyclic sites was shown to occur. It seems reasonable to believe that similar multiple pathways of metabolism for the other terpene hydrocarbons also exist and that the earlier studies, employing only a few animal species, have but scratched the surface of this subject.

The studies of Igimi et al. (1974) and Kodama et al. (1974, 1976) showed that urinary excretion of metabolites was far more important than faecal excretion and that no unchanged compound was detected in the urine in the species studied (rat, guinea pig, hamster, rabbit, dog and man). Most of the radioactivity appeared in the urine within 24 h and 75–95% was recovered in 2–3 days. Less than 10% of the dose was excreted in the faeces. These experiments employed rather high doses of 800 mg/kg p.o. in all species except dogs (400 mg/kg) and man (27 mg/kg). Igimi et al. (1974) found that loss of radioactivity as $^{14}CO_2$ was insignificant as less than 2% of the dose was recovered in this form in rats. Biliary excretion, in the rat at least, accounts for a fairly large portion of the radioactivity (25% of the dose), primarily as the glucuronide of uroterpenol (metabolite (22)).

The general patterns of d-limonene metabolism found with the ^{14}C-labelled compound are similar in all of the six species studied, however the quantitative relationships vary considerably among them. This quantitative data on metabolite excretion (Kodama et al., 1976) is summarized in Table 2.1. These values show that ring hydroxylation reactions at C2 to give metabolite (15) or at C6 to give metabolites (19) and (20) are usually minor reactions. However, Regan and Bjeldanes (1976) reported that about 16% of the unconjugated metabolites of d-limonene in rats corresponded to

TABLE 2.1

Species differences in $[^{14}C]$-d-limonene metabolism[a] (mean % of dose in 48 h urine)

Metabolite	Structure (see Fig. 2.1.)	Rat (3)[b]	Guinea pig (3)	Hamster (4)	Rabbit (3)	Dog (2)	Man (2)
p-Mentha-1,8-dien-10-ol	17		0.3	0.3	3.6	2.1	0.1
p-Mentha-1,8-dien-10-yl-β-D-glucopyranosiduronic acid	18	1.7	3.1	2.3	3.4	0.3	1.0
p-Mentha-1,8-dien-6-ol	19	0.9	0.1	0.1	0.4	0.2	—
p-Menth-1-ene-6,8,9-triol	20	tr[c]	tr	tr	0.1	0.2	3.4
p-Menth-1-ene-8,9-diol (uroterpenol)	21	3.5	1.3	11.3	9.6	28.9	1.3
8-Hydroxy-p-menth-1-en-9-yl-β-D-glucopyranosiduronic acid	22	3.2	27.7	7.9	6.8	2.4	27.4
Perillic acid	12	5.1	5.3	7.6	5.1	10.4	2.4
Perillylglycine	13	8.1	20.5	1.6	2.2	1.4	0.4
Perillyl-β-D-glucopyranosiduronic acid	14	1.8	4.2	22.4	3.5	0.2	2.9
2-Hydroxy-p-menth-8-en-7-oic acid	15	6.7	1.1	2.5	2.7	4.6	1.1
Perillic acid-8,9-diol	16	26.6	0.7	9.5	14.5	3.7	2.2
Total		56.3	64.3	65.5	51.9	54.4	42.2
Approx. % of urinary radioactivity identified		70	78	68	61	71	61

[a] Kodama et al. (1976).
[b] Number of animals.
[c] Trace amount.

metabolite (19). Furthermore, this metabolite consisted of a mixture of the 6-α-hydroxy and 6-β-hydroxy diastereoisomers. The reason for the appreciably higher excretion of 6-hydroxylated metabolites in this study is not known, however it may be related to the fact that the urine samples were collected from rats dosed daily for 10 days with d-limonene (400 mg/kg, p.o.), a regimen which may have led to induction of the hepatic mono-oxygenase enzyme systems. A further point of general interest found in Table 2.1 is that formation of one or more of the 8,9-diols, probably via the 8,9-epoxide intermediates, occurred to a much greater extent in all species than did ring oxidation. Pathways leading to or via perillic acid (12) accounted for as much as 85% (rat) and as little as about 20% (man) of the quantitated urinary metabolites. Regan and Bjeldanes (1976) characterized only one 8,9-diol, uroterpenol (21), and reported that about 5% of the unconjugated material was excreted in this form in rats. In addition to confirming the formation of metabolites (19) and (21), the latter study also showed that perillic acid (12) was excreted. A previously unreported metabolite, perillyl alcohol (23) which is an intermediate in the formation of perillic acid, was detected in small amounts. Two additional novel metabolites, a set of diastereoisomers of p-mentha-2,8-dien-1-ol (24), were the most abundant compounds identified. Regan and Bjeldanes proposed that the diastereoisomeric pairs of compounds (19) and (24) which are formed arise from the corresponding pair of the postulated 1,2-epoxide intermediate (25). However, it seems more reasonable to assume that the 6α- and 6β-hydroxy metabolites (19) are formed via allylic hydroxylation of d-limonene. This sequence is, in fact, analogous to that observed with Δ^1-tetrahydrocannabinol which undergoes allylic hydroxylation, giving the 6α- and 6β-hydroxy derivatives. Other clues to the pathways involved in d-limonene metabolism were obtained by Kodama et al. (1976) who administered some of the metabolites to rats or dogs. The results obtained suggest that formation of the 8,9-diol (metabolite (21)) leads to its conjugation with glucuronic acid rather than to further oxidation at C7 or in the ring. On the other hand, initial oxidation at the latter two sites may be followed by formation of the respective 8,9-diols. A general summary of these results is illustrated in Fig. 2.1 which shows the metabolites of d-limonene along with their possible routes of formation.

C. BICYCLIC TERPENE HYDROCARBONS

Several early investigations dealt with the metabolism of some bicyclic terpene hydrocarbons, however the analytical methods then available were inadequate to allow many firm conclusions to be drawn. Fortunately, a few

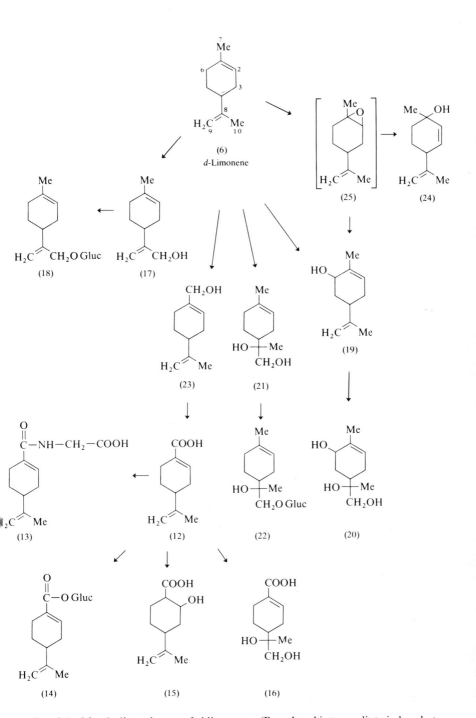

FIG. 2.1. Metabolic pathways of *d*-limonene. (Postulated intermediate in brackets; Gluc = glucuronic acid.)

recent reports on several of these compounds have added useful information, not the least of which being the evidence that the bicyclic structure can be metabolized to monocyclic derivatives.

(26)

α-Pinene

α-Pinene (2-pinene) (26) was reported by Fromm and Hildebrandt (1901) to undergo hydroxylation and conjugation with glucuronic acid in rabbits. Hämäläinen (1912) also found evidence for the hydroxylation of α-pinene in rabbits. Harvey (1942) was able to recover only 3–5% of the dose of α-pinene in the 24 h urine of sheep. This material was found in the neutral fraction and no acidic or phenolic products were detected. Wright (1945) repeated this experiment giving sheep two or three daily doses (10 g) of α-pinene but had similar difficulties. Little metabolic material was recovered, however α-pinene appeared to be metabolized to a glucuronide conjugate which was converted to p-cymene (7) upon heating with acid. Hämäläinen (1912) had previously noted the dehydration of an α-pinene metabolite to p-cymene and believed that the precursor might be a monocyclic triol. Newer data on this interesting reaction of conversion of a bicyclic terpene to a monocyclic derivative are equivocal in the case of α-pinene and, for the reason noted below, are included in the following summary of β-pinene metabolism. The hydroxylation of α-pinene was recently confirmed by Ishida et al. (1977) who demonstrated that it undergoes allylic hydroxylation when fed to rabbits. In this study, the (−)-isomer (firpene), (+)-isomer (australene) and (±)-form of α-pinene were each given in doses of 8–10 g and the neutral fraction of urine following hydrolysis with β-glucuronidase and removal of the acidic and phenolic fractions was studied. In each case the major metabolite was the monohydroxylated derivative 2-pinen-4-ol (verbenol) which results from an endocyclic allylic hydroxylation. Trace amounts of the corresponding exocyclic oxidation product, 2-pinen-10-ol (myrtenol) were also identified. It was also found that verbenol formation was appreciably higher from the (−)- than from the (+)-isomer of α-pinene.

β-Pinene (nopinene, 2(10)-pinene) (27) was included in the study of Hämäläinen (1912) who believed that it underwent hydroxylation at the 3- or 4-position in rabbits. In an experiment identical to that described above

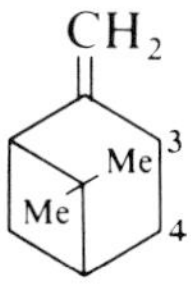

(27)   (28) 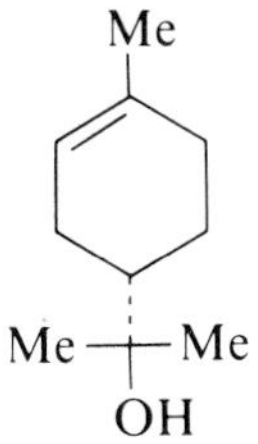   (29) 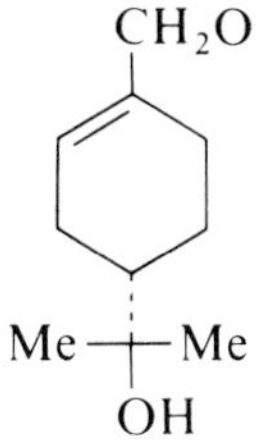   (30)

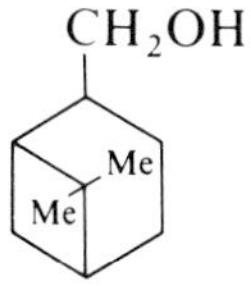

β-Pinene

with α-pinene, Ishida *et al.* (1977) showed that the alcohol formed is *trans*-pinocarveol (2(10)-pinen-3-ol) which is the product of endocyclic allylic hydroxylation. Three other alcohols were identified as urinary metabolites of (−)-β-pinene. The most abundant was *trans*-10-pinanol (28), however the monocyclic derivatives α-terpineol (29) and 1-*p*-menthen-7,8-diol (30) were also identified. The conversion of β-pinene to monocyclic derivatives was first suggested by Hämäläinen (1912) who reported that treatment of the urinary glucuronide material from rabbits with hot acid led to the formation of *p*-cymene. Further information on this subject was obtained by Southwell (1975) using the koala (*Phascolarctos cinereus*) which, however, is a marsupial rather than a mammal. The urine from these animals fed on eucalyptus leaves contained three monoterpene lactones. One of these was shown to be *o*-mentha-1,3-dien-1 → 8-olide (31) and the others were probably *p*-menth-1-en-8 → 3-olide (32) and *p*-menth-1-en-8 → 5-olide (33). The essential oil of eucalyptus leaves contains numerous terpenoid and related compounds, however it was noted that the major constituents are α- and β-pinene, 1,8-cineole and *p*-cymene. The structures of the three lactones show more resemblance to the pinenes than to other compounds and it was therefore suggested that they are metabolites of the pinenes. Whether the lactones arise from one or both of these bicyclic terpenes was not determined. However, the excretion of large amounts of ester glucuronides in koala urine was interpreted to indicate that the lactones identified may arise via cyclization of carboxylic acids which are formed by hydrolysis of urinary glucuronide conjugates.

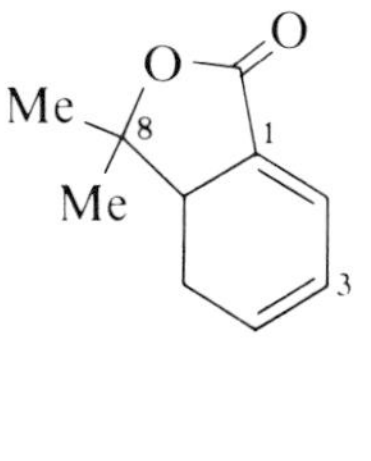

(31) 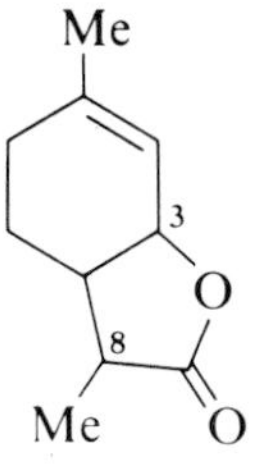   (32) 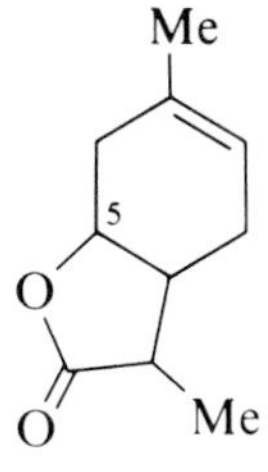   (33)

Hämäläinen (1912) also administered the fully saturated **bornane** (camphane) (34) to rabbits and found the glucuronide of borneol (2-hydroxybornane) in the urine. This was recently confirmed by Robertson and Hussain (1969) and Robertson and Solomon (1971) following the administration of 0·5–1·0 g doses of bornane to rabbits. However, most of the hydroxylated material was shown to be 3-hydroxybornane (epiborneol). While bornane can therefore be hydroxylated at either C2 or C3, Robertson and Solomon reported that hydroxylation at both sites to give the 2,3-diol does not occur. Compounds which would be formed from the diol via dehydrogenation (i.e. ketols) were also shown not to be excreted in the urine.

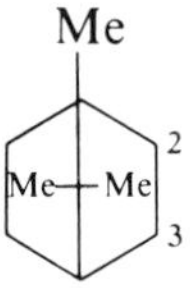

(34)	(35)	(36)
Bornane	Camphene	

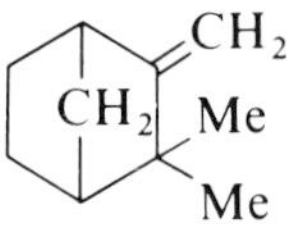

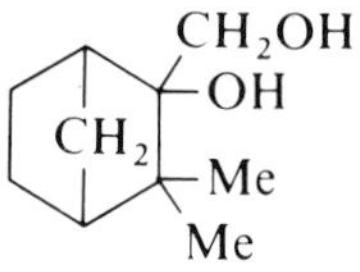

Camphene (35) was investigated by Fromm and Hildebrandt (1901) and by Hildebrandt (1902) who reported that it was excreted in rabbits as the glucuronide conjugate of a hydroxy derivative (camphenol). Subsequently, Fromm *et al.* (1903) reported that the monoglucuronide of camphene glycol (36) was also formed by rabbits.

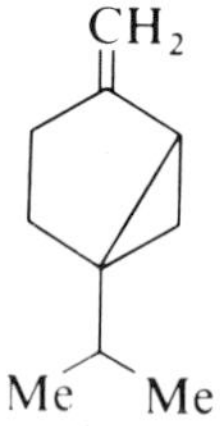

(37)

4(10)-Thujene

A similar result showing hydroxylation and excretion of this derivative as a glucuronide conjugate was reported with **4(10)-thujene** (sabinene) (37) (Fromm and Hildebrandt, 1901; Hildebrandt, 1902; Hämäläinen, 1912). Hämäläinen believed that oxidation occurred at one of the two sites adjacent to the isopropyl group.

The metabolism of (+)-**3-carene** (38) in rabbits was studied by Ishida *et al.* (1977) in an experiment identical to those mentioned above with α- and β-pinene. Three urinary alcohols were detected, the major metabolite

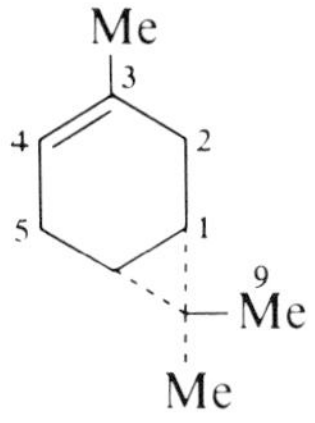

(38)
(+)-3-Carene

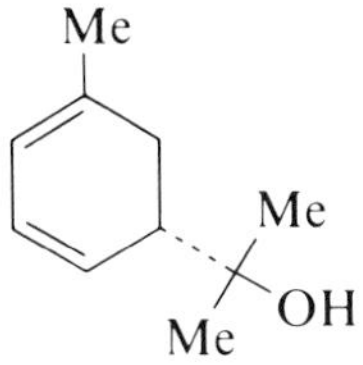

(39)

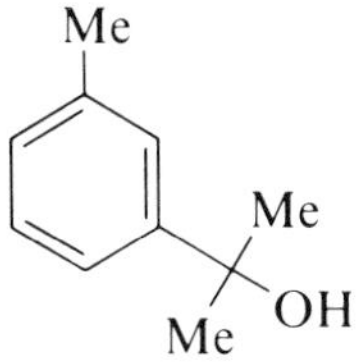

(40)

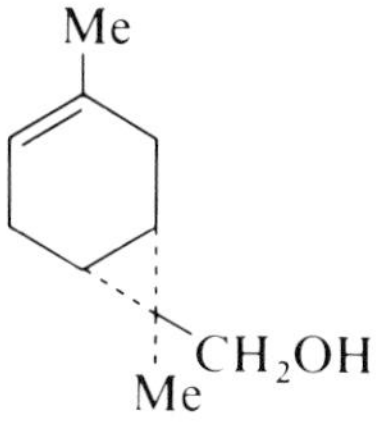

(41)

being *m*-mentha-4,6-dien-8-ol (39). Small amounts of another monocyclic derivative, *m*-cymen-8-ol (40), and of 3-caren-9-ol (41) were also identified. Thus, neither 3-caren-2-ol nor 3-caren-5-ol which would result from the allylic hydroxylation of 3-carene was detected, however it was believed that the rearrangement of either of these alcohols could give metabolite (39) which could subsequently be aromatized to the cymene derivative (40). The formation of 3-caren-9-ol furnishes an example of the stereoselective hydroxylation of the *gem*-dimethyl group.

III. Aromatic Hydrocarbons

Aromatic hydrocarbons are not found among the more typical representatives of plant compounds and, indeed, only a limited number of such compounds have been reported. Of course, fossil fuels such as coal and oil are partly derived from plant organisms, however the inclusion of the multitude of aromatic hydrocarbons from these sources in the present summary would hardly be in keeping with the general theme of this book. Therefore, the compounds that belong naturally to this section include *p*-cymene, styrene and a few hydrocarbons containing one or usually several unsaturated bonds (e.g. polyacetylenic compounds).

p-Cymene (7) shows close structural similarity to the monocyclic terpene hydrocarbons. *p*-Menthane is the fully reduced derivative and two compounds included above in Section II.B, *p*-menthene (4) and α-phellandrene (5), show intermediate degrees of unsaturation, having one or two ring double bonds, respectively. Studies of *p*-cymene metabolism in dogs were carried out nearly a century ago and indicated that the major pathway was oxidation of the methyl group giving cumic acid (43) followed by its conjugation with glycine to give cuminuric acid (44) (see Williams, 1959, p. 204). Other investigations in sheep also showed that cumic acid (Harvey, 1942) and its glycine conjugate (Wright, 1945) are urinary metabolites of *p*-cymene. These studies did not suggest that aromatic hydroxylation

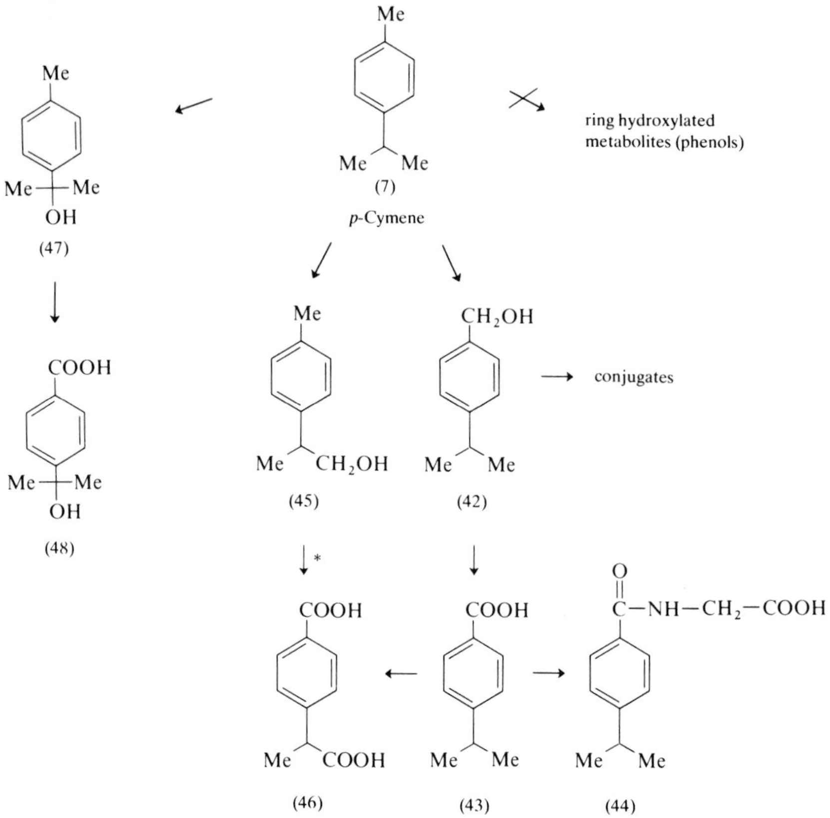

FIG. 2.2. Metabolic pathways of *p*-cymene (*see text).

occurs with *p*-cymene and this possibility was specifically checked by Bakke and Scheline (1970). Following acid or enzymic hydrolysis of the 48 h urines of rats given *p*-cymene (100 mg/kg, p.o.), neither of the possible monohydric phenols (thymol and carvacrol) was detected. This shows that only aliphatic oxidation occurs and small amounts of the intermediate *p*-isopropylbenzyl alcohol (42) were also detected. Earlier studies indicated that oxidation of the isopropyl group does not occur, however the presence of two unidentified metabolites in the *p*-cymene samples obtained by Bakke and Scheline suggested that the metabolic picture might be more complex than previously assumed. That this is indeed the case has been found by Ve and Scheline (1977) who identified several urinary metabolites with an oxidized isopropyl group when rats were given *p*-cymene (400 mg/kg). All of the three possible monohydric alcohols, compounds (42), (45) and (47), were present and the dicarboxylic acid (46) was also identified. Of the five possible monohydric monocarboxylic acids, three such compounds were also detected. One of these isomers is compound (48) in which the p-methyl group of metabolite (47) has been fully oxidized. Another metabolite is probably an intermediate in the reaction (45) → (46). p-Cymene metabolism is illustrated in Fig. 2.2.

Several studies of various aspects of the metabolism of **styrene** (49) have been made, with the result that its metabolic fate is reasonably well understood. Again, Williams (1959, pp. 199–200) summarized earlier work which showed that the ethylene moiety undergoes cleavage, giving rise to benzoic acid which is excreted as its glycine conjugate and to CO_2. El Masri et al. (1958) carried out an extensive investigation of styrene metabolism in rabbits. The major urinary metabolite was shown to be hippuric acid (58) which accounted for about 40% of the dose (520 mg/kg, p.o.). Mandelic acid (56) was also detected and James and White (1967), in similar experiments using a dose of about 145 mg/kg, reported that an average of 32% of the dose was excreted in this form. While these two metabolites account for the bulk of the administered styrene, El Masri et al. found that about 2% can be recovered unchanged in the respiratory air and a further 5% and 6% excreted in the urine as a mercapturic acid and as the monoglucuronide of 1-phenyl-1,2-ethanediol (phenylglycol) (55), respectively. James and White confirmed the excretion value of 5% for the mercapturic acid derivative and reported it to be N-acetyl-S-(2-hydroxy-2-phenylethyl)cysteine (54).

The above comments indicate that the metabolic fate of most of the styrene administered to rabbits can be accounted for. The same is true in humans and Bardodej and Bardodejova (1970), in experiments involving inhalation exposure to 22 p.p.m. for 8 h, found that 85% of the retained dose was excreted in the urine as mandelic acid (56) and 10% as phenyl-glyoxylic acid (57). The latter metabolite is also formed from styrene in rats (Ohtsuji and Ikeda, 1971). About 11% of the dose (455 mg/kg, i.p.) was excreted in the urine in 10 h as metabolite (57). Mandelic acid and hippuric acid accounted for about 8% each. These three metabolites were also reported by Braun et al. (1976) to be excreted in the urine of rats given styrene. James and White (1967) reported that 8·5% of the styrene (210 mg/kg) fed to rats was excreted as the mercapturic acid, metabolite (54). A subsequent investigation of styrene metabolism in rats by Bakke and Scheline (1970) was directed towards the identification of hydroxy-lated metabolites. After an oral dose of 100 mg/kg, the 48 h urine was found to contain 0·1% of the dose as p-vinylphenol (50). Interestingly, the previously identified diol (55) formed by rabbits was not detected in the rat urines. Instead, 1-phenylethanol (53) and a trace of 2-phenylethanol (52) were present. The absence of the diol as a urinary metabolite of styrene in rats is unexpected as Leibman and Ortiz (1969) showed that it is formed from styrene by both rat and rabbit liver microsomal preparations.

The pathways in the metabolism of styrene (49) are shown in Fig. 2.3. It now seems clear that epoxidation to styrene oxide (51) is the key step which then permits further metabolic change along several routes. Not unexpectedly, the epoxide itself has not been detected as a urinary metabolite of styrene in the studies cited above, however Leibman and Ortiz (1970) demonstrated its formation from styrene using rabbit liver microsomes. El Masri *et al.* (1958) and James and White (1967) administered styrene oxide to rabbits and found that the metabolites and their quantities were generally similar to those excreted when styrene is given. Ohtsuji and Ikeda (1971) reported that rats injected with the epoxide excreted phenylglyoxylic acid (57), mandelic acid (56) and hippuric acid (58) in the urine. These findings strongly support the role of the epoxide as the primary metabolic intermediate. The only discordant finding was that the expected glycol (55) and its glucuronide could not be detected (El Masri *et al.*, 1958). Smith *et al.* (1954) found that about 15% of the dose of styrene oxide (420 mg/kg, p.o.) is excreted by rabbits as unidentified material conjugated with glucuronic acid and similar experiments by James and White (1967) using a dose of about 200 mg/kg gave a value of 20%. This point should be clarified and it is possible that the conjugated material may be the glucuronide of 1-phenylethanol (53) rather than of the glycol. The *in vitro* metabolism of styrene oxide by preparations of epoxide hydratase from rats, guinea pigs and rabbits was studied by James *et al.* (1976). Highest activities were present in the liver and, when comparing a single tissue, small species differences were generally found.

The finding that a mercapturic acid derivative is a common metabolite of both styrene and its epoxide further supports the belief that the latter compound plays a key role in the metabolism of styrene. Boyland and Williams (1965) found that styrene oxide is a good substrate for the glutathione *S*-transferase which, by forming the glutathione conjugate, initiates the conversion of the epoxide to the mercapturic acid. Fjellstedt *et al.* (1973) described the preparation of this enzyme in purified form and James *et al.* (1976) investigated its properties in preparations of various tissues from rats, guinea pigs and rabbits using [^{14}C]-styrene oxide as a substrate. Ryan and Bend (1977) recently investigated the metabolism of the latter material in the isolated perfused rat liver. The excretion of radioactivity in the bile was 28–40% of the added styrene oxide (12 mg) in 90 min, after which time essentially no further excretion occurred. The perfusate contained 40–70% of the radioactivity at this time. The biliary radioactivity was due to a single compound which was identified as *S*-(2-hydroxy-1-phenylethyl)glutathione (59). It is noteworthy that this

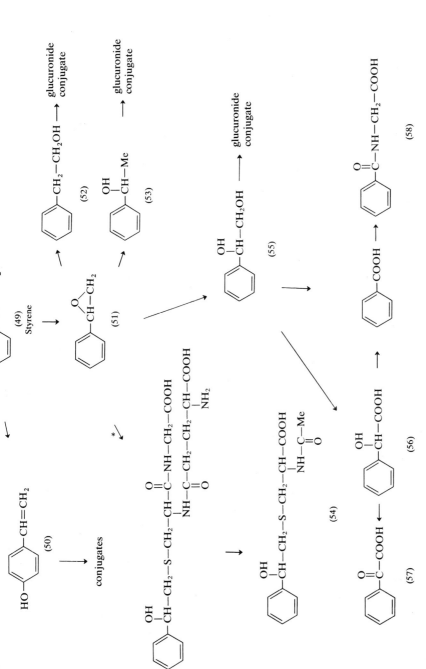

FIG. 2.3. Metabolic pathways of styrene. (* See text and Structure (59) for alternative structure of glutathione conjugate.)

metabolite is isomeric with the glutatione conjugate (S-(2-hydroxy-2-phenylethyl)glutathione) shown in Fig. 2.3 which is the precursor of the mercapturic acid derivative (54) previously reported. This newer finding strongly suggests that the mercapturic acid derivative formed from styrene and styrene oxide is N-acetyl-S-(2-hydroxy-1-phenylethyl)cysteine (60) rather than its isomer shown in structure (54). The perfusate was shown to contain approximately equal amounts of phenylglycol (55), mandelic acid (56) and the glutathione conjugate (59). Further support for the scheme shown in Fig. 2.3 comes from the finding that the glycol (55) when itself fed to rabbits, yields hippuric acid (58), mandelic acid (56) and phenylglycol glucuronide (El Masri et al., 1958). Similarly, Ohtsuji and Ikeda (1971) showed that the glycol was metabolized to hippuric acid, mandelic acid and phenylglyoxylic acid (57) in rats. Interestingly, administration of the latter compound to rats did not result in its conversion to hippuric acid or mandelic acid.

$$
\underset{(59)}{
\begin{array}{l}
\text{C}_6\text{H}_5\text{—CH—CH}_2\text{OH} \\
\qquad\quad | \\
\text{S—CH}_2\text{—CH—C(=O)—NH—CH}_2\text{—COOH} \\
\qquad\qquad\quad | \\
\qquad\qquad \text{NH—C(=O)—CH}_2\text{—CH}_2\text{—CH(NH}_2)\text{—COOH}
\end{array}}
$$

$$
\underset{(60)}{
\begin{array}{l}
\text{C}_6\text{H}_5\text{—CH—CH}_2\text{OH} \\
\qquad\quad | \\
\text{S—CH}_2\text{—CH—COOH} \\
\qquad\qquad\quad | \\
\qquad\qquad \text{NH—C(=O)—Me}
\end{array}}
$$

As noted above, a few aliphatic plant hydrocarbons are polyacetylenic compounds. While the metabolism of these naturally occurring compounds has not been reported, that of the simplest member of this type, phenyl-acetylene (61), was investigated and the results obtained may furnish some indication of the likely metabolism of the other derivatives. El Masri et al. (1958) found that a single oral dose (400 mg/kg) of phenylacetylene was slowly metabolized by rabbits and that 30–40% was eliminated unchanged in the respiratory air over three days. The urine did not contain metabolites conjugated with glucuronic acid but 5–15% of the dose was excreted daily during the first three days as phenaceturic acid (62). It was proposed that

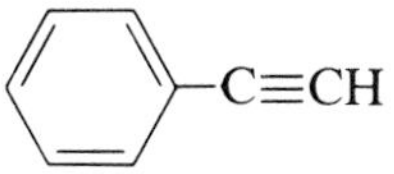

(61)

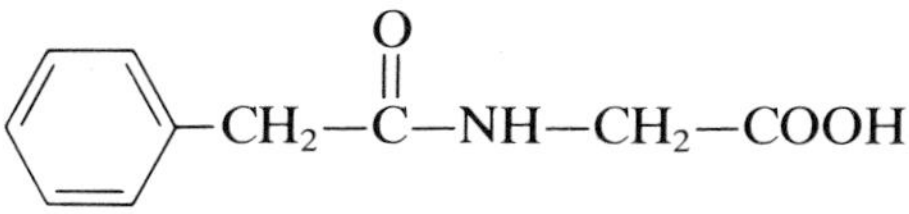

(62)

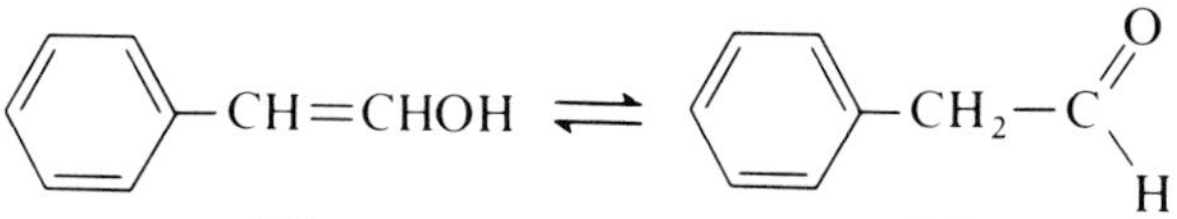

(63)

(64)

the latter metabolite arises via asymmetrical hydration of phenylacetylene to give compound (63) which is the enol form of phenylacetaldehyde (64). The latter compound will readily be oxidized to phenylacetic acid and conjugated with glycine to form the urinary metabolite (62). Hydration in the reverse manner would give phenylmethylketone which could undergo further metabolism to phenylmethylcarbinol, benzoic acid and mandelic acid (56). Of these possible metabolites only traces of the latter were detected and it can therefore be concluded that attachment of the oxygen function occurs nearly entirely at the ω-carbon.

References

Bakke, O. M. and Scheline, R. R. (1970). *Toxic appl. Pharmac.* **16**, 691–700.

Bardodej, Z. and Bardodejova, E. (1970). *Am. ind. Hyg. Ass. J.* **31**, 206–209.

Boyland, E. and Williams, K. (1965). *Biochem. J.* **94**, 190–197.

Braun, W. H., Madrid, E. O. and Karbowski, R. J. (1976). *Anal. Chem.* **48**, 2284–2285.

Das, M. L., Orrenius, S. and Ernster, L. (1968). *Eur. J. Biochem.* **4**, 519–523.

Dean, F. M., Price, A. W., Wade, A. P. and Wilkinson, G. S. (1967). *J. chem. Soc.* (C) 1893–1896.

Di Vincenzo, G. D., Kaplan, C. J. and Dedinas, J. (1976). *Toxic. appl. Pharmac.* **36**, 511–522.

El Masri, A. M., Smith, J. N. and Williams, R. T. (1958). *Biochem. J.* **68**, 199–204.

Fjellstedt, T. A., Allen, R. H., Duncan, B. K. and Jakoby, W. B. (1973). *J. biol. Chem.* **248**, 3702–3707.

Fromm, E. and Hildebrandt, H. (1901). *Hoppe-Seyler's Z. physiol. Chem.* **33**, 579–594.

Fromm, E., Hildebrandt, H. and Clemens, P. (1903). *Hoppe-Seyler's Z. physiol. Chem.* **37**, 189–202.

Frommer, U., Ullrich, V. and Staudinger, H. (1970). *Hoppe-Seyler's Z. physiol. Chem.* **351**, 903–912.

Frommer, U., Ullrich, V., Staudinger, H. J. and Orrenius, S. (1972). *Biochim. biophys. Acta* **280**, 487–494.

Frommer, U., Ullrich, V. and Orrenius, S. (1974). *FEBS Lett.* **41**, 14–16.
Hämäläinen, J. (1912). *Skand. Arch. Physiol.* **27**, 141–226.
Harvey, J. M. (1942). *Pap. Dep. Chem. Univ. Qd 1*, nr. 23.
Hildebrandt, H. (1902). *Hoppe-Seyler's Z. physiol. Chem.* **36**, 452–461.
Ichihara, K., Kusunose, E. and Kusunose, M. (1969). *Biochim. biophys. Acta* **176**, 713–719.
Igimi, H., Nishimura, M., Kodama, R. and Ide, H. (1974). *Xenobiotica* **4**, 77–84.
Ishida, T., Asakawa, Y., Okano, M. and Aratani, T. (1977). *Tetrahedron Lett.* 2437–2440.
James, M. O., Fouts, J. R. and Bend, J. R. (1976). *Biochem. Pharmac.* **25**, 187–193.
James, S. P. and White, D. A. (1967). *Biochem. J.* **104**, 914–921.
Kodama, R., Noda, K. and Ide, H. (1974). *Xenobiotica* **4**, 85–95.
Kodama, R., Yano, T., Furukawa, K., Noda, K. and Ide, H. (1976). *Xenobiotica* **6**, 377–389.
Kuhn, R., Köhler, F. and Köhler, L. (1936). *Hoppe-Seyler's Z. physiol. Chem.* **242**, 171–197.
Kusunose, M., Ichihara, K. and Kusunose, E. (1969). *Biochim. biophys. Acta* **176**, 679–681.
Leibman, K. C. and Ortiz, E. (1969). *Biochem. Pharmac.* **18**, 552–554.
Leibman, K. C. and Ortiz, E. (1970). *J. Pharmac. exp. Ther.* **173**, 242–246.
Lu, A. Y. H., Strobel, H. W. and Coon, M. J. (1970). *Molec. Pharmac.* **6**, 213–220.
McCarthy, R. D. (1964). *Biochim. biophys. Acta* **84**, 74–79.
Maynert, E. W., Foreman, R. L. and Watabe, T. (1970). *J. biol. Chem.* **245**, 5234–5238.
Mitchell, M. P. and Hübscher, G. (1968). *Eur. J. Biochem.* **7**, 90–95.
Neubauer, O. (1901). *Arch. exp. Path. Pharmak.* **46**, 133–154.
Ohtsuji, H. and Ikeda, M. (1971). *Toxic. appl. Pharmac.* **18**, 321–328.
Popović, M. (1970). *FEBS Lett.* **12**, 49–50.
Regan, J. W. and Bjeldanes, L. F. (1976). *J. agric. Fd Chem.* **24**, 377–380.
Robertson, J. S. and Hussain, M. (1969). *Biochem. J.* **113**, 57–65.
Robertson, J. S. and Solomon, E. (1971). *Biochem. J.* **121**, 503–509.
Ryan, A. J. and Bend, J. R. (1977). *Drug. Metab. Disposit.* **5**, 363–367.
Smith, J. N., Smithies, R. H. and Williams, R. T. (1954). *Biochem. J.* **56**, 320–324.
Smith, O. W., Wade, A. P. and Dean, F. M. (1969). *J. Endocr.* **45**, 17–28.
Southwell, I. A. (1975). *Tetrahedron Lett.* 1885–1888.
Ve, B. and Scheline, R. R. (1977). Unpublished observations.
Wade, A. P., Wilkinson, G. S., Dean, F. M. and Price, A. V. (1966). *Biochem. J.* **101**, 727–734.
Watabe, T. and Yamada, N. (1975). *Biochem. Pharmac.* **24**, 1051–1053.
Williams, R. T. (1959). "Detoxication Mechanisms". Chapman and Hall, London.
Wright, S. E. (1945). *Pap. Dep. Chem. Univ. Qd 1*, nr. 25.

3

METABOLISM OF ALCOHOLS, PHENOLS AND ETHERS

I. Alcohols

A summary of the general features of the oxidation of alcohols and of the enzyme systems involved is given in Chapter 1, Section I.A.6. Also, some aromatic alcohols undergo reduction by the gastrointestinal microflora (see Chapter 1, Section II.C.2.).

A. ALIPHATIC ALCOHOLS

A multitude of aliphatic alcohols have been isolated from plants, however these commonly occur in esterified forms and the amounts of free alcohols present may be quite small. Chemically these compounds span a wide range and include alcohols from C_1 to C_{34}. Both odd- and even-numbered examples are found and fully saturated derivatives are most common. Primary and secondary alcohols are encountered and, among the latter, the hydroxyl group is usually located at C2.

Two general metabolic pathways are available for alcohols: direct conjugation and oxidation. Conjugation of alcoholic hydroxyl groups occurs with glucuronic acid whereas the formation of sulphate esters, as seen with phenolic hydroxyl groups, is not seen. This lack of sulphate conjugation was demonstrated in rabbits by Kamil *et al.* (1953a) with several pentanols. This investigation also determined the extent of glucuronide formation in rabbits of a large number of alcohols, including many commonly found in plants. The data on the latter compounds are summarized in Table 3.1 which divides the compounds into normal alcohols, other primary alcohols and secondary alcohols. These findings are in general agreement with the early qualitative results of Neubauer (1901) who administered many of these alcohols to rabbits and dogs. It is evident that conjugation is a general reaction of these compounds, however this pathway is of minor importance with the lower homologues. Kamil *et al.* (1953c) subsequently reported the isolation of small amounts of both methyl and ethyl glucuronides from the urine of rabbits given the corresponding alcohols. Conjugation of the normal alcohols is low (10% or less of the dose) but branching increases it considerably. This is seen with the value given by 2-heptanol. Several other closely related 2- and 3-ols not listed in Table 3.1 gave conjugation values of between 50 and 70%. 2-Ethyl-1-hexanol is notable for its extensive conjugation, however the urine samples contained a reducing glucuronide which was shown by Kamil *et al.* (1953b) to be that of 2-ethylhexanoic acid rather than that of the administered alcohol.

The other main metabolic pathway of alcohols involves oxidation and in this case a fundamental difference exists between the primary and secondary compounds. With the former, oxidation proceeds via the aldehyde to the corresponding carboxylic acid. The latter may be oxidized completely to carbon dioxide and water or, if this is prevented, be conjugated with glucuronic acid. Extensive oxidation can be expected to occur with the normal alcohols which undergo chain shortening via β-oxidation, the fragments entering into the metabolic pool of 2-carbon compounds. Substitution in the α-position (2-alkyl alcohols) may greatly influence this oxidative pathway and whereas 2-methyl alcohols undergo extensive oxidation, this is not the case with 2-ethyl derivatives. This was demonstrated experimentally by Kamil *et al.* (1953a,b) and these effects were discussed by Williams (1959, pp. 47–48).

The oxidation of secondary alcohols leads to the formation of ketones rather than aldehydes and carboxylic acids. Further oxidation of the carbonyl group is therefore prevented, however the aliphatic moiety of this ketone may undergo oxidation. This is noted below with 2-hexanol. The latter reaction will increase the polarity of the molecule and also provide a site for conjugation.

In addition to these two general pathways of conjugation and oxidation, alcohols may also be lost unchanged from the body in the expired air. This occurs with volatile alcohols and, more importantly, with volatile ketones formed from several secondary alcohols. Kamil *et al.* (1953a) found small amounts of the ketones derived from 2-butanol and 2-heptanol in the expired air of rabbits given the alcohols. This point was studied in detail by Haggard *et al.* (1945) with seven of the eight isomers of pentanol. When rats were given 1-pentanol (1 g/kg, i.p.) about 0·9% of the dose was recovered unchanged in the expired air. The corresponding figure with 2-pentanol was 5·4%, however an additional 38% of the dose was lost by this route as methyl *n*-propyl ketone.

The following summary is devoted to the metabolism of particular aliphatic plant alcohols. The books of Williams (1959) and Browning (1965) cover the metabolism of numerous alcohols including many of plant origin and some of the following data were taken from these sources. It is felt that no useful purpose will be served by reviewing the voluminous

TABLE 3.1

Conjugation of some plant alcohols with glucuronic acid in rabbits[a]

Compound	Synonyms	Structure	Dose (mg/kg)	Conjugation (% of dose[b])
Normal alcohols				
Methanol		Me—OH	800	0[c]
Ethanol		Me—CH$_2$OH	765	0·5
1-Propanol		Me—CH$_2$—CH$_2$OH	800	0·9
1-Butanol		Me—(CH$_2$)$_2$—CH$_2$OH	400	1·8
1-Pentanol	n-Amyl alcohol	Me—(CH$_2$)$_3$—CH$_2$OH	735	7
1-Hexanol		Me—(CH$_2$)$_4$—CH$_2$OH	850	10
1-Heptanol		Me—(CH$_2$)$_5$—CH$_2$OH	965	5
1-Octanol	Caprylic alcohol	Me—(CH$_2$)$_6$—CH$_2$OH	1080	10
1-Nonanol		Me—(CH$_2$)$_7$—CH$_2$OH	1200	4
1-Decanol		Me—(CH$_2$)$_8$—CH$_2$OH	1320	4[d]
1-Octadecanol	Stearyl alcohol	Me—(CH$_2$)$_{16}$—CH$_2$OH	2250	5–10[d]
Other primary alcohols				
2-Methyl-1-propanol	Isobutyl alcohol	(Me)$_2$—CH—CH$_2$OH	615	4
2-Methyl-1-butanol		Me—CH$_2$—CH(Me)—CH$_2$OH	735	10
3-Methyl-1-butanol	Isoamyl alcohol	(Me)$_2$—CH—CH$_2$—CH$_2$OH	735	9
2-Ethyl-1-hexanol		Me—(CH$_2$)$_3$—CH(—CH$_2$Me)—CH$_2$OH	1080	87[c,e]
Secondary alcohols				
2-Butanol	sec.—Butyl alcohol	Me—CH$_2$—CH(OH)—Me	615	14
2-Pentanol	sec.—Amyl alcohol	Me—(CH$_2$)$_2$—CH(OH)—Me	735	45
2-Heptanol		Me—(CH$_2$)$_4$—CH(OH)—Me	965	55
2-Octanol		Me—(CH$_2$)$_5$—CH(OH)—Me	1080	16

[a] Kamil et al. (1953a).
[b] Usually average values from three animals.
[c] See text.
[d] Incomplete absorption, some alcohol recovered in faeces.
[e] Ester glucuronide of corresponding carboxylic acid.

literature on the metabolism of methanol and ethanol. These compounds have other areas of association by virtue of their toxic properties and the dangers involved in their social abuse. Accordingly, several recent reviews of the metabolism of these alcohols are available. Cornish (1975) summarized the major points in the metabolism of methanol and supplementary information on the enzyme systems involved in its metabolism in rats and monkeys has been reported by Tephly *et al.* (1964) and Makar *et al.* (1968), respectively. Notable recent reviews of the metabolism of ethanol are those by Lieber (1974) and the book by Majchrowicz (1975) which includes several chapters on various aspects of ethanol metabolism. Further information, especially regarding the roles of alcohol dehydrogenase, the microsomal ethanol-oxidizing system and catalase, is found in the articles by Mezey (1976) and Teschke *et al.* (1976).

Among the lower plant alcohols, both **1-propanol** and **1-butanol** are readily oxidized, little being conjugated with glucuronic acid (Table 3.1). Both alcohols are oxidized by liver alcohol dehydrogenase and, in fact, maximum activity among primary alcohols is found with 1-butanol (see McMahon, 1971). **1-Pentanol** was noted above to undergo a small amount of excretion unchanged in the expired air in rats. It is conjugated to a small extent with glucuronic acid in rabbits (Table 3.1) and undergoes rapid metabolism in rats following its injection (Haggard *et al.*, 1945). Following a dose of 1 g/kg, the alcohol had disappeared from the blood within 3·5 h. Hinson and Neal (1975) reported a kinetic study of the oxidation of **1-octanol** by horse liver alcohol dehydrogenase.

Of the higher normal alcohols, **1-hexadecanol** (cetyl alcohol, $Me-(CH_2)_{14}-CH_2OH$) was reported not to form a glucuronide conjugate in rabbits (Neubauer, 1901). It is oxidized to the corresponding fatty acid, palmitic acid, in rats (Stetten and Schoenheimer, 1940). However, the data summarized in Table 3.1 indicate that glucuronide formation occurs to a relatively small extent in rabbits with both lower and higher homologues of cetyl alcohol and the earlier negative report should perhaps be only tentatively accepted. Stetten and Schoenheimer also found that **1-octadecanol** was metabolized to the corresponding fatty acid. It was noted that some of the C_{16} alcohol was converted to the C_{18} acid and some of the C_{18} alcohol to the C_{16} acid.

Two common primary plant alcohols with branched chains are **2-methyl-1-propanol** (isobutyl alcohol) and **3-methyl-1-butanol** (isoamyl alcohol). Both are characterized by their rapid oxidation, however small amounts are conjugated with glucuronic acid (Table 3.1). Williams (1959, p. 59) summarized the main points in the metabolism of isobutyl alcohol. These involve oxidation to its aldehyde and then carboxylic acid derivatives, the latter compound then being decarboxylated to acetone and CO_2. The investigation of Haggard *et al.* (1945) showed that rats given isoamyl alcohol (1 g/kg, i.p.) excreted little (1%) unchanged in the expired air and even less in the urine. However, the blood levels declined to an undetectable amount within 5 h.

As noted above, the secondary alcohols are distinguished by higher degrees of conjugation with glucuronic acid (see Table 3.1) and by oxidation to ketones rather than to aldehydes and then carboxylic acids. The ketones formed are fairly volatile and may be lost in the expired air, as previously noted with **2-butanol**, **2-pentanol** and **2-heptanol**. The investigation of Haggard *et al.* (1945) also showed that 2-pentanol, and especially its ketone metabolite, were removed from the blood at a slower rate than were the primary pentanols. A recent metabolic investigation which included the homologue 2-hexanol demonstrated a previously unexplored metabolic route for these secondary alcohols. Di Vincenzo *et al.* (1976) determined the metabolites present in the serum of guinea pigs given the alcohol (450 mg/kg, i.p.). In addition to the expected hexan-2-one, several metabolites formed by ω-1 oxidation of the alkyl group were detected. These included 2,5-hexanediol, 2,5-hexanedione and 5-hydroxy-2-hexanone. There is no reason to believe that this pathway may not be involved in the metabolism of related secondary alcohols as well.

Sugar alcohols are polyhydric alcohols in which the aldehyde group of sugars is replaced by an alcohol group. They are of fairly common occurrence in plants and show many of the properties of the corresponding sugars. This close association includes their metabolism and it must be stressed that sugar–sugar alcohol interconversions occur in the pathways of mammalian carbohydrate metabolism. Thus, reversible pyridine nucleotide-linked dehydrogenations are involved in various sugar isomerizations which involve the polyols as intermediates. This is the case with the glucuronate-gulonate pathways taking place in liver mitochondria in which the C_5 polyol xylitol is an intermediate in the isomerization of L- and D-xylulose. In an analogous fashion, D-sorbitol is the intermediate in the glucopyranose-fructopyranose isomerization which is carried out by an NADP-specific D-hexitol dehydrogenase present in many mammalian tissues. These comments make it clear that the metabolism of polyhydric alcohols is to a large extent closely associated with normal carbohydrate metabolism and that much of the subject falls outside the scope of this book. Indeed, much of the interest in the metabolism of the polyols has centred on their caloric values and conversion to glycogen, factors of importance in the use of polyols as sweetening agents in dietary and diabetic foods. Carr and Krantz (1945) reviewed the subject of glycogen formation by polyhydric alcohols. Their metabolism was summarized briefly by Williams (1959, pp. 79–82) and more extensively by Touster and

Shaw (1962). The following account is based mainly on the former source and, where indicated, on additional or more recent findings.

Galactitol (dulcitol) (1) was shown earlier to be partly converted to glycogen in rats but more recent work (Weinstein and Segal, 1968) indicated that this occurs to a very limited extent. Using compounds labelled with ^{14}C in the 1-position, the rate of $^{14}CO_2$ formation from galactitol by liver preparations was found to be only 3% of that observed with sorbitol. When the polyol was injected in rats or humans nearly all was excreted unchanged in the urine and little or no $^{14}CO_2$ was detected in the expired air.

$$
\begin{array}{cc}
CH_2OH & CH_2OH \\
| & | \\
CHOH & HOCH \\
| & | \\
HOCH & HOCH \\
| & | \\
HOCH & CHOH \\
| & | \\
CHOH & CHOH \\
| & | \\
CH_2OH & CH_2OH \\
(1) & (2) \\
\text{Galactitol} & \text{Mannitol}
\end{array}
$$

The metabolism of **D-mannitol** (2) was conveniently summarized by Olmsted (1953) who noted that it is only slowly absorbed from the intestine and that 80% or more of this absorbed material is excreted unchanged in the urine in humans. Extensive urinary excretion of unchanged compound is also seen in other species (rats, rabbits, dogs and monkeys), however it is able to be converted to liver glycogen to a limited extent. More recently, Nasrallah and Iber (1969) studied the absorption and metabolism in man of mannitol uniformly labelled with ^{14}C. Following oral doses of 28–100 g, about 18% and 32% of the radioactivity was found in the urine and faeces, respectively, in 48 h. The urinary radioactivity was due entirely to unchanged compound. Some mannitol was metabolized in these experiments and as much as 18% of the dose was recovered in 12 h as respiratory CO_2.

D-Sorbitol (glucitol) (3) was noted above to be converted in mammalian tissues to both glucose and fructose and it is therefore understandable that this polyol more extensively enters into the pathways of carbohydrate metabolism than is the case with galactitol and mannitol. Although most of the administered D-sorbitol undergoes oxidation, some unchanged compound appears in the urine. Strack et al. (1965) reported a value of 12–18% of the dose (0·25–0·75 g/kg/h) in rabbits. They also found that

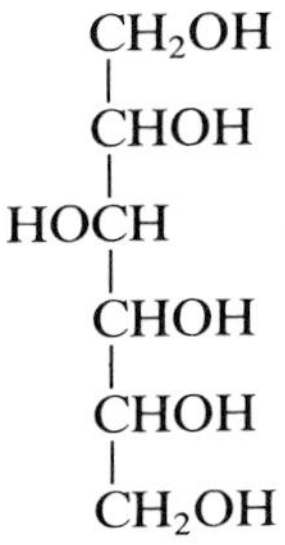

$$
\begin{array}{c}
CH_2OH \\
| \\
CHOH \\
| \\
HOCH \\
| \\
CHOH \\
| \\
CHOH \\
| \\
CH_2OH
\end{array}
$$

(3)

D-Sorbitol

blood fructose levels were increased whereas little effect on blood glucose was noted. The results of Maeda (1966), also obtained using rabbits, suggested that D-sorbitol is initially metabolized slowly to fructose and then rapidly to glycogen, glucose and CO_2. The delayed nature of the conversion of D-sorbitol to glucose and glycogen was also underlined by Olmsted (1953). A study of the blood and liver levels of sorbitol metabolites including C_3- and C_6-phosphorylated intermediates in rats was carried out by Heinz and Wittneben (1970).

Of the two naturally occurring heptitols, only **D-volemitol** (sedoheptitol, D-β-mannoheptitol) (4) appears to have been studied. Hiatt *et al.* (1938) reported that it, unlike mannitol, is not capable of serving as a glycogen precursor in rat liver. Maltitol, the alcohol of the disaccharide maltose, was shown by Rennhard and Bianchine (1976) to undergo rapid and extensive caloric utilization in animals.

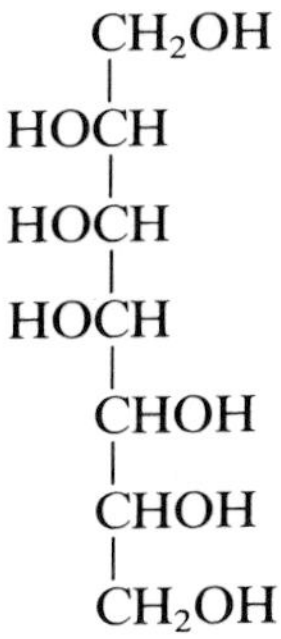

$$
\begin{array}{c}
CH_2OH \\
| \\
HOCH \\
| \\
HOCH \\
| \\
HOCH \\
| \\
CHOH \\
| \\
CHOH \\
| \\
CH_2OH
\end{array}
$$

(4)

D-Volemitol

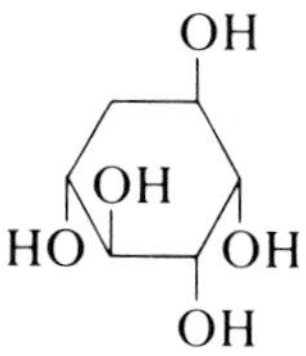

(5)

Quercitol

A closely related group of compounds is that of the carbocyclic polyols or cyclitols. Williams (1959, p. 118) referred to nineteenth century experiments which reported that **D-quercitol** (1,2,3,4,5-cyclohexanepentol, acorn sugar) (5) is not metabolized in the body but is largely excreted unchanged. The C_6-cyclitols are the inositols, of which myo-inositol (meso-inositol) is the best known. It is widely distributed in plants and animals, having physiologic functions, and a discussion of its metabolism therefore falls outside the scope of this book.

B. MONOTERPENOID ALCOHOLS

1. Acyclic Terpene Alcohols

The first metabolic studies in this area were carried out at the beginning of the present century when Hildebrandt (1901) and Neubauer (1901) administered **linalool** (3,7-dimethyl-1,6-octadien-3-ol) (6) and geraniol (trans-3,7-dimethyl-2,6-octadien-1-ol) (7), respectively, to rabbits. In the former study it was noted that treatement with acid resulted in the isomerization of linalool to geraniol. This conversion did not occur metabolically and linalool was reported to be excreted as its glucuronide conjugate. Parke et al. (1974) recently studied the absorption, distribution and excretion in rats of linalool labelled with ^{14}C in positions 1 and 2. The excretion of linalool as a glucuronide conjugate was confirmed. The nature of the urinary, faecal and biliary metabolites was not determined, however they were found to be mainly polar, ether-insoluble conjugates. Biliary excretion and enterohepatic circulation of metabolites was noted, with about 25% of the dose (500 mg/kg, i.p.) being excreted in the bile in 6–11 h and delayed faecal excretion of radioactivity arising due to enterohepatic circulation. A significant finding was that 23% of the dose was lost as respiratory $^{14}CO_2$. This indicates that linalool is more extensively metabolized than previously shown and suggests that it can to some extent enter into pathways of intermediary metabolism.

(6)   (7)   (8)
Linalool   Geraniol   Nerol

Neubauer (1901) did not detect a urinary glucuronide of **geraniol** (7) in rabbits and believed that it might undergo cyclization. However, Hildebrandt (1904) subsequently showed that geraniol was extensively oxidized in rabbits to a dicarboxylic acid derivative. **Nerol** (8), the *cis* isomer of geraniol, was excreted conjugated with glucuronic acid. Hildebrandt believed that the carboxyl groups were formed by oxidation of the alcohol group at C1 and the methyl group at C3. However, several subsequent studies with geraniol have shown that the allylic oxidation occurs instead at

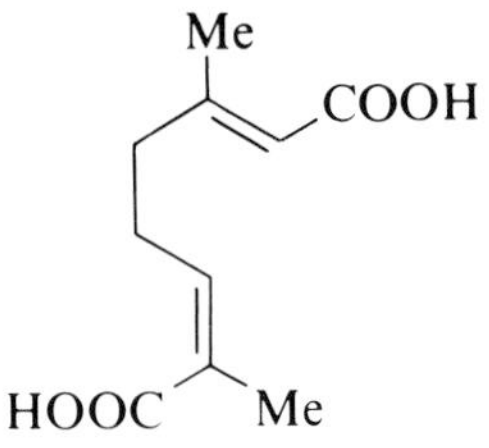

Me   Me

COOH

COOH

HOOC   Me   HOOC   Me

(9) 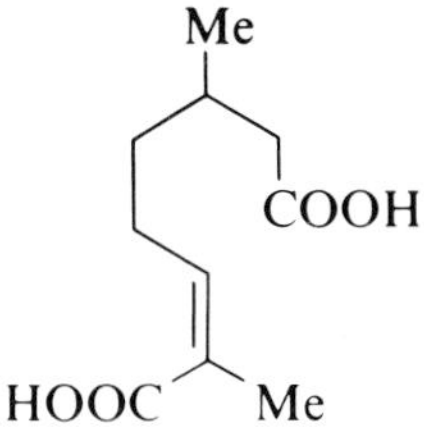   (10)

a C7 methyl group. This gives rise to metabolite (9), known as Hildebrandt acid (Kuhn *et al.*, 1936; Fischer and Bielig, 1940; Asano and Yamakawa, 1950). These three investigations also showed that the optically active metabolite (10), termed reduced Hildebrandt acid, was also excreted. Roughly 25–30% of the oral dose (17–55 g given in daily 2–10 g portions) was isolated as these two dicarboxylic acids in a ratio of about 2 : 1 between metabolites (10) and (9). Following i.p. dosage this ratio was reversed and the total recovery was reduced to about 15%. Recently, Licht and Jamroz (1977) showed that the initial step in the metabolism of geraniol and nerol, allylic hydroxylation at C8 to give the alcohol intermediate, took place *in vitro* with a mono-oxygenase system from rabbit liver microsomes.

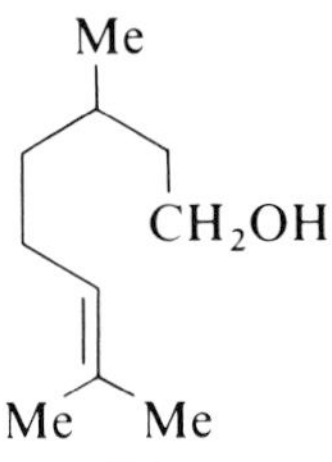

Me   Me

CH_2OH COOH

Me   Me HOH_2C Me

(11)   (12)

Citronellol

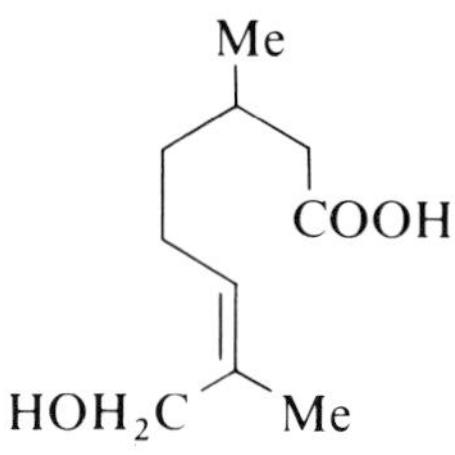

Fischer and Bielig (1940) also studied the metabolism in rabbits of **citronellol** (3,7-dimethyl-6-octen-1-ol) (11). This alcohol differs from geraniol or nerol by the lack of a double bond at C2,3. A small amount of the total dose of 30 g was recovered in the urine unchanged, but most of the identified material consisted of compounds formed by double ω-oxidation, i.e. oxidation of the alcohol group and allylic oxidation of the methyl group at C7. Both the intermediate alcohol, 7-hydroxymethyl-3-methyl-6-octenoic acid (12), and reduced Hildebrandt acid (10) were isolated to the extent of about 8% and 10%, respectively, of the dose.

The metabolism of the closely related sesquiterpenoid farnesol and the diterpenoid phytol is covered in Chapter 6, Sections I and II, respectively.

2. Monocyclic Terpene Alcohols

Studies on the metabolism of monocyclic terpene alcohols have been devoted nearly exclusively to their conjugation with glucuronic acid. This subject was first approached by Hämäläinen over 60 years ago, however the most detailed information available is of a later date and deals with the menthols, especially *l*-menthol.

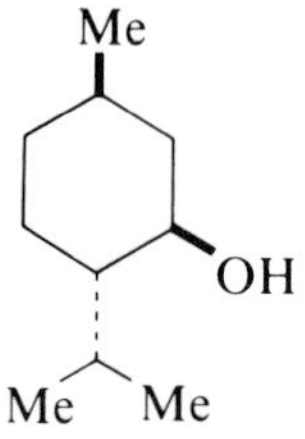

(13)
l-Menthol

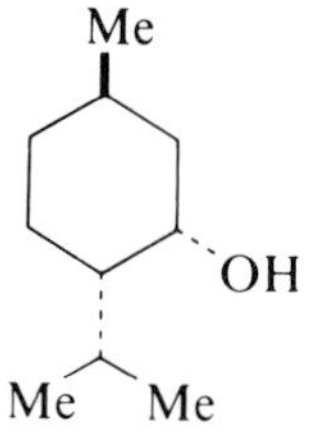

(14)
d-Neomenthol

The menthols (3-*p*-menthanols) contain three asymmetric carbon atoms and eight isomers are therefore possible. These are the *d* and *l* forms of menthol, isomenthol, neomenthol and neoisomenthol, however only **l-menthol** (13) and **d-neomenthol** (14) occur naturally. Quick (1924) administered *l*-menthol to rabbits and later to dogs and humans (Quick, 1928) and determined the extent of its conjugation. In the first case, oral doses of about 0·25 g/kg resulted in 48% being excreted in the urine as menthol glucuronide. The values decreased only slightly to 41–46% (0·5 g/kg) and 42–48% (1·0 g/kg) and then more rapidly to slightly under 20% as lethal doses of 2–2·5 g/kg were approached. Excretion of the conjugate was rapid, values of about 90% of the conjugated material being registered in 6 h following a dose of 1 g/kg. Interestingly, dogs fed menthol (5 g) excreted only about 5% of the dose as the glucuronide. In humans fed 1 g, which corresponds to a dose of only about 15 mg/kg, the value was much higher (79%). These experiments did not provide any information on the fate of the unconjugated menthol. Williams (1938b) studied the conjugation of menthols in detail and, in the case of *l*-menthol, found that rabbits fed 1 g/kg excreted an average of 48% of the dose as menthol glucuronide, thus confirming the result of Quick.

The fundamental point of interest in the studies of Williams on the glucuronic acid conjugation of the menthols was that of the effects of optical and geometrical isomerism on this metabolic pathway. When identical experiments to that noted above with *l*-menthol were carried out with *dl*-menthol, 59% of the dose was excreted conjugated with glucuronic acid (Williams, 1938b). From these values it can be calculated that the corresponding value for *d*-menthol must be 70%. Similarly, administration of *dl*-isomenthol and *d*-isomenthol gave values of 55% and 65%, respectively, which indicates that the *l*-form of this geometrical isomer must undergo conjugation only to the extent of 45% of the dose. In both cases the *d* isomer is more extensively conjugated and this fact was utilized to effect the resolution of *dl*- menthol (Williams, 1939) and *dl*-isomenthol (Williams, 1940a). A subsequent study using the naturally occurring *d*-neomenthol (14) showed that rabbits fed 1 g/kg excreted 67–68% of the dose combined with glucuronic acid (Williams, 1940b). Thus, the extent of conjugation of the *d* forms of menthol, isomenthol or neomenthol in rabbits is nearly identical.

Wright (1945) reported that sheep are comparable with rabbits in their ability to conjugate *l*-menthol with glucuronic acid. However, the question of the fate of the remainder of the administered *l*-menthol was not approached and this point remains a matter of speculation. Williams (1959) suggested that ring fission and degradation may occur, however it seems reasonable to assume that hydroxylation of the methyl and/or isopropyl groups may be involved in the metabolism of that portion of the dose previously unaccounted for. This possibility has now been studied using gas chromatographic-mass spectrometric techniques (Scheline, 1977). Rats were given *dl*-menthol (400 mg/kg, p.o.) and the 24 h urines hydrolysed using a β-glucuronidase-sulphatase preparation followed by separation into neutral and acidic fractions. In the neutral fraction a mixture of hydroxylated menthol derivatives was detected which appeared to account for nearly twice as much material as that due to menthol itself. Most of this consisted of monohydroxylated derivatives of menthol, of which three were recognized, however evidence was also obtained for the presence of small amounts of a di- and a trihydroxylated menthol derivative. The positions of the hydroxyl groups in these metabolites were not determined but the finding that the two major metabolites in the acidic fraction corresponded to monocarboxylated derivatives of menthol clearly indicates that both the methyl group and the isopropyl group undergo oxidation. It seems tempting to assume that the third monohydroxylated derivative arises from oxidation of the isopropyl group to give a tertiary carbinol which is, of course, unable to be oxidized further. The possible significance of this oxidative pathway in rabbits is not known, however the well-documented ability of rabbits to conjugate menthol directly with glucuronic acid suggests that, if present, it is probably less extensive than in rats.

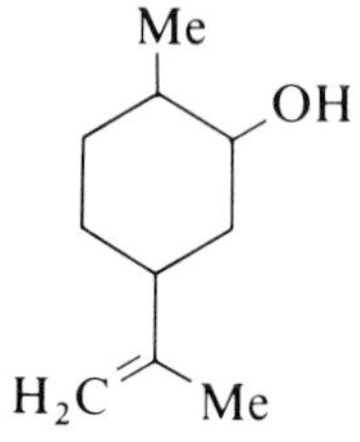

Me
OH

H₂C   Me
(15)
Dihydrocarveol

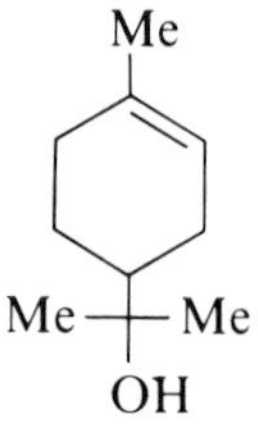

Me

Me —— Me
OH
(16)
α-Terpineol

Of the other monocyclic terpene alcohols **dihydrocarveol** (*p*-8(9)-menthen-2-ol) (15) was one of the many terpenoids studied by Hämäläinen (1912a) with regard to their ability to form conjugates with glucuronic acid when given to rabbits. Not unexpectedly, this reaction was found to occur with dihydrocarveol. In a similar study, Hämäläinen (1913) reported that the isomeric menthenol **α-terpineol** (*p*-1-menthen-8-ol) (16) formed a glucuronide conjugate in rabbits. Its metabolism in sheep was studied by Wright (1945) who found that terpin (17) could be isolated from the acid-hydrolysed urines. Recently, Horning *et al.* (1976) reported that the major neutral urinary metabolite of α-terpineol in man is *p*-menthan-1,2,8-triol (18). This metabolite was also excreted in rats and it was suggested that formation of the 1,2-diol moiety occurs via an epoxide intermediate which, however, was not detected in any of the urine extracts.

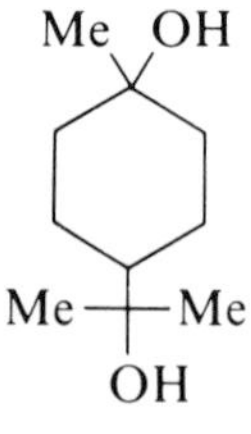

Me OH

Me —— Me
OH
(17)
Terpin

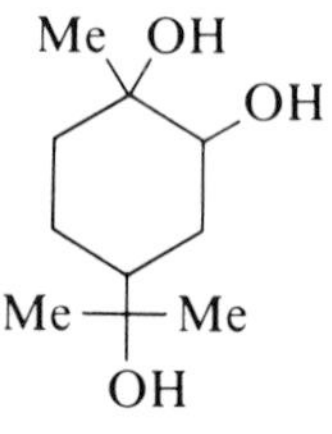

Me OH
OH

Me —— Me
OH
(18)

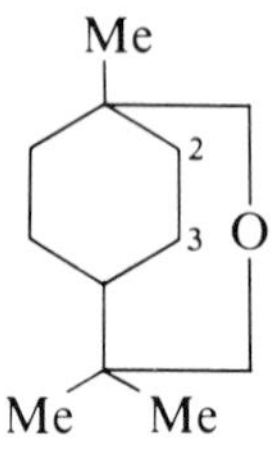

Me

^{2}
^{3} O

Me   Me
(19)
Cineole

Terpin (*p*-menthan-1,8-diol) (17), which usually occurs in the hydrated form, was noted above to be a metabolite of α-terpineol. When the diol itself is given to animals it is excreted in the urine conjugated with glucuronic acid, as shown by Hämäläinen (1912a) in rabbits and Wright (1945) in sheep. The anhydride of terpin is **cineole** (eucalyptol; 1,8-epoxy-*p*-menthane) (19) which was also included in the experiments of Hämäläinen (1911) and Wright (1945). In the former case, the compound was believed to undergo hydroxylation at the 2- or 3-position followed by conjugation of this alcohol with glucuronic acid. However, Wright found little metabolic material in the urine of sheep, even after giving large doses, and believed that cineole may be largely oxidized in this species.

3. Bicyclic Terpene Alcohols

It was noted in the preceding section on monocyclic derivatives that their conjugation with glucuronic acid was a major metabolic pathway. This is equally true of the bicyclic terpene alcohols. A further similarity between these two groups relates to the fact that a large number of both types were included in the extensive study carried out by Hämäläinen (1912a). Interestingly, many of the bicyclic terpene alcohols studied by Hämäläinen have never been reinvestigated and this early report on their conjugation with glucuronic acid remains the sole source of information on their metabolic fate. These alcohols include **α-santenol** (1,7-dimethyl-2-norbornanol) (20), **β-santenol** (santene hydrate; 2,3-dimethyl-2-norbornanol) (21), **camphenilol** (22), **fenchyl alcohol** (fenchol) (23), **isofenchyl alcohol** (24) and **thujyl alcohol** (25). In all cases the alcohols, when given to rabbits, were found to be excreted as their glucuronides. However, in the case of thujyl alcohol (25) another metabolite, the glucuronide of *p*-menthane-2,4-diol (26), was also reported. An analogous reaction was reported by

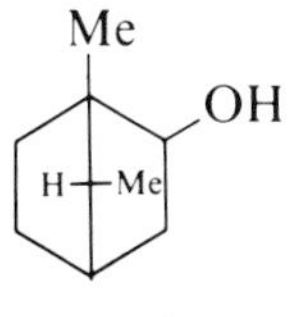

(20)
α-Santenol

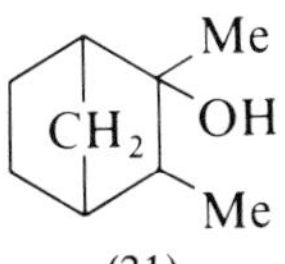

(21)
β-Santenol

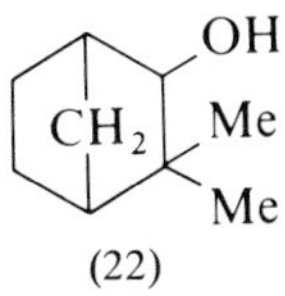

(22)
Camphenilol

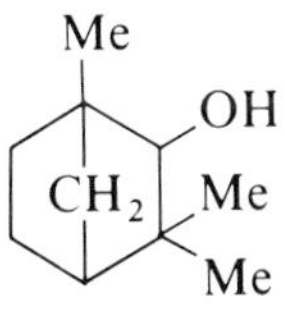

(23)
Fenchyl alcohol

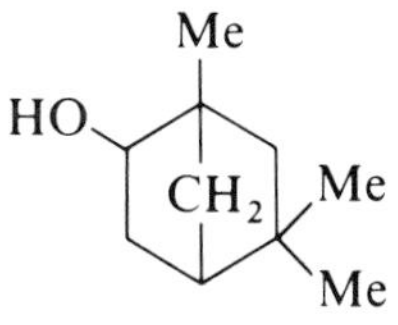

(24)
Isofenchyl alcohol

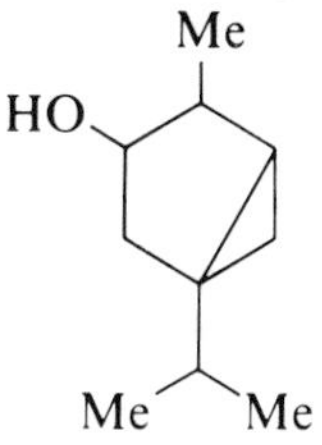

(25)
Thujyl alcohol

Hämäläinen to occur with the corresponding ketone, thujone (see Chapter 4, Section II.B.2). Recent findings with the terpenoid hydrocarbons β-pinene and 3-carene provide further examples of the metabolic conversion of bicyclic terpenoids to monocyclic derivatives (see Chapter 2, section II.C).

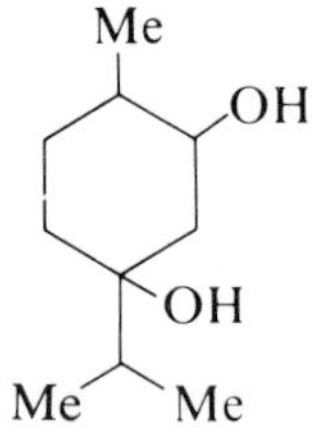

Me
OH

OH

Me   Me

(26)

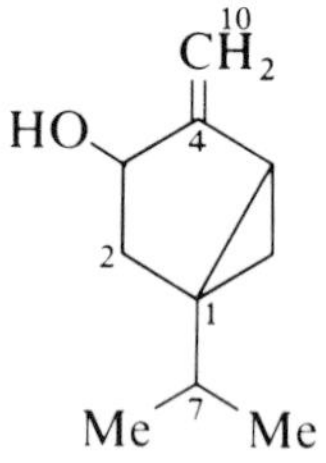

$\overset{10}{C}H_2$
HO
4
2
1
Me $\overset{7}{}$ Me

(27)
Sabinol

Another alcohol containing the thujane (sabinane) ring and therefore closely related to thujyl alcohol is **sabinol** (27). However, the presence of an unsaturated bond at C4,10 appears to protect the compound from cleavage of the bicyclic structure as noted above with thujyl alcohol (Hämäläinen, 1912a, 1912b). Instead, sabinol is itself excreted as a glucuronide conjugate, a finding which had been reported previously by Hildebrandt (1901) and Fromm and Hildebrandt (1901).

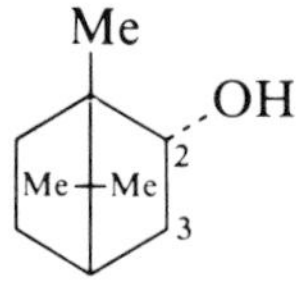

Me
OH
Me—Me
2
3

(28)
d-Borneol

Me
OH
Me—Me

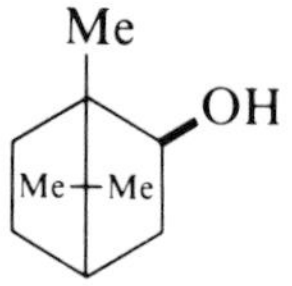

(29)
l-Borneol

The bicyclic terpene alcohol which has most often been the subject of metabolic studies is borneol (endo-2-hydroxycamphane). Again, the reports are of older date and are concerned with the ability of the alcohol to undergo glucuronide conjugation. Both **d-borneol** (Borneo camphor) (28) and **l-borneol** (29) were investigated and Magnus-Levy (1907) reported that rabbits excreted equal amounts of bornyl glucuronide after administration of 1 g of d- or l-borneol. Also, dogs given large, repeated doses excreted about 75% of both forms as the glucuronide. Likewise, Hämäläinen (1909) found that 22% of both d- and l-borneol were excreted in the urine conjugated with glucuronic acid when rabbits were given daily doses of 1·5–2 g. Nonetheless, Pryde and Williams (1934) repeated these experiments in dogs using a mixture of d- and l-borneol and found evidence for the preferential conjugation of the d-form. They isolated pure β-l-bornyl-d-glucuronide from the urine of dogs fed pure l-borneol. Further quantitative data on the extent of excretion of bornyl glucuronide following the administration of d-borneol to dogs and man or to man was obtained by Quick (1928) and by Pryde and Williams (1936), respectively. Dogs given an oral dose of 5 g excreted only about 50% of the dose as the conjugate, a value somewhat lower than that found earlier by Magnus-Levy. Quick obtained in man values of 69% (3·5 g, urine collected 6 h) and 81% (2·0 g, urine collected 10 h). The subsequent study of Pryde and Williams confirmed the rapid and extensive urinary excretion of bornyl glucuronide in man, reporting a value of about 80% of the dose in 12 h following administration of 2 g amounts to 24 subjects.

A compound showing close structural similarity to borneol is camphane-2,3-diol which contains an additional hydroxyl group at C3. Four racemic diols are possible, the hydroxyl groups at C2 and C3 being *endo-endo*, *exo-exo*, *endo-exo* or *exo-endo*, i.e. two *cis* and two *trans* forms. These diols were studied in connection with their possible formation from other camphane derivatives and are therefore mentioned in the summaries of the metabolism of camphane (Chapter 2, Section II.C) and camphorquinone (Chapter 4, Section II.B.2). However, Robertson and Solomon (1971) administered the racemic camphane-2,3-diols to rabbits and found that they are excreted in the urine partly free and partly combined with glucuronic acid. In addition, dehydrogenation of the diols to ketols occurred.

C. AROMATIC ALCOHOLS

(30)

Benzyl alcohol

The simplest member of this rather limited class of plant alcohols is **benzyl alcohol** (30) and, in fact, most of the metabolic studies in this area have dealt with this compound. Not unexpectedly, its metabolism has been shown to be straightforward, mainly involving oxidation to benzoic acid followed by conjugation of the latter with glycine to form hippuric acid. Snapper *et al.* (1925) reported that this pathway accounted for 80–90% of the dose (1·5 g) in man and Diack and Lewis (1928) recorded an average of 67% in rabbits fed a dose of 400 mg/kg of benzyl alcohol. Bray *et al.* (1951) also fed benzyl alcohol to rabbits and found that 76% of a dose of approx. 500 mg/kg was excreted as ether-soluble acids (mainly hippuric but also a small amount of benzoic). The remainder of the dose was excreted as benzoyl glucuronide, a finding which suggests that glucuronidation assumes quantitative importance when the amounts of benzoic acid being formed exceed those capable of being conjugated with glycine. This was substantiated in experiments in which glycine was fed concomitantly, in which case the formation of hippuric acid increased to 91–98% of the dose (approx. 250–800 mg/kg) and that of the glucuronide fell correspondingly. Teuchy et al. (1971) reported that rats given benzyl alcohol (44 mg/rat, i.p.) excreted an average of 58% of the dose as hippuric acid. It may be reasonably assumed that the remainder is conjugated with glucuronic acid, however other pathways may be operative. In this context it is interesting to note that Sloane (1965) found that guinea pig liver microsomes are able to convert benzyl alcohol to phenol.

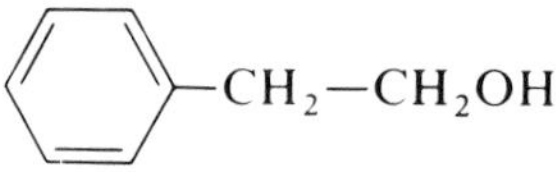

$$\langle \text{ring} \rangle - CH_2 - CH_2OH$$

(31)

2-Phenylethanol

The metabolism of **2-phenylethanol** (phenethyl alcohol) (31) in rabbits was studied by Smith et al. (1954) and by El Masri et al. (1956). In the former investigation, an average of 7% of an oral dose (460 mg/kg) was excreted combined with glucuronic acid while no increase in the ethereal sulphate output was detected. The other study indicated that the alcohol is oxidized to phenylacetic acid which is conjugated with glycine and excreted as phenaceturic acid. In addition, a small amount of hippuric acid was detected.

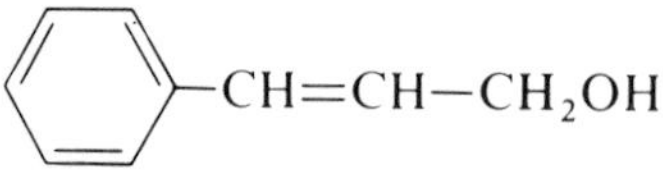

$$\langle \text{ring} \rangle - CH = CH - CH_2OH$$

(32)

Cinnamyl alcohol

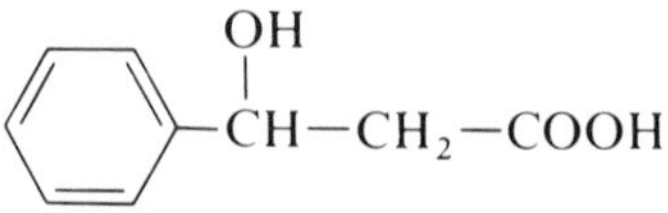

$$\langle \text{ring} \rangle - \overset{\displaystyle OH}{\underset{\displaystyle |}{CH}} - CH_2 - COOH$$

(33)

$$\langle \text{ring} \rangle - CH - CH - COOH$$

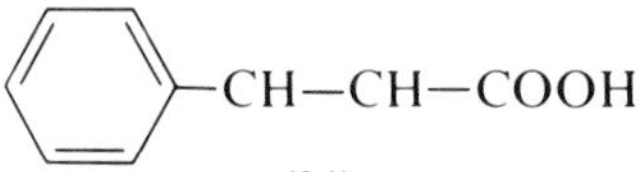

(34)

Another unsubstituted aromatic alcohol is **cinnamyl alcohol** (styryl alcohol) (32) which, when given to rabbits orally in repeated daily doses of

3–5 g, was shown by Fischer and Bielig (1940) to result in the urinary excretion of small amounts of unchanged compound, its reduced product dihydrocinnamyl alcohol (3-phenylpropanol) and perhaps some 1-phenylpropane-1,3-diol. The latter metabolite is probably an intermediate in the oxidation of the C_6—C_3 alcohol to C_6—C_1 compounds and 30–35% of the dose was recovered as benzoic acid. Recently, Peele and Oswald (1977) administered cinnamyl alcohol (274 mg/kg, i.p.) to rats and found that 3-hydroxy-3-phenylpropionic acid (33), the oxidation product of the diol noted above, was excreted in the urine. The major acidic metabolite was hippuric acid and cinnamyl alcohol itself was the sole neutral metabolite detected. Teuchy et al. (1971) showed earlier that cinnamyl alcohol is a precursor of benzoic acid in rats. They reported that an average of 10% of the dose (550 mg/rat, i.p.) was excreted in the urine as hippuric acid. Peele and Oswald did not detect cinnamic acid (34), the direct oxidation product of cinnamyl alcohol, as a urinary metabolite in rats, however Fischer and Bielig reported that it was the major metabolite in rabbits, accounting for 65–70% of the dose. They also detected trace amounts of the corresponding reduced acid, dihydrocinnamic acid.

The metabolism of **saligenin** (salicyl alcohol) (35a) has received little attention, however Williams (1959, p. 320) noted that it is oxidized to the expected salicylic acid. The rate of oxidation in rabbits may be somewhat slower than that seen with benzyl alcohol as some unchanged compound is excreted following a 1 g dose. Using about half this dose, Williams (1938a) found that 6–10% was excreted conjugated with sulphate.

(35)

a) Saligenin (R = H)
b) Salicin (R = β-D-glucose)

(36)

Coniferin

Salicin (35b) is the phenolic glucoside of saligenin. According to Drasaɪ and Hill (1974), this compound is hydrolysed by a number of intestinal bacteria including enterobacteria (*Klebsiella*), streptococci, staphylococci and non-sporing anaerobes (bifidobacteria). It seems evident that ingestion of salicin would lead to the intestinal liberation of saligenin which would then be absorbed and metabolized as noted above. **Coniferin** (36) is a closely related glycoside, being the 4-β-D-glucoside of coniferyl alcohol. Drasar and Hill reported that enterobacteria (*Proteus*, *Escherichia* and *Klebsiella*) are able to hydrolyse it. This will similarly bring about its absorption and metabolism, however the metabolic fate of coniferyl alcohol has not been studied.

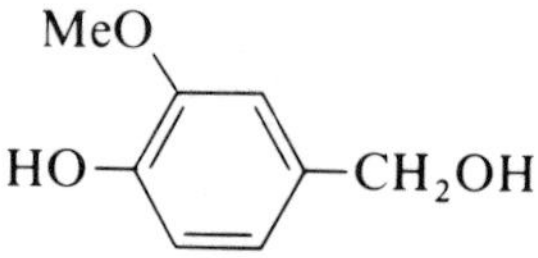

(37)

Vanillyl alcohol

Vanillyl alcohol (37) occurs in the vanilla fruit as a phenolic glucoside. Strand and Scheline (1975) dosed rats orally with the alcohol (100 or 300 mg/kg) and found that the pattern of metabolites excreted in the urine was essentially the same as that shown by the corresponding aldehyde, vanillin. The compounds identified were vanillyl alcohol itself, its oxidation products vanillin and vanillic acid, the latter being excreted to a small extent as its glycine conjugate. In addition, the O-demethylation, decarboxylation and alcohol reduction reactions shown by Scheline (1972) to be carried out with vanillyl alcohol and its metabolites by the rat intestinal microorganisms *in vitro* were also found to occur *in vivo*. These reactions accounted for the presence in the urine, in conjugated form, of catechol, 4-methylcatechol, guaiacol and 4-methylguaiacol. These metabolic pathways are illustrated in Fig. 4.1, Chapter 4. The O-methylation of vanillyl alcohol to 3,4-dimethoxybenzyl alcohol was reported by Friedhoff *et al.* (1972) to be carried out by a guaiacol-O-methyltransferase system found in the 100 000 g supernatant fraction from rat liver which utilized S-adenosylmethionine. This pathway is probably not quantitatively significant *in vivo*.

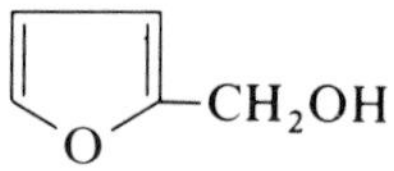

(38)

Furfuryl alcohol

The metabolism of **furfuryl alcohol** (38) appears to be straightforward, involving oxidation to furoic acid. Paul *et al.* (1949) found that the major urinary metabolite of the alcohol (approx. 50 mg/kg, p.o.) in rats is furoylglycine.

II. Phenols

Phenolic groups are a common feature of numerous classes of plant substances. Accordingly, phenolic compounds furnish a large and heterogenous group, examples of which are found in most of the chapters in this book. This distribution indicates that other structural features are usually considered to be more important or typical for the compound in question. We find that compounds in which the phenolic group is the sole or most prominent functional group are relatively rare in plants. It is with this rather restricted group of compounds that the present section is concerned.

Mono-, di- and trihydric phenols of the simpler types generally seem to be found in relatively few plants. The most widely distributed of these compounds is probably hydroquinone, however even the simplest member, phenol, has been occasionally reported to occur in small amounts in a few plant sources. In view of the rather limited amount of metabolic data available on most of the other members of this group, it seems justified to begin this section with a summary of the metabolic fate of phenol itself.

A general understanding of the main metabolic pathways of **phenol** has been available for some time, the earliest investigations dating from the last century. In brief, phenol may undergo conjugation reactions directly by virtue of the presence of a free hydroxyl group or it may undergo oxidation which gives rise to hydroxylated metabolites, those in turn being subject to conjugation. The former pathways largely involve the formation of glucuronide or ethereal sulphate conjugates, however a phosphate conjugate of phenol has also been identified as a urinary metabolite of phenol in cats (Capel *et al.* 1974). The oxidative pathway furnishes mainly hydroquinone, the *p*-hydroxylated product, although some *o*-hydroxylation to catechol may sometimes be detected. Not unexpectedly, the conjugative pathways are quantitatively far more important than is the oxidative pathway. Differences in experimental procedures make it difficult to compare the results of earlier studies with those obtained more recently which have made use of ^{14}C-labelled phenol. The former investigations were carried out in rabbits and measured primarily the urinary excretion of glucuronide and sulphate conjugates. These are largely conjugates of phenol itself, but the contribution of small amounts of oxidized metabolites to these values was usually not ascertained. In any case, Williams (1938a) showed that the extent of phenol conjugation with sulphate in rabbits is dose dependent, oral doses of 25 or 50 mg/kg giving values of 41 and 37% of the dose, respectively, whereas doses of 100–250 mg/kg gave values of 18–19%. The latter values are in close agreement with those of Bray *et al.* (1952a) who reported 15–16% using a dose of 250 mg/kg in similar experiments. Conjugation of phenol with glucuronic acid at these higher dose levels (125–250 mg/kg, p.o.) accounts for about 70% of the dose (Porteous and Williams, 1949; Bray *et al.*, 1952b). Porteous and Williams also found that approx. 14% of the dose was oxidized to other phenols, mainly hydroquinone.

The availablility of ^{14}C-labelled phenol has resulted in a much more comprehensive understanding of the pathways of phenol metabolism, especially in regard to the oxidized metabolites. The initial experiments of this type were carried out by Parke and Williams (1953) who found that orally administered phenol (50–60 mg/kg) was about equally excreted in the urine of rabbits as the glucuronide and sulphate conjugates (approx. 45% of dose, each). The value for the latter conjugate is similar to that noted above at this dose level. In addition, about 10% was recovered as hydroquinone, 0.5–1% as catechol and a trace as hydroxyquinol (1,2,4-trihydroxybenzene). No evidence was found for the formation of phenylmercapturic acid or any of the isomers of muconic acid (HOOC—CH=CH—CH=CH—COOH) which might have arisen by ring cleavage. Another development facilitated by the use of ^{14}C-phenol has been the study of the species differences in its metabolism. These studies have been carried out in a large number of species by Capel et al. (1972) and French et al. (1974). Some of the major findings from these and a few related investigations are summarized in Table 3.2. Free phenol is not excreted following oral dosage. These results show that the formation of both glucuronide and sulphate conjugates is common, however in cats and the closely related genet, civet and lion only sulphate conjugation is seen, except for a trace of phenyl glucuronide detected in cats. It is of interest to note that conjugation with phosphate also occurs in cats and, in experiments similar to those summarized in Table 3.2 but using i.p. dosage, as much as 12% of the dose was excreted as monophenyl phosphate ($C_6H_5OPO(OH)_2$). In regard to the glucuronide : sulphate conjugation ratio, pigs were found to represent the other extreme with nearly all the excreted material being the glucuronide. Oehme and Davis (1969) also reported that pigs excrete phenol conjugated primarily with glucuronic acid. Goats, a species not included in Table 3.2, were found to excrete mainly phenyl sulphate.

Another recent and significant finding concerning the metabolism of phenol was reported by Powell et al. (1974) who showed that conjugation of orally administered phenol in rats occurs largely in the gastrointestinal tract prior to its uptake into the portal circulation and transport to the liver. A novel but minor metabolite of phenol in rats is the o-methoxylated derivative guaiacol (Bakke, 1970). This metabolite, found consistently but only in trace amounts following oral doses of 100 mg/kg, is formed via catechol which can then be O-methylated.

As noted above, the simple dihydric phenols are not common plant compounds. Catechol (1,2-dihydroxybenzene) and especially hydroquinone (quinol; 1,4-dihydroxybenzene) occur naturally, however the resorcinol (1,3-dihydroxybenzene) structure appears to be found mainly in

Urinary metabolites of phenol in various species

Species	Dose mg/kg, p.o.	% of dose excreted in 24 h	% excreted in 24 h as:					Reference
			Phenyl glucuronide	Phenyl sulphate	Hydro-quinone glucuronide	Hydro-quinone sulphate	Other	
Rodents								
Mouse	25	66	35	46	15	5		a
Rat	25	95	42	54	2	1		a
Egyptian jerboa	25	47	26	61	1	12		a
Gerbil	25	55	35	42	1	19		a
Hamster	25	73, 78	44, 41	27, 24	28, 27	1, trace		a
Lemming	25	40	39	35	15	10		a
Guinea pig	25	64	82, 73	13, 22	5	trace		a
Other								
Indian fruitbat	25	50, 58	91, 89	9, 11	—	—		a
Rabbit	25	48	46	45	—	9		a
European hedgehog	20	34, 43	20, 10	63, 86	—	17, 4		a
Pig	25	76, 95	96, 99	4, 1	—			a
Elephant	10	49	25	73	—	1		b
Carnivores								
Ferret	25	51	40	28	—	30		a
Cat	25	49	0·2	88	—	9		a,c
Cat	20[d]		—	80	—	20	Monophenyl phosphate, 2·5	e
Cat	40[d]		—	78	—	16	Phenol,2; hydroquinone,4	e
Dog	25	53, 62	24, 12	33, 68	—	43, 20		a
Forest genet	10	37, 58	—	100, 98				f
African civet	10	60	—	97				f
Hyaena	10	15, 47	0	93, 86		6, 13		b
Lion	10	77	—	99				f
Primates								
Capuchin	25	73	65	14	21	—		a
Squirrel monkey	25	31	68	7	25	—		a
Rhesus monkey	50	37, 49	40, 30	60, 70	—	—		a
Man	0·01	90	16	77	trace	1		a

[a] Capel et al. (1972).
[b] Caldwell et al. (1975).
[c] Capel et al. (1974).
[d] Given i.v., urine collected for 6 h.
[e] Miller et al. (1973, 1976).
[f] French et al. (1974).

various 5-alkyl derivatives. The simplest of these is **orcinol** (5-methyl-resorcinol) which is generally found in lower plants (e.g. lichens) but may also occur as glycosides in some heathers. According to early reports summarized by Williams (1959, p. 306), orcinol undergoes the expected metabolic reactions of conjugate formation with one of the phenolic groups. Martin *et al.* (1975) found that when orcinol (300 mg/day) was administered to sheep by continuous ruminal infusion, approximately 90–100% of the dose was recovered as urinary orcinol. However, it was not stated if this was as free or conjugated material. Corresponding urinary recoveries of quinol and catechol were 55–88% and 22–44%, respectively.

Not unexpectedly, formation of mono-glucuronide and mono-sulphate conjugates are the main routes of metabolism with the simple dihydric phenols and, again, Williams (1959, pp. 303–306) has conveniently summarized the earlier literature which furnishes most of the data on the conjugation of these compounds. It seems reasonable to assume that the conjugation of catechol and hydroquinone is very similar to that noted above with phenol, however the data obtained from various animal species are quite limited. In rabbits, **catechol** (100–200 mg/kg, p.o.) is excreted in the urine as glucuronide (70–75% of dose) and sulphate (18%) conjugates (Garton and Williams, 1948; Bray *et al.*, 1952b). In the former study free catechol (2% of dose) was isolated from the urine and traces of the ethereal sulphate of hydroxyquinol (1,2,4-trihydroxybenzene) were detected. With **hydroquinone** in similar experiments, Garton and Williams (1949) reported values of 43% and 30% for glucuronide and sulphate conjugation, respectively, while Bray *et al.* (1952b) reported corresponding values of 78% and 18%. More recently, Miller *et al.* (1973, 1976) found that the radioactivity in the urine from cats given [^{14}C]-hydroquinone (20 mg/kg, i.v.) consisted mainly of its sulphate conjugate (87%) together with a glucuronide conjugate (3%) and free hydroquinone (10%).

In addition to the conjugation reactions noted above, polyhydric phenols may undergo other routes of metabolism not shown by phenol or related monohydric derivatives. These reactions include oxidation to quinone derivatives and, in the case of *o*-dihydric compounds (catechols), *O*-methylation. Oxidation to quinones may proceed further to polymerization products which give the urine a dark appearance (see Miller *et al.*, 1973). Catechol is known to be an excellent substrate of the *O*-methyl transferase system in rat liver (Axelrod and Tomchick, 1958). Bakke (1970) studied this reaction *in vivo* in rats. Following the oral administration of catechol (100 mg/kg, p.o.), the 48 h urines contained, in conjugated form, roughly 7% of the dose as guaiacol. Di-*O*-methylation, which would form veratrol, was not detected. As expected, resorcinol and hydroquinone did not form *O*-methylated metabolites.

Arbutin is the monoglucoside of hydroquinone. It seems likely that this compound, if fed to animals, would be poorly absorbed from the gastrointestinal tract and therefore come into contact with the β-glucosidases produced by the gut microflora. This should result in its hydrolysis, after which the aglycone may be absorbed and metabolized as described above. Drasar and Hill (1974) reported that arbutin is hydrolysed by enterobacteria (*Escherichia* sp.).

Both **guaiacol**, the monomethyl ether of catechol, and **hydroquinone monomethyl ether** occur naturally. Their metabolism to glucuronide and sulphate conjugates appears to be very similar to that shown by phenol and Williams (1938a) found that similar amounts (roughly 20%) of the three phenols given to rabbits in equivalent doses (250 mg/kg of phenol) were excreted as ethereal sulphates. Bray *et al.* (1952b), in similar experiments with hydroquinone monomethyl ether using a dose of 380 mg/kg, reported respective values of 69% and 13% for glucuronide and sulphate excretion. Grischkanski (1941) reported that some unchanged guaiacol is excreted in the respiratory air of rabbits given large oral doses (1 g/kg). Another possible metabolic pathway of these phenolic ethers is O-demethylation and Wong and Sourkes (1966) found that the urine of rats given guaiacol (50 mg/kg, i.p.) contained the glucuronide and/or sulphate conjugates of catechol. Bray *et al.* (1955) estimated that about 3% of a dose of hydroquinone monomethyl ether (approx. 250 mg/kg) was excreted in the urine of rabbits as demethylated product (hydroquinone). Similar experiments with **hydroquinone dimethyl ether** also showed O-demethylation. About a third of the dose was excreted as the monomethyl ether and some hydroquinone was also formed.

The metabolism of the trihydric phenols pyrogallol and phloroglucinol closely follows the general patterns outlined above, involving formation of glucuronides and ethereal sulphates (Williams, 1959, p. 306). **Pyrogallol** (1,2,3-trihydroxybenzene), being a catecholic compound, would be expected to undergo O-methylation and this reaction has been shown to take place in both *in vitro* and *in vivo* experiments. Some confusion as to the nature of the product or products formed has occurred. Archer *et al.* (1960) believed that catechol-O-methyl transferase catalysed the stepwise methylation of pyrogallol, first to 1-O-methylpyrogallol and then to 1,2-di-O-methylpyrogallol. However, similar experiments carried out by Masri *et al.* (1964) indicated the selective O-methylation of the middle hydroxyl group, forming 2-O-methylpyrogallol. It was noted that pyrogallol is a relatively poor substrate for catechol-O-methyltransferase and Masri *et al.* (1962) had earlier reported that only a trace of the 2-methoxy metabolite could be detected in the urine of rats given pyrogallol (50 mg, p.o.), although it is not clear if this compound was also present in the large amount of conjugated material that was excreted. A subsequent *in vivo* study in rats by Scheline (1966) clearly indicated that a moderate amount of pyrogallol was converted to 2-*O*-methylpyrogallol, which was then excreted entirely in conjugated form. This finding was confirmed by Bakke (1970) who showed that about 6% of the dose (100 mg/kg, p.o.) underwent this metabolic pathway. Another reaction with pyrogallol is its dehydroxylation to resorcinol (1,3-dihydroxybenzene) (Scheline, 1966). However, this dehydroxylation reaction, which is of minor quantitative importance, is carried out not by the tissues but by the intestinal bacteria.

Phloroglucinol (1,3,5-trihydroxybenzene), which lacks the catechol structure, is not a substrate of catechol-*O*-methyltransferase and no methylated derivatives were detected in the urine of rats given oral doses of 100 mg/kg of the phenol (Bakke, 1970). Takaji *et al.* (1971) gave phloroglucinol i.v. to humans and found that the urine contained unchanged compound (1–2%), phloroglucinol glucuronide (37–50%) and other conjugates (12–14%). The distribution and excretion of [³H]— phloroglucinol (50 mg/kg, i.v.) in rats was studied by Fujie and Ito (1972). It is of interest that more than a fifth of the injected dose was excreted in the faeces, a finding no doubt related to the often demonstrated ability of the rat to excrete phenolic compounds in the bile as their glucuronide conjugates.

Of the *O*-methylated derivatives of pyrogallol known to occur in plants, the 1,3-dimethyl ether **(2,6-dimethoxyphenol)** was shown by Miller *et al.* (1973, 1976) to be metabolized in cats mainly by hydroxylation and ethereal sulphate formation to 2,6-dimethoxyquinol disulphate. Over 90% of the urinary radioactivity was due to this metabolite when the ¹⁴C-labelled parent phenol was administered (20 or 40 mg/kg, i.v.). Small amounts of the unchanged compound and a glucuronide conjugate were detected and, at the higher dosage, a few percent of the urinary radioactivity was due to free 2,6-dimethoxyquinol. The urine samples obtained using the higher dosage were characteristically dark in colour.

(39)   (40)   (41)
Thymol   Carvacrol

Two isomeric methyl isopropyl phenols are **thymol** (39) and **carvacrol** (40). No new data on the metabolism of these compounds have appeared since Williams (1959, pp. 300–301) reviewed the earlier reports which showed that thymol is excreted as glucuronide and ethereal sulphate conjugates in rabbits, dogs and man. Robbins (1934) reported that dogs and man did not excrete unchanged thymol in the faeces but that roughly a third of the dose (1 g) was excreted in the urine as free thymol. Williams also stated that a small amount of oxidation, giving thymoquinol (41), also occurs, however our present understanding of the patterns of metabolism of this antiseptic phenol is scanty. Scheline (1977), using gas chromatographic-mass spectrometric techniques, found that rats given thymol (400 mg/kg, p.o.) excrete it in the urine partly as hydroxylated phenols. Following hydrolysis of the urine with a β-glucuronidase-sulphatase preparation, thymol itself was the most prominent metabolite. Two monohydroxylated derivatives of thymol were detected which were together present to the extent of about 15% of that found with thymol. This preliminary study did not indicate at which of the six available positions (three aliphatic and three aromatic) oxidation had taken place. The lack of metabolic data in the case of carvacrol (40) is even more pronounced and little useful information is available. Schröder and Vollmer (1932) found that carvacrol is rapidly excreted in the urine in rats and rabbits following its absorption with very little appearing in the faeces or expired air.

The stilbenes furnish an interesting but relatively limited group of phenolic compounds which possess fungicidal activity. The majority of these $C_6-C_2-C_6$ compounds have substituents (hydroxyl and/or methoxyl) in the 3- and 5-positions. None of these has been studied metabolically. In addition, a few naturally occurring stilbenes contain a single substituent in the 4-position and the metabolism of one of these, **4-hydroxystilbene** (42), is partially understood by virtue of the fact that it is the primary hydroxylation product of stilbene, an aromatic hydrocarbon which has been the subject of several metabolic investigations (Sinsheimer and Smith, 1968, 1969; Scheline, 1974; Tay and Sinsheimer, 1976). Several interesting findings emerge from these studies, perhaps the most notable being the discovery of several tri- and tetrahydroxylated metabolites. Furthermore, considerable reduction of the double bond of the stilbene metabolites to give bibenzyl derivatives was noted in rats. This reaction is carried out by the intestinal bacteria following the biliary excretion of glucuronide conjugates of the hydroxylated stilbenes. While many of these results using stilbene fall outside the scope of this book, the findings also suggest the probable metabolic pathways of 4-hydroxystilbene (Fig. 3.1).

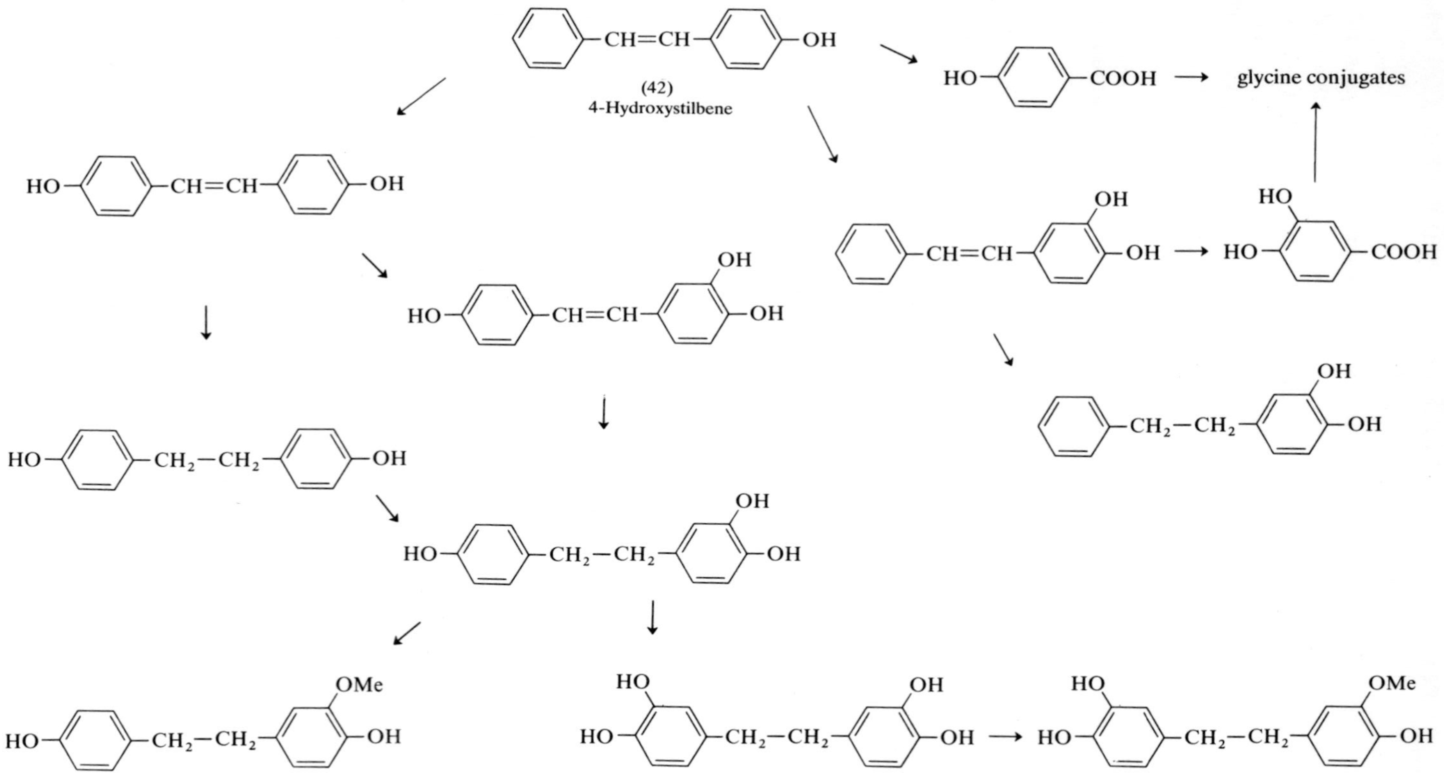

FIG. 3.1. Probable metabolic pathways of 4-hydroxystilbene in rats.

It is noteworthy that rats metabolize 4-hydroxystilbene more extensively than do rabbits and that they also reduce a larger portion of the dose to bibenzyl derivatives. The only polyhydroxylated metabolites found in rabbit urine were dihydroxylated (3,4- or 4,4'-) derivatives, however cleavage to C_6—C_1 (benzoic acid) derivatives is more pronounced in rabbits than in rats (Tay and Sinsheimer, 1976). Scheline (1977), in experiments similar to those carried out earlier with stilbene, administered 4-hydroxystilbene to rats and confirmed the pattern of phenolic metabolites found in earlier studies. The most abundant of these metabolites were the trihydroxylated derivatives. Small amounts of O-methylated derivatives of the tri- and tetrahydroxy compounds were detected in this newer study.

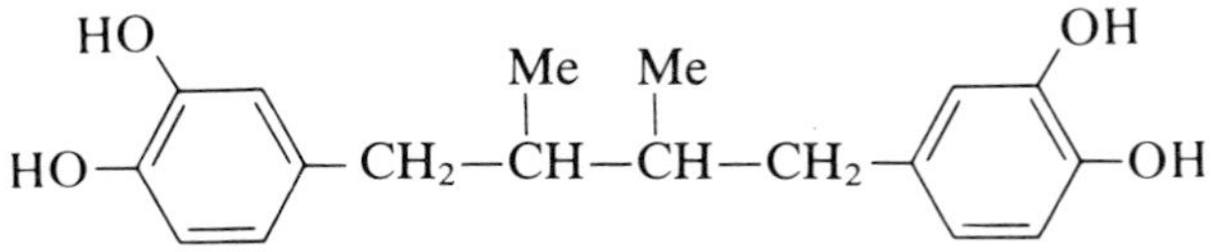

(43)

Nordihydroguaiaretic acid

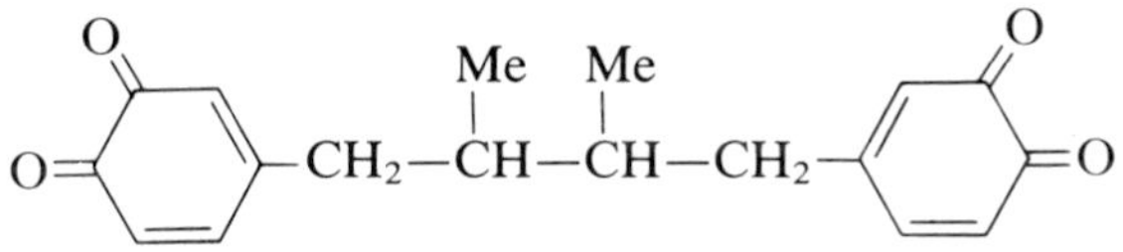

(44)

Nordihydroguaiaretic acid (43) is a plant phenol possessing antioxidant properties. Its metabolic fate is poorly understood, however Grice et al. (1968) reported that the corresponding o-quinone derivative (44) was present in the kidneys of rats fed for four weeks or more on a diet containing 2% of the phenol. Urine samples from these animals contained traces of the o-quinone but no free phenol. Interestingly, when rats were given single 250 mg doses of nordihydroguaiaretic acid, the ileum and caecum contained about 1% and 0.6%, respectively, of the dose as the o-quinone after 7·5 h. This finding is somewhat surprising as the reducing environment of the lower intestine would be expected to favour the phenolic rather than the quinoid form (see Chapter 4, Section III.B), thus making it unlikely that the lower intestine is the site of the o-quinone formation as suggested by Grice et al. Recently, Ve and Scheline (1976) found that rats given oral doses of nordihydroguaiaretic acid (100 mg) excreted in the urine small amounts of glucuronide and/or sulphate conjugates of the unchanged compound and its monomethyl ether.

III. Ethers

In the previous section it was noted that the phenolic group is a common feature of many classes of plant compounds. The same is equally true of the ether moiety and examples of the mammalian metabolism of many ethers are accordingly found in most sections of this book. Additionally, the general features of the metabolism of ethers, both that carried out by tissue enzymes and by the intestinal microflora, are reviewed in Chapter 1. The present section will therefore be limited to a review of the metabolism of a group of closely related ethers based on the alkenebenzene structure. These compounds can be divided into two types, derivatives of allylbenzene (45) or propenylbenzene (46), the latter compounds often having the prefix iso added to their common names. These compounds usually contain one or more substituents (hydroxyl, methoxyl or methylenedioxyl) in the 3-, 4- and 5-positions, although 2-substitution may also occur.

(45)   (46)

The simplest alkenebenzene ethers which have been studied metabolically are the 4-methoxyl derivatives. A detailed investigation of the fate of both **estragole** (4-allylanisole) (47) and **anethole** (4-propenylanisole) (56) in rats was carried out by Solheim and Scheline (1973). A large number of urinary and biliary metabolites was identified in both cases and the major routes of metabolism were also ascertained. The major urinary metabolites of estragole and anethole are shown in Figs 3.2 and 3.3, respectively, together with the approximate amounts excreted following doses of 100 mg/kg (p.o. or i.p.). Metabolites containing alcoholic or phenolic hydroxyl groups were excreted largely as glucuronide and/or sulphate conjugates whereas acidic metabolites were excreted mainly free. A major metabolic reaction with both of these compounds is O-demethylation which forms the corresponding p-hydroxy derivatives (48) and (57). The demethylation reaction was shown by Axelrod (1956) to be carried out with anethole by the liver microsomal O-demethylating system. Another common metabolite shown in Figs 3.2 and 3.3 is 4-methoxyhippuric acid (52). This compound is the most prominent urinary metabolite of anethole and the values found agreed well with those of Le Bourhis (1970) who reported that 35–50% of the anethole given to rats, rabbits and humans was excreted as conjugated material which included 4-methoxyhippuric acid. These results show that metabolism via cinnamoyl derivatives, e.g.

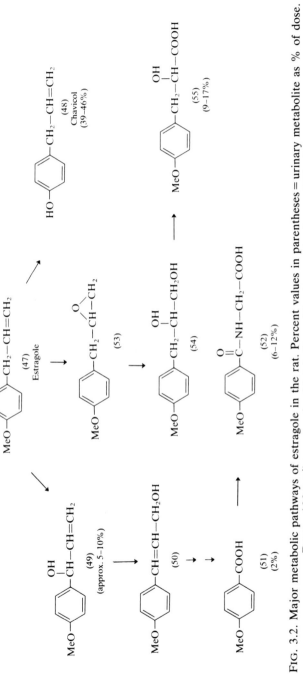

FIG. 3.2. Major metabolic pathways of estragole in the rat. Percent values in parentheses = urinary metabolite as % of dose. Dose = 100 mg/kg, p.o. or i.p. Metabolites without percent values = <1% of dose.

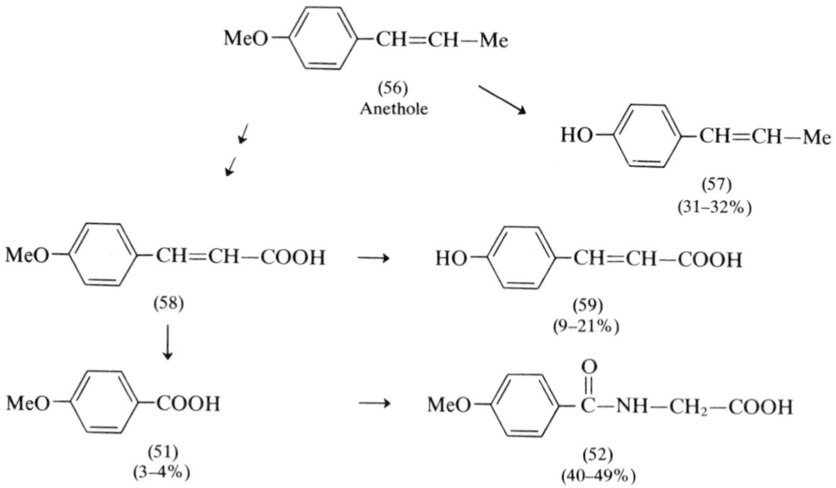

FIG. 3.3. Major metabolic pathways of anethole in the rat. Percent values in parentheses = urinary metabolite as % of dose. Dose = 100 mg/kg, p.o. or i.p. Metabolites without percent values = <1% of dose.

compounds (50) and (58), is more extensive with anethole than with estragole. As seen in Fig. 3.3, as much as 75% of the dose of anethole was excreted as metabolites (51), (52) and (59). This is perhaps to be expected as this pathway does not require migration of the double bond with anethole as with estragole. Conversely, epoxidation of the double bond was more pronounced with estragole than with anethole. This pathway, proceeding via metabolites (53) and (54), led to the formation of an α-hydroxy acid (55) which was a major urinary metabolite of the allyl derivative. Interestingly, no sulphur-containing metabolites were detected in the urine, suggesting that estragole epoxide is not conjugated with glutathione by the glutathione S-transferases. The epoxide-diol pathway appears to be of minor quantitative importance with anethole and only trace amounts of the isomeric 1',2'-diol (not shown in Fig. 3.3) were detected.

An interesting reaction with estragole is allylic hydroxylation to give the allyl alcohol derivative (49). This labile compound is excreted partly as a glucuronide conjugate, but it also isomerizes to alcohol (50) and may be oxidized to the keto derivative of metabolite (49) (not shown in Fig. 3.2). The latter compound can react with endogenous amines to form tertiary aminopropiophenones and evidence was obtained for the formation of small amounts of these metabolites. A more extensive discussion of these nitrogen-containing metabolites of allylbenzene derivatives is found below n the summary of the metabolism of safrole.

An alkenbenzene derivative closely related to estragole is **chavicol** (4-allylphenol) (48). While it has not itself been studied metabolically, the results obtained with estragole suggest that formation of glucuronide and/or sulphate conjugates is the major metabolic pathway. Nonetheless, administration of chavicol itself rather than its endogenous formation from estragole would likely lead also to the formation of some derivatives in which the allyl moiety had been oxidized.

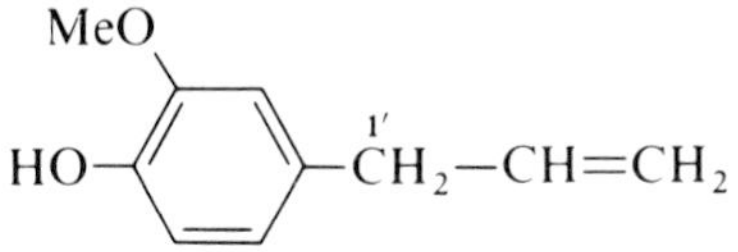

MeO

$$HO-\langle\rangle-\overset{1'}{C}H_2-CH=CH_2$$

(60)

Eugenol

The recent expansion of knowledge on the metabolism of alkeneben-zenes has not extended to **eugenol** (4-allyl-2-methoxyphenol) (60), about which only a few fragmentary results are available. Yuasa (1974) noted the effectiveness of eugenol in increasing the urinary excretion of ether glucuronides in rats. Preparations of liver microsomes from these animals showed elevated levels of UDP-glucuronyltransferase activity. Wineberg *et al.* (1972) administered [*methoxy*-^{14}C]-eugenol (450 mg/kg, i.p.) to rats and found that the tissue radioactivity declined rapidly, although traces could be detected after four days. The bulk of the tissue radioactivity was due to unchanged compound, but some unidentified, water-soluble, radioactive material was also usually present. Both ether-soluble and water-soluble metabolites were detected in the excreta, however the nature of these was not determined. Demethylation of eugenol occurred and 0·2–1% of the dose was recovered as respiratory $^{14}CO_2$. Another report on the metabolism of eugenol was made by Oswald *et al.* (1971b) who found that the same type of nitrogen-containing metabolites is formed as those seen when safrole or myristicin are administered to rats or guinea pigs (see below). This finding indicates that eugenol also undergoes allylic hydroxylation at the 1'-position. This suggests that the metabolic pathways shown in Fig. 3.4 for eugenol methyl ether (61) which proceed via metabolite (70) also occur with eugenol. A subsequent report by Oswald *et al.* (1972b) dealt with the identification of the highly labile tertiary amino-propiophenone metabolites of eugenol. These were found to be the *N,N*-dimethylamino, pyrrolidinyl and piperidyl derivatives similar to those formed from safrole which are illustrated below in structures (81), (82) and (83) of Fig. 3.6. A further indication that eugenol metabolism follows the general pathway shown with other allylbenzene derivatives is the finding of

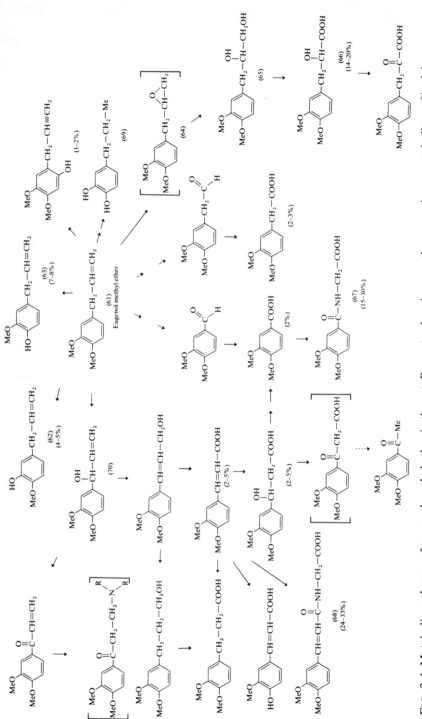

Fig. 3.4. Metabolic pathways of eugenol methyl ether in the rat. Percent values in parentheses = urinary metabolite as % of dose. Dose = 200 mg/kg, p.o. or i.p. Metabolites without percent values = <1% of dose. Postulated metabolites in brackets. Artifact formation indicated by broken arrow.

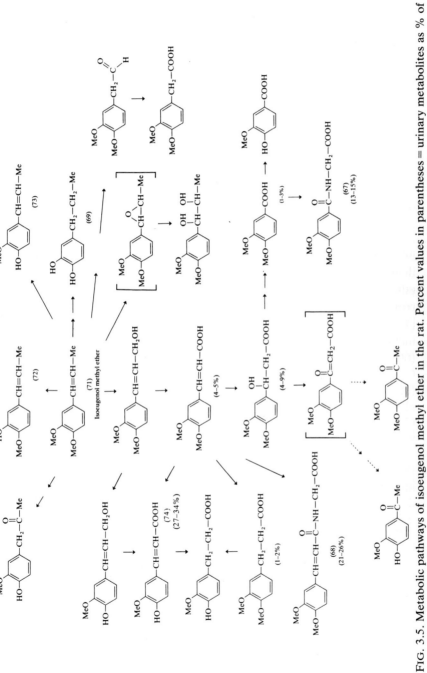

FIG. 3.5. Metabolic pathways of isoeugenol methyl ether in the rat. Percent values in parentheses = urinary metabolites as % of dose. Dose = 200 mg/kg, p.o. or i.p. Metabolites without percent values = <1% of dose. Postulated metabolites in brackets. Artifact formation indicated by broken arrows.

Padieu and Maume (1977) that rat liver cell lines convert it to its epoxide and then to the corresponding diol.

Although the metabolism of **eugenol methyl ether** (4-allylveratrole) (61) and **isoeugenol methyl ether** (4-propenylveratrole) (71) has been the subject of but a single investigation in rats (Solheim and Scheline, 1976), the use of gas chromatographic and gas chromatographic-mass spectrometric techniques resulted in the identification and quantitation of a very large number of both the major and minor urinary metabolites of these compounds. Their proposed routes of metabolism are shown in Fig. 3.4 (eugenol methyl ether) and Fig. 3.5 (isoeugenol methyl ether). These figures are much more detailed than those shown above for the monomethoxy derivatives (Figs 3.2 and 3.3) due to the inclusion of all identified minor metabolites and postulated intermediates. Nonetheless, common metabolic pathways are observed with the mono- and dimethoxy derivatives of each set of compounds. Quantitatively, O-demethylation is much less pronounced with the dimethoxy compounds than with the monomethoxy compounds. A total of only 11–13% due to the O-demethylated compounds (62) and (63) was found with eugenol methyl ether compared with the 39–46% noted with estragole (Fig. 3.2). With the corresponding propenyl derivatives these values were 1% (metabolites (72) and (73)), and 31–32% (Fig. 3.3), respectively. In summary, the major metabolic pathways of eugenol methyl ether give either the α-hydroxy acid (66), presumably via an epoxide (64) and diol (65) or the C_6—C_1 and C_6—C_3 acids which are also excreted as their glycine conjugates (67) and (68). With isoeugenol methyl ether the major metabolites were these glycine conjugates of the C_6—C_1 and C_6—C_3 acids and a free C_6—C_3 acid (74), indicating that the cinnamoyl pathway is the major metabolic route with the propenyl derivative. Further points of interest include the finding that the fully reduced and demethylated metabolite (69) is formed by the intestinal bacteria. Biliary metabolites, which in both numbers and amounts were greater with the allyl derivative than with the propenyl derivative, included most of the compounds found in the urine.

Safrole (75) is structurally closely related to eugenol methyl ether (61), differing only by the presence of a methylenedioxy group rather than two methoxy groups in the 3,4-position. Because of the discovery of the hepatotoxic and weak hepatocarcinogenic activities of safrole in animals, the study of its metabolic fate has recently assumed increased importance. This alkenebenzene derivative is therefore the most widely studied compound in this group and a general outline of its metabolism is now reasonably clear, although the quantitative assessments of the metabolites formed are somewhat deficient. Interestingly, metabolism studies with safrole have a long history. Heffter (1895) administered it orally to rabbits and dogs and was able to show small amounts of piperonylic acid (78) in the urine. This was the only metabolite found, however, and it was believed that much of the dose was excreted unchanged in the expired air. Several recent investigations clearly indicate, on the other hand, that safrole undergoes extensive metabolism which shows a close similarity to that described above with the related allylbenzene derivatives, estragole and eugenol methyl ether.

Cleavage of the methylenedioxy moiety appears to be a prominent metabolic reaction with safrole and this leads to the formation of catechol (3,4-dihydroxyphenyl) derivatives. This subject was studied by Casida *et al.* (1966) and Kamienski and Casida (1970) who administered safrole (0·8 mg/kg, p.o.) labelled with ^{14}C in the methylenedioxy position to mice. About 61% of the radioactivity was recovered as $^{14}CO_2$ in the expired air in 48 h (nearly entirely during the first 12 h). Urinary radioactivity amounted to 23% of the dose, mainly as water-soluble metabolites, and the faeces contained 3%. Five metabolites were detected chromatographically in the ether extract of the urine and one of these, piperonylic acid (78), was identified. This fraction did not contain piperonylglycine (79). *In vitro* experiments using the mouse liver microsome-NADPH system showed that safrole was demethylenated, yielding [^{14}C]-formate and a catechol derivative which, however, was not identified. It seems reasonable to assume that the catechol detected is the expected compound (84) as it, mainly in conjugated form, was found to be a major urinary metabolite of safrole in rats and guinea pigs (Stillwell *et al.*, 1974). Similarly, Strolin Benedetti *et al.* (1977) reported that this conjugate was the main urinary metabolite of [1'-^{14}C]-safrole in rats and man. The proposed metabolic pathways of safrole are shown in Fig. 3.6 which is largely based on the findings of these two investigations. The intraperitoneal doses employed by Stillwell *et al.* were 125 mg/kg (rats) and 50 mg/kg (guinea pigs). Oral doses of 0·8 or 750 mg/kg (rats) and approx. 0·16 or 1·7 mg (man) were given in the investigation of Strolin Benedetti *et al.* who found that nearly quantitative excretion of radioactivity in the urine occurred in 24 h when small amounts of safrole were given. However, only about 25% of the radioactivity was excreted in this period by rats which had received the larger dose.

The metabolic pathways of safrole shown in Fig. 3.6 indicate that allylic hydroxylation, giving 1'-hydroxysafrole (76) occurs. Stillwell *et al.* (1974) found that this compound was excreted by both rats and guinea pigs and also that the ultimate product via this route, piperonylglycine (79), is a major urinary metabolite. A further point of interest with this pathway is the finding that the isomeric allyl alcohol (77) arises from compound (76) by a chemical rearrangement reaction. This was previously reported by

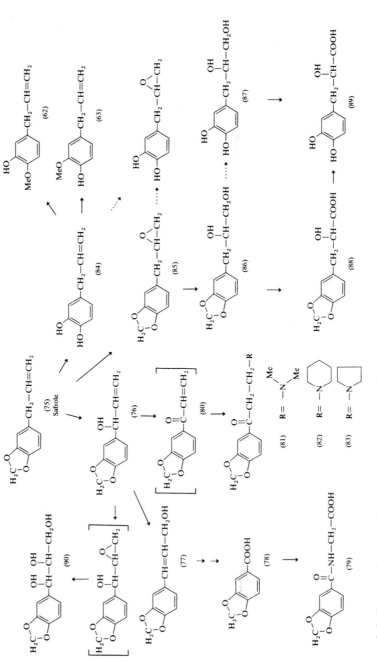

FIG. 3.6. Proposed metabolic pathways of safrole. Postulated metabolites in brackets. Broken arrows denote alternative pathways. Phenolic and alcoholic metabolites excreted mainly as conjugates.

Borchert *et al.* (1973a) who directed their attention to the allylic hydroxylation of safrole to form metabolite (76) in mice, rats, hamsters and guinea pigs. They found that as much as a third of the dose (300 mg/kg, i.p.) was excreted as the glucuronide conjugate of (76) in mice whereas the corresponding values in the other species were only 1–3%. Strolin Benedetti *et al.* (1977) found that metabolite (76) and its isomer, 3'-hydroxyisosafrole (77), were excreted in conjugated form in the urine of rats but not humans. The role of 1'-hydroxysafrole (76) in relation to the carcinogenic properties of safrole has been studied by Borchert *et al.* (1973a,b).

Another consequence of the formation of 1'-hydroxysafrole is its oxidation to 1'-ketosafrole (80). While Stillwell *et al.* (1974) and Borchert *et al.* (1973b) reported that this compound was not detected in the urines of safrole-treated rats, Borchert *et al.* (1973a) believed that it is formed *in vivo.* This interpretation is also strongly supported by the finding that condensation products of 1'-ketosafrole with several endogenous amines to give tertiary aminopropiophenone derivatives has been reported. These metabolites, found as urinary metabolites, were first described by Oswald *et al.* (1969) and subsequently characterized (Oswald *et al.*, 1971a). The tertiary aminopropiophenone derivative excreted by the guinea pig was shown to be the *N,N*-dimethyl derivative (81) whereas in rats the major component was the piperidyl derivative (82) with some *N,N*-dimethyl derivative and trace amounts of the pyrrolidinyl derivative (83) also being present. Peele (1976) reported that administration of synthetic 1'-hydroxysafrole to rats resulted in the urinary excretion of unchanged compound, tertiary aminopropiophenone derivatives and 3,4-methylenedioxyhippuric acid (79). The basic metabolites were not formed when 3'-hydroxyisosafrole (77) was given. A chemical study of the formation of the tertiary aminopropiophenone metabolites of safrole was reported by McKinney *et al.* (1972).

Another important metabolic pathway of safrole shown in Fig. 3.6 is that which leads to the formation of several postulated or detected epoxides, their diols and, ultimately, the α-hydroxy acids (88) and (89). The formation of relatively large amounts of α-hydroxy acids was noted above with related allylbenzenes and it is therefore not surprising that Stillwell *et al.* (1974) reported that compound (88) is a major urinary metabolite of safrole in rats and guinea pigs. The excretion of the 2',3'-diols (86) and (87) and the α-hydroxy acids following administration of safrole epoxide (85) supports this concept of the epoxide-diol pathway. Watabe and Akamatsu (1974) demonstrated the conversion of safrole oxide (85) to the diol (86) by hepatic microsomal epoxide hydratase. The triol (90) is an interesting polyhydroxylated metabolite which is excreted in small amounts by rats but not guinea pigs. Padieu and Maume (1977) recently reported the results of experiments designed to identify the metabolites of safrole following its administration to rats (dose = 1200 mg/kg, p.o.) and its incubation with rat liver homogenates and cell lines. These studies confirmed the reactions of demethylenation, 1'-hydroxylation, epoxidation and hydration of the epoxides to diols. Numerous unidentified biliary and urinary metabolites of safrole given intraperitoneally to rats were detected by Fishbein *et al.* (1967).

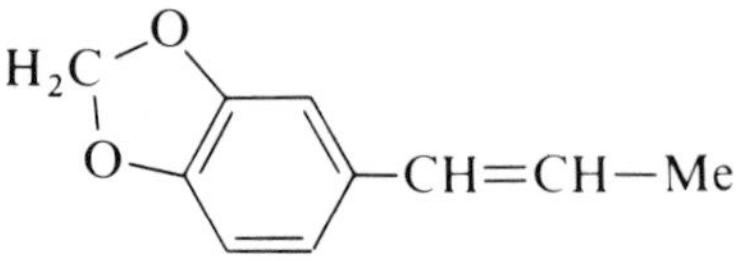

(91)

Isosafrole

Although it seems reasonable to assume that the metabolic fate of **isosafrole** (91) parallels that described above with related propenylbenzene derivatives, relatively little information is available on this subject. Heffter (1895) was unable to detect any piperonylic acid in the urine of isosafrole-treated rabbits and believed that most of the dose was lost unchanged in the expired air. That this situation is probably not the case was demonstrated by Fishbein *et al.* (1967) who detected chromatographically several metabolites of isosafrole in rat bile and urine. Padieu and Maume (1977) found that isosafrole was converted to its epoxide and then to the corresponding diol when incubated with rat liver cell lines. This epoxide-diol pathway is postulated to give rise to the corresponding diol identified as a metabolite of the closely related isoeugenol methyl ether (see Fig. 3.5). Oswald *et al.* (1969) reported that both safrole and isosafrole behaved similarly with respect to the formation of ketone derivatives and nitrogen-containing metabolites. However, allylic hydroxylation of isosafrole would form the 3-phenylallyl alcohol derivative (77) rather than the 1-phenylallyl alcohol derivative (76) which is probably the precursor of the postulated ketone intermediate. Therefore, the significance of this finding and the actual nature of the metabolites detected remains unclear.

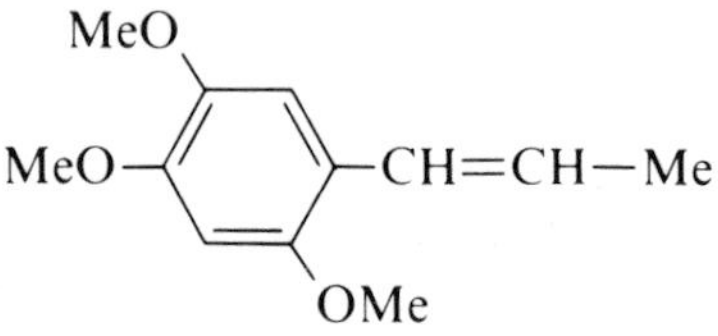

(92)

Asarone (*trans*)   β-Asarone (*cis*)

Asarone (*trans* isomer of (92)) and **β-asarone** (*cis* isomer of (92)) are structural isomers of isoelemicin (94). The only report dealing with their metabolic fate is that of Oswald *et al.* (1969) who studied the urinary excretion of nitrogen-containing metabolites of alkenebenzene derivatives. With the asarones the effect of *cis* and *trans* double bonds on the formation of these metabolites could be studied and it was found that the *trans* isomer underwent this reaction (producing ninhydrin-positive material) more rapidly and more extensively than did the *cis* isomer. Asarone gave rise to three such metabolites, however their identities have not been ascertained.

MeO—, MeO—, MeO— $-CH_2-CH=CH_2$

(93)

Elemicin

MeO—, MeO—, MeO— $-CH=CH-Me$

(94)

Isoelemicin

Little information has been published on the metabolic fate of **elemicin** (93). Oswald *et al.* (1972a) reported that it is converted to a small extent to tertiary aminopropiophenone derivatives. These metabolites are the 3-*N,N*-dimethylamino-, 3-piperidyl- and 3-pyrrolidinyl-1-(3′,4′,5′-trimethoxyphenyl)-1-propanones which are analogous to the safrole metabolites (81), (82) and (83), respectively (see Fig. 3.6). Recently, Solheim and Scheline (1977) administered elemicin and **isoelemicin** (94) orally (400 mg/kg) to rats and, similarly to that carried out previously with the mono- and dimethoxy derivatives noted above, identified the urinary metabolites. These experiments showed that the same general pathways of metabolism are utilized in all three sets of compounds, however the total amounts of identified metabolites were much smaller with the trimethoxy derivatives. Thus, total recoveries reached only 15–20% of the dose and most metabolites were detected in amounts ranging from traces to about 1% of the dose. Metabolite excretion occurred mainly during the first 24 h and only traces of the major metabolite were subsequently detected. 3,4,5-Trimethoxyphenylpropionic acid (95) was the major urinary metabolite of both compounds and accounted for approx. 7–10% of the dose. O-Demethylation of elemicin and isoelemicin did not occur to a large extent and the degradation to benzoic acid derivatives seen earlier with the mono- and dimethoxy compounds was not encountered. However, some of the metabolites of elemicin clearly demonstrated the presence of the epoxide-diol pathway. These included a triol (96) analogous to the safrole metabolite (90) shown in Fig. 3.6 and an epoxide derivative (97).

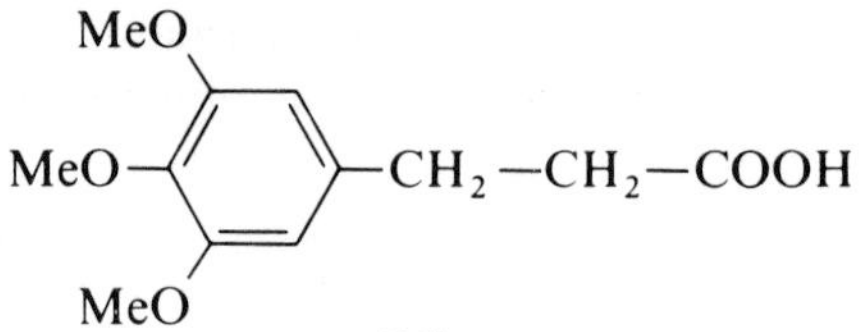

MeO
MeO—⟨benzene ring⟩—CH$_2$—CH$_2$—COOH
MeO

(95)

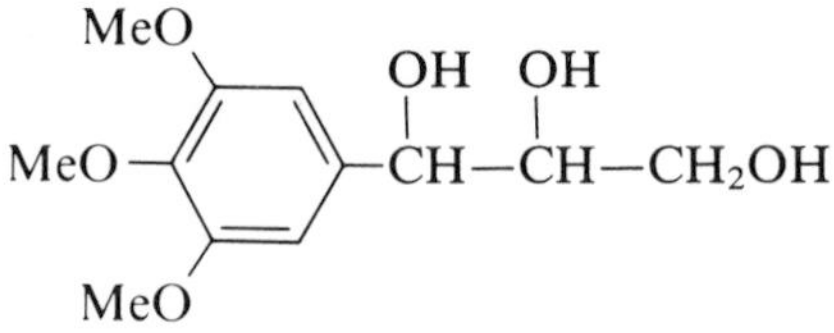

MeO
                    OH OH
                    | |
MeO—⟨benzene ring⟩—CH—CH—CH$_2$OH
MeO

(96)

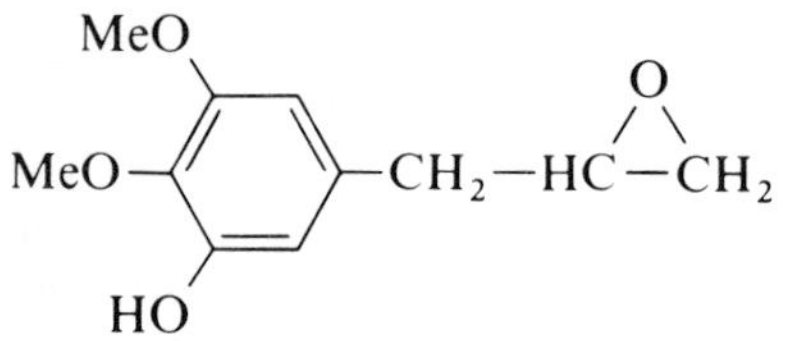

MeO
MeO—⟨benzene ring⟩—CH$_2$—HC—CH$_2$ (epoxide O)
HO

(97)

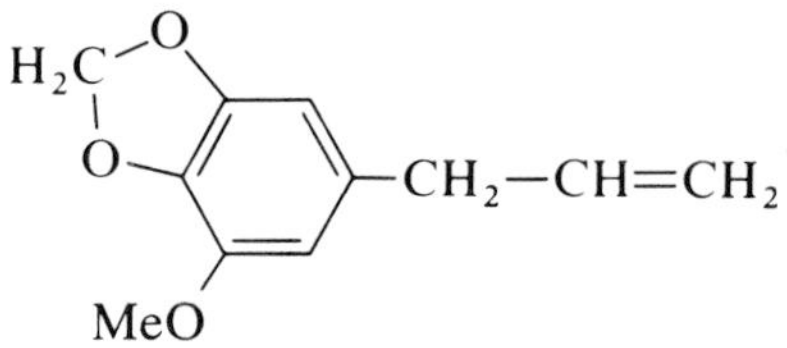

H$_2$C—O
        ⟨benzene ring⟩—CH$_2$—CH=CH$_2$
O
MeO

(98)
Myristicin

H$_2$C—O
        ⟨benzene ring⟩—CH$_2$—CH—NH$_2$
O |
MeO   Me

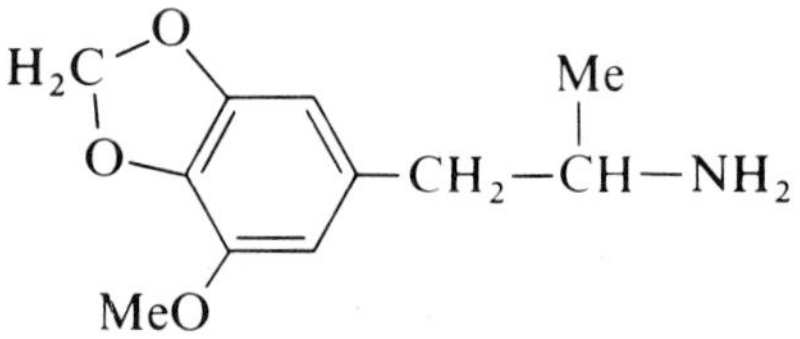

(99)

Several possible routes of metabolism of **myristicin** (98) were discussed by Shulgin (1966) in relationship to the possible psychotropic activity of this allylbenzene derivative. It was suggested that addition of ammonia to the allyl group may occur to give the amphetamine derivative (99). This hypothesis was investigated experimentally by Braun and Kalbhen (1973) using the isolated perfused rat liver or rat liver homogenates. Their results showed that myristicin was converted to 3-methoxy-4,5-methylene-dioxyamphetamine (99) by both preparations. Oxygenation of the liver homogenates during incubation markedly increased the yield of the metabolite and this was felt to indicate that an oxidation relation precedes introduction of the amino group. However, the quantitative importance of this pathway *in vivo* remains to be assessed.

Peele (1976) recently reported on the metabolism of myristicin, also labelled with ^{14}C in the methylenedioxy or methoxy groups, in rats. With the ^{14}C-labelled compounds, excretion of radioactivity was nearly complete in 48 h and occurred largely by the urinary route. Only about 2% of the radioactivity was lost as respiratory $^{14}CO_2$ when the compound labelled in the methoxy group was given, however this figure increased to 25% with the methylenedioxy-^{14}C material. Casida *et al.* (1966) and Kamienski and Casida (1970) similarly showed that myristicin undergoes extensive demethylenation *in vivo*. When mice were given a small dose (approx. 1 mg/kg, p.o.) of myristicin labelled with ^{14}C in the methylenedioxy moiety, 73% of the radioactivity was lost in 48 h as $^{14}CO_2$. Most of this excretion occurred during the first 12 h and the 48 h recoveries of urinary and faecal radioactivity were 15% and 3%, respectively. *In vitro* experiments using the mouse liver microsome-NADPH system indicated demethylenation to give formate and the corresponding catechol, however the two metabolites which were detected were not identified.

Peele (1976) also identified numerous urinary metabolites of myristicin in rats. The results obtained from the 24 h urine samples hydrolysed with β-glucuronidase showed similarities to that illustrated in Fig. 3.6 for safrole. Thus, 5-hydroxyeugenol, the catechol metabolite arising from demethylenation of myristicin, was identified. This metabolite is analogous to metabolite (84) produced from safrole. Additional metabolites were myristicin glycol (analogous to metabolite (86)), 1'-hydroxymyristicin (analogous to metabolite (76)) and 3'-hydroxyisomyristicin (analogous to metabolite (77)). Acidic metabolites included the methoxymethylenedioxy derivatives of benzoic and phenyllactic acids analogous to metabolites (78) and (88), respectively. Several additional metabolites were identified including the similarly substituted phenylacetic acid derivative and myristicin acetophenone. These pathways are known to occur with eugenol methyl ether (see Fig. 3.4) and the acetophenone derivative is probably an artifact produced from decarboxylation of the corresponding β-keto acid. Another metabolite of myristicin was tentatively identified as dihydromyristicin which is formed by the reduction of the allylic double bond.

In addition to these neutral and acidic metabolites of myristicin, Peele (1976) also observed the presence of Mannich base derivatives. The

(100)

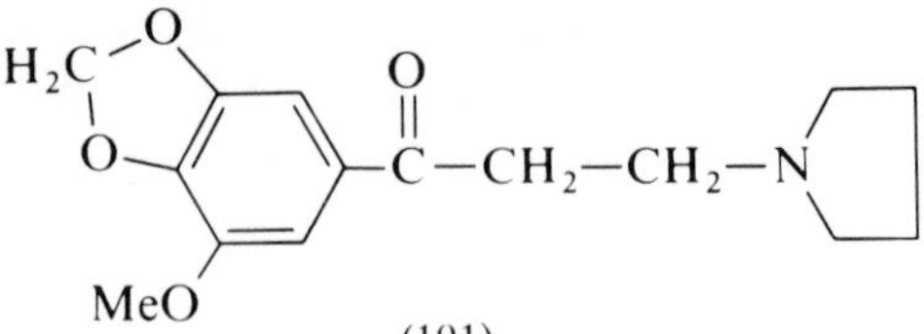

(101)

urinary excretion of these nitrogen-containing metabolites of myristicin and several related alkenbenzene derivatives was first reported by Oswald *et al.* (1969) and further investigated by Oswald *et al.* (1971b) who showed that these compounds are, in the rat, mainly 3-piperidyl-1-(3'-methoxy-4',5'-methylenedioxyphenyl)-1-propanone (100) and, in the guinea pig, mainly the corresponding 3-pyrrolidinyl derivative (101). Trace amounts of the other of these tertiary aminopropiophenone derivatives were also excreted by these two species, however a third type, the *N,N*-dimethyl-amino derivative analogous to that formed from safrole (Fig. 3.6, compound (81)), was not detected.

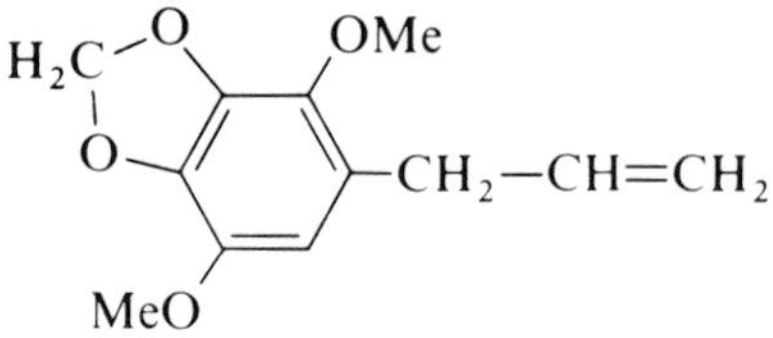

(102)

Apiol

Apiol (102) is 2-methoxymyristicin and Heffter (1895) reported that it is extensively oxidized in the body. An unidentified metabolite was found to be excreted in the urine of both rabbits and dogs. It seems reasonable to assume that apiol is metabolized along the pathways described above for related allylbenzenes, however the presence of the 2-methoxyl group may possibly reduce the extent of allylic hydroxylation and thus the production of the key intermediate, 1'-hydroxyapiol.

References

Archer, S., Arnold, A., Kullnig, R. K. and Wylie, D. W. (1960). *Archs Biochem. Biophys.* **87**, 153–154.
Asano, M. and Yamakawa, T. (1950). *J. Biochem.* Tokyo **37**, 321–327.
Axelrod, J. (1956). *Biochem. J.* **63**, 634–639.
Axelrod, J. and Tomchick, R. (1958). *J. biol. Chem.* **233**, 702–705.
Bakke, O. M. (1970). *Acta pharmac. tox.* **28**, 28–38.

Borchert, P., Wislocki, P. G., Miller, J. A. and Miller, E. C. (1973a). *Cancer Res.* **33**, 575–589.

Borchert, P., Miller, J. A., Miller, E. C. and Shires, T. K. (1973b). *Cancer Res.* **33**, 590–600.

Braun, U. and Kalbhen, D. A. (1973). *Pharmacology* **9**, 312–316.

Bray, H. G., Thorpe, W. V. and White, K. (1951). *Biochem. J.* **48**, 88–96.

Bray, H. G., Humphris, B. G., Thorpe, W. V., White, K. and Wood, P. B. (1952a). *Biochem J.* **52**, 419–423.

Bray, H. G., Thorpe, W. V. and White, K. (1952b). *Biochem. J.* **52**, 423–430.

Bray, H. G., Craddock, V. M. and Thorpe, W. V. (1955). *Biochem. J.* **60**, 225–232.

Browning, E. (1965). "Toxicity and Metabolism of Industrial Solvents". Elsevier, Amsterdam, London and New York.

Caldwell, J., French, M. R., Idle, J. R., Renwick, A. G., Bassir, O. and Williams, R. T. (1975). *FEBS Lett.* **60**, 391–395.

Capel, I. D., French, M. R., Millburn, P., Smith, R. L. and Williams, R. T. (1972). *Xenobiotica* **2**, 25–34.

Capel, I. D., Millburn, P. and Williams, R. T. (1974). *Biochem. Soc. Trans.* **2**, 305–306.

Carr, C. J. and Krantz, J. C. (1945). *Adv. Carbohyd. Chem.* **1**, 175–192.

Casida, J. E., Engel, J. L., Essac, E. G., Kamienski, F. X. and Kuwatsuka, S. (1966). *Science* N.Y. **153**, 1130–1133.

Cornish, H. H. (1975). *In* "Toxicology, The Basic Science of Poisons" (L. J. Casarett and J. Doull, Eds), pp. 503–526. Macmillan, New York.

Diack, S. L. and Lewis, H. B. (1928). *J. biol Chem.* **77**, 89–95.

Di Vincenzo, G. D., Kaplan, C. J. and Dedinas, J. (1976). *Toxic. appl. Pharmac.* **36**, 511–522.

Drasar, B. S. and Hill, M. J. (1974). "Human Intestinal Flora" p. 63. Academic Press, London, New York, San Francisco.

El Masri, A. M., Smith, J. N. and Williams, R. T. (1956). *Biochem. J.* **64**, 50–56.

Fischer, F. G. and Bielig, H.-J. (1940). *Hoppe-Seÿler's Z. physiol. Chem.* **266**, 73–98.

Fishbein, L., Fawkes, J., Falk, H. L. and Thompson, S. (1967). *J. Chromat.* **29**, 267–273.

French, M. R., Bababunmi, E. A., Golding, R. R., Bassir, O., Caldwell, J., Smith, R. L. and Williams, R. T. (1974). *FEBS Lett.* **46**, 134–137.

Friedhoff, A. J., Schweitzer, J. W., Miller, J. and Van Winkle, E. (1972). *Experientia* **28**, 517–519.

Fromm, E. and Hildebrandt, H. (1901). *Hoppe-Seyler's Z. physiol. Chem.* **33**, 579–594.

Fujie, K. and Ito, H. (1972). *Arzneimittel-Forsch.* **22**, 777–780.

Garton, G. A. and Williams, R. T. (1948). *Biochem. J.* **43**, 206–211.

Garton, G. A. and Williams, R. T. (1949). *Biochem. J.* **44**, 234–238.

Grice, H. C., Becking, G. and Goodman, T. (1968). *Fd Cosmet. Toxicol.* **6**, 155–161.

Grischkanski, A. (1941). *Naunyn-Schmiedebergs Arch. exp. Path. Pharmak.* **65**, 283–296.

Haggard, H. W., Miller, D. P. and Greenberg, L. A. (1945). *J. ind. Hyg. Toxicol.* **27**, 1–14.

Hämäläinen, J. (1909). *Skand. Arch. Physiol.* **23**, 86–98.

Hämäläinen, J. (1911). *Skand. Arch. Physiol.* **24**, 1–12.

Hämäläinen, J. (1912a). *Skand. Arch. Physiol.* **27**, 141–226.
Hämäläinen, J. (1912b). *Biochem. Z.* **41**, 241–246.
Hämäläinen, J. (1913). *Biochem. Z.* **50**, 220–222.
Heffter, A. (1895). *Arch. exp. Path. Pharmak.* **35**, 342–374.
Heinz, F. and Wittneben, H. E. (1970). *Hoppe-Seyler's Z. physiol. Chem.* **351**, 1215–1220.
Hiatt, E. P., Carr, C. J., Evans, W. E. and Krantz, J. C. (1938). *Proc. Soc. exp. Biol. Med.* **38**, 356–357.
Hildebrandt, H. (1901). *Arch. exp. Path. Pharmak.* **45**, 110–129.
Hildebrandt, H. (1904). *Beitr. chem. Physiol. Path.* **4**, 251–253.
Hinson, J. A. and Neal, R. A. (1975). *Biochim. biophys. Acta* **384**, 1–11.
Horning, M. G., Butler, C. M., Stafford, M., Stillwell, R. N., Hill, R. M., Zion, T. E., Harvey, D. J. and Stillwell, W. G. (1976). *In* "Advances in Mass Spectrometry in Biochemistry and Medicine" (A. Frigerio and N. Castagnoli, Eds), Vol. I, pp. 91–108. Spectrum Publications, New York.
Kamienski, F. X. and Casida, J. E. (1970). *Biochem. Pharmac.* **19**, 91–112.
Kamil, I. A., Smith, J. N. and Williams, R. T. (1953a). *Biochem. J.* **53**, 129–136.
Kamil, I. A., Smith, J. N. and Williams, R. T. (1953b). *Biochem. J.* **53**, 137–140.
Kamil, I. A., Smith, J. N. and Williams, R. T. (1953c). *Biochem. J.* **54**, 390–392.
Kuhn, R., Köhler, F. and Köhler, L. (1936). *Hoppe-Seyler's Z. physiol. Chem.* **242**, 171–197.
Le Bourhis, B. (1970). *Annls Pharm. Fr.* **28**, 355–361.
Licht, H. J. and Jamroz, G. (1977). *Fedn Proc. Fedn. Am. Socs exp. Biol.* **36**, 832.
Lieber, C. S. (1974). *In* "Chemical and Biological Aspects of Drug Dependence" (S. J. Mulé and H. Brill, Eds), pp. 135–161. CRC Press, Cleveland, Ohio.
McKinney, J. D., Oswald, E., Fishbein, L. and Walker, M. (1972). *Bull. Environ. Contam. Toxicol.* **7**, 305–310.
McMahon, R. E. (1971). *In* "Concepts in Biochemical Pharmacology" (B. B. Brodie and J. R. Gillette, Eds), Part 2, pp. 500–517. Springer, Berlin, Heidelberg, New York.
Maeda, K. (1966). *Sapporo Igaku Zasshi* **29**, 144–152 (*Chem. Abstr.* (1968) **69**, 50666s).
Magnus-Levy, A. (1907). *Biochem. Z.* **2**, 319–331.
Majchrowicz, E. (1975). "Biochemical Pharmacology of Ethanol". Plenum Press, New York and London.
Makar, A. B., Tephly, T. R. and Mannering, G. J. (1968). *Molec. Pharmac.* **4**, 471–483.
Martin, A. K., Milne, J. A. and Moberley, P. (1975). *Proc. Nutr. Soc.* **34**, 70A–71A.
Masri, M. S., Booth, A. N. and DeEds, F. (1962). *Biochim. biophys. Acta* **65**, 495–505.
Masri, M. S., Robbins, D. J., Emerson, O. H. and DeEds, F. (1964). *Nature Lond.* **202**, 878–879.
Mezey, E. (1976). *Biochem. Pharmac.* **25**, 869–875.
Miller, J. J., Powell, G. M., Olavesen, A. H. and Curtis, C. G. (1973). *Biochem. Soc. Trans.* **1**, 1163–1165.
Miller, J. J., Powell, G. M., Olavesen, A. H. and Curtis, C. G. (1976). *Toxic. appl. Pharmac.* **38**, 47–57.
Nasrallah, S. M. and Iber, F. L. (1969). *Am. J. med. Sci.* **258**, 80–88.
Neubauer, O. (1901). *Arch. exp. Path. Pharmak.* **46**, 133–154.
Oehme, F. W. and Davis, L. E. (1969). *Pharmacologist* **11**, 241.

Olmsted, W. H. (1953). *Diabetes* **2**, 132–137.

Oswald, E. O., Fishbein, L. and Corbett, B. J. (1969). *J. Chromat.* **45**, 437–445.

Oswald, E. O., Fishbein, L., Corbett, B. J. and Walker, M. P. (1971a). *Biochim. biophys. Acta* **230**, 237–247.

Oswald, E. O., Fishbein, L., Corbett, B. J. and Walker, M. P. (1971b). *Biochim. biophys. Acta* **244**, 322–328.

Oswald, E. O., Fishbein, L., Corbett, B. J. and Walker, M. P. (1972a). *J. Chromat.* **73**, 43–57.

Oswald, E. O., Fishbein, L., Corbett, B. J. and Walker, M. P. (1972b). *J. Chromat.* **73**, 59–72.

Padieu, P. and Maume, B. F. (1977). *In* "Quantitative Mass Spectrometry in Life Sciences" (A. P. de Leenheer and R. R. Roncucci, Eds), pp. 49–82. Elsevier Scientific Publishing Co., Amsterdam, Oxford, New York.

Parke, D. V. and Williams, R. T. (1953). *Biochem. J.* **55**, 337–340.

Parke, D. V., Rahman, K. M. Q. and Walker, R. (1974). *Biochem. Soc. Trans.* **2**, 612–615.

Paul, H. E., Austin, F. L., Paul, M. F. and Ells, V. R. (1949). *J. biol. Chem.* **180**, 345–363.

Peele, J. D. (1976). *Diss. Abstr. Int.* **37B**, 1234.

Peele, J. D. and Oswald, E. O. (1977). *Biochim. biophys. Acta* **497**, 598–607.

Porteous, J. W. and Williams, R. T. (1949). *Biochem. J.* **44**, 46–55.

Powell, G. M., Miller, J. J., Olavesen, A. H. and Curtis, C. G. (1974). *Nature Lond.* **252**, 234–235.

Pryde, J. and Williams, R. T. (1934). *Biochem. J.* **28**, 131–135.

Pryde, J. and Williams, R. T. (1936). *Biochem. J.* **30**, 799–800.

Quick, A. J. (1924). *J. biol. Chem.* **61**, 679–683.

Quick, A. J. (1928). *J. biol. Chem.* **80**, 535–541.

Rennhard, H. H. and Bianchine, J. R. (1976). *J. agric. Fd Chem.* **24**, 287–291.

Robbins, B. H. (1934). *J. Pharmac. exp. Ther.* **52**, 54–60.

Robertson, J. S. and Solomon, E. (1971). *Biochem. J.* **121**, 503–509.

Scheline, R. R. (1966). *Acta pharmac. tox.* **24**, 275–285.

Scheline, R. R. (1972). *Xenobiotica* **2**, 227–236.

Scheline, R. R. (1974). *Experientia* **30**, 880–881.

Scheline, R. R. (1977). Unpublished observations.

Schröder, V. and Vollmer, H. (1932). *Naunyn-Schmiedebergs Arch. exp. Path. Pharmak.* **168**, 331–353.

Shulgin, A. (1966). *Nature Lond.* **210**, 380–384.

Sinsheimer, J. E. and Smith, R. V. (1968). *J. pharm. Sci.* **57**, 713–714.

Sinsheimer, J. E. and Smith, R. V. (1969). *Biochem. J.* **111**, 35–41.

Sloane, N. H. (1965). *Biochim. biophys. Acta* **107**, 599–602.

Smith, J. N., Smithies, R. H. and Williams, R. T. (1954). *Biochem. J.* **56**, 320–324.

Snapper, J., Grünbaum, A. and Sturkop, S. (1925). *Biochem. Z.* **155**, 163–173.

Solheim, E. and Scheline, R. R. (1973). *Xenobiotica* **3**, 493–510.

Solheim, E. and Scheline, R. R. (1976). *Xenobiotica* **6**, 137–150.

Solheim, E. and Scheline, R. R. (1977). Unpublished observations.

Stetten, D. and Schoenheimer, R. (1940). *J. biol. Chem.* **133**, 347–357.

Stillwell, W. G., Carman, M. J., Bell, L. and Horning, M. G. (1974). *Drug. Metab. Disposit.* **2**, 489–498.

Strack, E. Kukfahl, E., Mueller, F. and Beyreiss, K. (1965). *Z. ges. exp. Med.* **139**, 23–32 (*Chem. Abstr.* (1966) **64**, 1120f).

Strand, L. P. and Scheline, R. R. (1975). *Xenobiotica* **5**, 49–63.
Strolin Benedetti, M., Malnoë, A. and Broillet, A. L. (1977). *Toxicology* **7**, 69–83.
Takaji, K., Suzuki, T., Saitoh, Y. and Nishihara, K. (1971). *Yakuzaigaku* **31**, 213–219 (*Chem. Abstr.* (1974) **80**, 78285t).
Tay, L. K. and Sinsheimer, J. E. (1976). *Drug. Metab. Disposit.* **4**, 154–163.
Tephly, T. R., Parks, R. E. and Mannering, G. T. (1964). *J. Pharmac. exp. Ther.* **143**, 292–300.
Teschke, R., Hasumura, Y. and Lieber, C. S. (1976). *Archs Biochem. Biophys.* **175**, 635–643.
Teuchy, H., Quatacker, J., Wolf, G. and Van Sumere, C. F. (1971). *Archs. int. Physiol. Biochim.* **79**, 573–587.
Touster, O. and Shaw, D. R. D. (1962). *Physiol. Rev.* **42**, 181–225.
Ve, B. and Scheline, R. R. (1976). Unpublished observations.
Watabe, T. and Akamatsu, K. (1974). *Biochem. Pharmac.* **23**, 2839–2844.
Weinberg, J. E., Rabinowitz, J. L., Zanger, M. and Gennaro, A. R. (1972). *J. Dent. Res.* **51**, 1055–1061.
Weinstein, A. N. and Segal, S. (1968). *Biochim. biophys. Acta* **156**, 9–16.
Williams, R. T. (1938a). *Biochem. J.* **32**, 878–887.
Williams, R. T. (1938b). *Biochem. J.* **32**, 1849–1855.
Williams, R. T. (1939). *Biochem. J.* **33**, 1519–1524.
Williams, R. T. (1940a). *Biochem. J.* **34**, 48–50.
Williams, R. T. (1940b). *Biochem. J.* **34**, 690–697.
Williams, R. T. (1959). "Detoxication Mechanisms". Chapman and Hall, London.
Wong, K. P. and Sourkes, T. L. (1966). *Can. J. Biochem.* **44**, 635–644.
Wright, S. E. (1945). *Pap. Dep. Chem. Univ. Qd* **1**, nr. 25.
Yuasa, A. (1974). *Jap. J. vet. Sci.* **36**, 427–432.

4

METABOLISM OF ALDEHYDES, KETONES AND QUINONES

I. Aldehydes

A guide to the general pathways of metabolism of aldehydes and to the enzyme systems involved is given in Chapter 1, Sections I.A.7 and I.B.2. In addition, some aromatic aldehydes have been found to undergo metabolism by the intestinal bacteria (Chapter 1, Section II.C.3).

A. ALIPHATIC ALDEHYDES

The existence of a profusion of plant aldehydes is well documented. These include straight-chain compounds from the simplest type, methanal (formaldehyde), to the 18-carbon aldehyde, n-octadecanal, as well as numerous branched-chain examples. As a rule, they are found in low concentrations, often in fruits to which they impart characteristic aromas. Among the simpler homologues, acetaldehyde (ethanal) is fairly frequently encountered but aldehydes containing from six to 12 carbon atoms are generally those which one commonly associates with this group of plant compounds. Decanal is the most widespread of the higher homologues, being found in citrus fruits and numerous aromatic oils.

Most of the data on the metabolic fate of aliphatic aldehydes deals with the simpler members of the group, **formaldehyde** and **acetaldehyde**. This results naturally from the long-standing interest in the toxicity of the former compound, a metabolite of methanol, and the formation of the latter in ethanol metabolism. Furthermore, these compounds enter into the normal pathways of one- and two-carbon metabolism. Formaldehyde metabolism has recently been summarized by Hathway (1975), whereas Williams (1959, pp. 89–90) reviewed the metabolism of both it and acetaldehyde. In both cases their metabolism is straightforward and involves conversion to the corresponding acid followed by oxidation to CO_2.

Williams (1959, p. 92) stated that little is known about the metabolism of the higher aliphatic aldehydes but assumed that these also undergo oxidation to the corresponding acids which then enter into the normal pathways of fatty acid metabolism. The general metabolic pathway of these compounds is chain-shortening due to β-oxidation. The ready oxidation of these aldehydes to CO_2 and water ensures that little or no urinary excretion of intermediate products occurs. This general understanding has subsequently been neither confirmed nor contradicted and no new metabolic results were presented by Opdyke (1973a,b) in reviews of biological data on several naturally-occurring saturated aliphatic aldehydes containing from six to 14 carbon atoms. Boyland (1940) administered *n*-heptanal (oenanthal, heptaldehyde) orally to rats and rabbits and was unable to detect any acidic metabolites including the ω-oxidation product, pimelic acid, in the urine. It was concluded that the aldehyde undergoes complete oxidation in the body. In an early study Neubauer (1901) administered heptaldehyde and also isobutyraldehyde and iso-valeraldehyde to rabbits, however no useful conclusions could be drawn from the results. Hinson and Neal (1972, 1975) carried out kinetic studies of the oxidation of *n*-butanal (butyraldehyde), *n*-hexanal and *n*-octanal (octylaldehyde, caprylaldehyde) by horse liver alcohol dehydrogenase and showed that they were oxidized to the corresponding carboxylic acids.

A few aliphatic plant aldehydes contain double bonds. A common example of this group is *trans*-hex-2-en-1-al (leaf aldehyde). It seems likely that this compound may be partly metabolized to a mercapturic acid derivative. This is suggested by the finding that its administration to rats gives rise to a moderate fall in the liver glutathione content in rats (Boyland and Chasseaud, 1970). Also, the aldehyde is an excellent substrate for the glutathione S-transferase activity in rat liver which forms glutathione conjugates with a number of α,β-unsaturated carbonyl compounds (Boyland and Chasseaud, 1967, 1968).

B. MONOTERPENOID ALDEHYDES

1. *Acyclic Terpene Aldehydes*

Metabolic data on this group are limited to two compounds, citral (geranial, neral) (1) and citronellal (2). The former compound is 3,7-

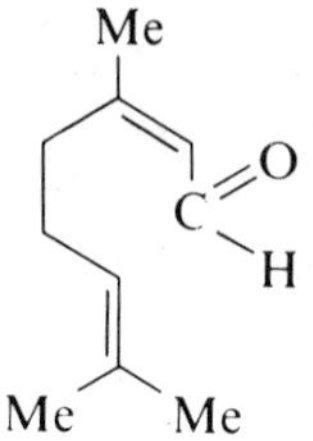

(1)
Citral

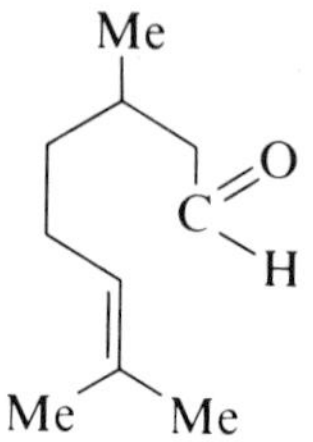

(2)
Citronellal

dimethyl-2,6-octadienal and differs from the latter, 3,7-dimethyl-6-octenal, only by the presence of a double bond at C2,3.

The metabolism of **citral** (1) was first studied by Hildebrandt (1901) in rabbits. Two urinary metabolites were detected, one of which was geranic acid (3) formed by the oxidation of the aldehyde group. The other compound was a dicarboxylic acid which was then believed to be a geranic acid derivative with a carboxyl rather than a methyl group at C3. This problem was reinvestigated by Kuhn *et al.* (1936) who administered large, repeated doses (2×5 g daily, p.o.) to rabbits. They found that this dicarboxylic acid, now known as "Hildebrandt acid", has structure (4). In an earlier brief report, Kuhn and Livada (1933) found that about 10% of the citral (2–4 g/day, i.p.) given to dogs was excreted in the urine as metabolite (4). A second dicarboxylic acid metabolite was detected in which the double bond at C2,3 was reduced. The latter compound (5), reduced Hildebrandt acid, was shown to be optically active (dextrorotatory) and it therefore results from an asymmetric reduction of the double bond. Neubauer (1901) detected a glucuronide conjugate in the urine of rabbits given citral (1 g/kg, p.o.). Parke and Rahman (1969) found that the administration of citral for three days to rats resulted in moderate (about 25%) increases in the activities of several liver microsomal enzyme systems.

Hildebrandt (1901) also investigated the metabolic fate of **citronellal** (2) in rabbits. In view of the close structural similarity of compounds (1) and (2), it was considered likely that their pathways of metabolism would be similar. However, none of the expected carboxylic acid derivatives was detected in the urine. This anomalous finding was subsequently clarified by Kuhn and Löw (1938) who reported that citronellal is cyclized to *p*-menthane-3,8-diol (6). This reaction takes place readily under acidic conditions and they believed that metabolite (6) is formed in the stomach. Following its absorption it is conjugated with glucuronic acid at position C3. About 25% of a total dose of 50 g of citronellal fed to rabbits was isolated from the urine as this glucuronide. That this is not the sole metabolic pathway of citronellal, however, was shown by Asano and Yamakawa (1950) who, unlike previous workers, were able to isolate the reduced dicarboxylic acid (5) from the urine of rabbits given the aldehyde by subcutaneous injection.

C. Aromatic Aldehydes

The metabolism of **benzaldehyde** (7) is straightforward and involves oxidation to benzoic acid (8) which may be excreted as such or as hippuric acid (9) following conjugation of (8) with glycine. This was shown by Friedmann and Türk (1913) who gave two dogs a total of 10 g each of benzaldehyde by i.p. injection over a period of five days. They isolated from the urine about 2% and 13% of the dose as benzoic acid and 72% and 39% as hippuric acid. Bray *et al.* (1951) carried out a kinetic study of the formation of benzoic acid in rabbits from several precursors including benzaldehyde. They found that the combined urinary benzoic acid and hippuric acid fraction accounted for about 90% of the administered dose of benzaldehyde (approx. 250–750 mg/kg, p.o.). No excretion of benzoyl glucuronide occurred in three animals while 1% and 5% of the dose was found with two others. The extent of formation of the latter metabolite was noted to be dependent on the amount of benzoic acid present in the body. With benzaldehyde, the rate of formation of the acid under the experimental conditions was not sufficient to appreciably exceed the capability of conjugation with glycine. Teuchy *et al.* (1971) found that about 30% of the dose of benzaldehyde (44 mg/animal, i.p.) was excreted as urinary hippuric acid by rats. This suggests that conjugation of the metabolically formed benzoic acid with glycine may be less extensive in rats than in rabbits, however this study did not measure the extent of urinary benzoic acid excretion by rats. The possibility that benzaldehyde may be partly reduced to benzyl alcohol has not been investigated, however Scheline (1972b) noted that this reaction occurred *in vitro* when the aldehyde was incubated with a mixed culture of rat caecal microorganisms. Whether or not this reduction has significance *in vivo* will depend on the extent of absorption of the administered aldehyde in the upper regions of the intestine. It seems reasonable to assume that absorption will be essentially complete before the compound reaches the abundant microflora of the lower intestine and that little microbial reduction of benzaldehyde will occur.

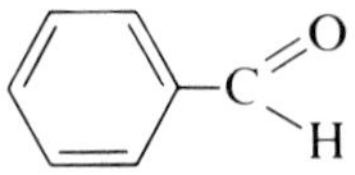

(7)

Benzaldehyde

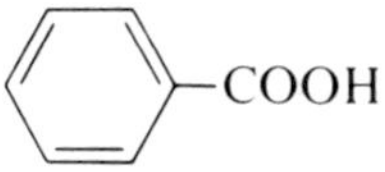

(8)

Benzoic acid

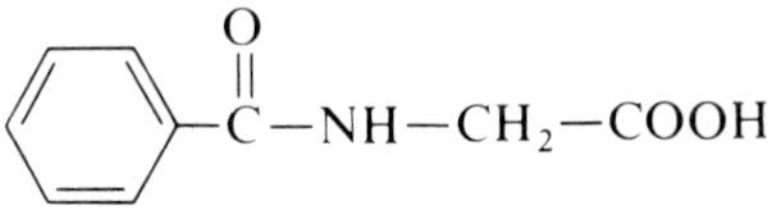

(9)

Hippuric acid

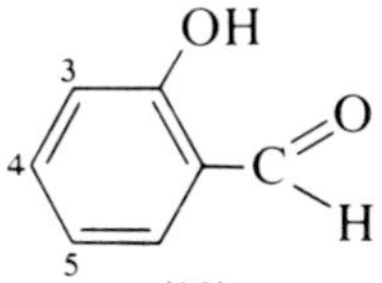

(10)

Salicylaldehyde

Information on the metabolism of **salicylaldehyde** (2-hydroxy-benzaldehyde) (10), obtained both from the earlier literature and from unpublished data, was summarized by Williams (1959, pp. 333–334). Salicylic acid was found to be the major metabolite in rabbits, cats and dogs. Excretion of the unchanged aldehyde, both free and conjugated with glucuronic acid and sulphate, has been detected in rabbits. Williams (1938) recorded a value of about 6% of the dose (325 mg/kg, p.o.) for the latter metabolite. Quantitative data were also obtained by Bray et al. (1952) who fed salicylaldehyde (0·4 g/kg) to rabbits and found that 75% was excreted as the corresponding acid. Only about 3% was due to the ethereal sulphate and the remainder, in a 2:1 ratio, consisted of ester and ether glucuronides. Sato et al. (1956b) studied the metabolism of salicylaldehyde using rat liver preparations, mainly in order to determine the ability of the phenolic group to undergo sulphate conjugation. Two metabolites were detected when the supernatant fraction of liver homogenates was used. One was the sulphate conjugate of salicylic acid and the other probably the corresponding conjugate of salicylaldehyde. These metabolites were also detected when liver slices were used and two additional compounds were also noted. These were tentatively identified as the glycine conjugate of salicylic acid and the sulphate conjugate of 2,5-dihydroxybenzoic acid. Interestingly, the sulphate conjugate of salicylic acid found in these experiments was not formed when salicylic acid itself was used as the substrate. It was therefore assumed that the conjugation reaction precedes oxidation of the aldehyde group. However, it is not clear if the latter step occurred enzymically or merely as a result of chemical oxidation. Salicylaldehyde is metabolized *in vitro* both to salicylic acid and to salicyl alcohol when incubated with rat caecal microorganisms (Scheline, 1972a).

Hartles and Williams (1948) studied the metabolism of **4-hydroxy-benzaldehyde** in rabbits. Using oral doses of about 300 mg/kg it was found that slightly more than 40% was excreted as 4-conjugates, the glucuronide:ethereal sulphate ratio being about 3·5:1. Nearly identical experiments were carried out by Bray *et al.* (1952) who reported a value of 25% of the dose (400 mg/kg) for 4-conjugates, the glucuronide to sulphate ratio being about 2:1. A small amount (4%) of ester glucuronide was excreted and the main metabolite was 4-hydroxybenzoic acid (67%). The conjugates reported by Hartles and Williams were mainly of the hydroxy acid rather than of the hydroxy aldehyde, however about 2–3% of the dose was found to be excreted as the 4-glucuronide of the latter compound. These *in vivo* experiments showed that the extent of glucuronide excretion was higher with the aldehyde than with the corresponding acid and it therefore seems likely that some conjugation occurs prior to aldehyde oxidation. This view was also held by Quick (1932) who also found evidence for the formation in dogs of a diglucuronide of 4-hydroxybenzoic acid following administration of the aldehyde. Sato *et al.* (1956a), in experiments similar to those described above with salicylaldehyde, studied the metabolism of 4-hydroxybenzaldehyde *in vitro* by rat liver preparations. The ethereal sulphates of the substrate and its carboxylic acid derivative were formed both by supernatant and slices. Smaller amounts of two other metabolites were also formed by liver slices, these being possibly the sulphate conjugate of 3,4-dihydroxybenzaldehyde and a hippuric acid derivative. Metabolism of 4-hydroxybenzaldehyde by rat caecal microorganisms leads to both the corresponding acid and alcohol (Scheline, 1972a). Interestingly, reduction can proceed further in this case by converting the alcohol moiety to a methyl group, forming *p*-cresol.

The metabolism of **protocatechualdehyde** (3,4-dihydroxybenzaldehyde) follows a similar pattern to that noted above with other phenolic aldehydes. Dodgson and Williams (1949) reported that about 40% of the dose (250 mg/kg, p.o.) was excreted in the urine by rabbits as two glucuronide conjugates, apparently of the aldehyde and of protocatechuic acid, and as an ethereal sulphate. The glucuronide:ethereal sulphate ratio was nearly 2:1. Large amounts of free protocatechuic acid were also excreted in the urine and this metabolite probably accounted for much of the remainder of the dose. Wong and Sourkes (1966) studied the urinary metabolites of protochualdehyde (50 mg/kg, i.p.) in rats and found that it was *O*-methylated to vanillin and ultimately converted to catechol. These pathways are discussed in the following summary of the metabolism of vanillin. The metabolism of protocatechualdehyde by rat caecal microorganisms is similar to that found with 4-hydroxybenzaldehyde (Scheline, 1972a). Some oxidation to protocatechuic acid takes place but the main reaction is complete reduction of the aldehyde group, giving 4-methylcatechol.

The aromatic aldehyde which has received the largest amount of attention from a metabolic viewpoint is **vanillin** (4-hydroxy-3-methoxybenzaldehyde) (11). A few very early studies, the results of which were summarized by Sammons and Williams (1941), delineated the major pathways in its metabolism including oxidation to vanillic acid (12), conjugation of some of this metabolite with glucuronic acid, an increase in the urinary output of ethereal sulphates and, possibly, minor excretion of unchanged compound. With this background, Sammons and Williams (1941) set out to obtain a quantitative assessment of the metabolism of vanillin (1 g/kg, p.o.) in rabbits. They accounted for about 83% of the dose, 69% being excreted as vanillic acid, either free (44%) or conjugated (25%). The latter material consisted mainly of the ether glucuronide of vanillic acid. About 8% of the dose was recovered as ethereal sulphates of vanillin and/or vanillic acid. The remaining 14% was associated with the 4-glucuronide of vanillin. No demethylation of vanillin was detected in this study.

The most abundant data on the metabolic pathways of vanillin (Fig. 4.1) and the amounts of metabolites formed have been obtained in studies using rats. Using a dose of 100 mg/kg (i.p.), Wong and Sourkes (1966) found 17% in the urine as free vanillic acid and 24% as its glucuronide conjugate. Strand and Scheline (1975), in similar experiments, reported a value of 47% for total urinary vanillic acid with a further 10% being accounted for as vanilloylglycine (13). A small amount of protocatechuic acid (14) was detected and the simple phenols, guaiacol (15) and catechol (16), formed via decarboxylation of the benzoic acid derivatives, accounted for 8–9% of the dose. The oxidative pathway therefore made up 65–70% of the dose in these experiments. A further 7% was excreted as vanillin and 19% was recovered as the reduction product vanillyl alcohol (17). As previously demonstrated in *in vitro* experiments with rat caecal microorganisms (Scheline, 1972a), further reduction of vanillin or vanillyl alcohol takes place and 2–3% of the dose was found in the urine as 4-methylguaiacol (18) and 4-methylcatechol (19). With the exception of the glycine conjugate and some of the vanillic acid, the metabolites were excreted in the urine as glucuronide and/or sulphate conjugates. The reductive pathway to give vanillyl alcohol was reported earlier by Wong and Sourkes (1966) who gave a value of 10% for this conversion . They also showed that catechol (4% of the dose) was excreted in the urine. The study of Strand and Scheline (1975) also ascertained the biliary metabolites of vanillin in the rat, the effect of prevention of biliary excretion on the metabolic pattern, the effects of suppression of the intestinal microorganisms and the effects of inhibition of the bacterial β- glucuronidase in the intestine. It was found that much of the extensive metabolism of vanillin in rats is dependent on the biliary excretion of the glucuronides of vanillin and its primary

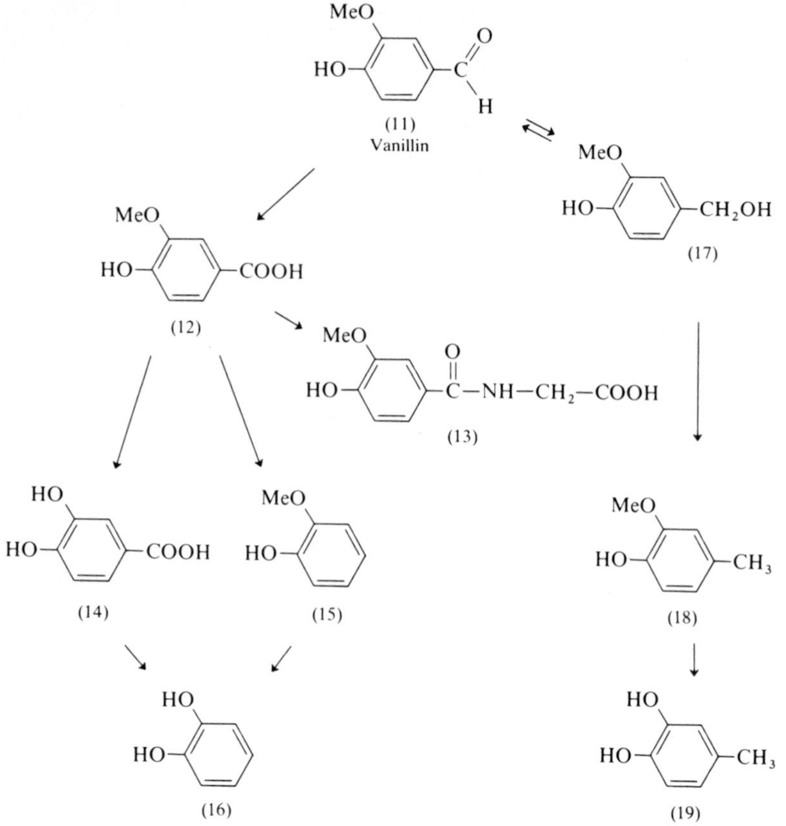

FIG. 4.1. Metabolic pathways of vanillin in the rat. (Metabolites also excreted in conjugated form. See text.)

metabolites, vanillic acid and vanillyl alcohol. These conjugates are substrates for bacterial enzymes which produce the decarboxylated products and toluene derivatives.

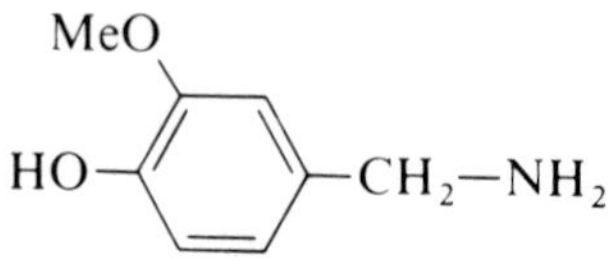

Little study of the metabolism of vanillin in man has been made although Grebennik *et al.* (1963) reported that 2% was excreted in the urine unchanged and 73% as vanillic acid. A previously unreported metabolite of vanillin, 3-methoxy-4-hydroxybenzylamine (20), was reported to be excreted in the urine of human subjects who ingested vanilla extract (Perry et al., 1965). The quantitative aspects of this conversion were not ascertained but based on the likely content of 50–100 mg of vanillin in the 60 ml of vanilla extract consumed and a normal urinary creatine excretion of 2 g/day, the data presented suggest a value of roughly 1% of the administered vanillin for this metabolite. This appears to be the only study in which the formation of a basic metabolite of vanillin has been investigated and this finding should be given further attention.

Other studies of vanillin metabolism have considered its conversion *in vitro*. Dirscherl and Brisse (1966) studied the formation of vanillic acid from several precursors including vanillin and found that 81% of this substrate was converted by rat liver homogenates and 12–70% by human liver homogenates under the experimental conditions employed. Friedhoff *et al.* (1972) reported that an enzyme found in the supernatant fraction of rat liver preparations and referred to as guaiacol-*O*-methyltransferase showed weak activity in transferring the methyl group of *S*-adenosylmethionine to vanillin to form 3,4-dimethoxybenzaldehyde.

The metabolism of isovanillin (3-hydroxy-4-methoxybenzaldehyde), which unlike its isomer vanillin does not occur naturally, was studied in rats by Strand and Scheline (1975). A major aim of this investigation was to assess the significance of the presence or absence of a free hydroxyl group in the 4-position on the patterns of metabolism of the two aldehydes. As noted above with vanillin, complete reduction of the aldehyde group to a toluene derivative is carried out by the intestinal microflora and this reaction is characteristic of the 4-hydroxy aldehydes (Scheline, 1972a). Isovanillin lacks this group and was therefore not fully reduced in the *in vitro* experiments. This difference in metabolism was also seen in the *in vivo* experiments as a methylguaiacol derivative was not detected in the urine from rats given isovanillin, in contrast to that seen with vanillin. In most other respects, however, the metabolism of the two aldehydes showed a similar picture. Following oral doses of isovanillin (100 mg/kg) the urinary metabolites and their percentages were: isovanillin, 19; isovanillyl alcohol, 10; isovanillic acid, 22; vanillic acid, 11; isovanilloylglycine, 19; catechol, 7; 4-methylcatechol, 1 and a small amount of protocatechuic acid. An interesting feature of these results is finding that considerable demethylation to catechol derivatives occurs. This reaction liberates the 4-hydroxyl group and thereby furnishes substrates for the bacterial decarboxylation and reduction reactions that form catechol and 4-methylcatechol, respectively, as well as for the *O*-methylation reaction which forms vanillic acid.

Anisaldehyde (4-methoxybenzaldehyde) was found by Sammons and Williams (1946) to form an ester glucuronide when fed to rabbits

(2 g/animal). This aldehyde is also metabolized to the corresponding acid and alcohol when incubated *in vitro* with rat caecal microorganisms (Scheline, 1972a). No information is available on the urinary excretion of that portion of the anisic acid which is not conjugated with glucuronic acid. However, it is reasonable to assume that excretion of the free acid occurs as Sammons and Williams (1946) also showed that the closely related **veratraldehyde** (3,4-dimethoxybenzaldehyde) was excreted as veratric acid to an extent of about 28% of the dose in similar experiments in rabbits. The amount of veratroylglucuronide excreted accounted for a further 38%. The only other metabolite detected in these experiments with veratraldehyde was catechol, which was excreted to a minor extent and in conjugated form. This reaction sequence, involving both O-demethylation and decarboxylation, is no doubt dependent on the intestinal microflora, as shown *in vitro* with rat caecal microorganisms (Scheline, 1972a). Additionally, O-demethylation can be carried out in the liver. Müller-Enoch *et al.* (1974) found that the perfused rat liver oxidized veratraldehyde and then O-demethylated the veratric acid to both vanillic acid and isovanillic acid. It was shown that 4-O-demethylation occurred more extensively as the ratio of the vanillic and isovanillic acids formed was 15:1.

Heffter (1895) administered **piperonal** (3,4-methylenedioxybenzaldehyde) orally to a rabbit and isolated from the urine about 28% of the dose (4 g) as piperonylic acid. Williams (1959, p. 334) reported that some of the latter metabolite is excreted as the glucuronide conjugate. More recently, Kamienski and Casida (1970) studied the metabolic fate of piperonal labelled with ^{14}C in the methylenedioxy moiety. About 1% of the oral dose (0·75 mg/kg) was recovered as $^{14}CO_2$ in 48 h indicating that this compound was fairly resistant to demethylenation. Most of the radioactivity (89%) was recovered in the urine during this period and the major metabolite was piperonylglycine. Some free piperonylic acid was also detected in addition to two unidentified metabolites. These unidentified metabolites were also found when piperonyl alcohol but not piperonylic acid was administered. *In vitro* experiments using mouse liver preparations showed some reduction of piperonal to piperonyl alcohol but mainly oxidation to the acid and its glycine derivative. A microsome-NADPH system showed low demethylenation activity, converting some piperonal to 3,4-dihydroxybenzoic acid.

(21)

Cinnamaldehyde

The metabolism of **cinnamaldehyde** (cinnamic aldehyde) (21) follows the expected pathway of oxidation to the corresponding acid, although the presence of the three-carbon side-chain allows for shortening by β-oxidation. This was first reported by Friedmann and Mai (1931) who administered the aldehyde (150 mg/kg, i.p.) to rabbits and found benzoic acid, hippuric acid and the glycine conjugate of cinnamic acid in the urine. Teuchy et al. (1971) recovered nearly 30% of the dose (30–40 mg/animal, i.p.) as urinary hippuric acid when cinnamaldehyde was given to rats. It seems probable that cinnamaldehyde may also be metabolized to a mercapturic acid derivative. Boyland and Chasseaud (1967, 1968) reported that the first step in this pathway, conjugation with glutathione, is carried out by the glutathione S-transferase activity present in the livers of numerous mammalian species. Accordingly, the administration of cinnamaldehyde to rats causes an appreciable reduction in the liver glutathione level in rats (Boyland and Chasseaud, 1970).

(22)

Furfural

(23)

Furfural (22) contains a heterocyclic rather than a carbocyclic ring but its metabolism follows the expected pathway of oxidation to the corresponding acid (furan-2-carboxylic acid, α-furoic acid) which is also conjugated with glycine to give furoylglycine. These metabolites were detected in the urine of rabbits and dogs by Jaffé and Cohn (1887) and Paul et al. (1949) identified the glycine conjugate as the major urinary metabolite of furfural (50–60 mg/kg, p.o.) in rats. In the former study a third metabolite, furfuracryluric acid (furylacryloyl glycine) (23), was also isolated. The formation of an acrylic acid derivative, in essence an aldol-like condensation of the aldehyde with acetate, is an unusual reaction which has not been observed with other aldehydes. The results of Paul et al. (1949) do not indicate whether or not compound (23) is also formed in rats.

Gossypol (24) is a toxic polyphenolic aldehyde which is found in cottonseed. Its presence in meal made from cottonseed limits the usefulness of this product in livestock feeding, except in ruminant animals which are resistant to the toxic effects. Since cottonseed meal has been used as a feedstuff for various species of animals, it is not surprising that metabolic data on gossypol are available from both mammalian and non-mammalian species. The following account of gossypol metabolism will not include the results obtained with fish and chickens, however these subjects are

(24)

Gossypol

included in the review of Abou-Donia (1976) on the physiological effects and metabolism of gossypol.

Information on the mammalian metabolism of gossypol has been obtained mainly in rats and swine, however Bressani *et al.* (1964) fed dogs on a diet containing cottonseed meal and studied the excretion of free and bound gossypol. They found that the bulk of the gossypol was eliminated in the faeces and that the amount of free gossypol excreted was several times higher than that present in the food. They believed that this indicates a capacity for intestinal hydrolysis of the bound forms of gossypol which occur in the cottonseed meal. The biological disposition in rats of gossypol labelled with ^{14}C in the aldehyde groups was investigated by Abou-Donia *et al.* (1970). When given orally at a dose level of 25 mg/kg, a relatively slow excretion of radioactivity was observed in respiratory CO_2, urine and faeces. Respiratory $^{14}CO_2$ accounted for about 12% of the radioactivity when the animals were given a basal diet and nearly 21% when this diet was supplemented with ferrous sulphate. Urinary excretion of radioactivity accounted for a mere 3% of the dose whereas 77% was recovered in about a week in the faeces. Similar results were obtained by Skutches and Smith (1974) who employed two labelled forms of biosynthetic [^{14}C]-gossypol. Material labelled in the formyl groups resulted in the excretion of $^{14}CO_2$ in the expired air whereas negligible amounts of $^{14}CO_2$ were expired when the rats were given material orally in which the ^{14}C was located only in the naphthalene rings or in the 12- and 12'-positions. These results show that the carbonyl groups are partly converted to CO_2 but that degradation of the naphthalene rings does not occur. In view of the patterns of tissue radioactivity obtained and the finding that about 90% of the radioactivity was recovered in three days in the faeces, it was concluded that gossypol or gossypol metabolites are excreted in the bile.

The most extensive studies of gossypol metabolism have been carried out in swine and the above findings pointing to the importance of the biliary excretion of metabolites in rats have their parallels in swine (Smith and Clawson, 1965; Albrecht *et al.*, 1972; Skutches *et al.*, 1973). The nature of the biliary metabolites of gossypol in swine was studied by Abou-Donia and Dieckert (1974) who found that this material consisted mainly of glucuronide (33%) and sulphate (22%) conjugates. Only about 7% was due to non-conjugated material and, of the remainder, about 16% and 22% were due to further conjugated material hydrolysed by hot-acid treatment and to the remaining water-soluble metabolites, respectively. The radioactive material excreted in the urine consisted of the same components that were found in the bile, however in different proportions. About 43% was due to the water-soluble metabolites and 27% to non-conjugated material. Abou-Donia and Dieckert (1975) subsequently reported the fate in swine of a single oral dose (6·7 mg/kg) of gossypol labelled with ^{14}C in the aldehyde groups. Faecal excretion of radioactivity was extensive and amounted to 7% in one day, 36% in two days, 56% in three days, 74% in four days and 95% in 20 days. The loss as expired $^{14}CO_2$ was 2·1% during this period and the urinary excretion of radioactivity amounted to only 0·7%. The levels of radioactivity were highest in muscle and liver and especially those in the liver showed a slow decline. The nature of the non-conjugated metabolites in the liver two days after dosing was studied and evidence was obtained for the presence of four radioactive compounds. The major metabolite was shown to be unchanged gossypol and good evidence for the formation of the oxidized product, gossypolone (25a), was also obtained. The remaining metabolites were tentatively identified as gossypolonic acid (25b) and demethylated gossic acid (26). Apogossypol is the compound formed by loss of the two aldehyde groups from gossypol and it was assumed that this non-radioactive metabolite was also formed. However, the loss of only 2·1% of the dose as $^{14}CO_2$ indicates that this pathway is of minor importance in pigs.

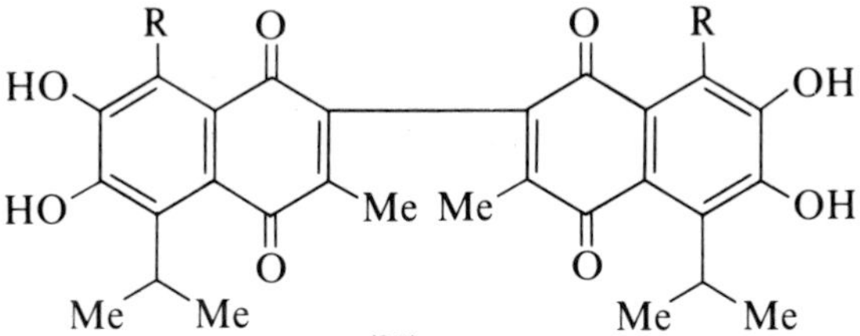

(25)
(a) R = CHO
(b) R = COOH

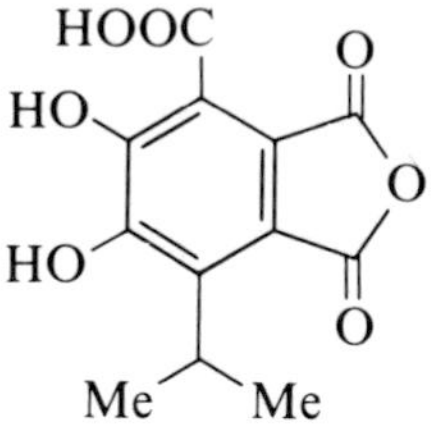

(26)

II. Ketones

A guide to the general pathways of metabolism of ketones and to the enzyme systems involved is given in Chapter 1, Section I.B.3.

A. Aliphatic Ketones

Williams (1959, p. 95) noted that the metabolism of very few aliphatic ketones had been investigated in detail. This statement is equally true today. The aliphatic ketones occurring naturally are very often methyl ketones. These have the general formula Me—CO—R and R varies from methyl (acetone) to n-undecyl (C_{11}). In addition to these 2-ones, some 3-ones and other ketones with branched chains are also found.

The lower aliphatic ketones have relatively low boiling points and this volatility influences their biological disposition in that elimination may occur in the expired air. Schwarz (1898) reported that 50–60% of the acetone (b.p. 56°) given to dogs at a dose level of 0·5–0·6 g/kg was eliminated in the expired air. Corresponding values for 2-pentanone (b.p. 102°, 0·36 g/kg) and 3-pentanone (b.p. 101°, 0·33 g/kg) were about 24% and 9%, respectively. Haggard et al. (1945) found that 2-pentanol (1 g/kg) was extensively metabolized in rats to 2-pentanone and that over 35% of the dose was recovered as the ketone in the expired air. If, however, the dose levels are reduced, the amounts of expired ketone are also lowered. Schwarz (1898) also used a smaller dose (3·5 mg/kg) and found that 18% of the acetone was lost by the respiratory route. In rats, only about 7% was lost by this route when this ketone (1 mg/g) was administered (Price and Rittenberg, 1950).

The general and major metabolic change with aliphatic ketones is reduction to the corresponding secondary alcohol followed by conjugation of the latter with glucuronic acid. This pathway was detected by Neubauer (1901) with numerous ketones ncluding the naturally occurring **acetone, 2-pentanone, 3-pentanone, 2-hexanone** and **2-octanone**. The compounds were given orally to rabbits (approx. 1–4 g/kg) and in some cases to dogs and the results indicated that at least part of the dose was reduced and conjugated. A similar finding was made by Kamil et al. (1953) with the closely related **2-heptanone**. When given to rabbits orally, this ketone (0·95 g/kg) was reduced to the alcohol which was excreted as the glucuronide conjugate to the extent of slightly over 40% of the dose.

As noted above, acetone is partly reduced and excreted as a glucuronide, however this pathway is less important with this compound than with higher ketones. Neubauer (1901) found the glucuronide after administering acetone to rabbits but not to dogs. When relatively small doses of acetone are administered, oxidation may be considerable. Price and Rittenberg (1950) found that at least half of the radioactivity was lost as $^{14}CO_2$ when methyl-labelled acetone (1–7 mg/kg) was given orally to rats. However, this process takes place relatively slowly. This study also suggested that acetone may be metabolized to one- and two-carbon fragments which then enter normal pathways of metabolism. Sakami and Lafaye (1951) obtained evidence indicating that acetone may also be converted to a three-carbon intermediate of glycolysis.

Recently, Di Vincenzo et al. (1976) studied the metabolism of 2-hexanone in guinea pigs. Following a dose of 450 mg/kg (i.p.) the serum was found to contain some unchanged compound and the expected reduction product, 2-hexanol. However ω-1 oxidation was also shown to occur and both 5-hydroxy-2-hexanone and 2,5-hexanedione were detected. In fact, the latter compound was the principal metabolite of 2-hexanone in guinea pig serum. It seems reasonable to assume that ω-1 oxidation is involved in the metabolism of related plant ketones as well since this reaction was also found to occur with a lower homologue, 2-butanone.

Alicyclic ketones are covered in the following section on terpenoid ketones, however a few miscellaneous alicyclic plant ketones not of the latter type are also known and these may conveniently be included here. These are commonly cyclopentanone and cyclohexanone derivatives containing various alkyl, alkenyl or acyl groups. Furthermore, a ring double bond is common, as seen with jasmone (3-methyl-2-(2-pentenyl)-2-cyclopenten-1-one). The metabolism of the substituted derivatives has not been studied, however some information is available on the parent compounds, cyclopentanone and cyclohexanone, and this may offer useful insights into the possible routes of metabolism of the substituted derivatives. James and Waring (1971) administered oral doses (approx. 190 mg/kg) of the two ketones to rats and rabbits. In rabbits, most of the material was found in the urine conjugated with glucuronic acid. This accounted for approx. 40–60% of the dose with the C_5-compound and 50–85% with the C_6-compound. The latter range agreed well with the value of 66% reported by Elliott et al. (1959) in similar experiments (dose = 250 mg/kg). This material consists of cyclopentyl and cyclohexyl glucuronide, respectively, and their excretion indicates that the ketones undergo appreciable reduction to the corresponding alcohols. No cyclopentane-1,2-diol was excreted, however a few percent of the dose was accounted for as sulphur-containing metabolites. These were identified as 2-hydroxycyclopentylmercapturic acid (27a), which was excreted only in trace amounts as both the cis and trans isomers, and a metabolite accounting for about 2·5% of the dose which is probably the corresponding ethereal sulphate derivative (27b). Similar results showing a trace of the 2-hydroxycycloalkylmercapturic acid and approx. 2–3% of the sulphate metabolite were seen with cyclohexanone following its administration to rabbits or after administration of both ketones to rats. Boyland and Chasseaud (1970) found that the administration of cyclohexanone and its unsaturated derivative, cyclohex-2-en-1-one, to rats led to a moderate fall

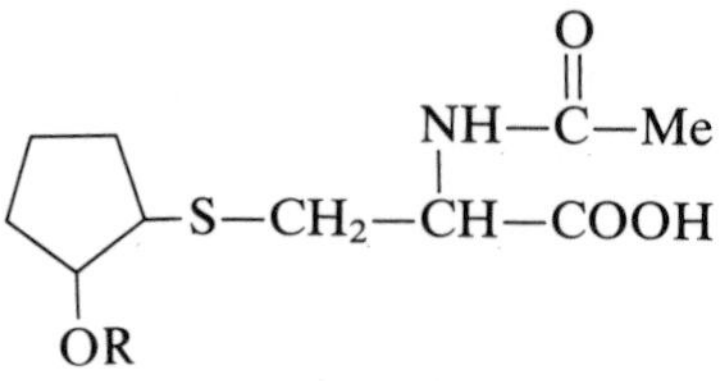

$$NH-\overset{\overset{\displaystyle O}{\|}}{C}-Me$$

$$\text{—S—CH}_2\text{—}\overset{|}{\text{CH}}\text{—COOH}$$

OR

(27)

(a) R = H
(b) R = SO$_3$H

in the liver glutathione levels, a change which generally points to mercapturic acid formation.

B. MONOTERPENOID KETONES

1. *Monocyclic Terpene Ketones*

Information on this group of compounds is limited to five closely related ketones based on the *p*-menthane (1-isopropyl-4-methylcyclohexane) (28) structure and to the ionones. The published investigations on the menthane derivatives are, without exception, of older vintage and it seems safe to conclude that the metabolism of these compounds is much more complex than presently documented. For example, Meyer and Scheline (1973) detected by gas chromatography-mass spectrometry at least six metabolites of piperitone (35) in the neutral fraction from the β-glucuronidase-hydrolysed urines of rats given this compound (100 mg/kg, p.o.). These metabolites included dehydrogenated and oxygenated derivatives. It seems likely that this subject offers a rich although complex field for study using modern and sensitive analytical methods. Williams (1959, pp. 524–527) reviewed the earlier literature on the metabolism of monocyclic terpene ketones.

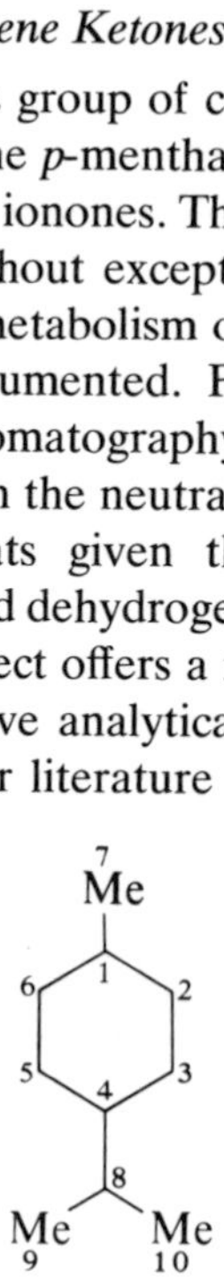

(28)

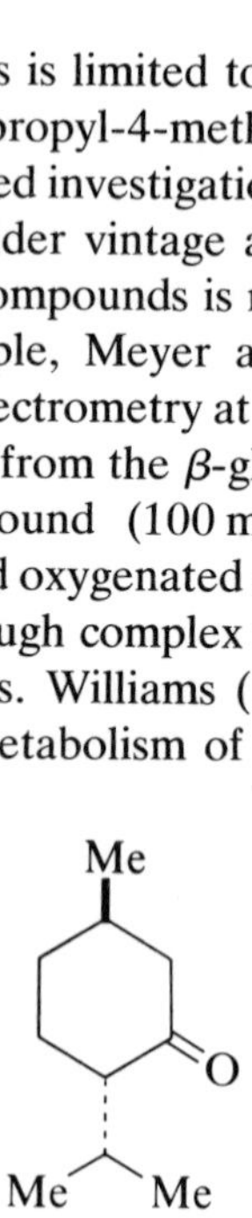

(29)

l-Menthone

The simplest of these ketones is *l*-menthone (*p*-menthan-3-one) (29). The earliest work on this compound indicated that it is excreted by rabbits as a glucuronide conjugate (Neubauer, 1901). This metabolite could be formed either following the reduction of the keto group or oxidation at some other site in the molecule. The latter possibility was suggested by Hildebrandt (1902b) and Hämäläinen (1912) believed that hydroxylation of the tertiary carbon at C4 occurred to give 4-hydroxymenthone. This suggestion was based on the finding that the metabolite underwent loss of water when warmed in 5% H_2SO_4, giving pulegone (39) or its isomer, p-menth-4(5)-en-3-one. This subject was place on a firmer footing by Williams (1940) who showed that at least part (10–15%) of the l-menthone undergoes reduction of the keto group in rabbits to give d-neomenthol (30). Interestingly, the other possible reduction product, l-menthol (31), was not detected although the experimental methods employed would have easily shown the presence of its glucuronide had it been excreted. These results were best explained by assuming that l-menthone undergoes an asymmetric reduction. Metabolites other than d-neomenthol glucuronide were undoubtedly formed in these experiments as quantitative measurements of the urinary glucuronic acid output showed that about 30–40% of the l-menthone fed was converted to glucuronides of hydroxy derivatives.

(30)   (31)

The metabolism of the isomeric **d-isomenthone** (32) was also investigated by Williams (1940). The possible reduction products in this case are d-isomenthol (33) and d-neoisomenthol (34) and, of these, only the former was definitely identified in the menthanol glucuronide fraction obtained

(32)   (33)   (34)
d-Isomenthone

from rabbit urine. However, this compound seemed to be the major if not the only metabolite present.

The metabolic fate of **piperitone** (*p*-menth-1-en-3-one) (35) is very poorly understood. It is noted above that rats excrete at least six unidentified metabolites in the urine, however only a single report is available on the nature of the piperitone metabolites (Harvey, 1942). When the ketone (9·3 g) was given orally to sheep, no unchanged compound was recovered in the faeces. The urine was subjected to acid hydrolysis and steam distillation. No acidic metabolite or neutral hydrocarbons were detected in the distillate but thymol (36) and perhaps diosphenol (37) were identified. Some unchanged piperitone was also present as well as another ketone, carvotanacetone (38). Formation of the latter compound involves an unexpected migration of oxygen and this point especially would benefit by reinvestigation to see if it is a metabolite or an artefact produced by the dehydration of an alcohol under acidic conditions.

(35)   (36)   (37)   (38)
Piperitone

d-**Pulegone** (*p*-menth-4(8)-en-3-one) (39) is similar to piperitone both structurally and by virtue of the paucity of data on its metabolic fate. Hildebrandt (1902b) reported that it was excreted conjugated with glucuronic acid and Teppati (1937) found that the metabolites formed are the glucuronides of pulegol and, apparently, the fully reduced compound, menthol.

(39)   (40)   (41)
Pulegone   Carvone

Carvone (*p*-mentha-6,8-dien-2-one) (40) was reported by Hildebrandt (1902b) to undergo conjugation with glucuronic acid in rabbits. Also, the methyl group adjacent to the ketone moiety was reported to be oxidized to a carboxyl group. Hildebrandt (1902a) identified the glucuronide of hydroxycarvone (41) in the urine of carvone-treated rabbits. Fischer and Bielig (1940) stated that unsaturated ketones which escape complete oxidation are excreted chiefly as optically active carbinols in which one double bond has been reduced. This general statement has been assumed to apply to carvone, which was one of the numerous test compounds given to rabbits. However, due to the high toxicity shown by carvone at the dose employed (2×2 g/kg, p.o.) no metabolic information was obtained and the applicability of this conclusion to carvone can only be guessed at.

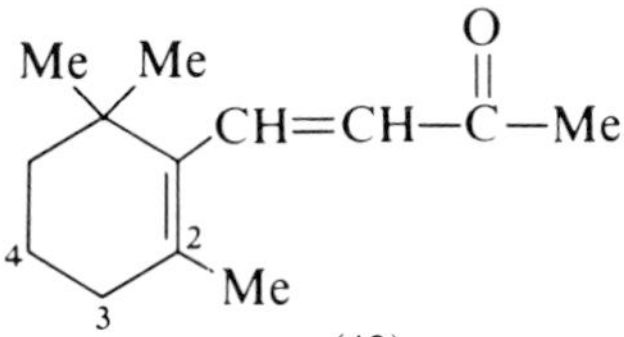

(42)
β-Ionone

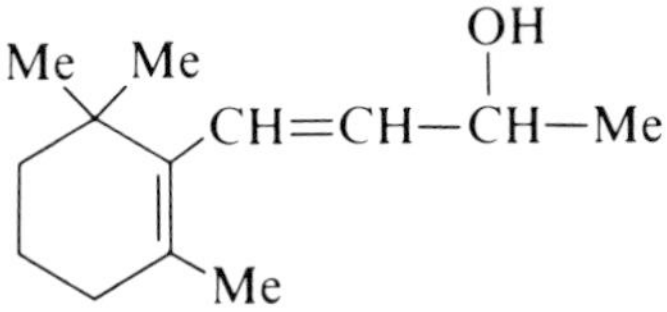

(43)

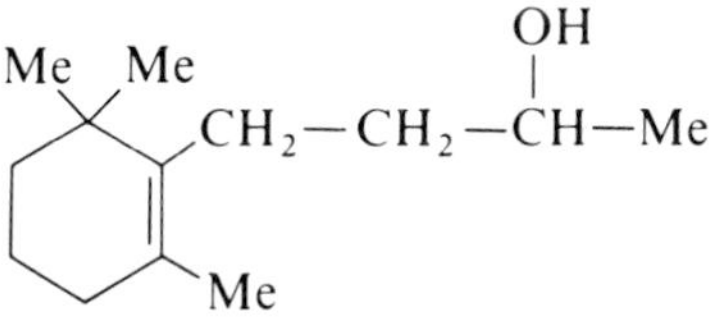

(44)

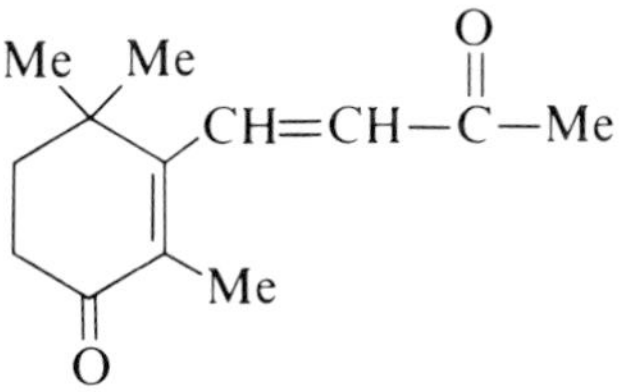

(45)

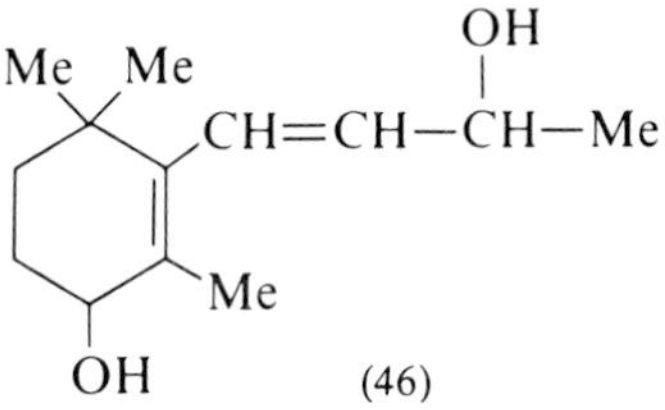

(46)

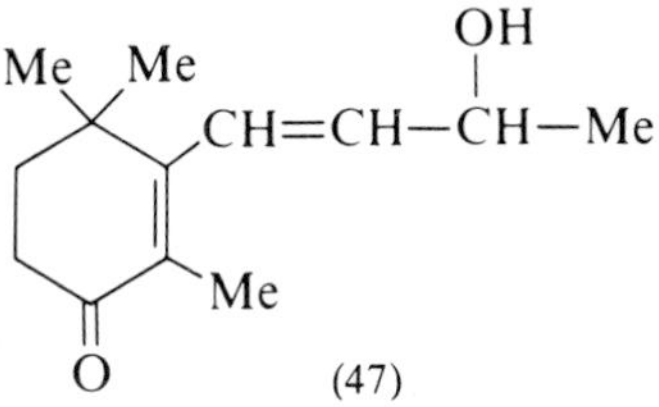

(47)

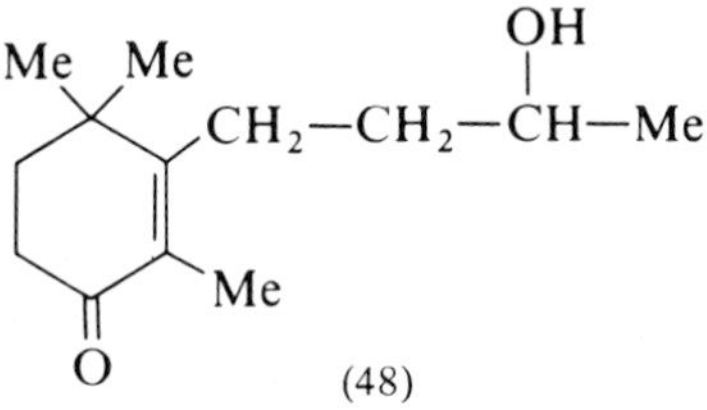

(48)

Several investigations have dealt with the metabolism of **β-ionone** (42) in rabbits. Three major metabolic routes are suggested by its structure (reduction of the keto group, double bond reduction and oxidation of the alicyclic ring) and the available data demonstrate that all of these pathways are utilized. Additionally, some unchanged compound was detected in the urine (Bielig and Hayasida, 1940; Ide and Toki, 1970), although the latter investigation indicated that the amount excreted is quite small. Ketone reduction of β-ionone was first reported by Bielig and Hayasida (1940) who detected the presence of the corresponding carbinol, β-ionol (43), in urine together with dihydro-β-ionol (44). Furthermore, four oxygenated derivatives of β-ionone, two derived from (43) and two from (44), were also detected. These compounds were believed to be derivatives in which the methyl group at C2 was oxidized to a hydroxymethyl group and it was speculated that further oxidation to the corresponding aldehydes and carboxylic acids was likely. This study showed that the carbinol metabolites are optically active and that the ring double bond is not reduced. This subject was reinvestigated by Prelog and Meier (1950) who detected three oxygenated metabolites in the urine of rabbits fed β-ionone. However, these were shown to be substituted at the 3-position and identified as 3-oxo-β-ionone (45), 3-hydroxy-β-ionol (46) and, probably, 3-oxy-β-ionol (47). This work was done with acid-hydrolysed urine and the above metabolites were found in the neutral fraction which showed roughly a 2 : 1 ratio between the ketonic and non-ketonic components. A third investigation, very similar to the previous two, was carried out by Ide and Toki (1970). All of the urinary metabolites detected were compounds oxygenated at the 3-position. They confirmed the excretion of metabolites (45), (46) and (47), the latter being a major metabolite in the free fraction. Another major metabolite in this fraction was found to be dihydro-3-oxo-β-ionol (48). The glucuronides of metabolites (47) and (48) were detected in the urine.

The data summarized above indicate that β-ionone can undergo fairly extensive metabolism and the results of Ide and Toki (1970) suggest that the major urinary metabolites are relatively highly metabolized. Nonetheless, our understanding of this subject is somewhat less than satisfactory, largely because of experimental inadequacies in the published work. All of the studies reported were done using rabbits which received oral doses of 20–50 g of β-ionone over periods ranging from seven to 18 days. Parke and Rahman (1969) showed that treatment of rats for three days with β-ionone gave 50–75% increases in the activities of several hepatic enzyme systems including that which carries out aromatic hydroxylation. Results following a more conventional dosage approach would therefore be of interest as would a closer investigation of the conjugated metabolites of β-ionone. The free metabolites isolated by Ide and Toki accounted for only a relatively small fraction of the administered dose and it seems logical to assume that the bulk is excreted as glucuronide conjugates. An approach utilizing enzyme hydrolysis of the urine samples followed by analysis using gas chromatography-mass spectrometry is warranted. The evidence concerning β-ionone oxidation suggests that allylic hydroxylation is involved, giving rise to 3-hydroxy derivatives which may then undergo further oxidation to keto compounds . As noted above, 3-hydroxy-β-ionol (46) has been identified as a urinary metabolite of β-ionone. 3-Hydroxy-β-ionone has not been detected although Ide and Toki (1970) reported that this metabolite is produced from β-ionone by the rabbit liver microsomal enzyme system. Allylic hydroxylation may also afford metabolites containing the hydroxymethyl group at C2. In view of the suggestions by Bielig and Hayasida (1940) noted above concerning the formation of several metabolites of this type, the possibility of this alternative pathway must be taken into consideration. An additional point is that the carbinols formed may be oxidized to ketones. Takenoshita and Toki (1974) purified a dehydrogenase from the soluble fraction of rabbit liver which dehydrogenated several alicyclic and acyclic alcohols including β-ionol (43) and 3-hydroxy-β-ionol (46).

(49)

α-Ionone

The metabolism of **α-ionone** (49) in rabbits was studied by Prelog and Würsch (1951). They obtained evidence indicating that allylic hydroxylation also occurs with this compound, resulting in the 4-hydroxy derivative.

2. Bicyclic Terpene Ketones

The ketones in this group are derivatives of three types of bicyclic terpenoid structures based on thujane (sabinane) (50), fenchane (51) and bornane (camphane) (52). Metabolic data on the first two types are derived entirely from the older literature and are limited to single representatives, thujone (53) and fenchone (54), respectively. Hildebrandt (1901) gave **thujone** (53) to rabbits and found that an optically active glucuronide was excreted in the urine. It was suggested that thujone is hydroxylated prior to conjugation. This was corroborated by Hämäläinen (1912) in similar

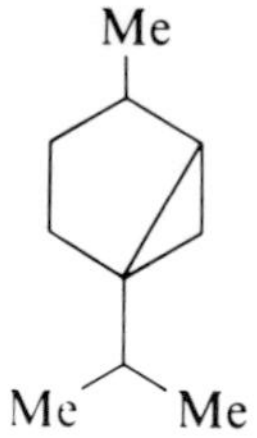

Me

Me

Me

(50)

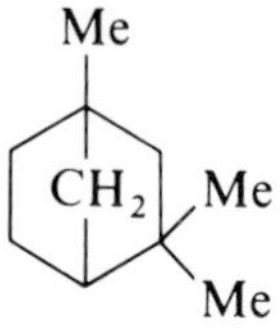

(51)

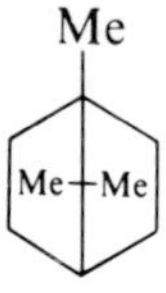

(52)

experiments which showed that the glucuronide conjugate of 4-*p*-menthanol-2-one (55) was a urinary metabolite of thujone. The identity of the metabolite was ascertained by virtue of its ability to undergo dehydration to carvenone (56) when treated with 5% H_2SO_4 and degradation to ω-dimethyllevulinic acid (57) when treated with an alkaline solution of bromine. A similar reaction involving degradation of the thujane skeleton to a monocyclic structure was also reported by Hämäläinen with the closely related thujyl alcohol (see Chapter 3, Section I.B.3). Conversion of bicyclic terpenoids to monocyclic derivatives is an interesting reaction about which little is known. However, analogous reactions have recently been demonstrated with the terpenoid hydrocarbons β-pinene and 3-carene (see Chapter 2, Section II.C).

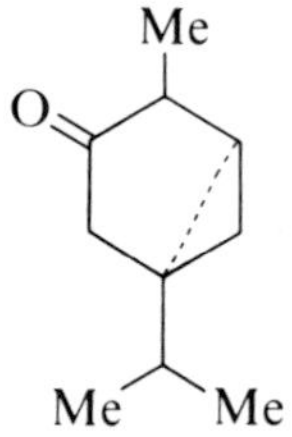

(53)
Thujone

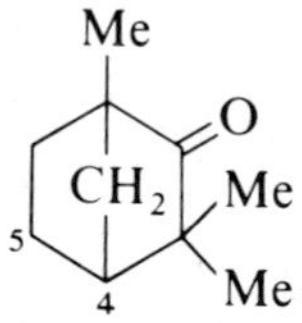

(54)
Fenchone

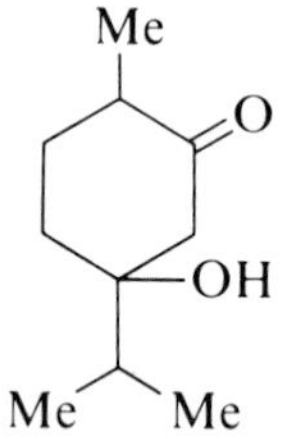

(55)

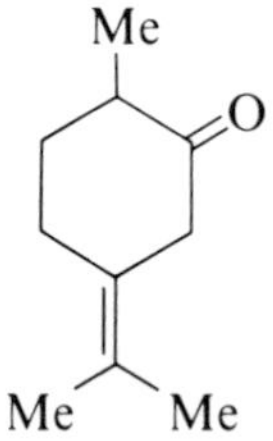

(56)

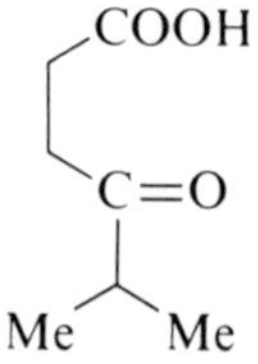

(57)

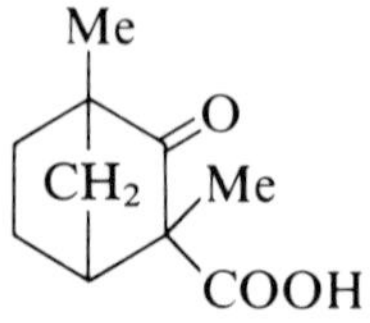

(58)

The earliest reports on the metabolism of (+)-**fenchone** (54) in dogs showed that it is probably oxidized to its 4-hydroxy derivative (Rimini, 1909). Hämäläinen (1912) also recorded the conversion of fenchone to a hydroxylated derivative in rabbits. Reinartz and Zanke (1936) administered fenchone to dogs and obtained evidence which indicated that besides 4-hydroxyfenchone, 5-hydroxyfenchone and Π-apofenchone-3-carboxylic acid (58), all as their glucuronides, were excreted in the urine.

(59)
Camphor

(60)
(+)-Camphor

Ordinary camphor, (+)-**camphor**, is also known under several other names including Japan camphor, Formosa camphor, 2-bornanone and 2-camphanone. Its structure is commonly shown as illustrated in (59), however structure (60) gives a more informative representation of (+)-camphor, especially with regard to the orientation of the *endo* bonds (downward in the illustration) and *exo* bonds (towards the convex side of the ring). Substituents 8 and 9 are often designated Π, these being termed either *cis* or *trans*, respectively, because of their orientation to the keto group. Studies on the metabolic fate of (+)-camphor began a century ago when Wiedemann (1877) obtained a glycosidic material from the urine of camphor-treated dogs. Schmiedeberg and Mayer (1879) soon showed that this material consisted of conjugates which liberated hydroxycamphor upon hydrolysis. This subject received no further attention for a period of 50 years until new investigations were initiated by Japanese workers. Asahina and Ishidate (1928) found that the hydroxycamphor (termed camphorol) fraction obtained from the camphoglucuronic acid mixture excreted by camphor-treated dogs consisted of both the 3- and 5-hydroxy derivatives. Later reports by Asahina and Ishidate (1933, 1934, 1935) indicated that oxidation at the Π-positions also occurred, giving both the *cis* and *trans* isomers of Π-hydroxycamphor. Small amounts of another metabolite, *trans*-Π-apocamphor-7-carboxylic acid (61), were also found in the urine. Greater amounts of the latter metabolite were excreted when a mixture of the Π-hydroxycamphors was administered. However, Shimamoto (1934a) found no evidence for the formation of *cis*-Π-hydroxycamphor in similar experiments. The relative values obtained for camphor metabolites in a 1 l portion of urine from dogs given daily doses (5 g) of camphor were: 3-hydroxycamphor, 15%; 5-hydroxycamphor, 55%; *trans*-Π-hydroxycamphor, 20%; losses during work-up, 10%.

Shimamoto (1934b) carried out similar experiments in rabbits given camphor (0·4 g/kg). The relative amount of hydroxycamphor excreted was only about 10% of that found previously in dogs and changes in the proportions of these metabolites were noted. This was especially true of the Π-hydroxy derivative which was no longer detected. The values for the 3- and 5-hydroxy compounds were 25% and 60%, respectively. Kawabata (1943) carried out metabolic experiments in which dogs were given chronic doses (i.p. or p.o.) of camphor. The Π-hydroxy derivative was reported to be a major metabolite and small amounts of the carboxylic acid derivative (61) were also detected. These studies suggest that, in dogs at least, camphor can also be oxidized at C9 to give the *trans*-Π-hydroxy derivative which may in part be further oxidized to the carboxylic acid derivative. Nonetheless, the investigation of Leibman and Ortiz (1973), using a gas chromatographic method of analysis, did not detect any Π-oxygenated metabolites of camphor in the acid-hydrolysed urines of dogs given camphor.

(61)   (62)   (63)

As noted above, the main pathways in the metabolism of (+)-camphor involve oxidation at the C3- and especially the C5-positions. These reactions have been the main points of interest in investigations of camphor metabolism by Robertson and Hussain (1969) and Leibman and Ortiz (1973). Additionally, these newer studies have demonstrated that the keto group is subject to metabolic reduction, a reaction which was not detected in earlier studies. Robertson and Hussain found three metabolites of (+)-camphor in the acid-hydrolysed urine of rabbits given the compound (290–560 mg/kg, p.o.). One of these, accounting for slightly more than 20% of the total, was shown to be (+)-borneol (62). The same qualitative result was obtained in this species by Leibman and Ortiz. The latter workers also studied this reduction *in vitro* using rat and rabbit liver preparations. With the former species, the 9000 g supernatant fraction was without effect whereas this fraction from rabbit liver brought about appreciable reduction. The major reduction product was (+)-borneol (2-*endo*-hydroxybornane) (62), but about 1% of the total consisted of isoborneol (2-*exo*-hydroxybornane) (63). Further experiments showed that reduction was much more readily carried out by the cytosol than by the microsomal fraction.

(64)   (65)   (66)   (67)

The two other urinary metabolites of (+)-camphor in rabbits described by Robertson and Hussain (1969) were found to be conjugates of (+)-3-endo-hydroxycamphor (64) and (+)-5-endo-hydroxycamphor (65). These accounted for about 20% and 60%, respectively, of the identified material and the data thus harmonize with earlier results. Leibman and Ortiz (1973) found also that camphor was excreted in the urine of dogs and rabbits as conjugates of both 3- and 5-hydroxycamphor. The former metabolite was most likely the endo isomer, however in the latter case both the endo (65) and exo (66) isomers were detected. Both of these 5-hydroxy derivatives were formed in vitro when (+)-camphor was incubated with rat liver microsomes or rabbit liver 9000 g supernatant fraction. This makes it unlikely that the in vivo results may have been influenced by isomerization of the hydroxycamphors during hydrolysis of the urinary glycosides. An important related point is that a small amount of the corresponding 5-oxo derivative (67) was detected in the urine of camphor-treated rabbits as well as in the incubates in the liver microsome experiments. Furthermore, reduction of the 5-keto group to a hydroxyl group takes place readily in the liver cytosol and the possibility arose that interconversion of the diketo derivative might be responsible for the proportions of the two 5-hydroxy isomers found. However, the data indicated that this was not the case and that both the 5-endo- and 5-exo-hydroxycamphors are primary hydroxylation products of (+)-camphor.

As noted above in the discussion of the metabolism of (+)-camphor, hydroxylated derivatives may be converted to ketocamphors (oxocamphors). This is true of the 5-hydroxy derivative which forms camphane-2,5-dione (5-oxocamphor) (67) and of Π-hydroxycamphor and its keto derivative Π-oxocamphor (68). The latter compound is, strictly speaking, a ketoaldehyde rather than a diketone and this fact has bearing on its metabolic fate. Nonetheless, its metabolism can conveniently be summarized at this point. In addition to the two compounds noted above, camphorquinone (3-oxocamphor) (69) has also been studied metabolically. When the latter compound, in the (±)-form, was given to rabbits (dose range of 140–440 mg/kg, p.o.), Robertson and Hussain (1969) found that it was excreted in the urine as conjugates of 3-hydroxycamphor (70) and

2-hydroxy-*epi*-camphor (71), both in the *endo* configurations. These results thereby confirmed those of Reinarz and Zanke (1934a) who reported that the glucuronides of metabolites (70) and (71) were excreted in the urine of dogs treated with camphorquinone. The data of Robertson and Hussain also showed that nearly 40% of the dose could be accounted for as urinary glucuronides. Subsequently, Robertson and Solomon (1971) reported that the urine extract obtained in the earlier study contained small amounts of the *cis* and *trans* isomers of the diol (camphane-2,3-diol). These *in vivo* results showing mainly reduction of a single keto group at either C2 or C3 differ from those obtained using rat liver microsomes or cytosol (Leibman and Ortiz, 1973). In the latter investigation, camphorquinone was reduced equally well in both liver preparations but only at C3 as 3-hydroxycamphor (70), presumably the *endo* isomer, was the sole reported metabolite.

The initial study on the metabolism of 5-oxocamphor (67) suggested that this compound is metabolized rather differently than is its 3-oxo isomer (Reinarz *et al.*, 1934). Whereas the latter compound is reduced *in vivo* at either C2 or C3 to hydroxyketones, 5-oxocamphor was reported to be oxidized at C4 to 4-hydroxy-5-oxocamphor and at C8 or C9 to Π-hydroxy-5-oxocamphor. In addition, some reduction of the 5-keto group giving 5-*endo*-hydroxycamphor (65) was noted and a fourth metabolite, probably β-cyclocamphanone-Π-carboxylic acid, was reported. Formation of the latter compound, a tricyclic terpene, seems implausible as it involves the creation of a carbon–carbon bond between C3 and C5 concomitant with loss of the oxygen atom at C5. This problem was therefore reinvestigated by Ishidate *et al.* (1941) who administered a total of 330 g of 5-oxocamphor to dogs in 5 g daily portions. Neither the 4-hydroxy diketone nor the cyclocamphanone derivative was found in the urine. Instead, the hydrolysis product of the glucuronide excreted consisted nearly entirely of 5-hydroxycamphor. Robertson and Hussain (1969) fed 5-oxocamphor (30–35 mg/kg) to rabbits and found a single urinary metabolite in the acid-hydrolysed, 24 h urines. This compound was shown to be 5-*endo*-hydroxycamphor (65) but, curiously, no increase in the urinary output of glucuronides was registered in these experiments. Leibman and Ortiz (1973) investigated the interconversions between the 5-oxo- and 5-hydroxycamphors *in vitro* using rat liver fractions. The main finding was that the cytosol or 9000 g supernatant fractions, but not the microsomal fraction, extensively reduced the 5-oxo compound to 5-*endo*-hydroxycamphor.

The third oxocamphor to have been studied metabolically is Π-oxocamphor (Π-apocamphor-7-aldehyde) (68). This compound also represents an intermediate stage in the metabolism of Π-hydroxycamphor to Π-apocamphor-7-carboxylic acid, a reaction discussed above. However, the present summary deals only with data obtained in experiments in which Π-oxocamphor itself was administered. Kawabata (1943) administered chronically 120 g of Π-oxocamphor to dogs and recovered the corresponding Π-hydroxy compound and carboxylic acid (isoketopinic acid; *trans*-Π-apocamphor-7-carboxylic acid) (61) in the urine. Approximately 20% and < 1%, respectively, of the total dose were obtained as these metabolites. Tamura and Imanari (1964, 1970) identified the urinary metabolites of Π-oxocamphor in humans following its p.o. or s.c. administration. After oral dosage, about 70% was recovered as the glucuronide of the Π-hydroxy metabolite with a further 5% and 25% excreted as the carboxylic acid and its ester glucuronide, respectively. Following injection (20 mg), however, only a trace of the alcohol glucuronide was excreted and the acid and its conjugate accounted for 30% and 70%, respectively, of the dose.

Early studies on the metabolism of (−)-**camphor** in dogs (Magnus-Levy, 1907; Mayer, 1908) and in rabbits (Hämäläinen, 1909) indicated that it was converted to a hydroxycamphor, thought to be the 3-hydroxy derivative, which was excreted in the urine both free and conjugated with glucuronic acid. This isomer of camphor was also included in the investigations of Robertson and Hussain (1969) and Leibman and Ortiz (1973). In the former study, rabbits were given oral doses (85–280 mg/kg) of (−)-camphor. The acid hydrolysed urines contained four metabolites, three of which were identified as borneol, 3-*endo*-hydroxycamphor and 5-*endo*-hydroxycamphor. The relative proportions of these metabolites were 1·0:2·4:5·0, respectively. These results demonstrate that (−)-camphor is metabolized in rabbits qualitatively in the same way as noted above with (+)-camphor and that both reduction of the keto group and ring oxidation at C3 and C5 occur. Leibman and Ortiz (1973) studied the *in vitro* metabolism of (−)-camphor using the 9000 g supernatant fraction of rabbit liver. Reduction of the keto group was noted but both the *endo* isomer (borneol) and the *exo* isomer (isoborneol) were identified, the amounts formed being roughly 1:3. Both isomers were formed when a similar preparation from rat liver was used, however the relative amounts of these products were now about 6:1. These *in vitro* studies using the rat liver preparation also showed that (−)-camphor was oxidized in a similar fashion to that seen with (+)-camphor. Thus, oxidation to 3-hydroxy-camphor and, to a somewhat greater extent, 5-hydroxycamphor was observed and the latter metabolite was present in both the *endo* and *exo* configurations.

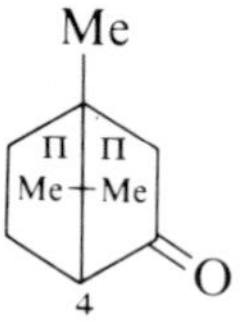

(72)
Epi-camphor

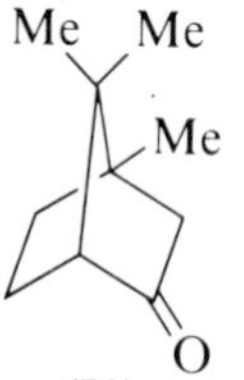

(73)
(+)-*Epi*-camphor

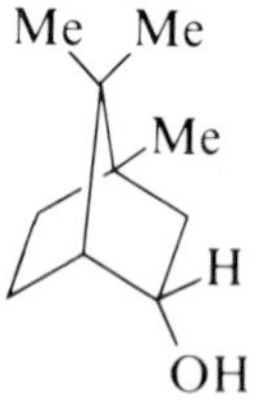

(74)

epi-Camphor (β-camphor) (72) differs from ordinary camphor in that the keto group is located at C3 rather than C2. Reinartz and Zanke (1934b) investigated the urinary metabolites of this compound in a dog which had received a total dose of 108 g divided equally over a 13-day period. The main metabolic reaction found was oxidation *ortho* to the keto group to give 4-hydroxy-*epi*-camphor which was excreted conjugated with glucuronic acid. Small amounts of the *cis* and *trans* isomers of Π-hydroxy-*epi*-camphor were also formed. More recently Robertson and Hussain (1969) gave oral doses (200–400 mg/kg) of (+)-*epi*-camphor (73) to rabbits. Three metabolites were detected in the hydrolysed urines but only one of these, (+)-*epi*-borneol (74), was identified. It is tempting to speculate that the other metabolites may be hydroxylated ketones similar to or identical with those described by Reinartz and Zanke, however this possibility was not clarified.

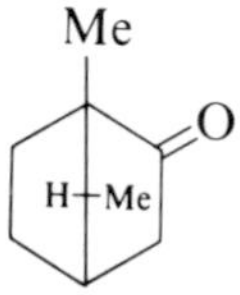

(75)
Santenone

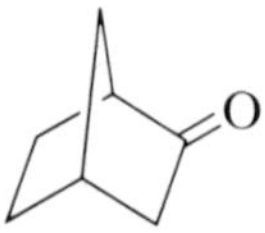

(76)
Norcamphor

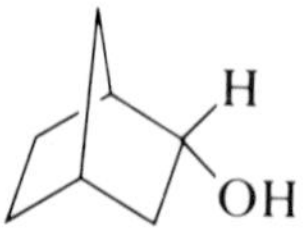

(77)

Santenone (1,7-dimethyl-2-norbornanone) (75), which was earlier known as Π-norcamphor, was one of the many alicyclic compounds studied by Hämäläinen (1912) in regard to their ability to undergo glucuronide formation in rabbits. The glucuronide obtained from santenone was of a ketoalcohol, santenonol. The position of the hydroxyl group was not determined but the results show that the keto group did not undergo reduction in this case. The latter reaction is, however, prominent in the metabolism of (±)-**norcamphor** (2-norbornanone) (76). Robertson and Hussain (1969) gave oral doses of roughly 200–400 mg/kg of this compound to rabbits and found that 30% of the dose could be accounted for as urinary glucuronides. Hydrolysis of the urines showed the presence of a single aglycone which was isolated and identified as (+)-*endo*-norborneol (77). The reduction of norcamphor to norborneol in rabbits was also reported by Krieger (1962). However, about equal quantities of both the *endo* and the *exo* isomers of norborneol were found to be excreted in the urine, chiefly in conjugated form.

C. Aromatic Ketones

Aromatic ketones do not furnish a large group of plant compounds, nonetheless it is convenient to divide these into several types according to their chemical features. True aromatic ketones are derivatives of benzo-phenone (diphenyl ketone) and several compounds, usually with hydroxyl and/or methoxyl groups in one or both rings, occur naturally. However, no metabolic data on these compounds are available. The remaining aromatic ketones are mixed ketones in which the aromatic ring is either adjacent to the keto group (phenones) or at another position. Of the former type, **acetophenone** (methyl phenyl ketone) (78) is the simplest example. The earliest studies of its metabolism were carried out a century ago and Williams (1959, pp. 336–338) has summarized the earlier work. In brief, acetophenone was shown to be metabolized in rabbits and dogs both by reduction and oxidation. The former reaction gave methyl phenyl carbinol (79) whereas both benzoic acid and a small amount (1–2%) of mandelic acid (80) were found as oxidized metabolites. Smith *et al.* (1954a,b) gave acetophenone (450 mg/kg, p.o.) to rabbits and found that nearly half of

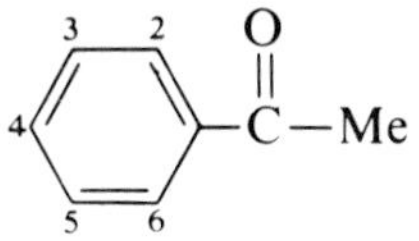

(78)
Acetophenone

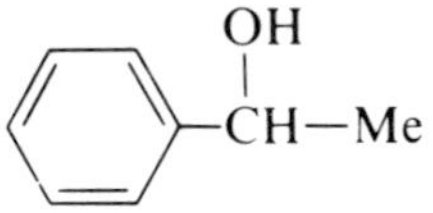

(79)

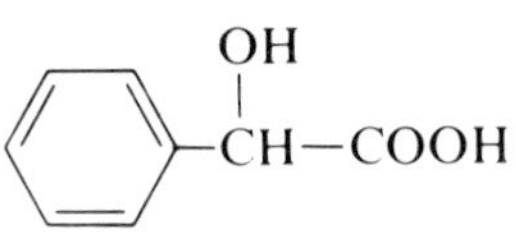

(80)

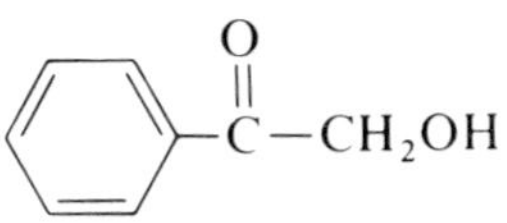

(81)

the dose was excreted in the urine as the glucuronide of (−)-methyl-phenylcarbinol. A small amount (3%) of ethereal sulphate was also detected. El Masry *et al.* (1956), in similar experiments but with a dose of 240 mg/kg, reported that 19% of the dose was found as urinary hippuric acid. The initial oxidation product in the metabolic sequence leading to hippuric acid is ω-hydroxyacetophenone (81) and Kiese and Lenk (1974) showed that rabbits excreted small amounts (0·5–1%) of it in the urine in conjugated form following an intraperitoneal dose of about 350 mg/kg of the ketone. Kiese and Lenk (1973) found that microsomal preparations from rabbit liver readily produced the ω-hydroxylated product of acetophenone. Additional urinary metabolites of acetophenone in rabbits include the 3- and 4-hydroxylated derivatives, however these phenols accounted together for only about 0·5% of the dose (Kiese and Lenk, 1974).

A number of derivatives of acetophenone have been found in plants and, like that noted above with the true aromatic ketones, these generally contain hydroxyl and/or methoxyl groups. The metabolic data on these naturally occurring acetophenones are extremely limited. **Acetovanillone** (4-hydroxy-3-methoxyacetophenone) is mainly excreted in the urine as the corresponding glucuronide conjugate. Daly *et al.* (1960) recorded a value of about 80% for this metabolite in 20 h following a dose of 120 mg/kg in rats. No urinary 3,4-dihydroxyacetophenone was detected, however the presence in the urine of a small amount (0·5%) of acetoisovanillone (3-hydroxy-4-methoxyacetophenone) indicates that some demethylation (followed by *p*-*O*-methylation) must have occurred. *In vitro* experiments indicated that acetovanillone is *O*-demethylated by microsomal preparations from both rat and guinea pig liver. However, acetovanillone can also be demethylated by the intestinal bacteria (Scheline, 1970) and this possibility, which would be quite likely in the event that acetovanillone glucuronide is excreted in the bile, must be kept in mind.

In addition, a few reports have appeared which dealt with the metabolism of acetophenone derivatives closely related to those occurring naturally. Because of this close structural similarity, it seems worthwhile to briefly review the metabolism of these compounds. Williams (1959, pp. 338–339) summarized the older literature which dealt with the metabolism of 2,4-dihydroxyacetophenone (resacetophenone), 2,3,4-trihydroxy-acetophenone (gallacetophenone) and 2,4,6-trihydroxyacetophenone (phloroacetophenone) in rats, rabbits and dogs. These compounds are apparently not metabolized by reduction of the keto group or by oxidation of the methyl group. Instead, the availability of the phenolic hydroxyl groups results in direct conjugation giving glucuronides and ethereal sulphates. Dodgson (1950) showed that, in the case of resacetophenone, glucuronic acid conjugation occurs at the 4-position. Haley and Bassin (1952) administered phloroacetophenone (100 mg/kg, s.c.) to rats and found that the excretion of metabolites in the urine was complete within 24 h. Nearly 80% of the dose was accounted for and the free unchanged phenol and glucuronide fractions each amounted to slightly over a fifth and the ethereal sulphate fraction a third of the dose.

The metabolism of 2,6-dihydroxyacetophenone, 2-hydroxy-6-methoxyacetopheneone and 2,6-dimethoxyacetophenone was studied in rats by Bobik *et al.* (1975). Both unlabelled material and material labelled with ^{14}C in the keto group were given by i.p. injection at doses of roughly 60–80 mg/kg. About 80% of the radioactivity was excreted in the urine in 24 h and this was extracted nearly quantitatively into ether following acid hydrolysis. In the case of the 2,6-dihydroxy derivative, this material consisted solely of the unchanged compound. With 2-hydroxy-6-methoxyacetophenone, most of the hydrolysed material was unchanged compound but 4% was shown to be two hydroxylated metabolites. These were identified as 2,3-dihydroxy-6-methoxyacetophenone and 2,5-dihydroxy-6-methoxyacetophenone. When 2,6-dimethoxyacetophenone was given no unchanged compound was excreted and the 2-hydroxy-6-methoxy derivative was the major (66%) metabolite. Aromatic hydroxylation at the 3-position was also detected in this case and 2,6-dimethoxy-3-hydroxyacetophenone (26%) and 2,3-dihydroxy-6-methoxyacetophenone (1%) were found. The remainder (7%) of the radioactivity was present as unidentified polar material. No aromatic carboxylic acids were detected as metabolites in any of the experiments. These results therefore substantiate the earlier findings which indicated that hydroxylated acetophenones are neither reduced to carbinols nor oxidized to aromatic acids. Their major metabolic pathway is direct conjugation but, in some cases, aromatic hydroxylation may also occur.

Another type of phenone includes compounds in which the methyl group is replaced by various other substituents. Typical of this group are the chalcones and dihydrochalcones. However, these compounds show both biosynthetic and metabolic similarities to the flavonoids and their metabolic fate is therefore summarized in Chapter 7, Section IV.F.

The final type of aromatic mixed ketone includes a relatively small number of compounds in which the aromatic ring is not situated adjacent to the keto group. As with most other groups of ketones, the amount of metabolic data available on this type is limited. In fact, only a single compound, **zingerone** (4-(4-hydroxy-3-methoxyphenyl)butan-2-one) (82), has been studied. Monge *et al.* (1976) found that zingerone, a pungent principle of ginger, was largely excreted in the urine of rats within 24 h as glucuronide and/or sulphate conjugates. About 50–55% of the dose

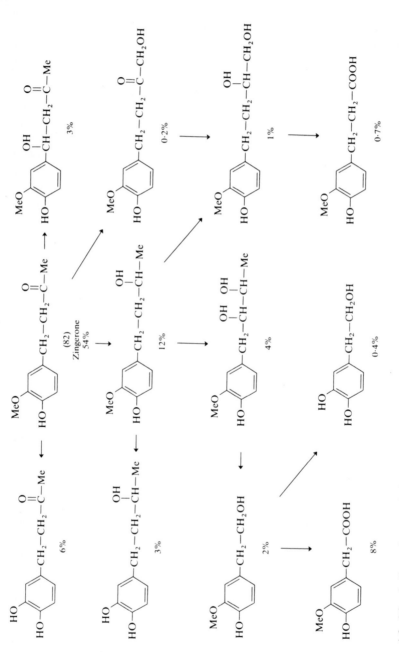

FIG. 4.2. Metabolic pathways of zingerone in rats. Percent values in parentheses = urinary metabolite as % of dose. Dose = 100 mg/kg, p.o. or i.p.

(100 mg/kg, p.o. or i.p.) was excreted as conjugates of zingerone itself and reduction to the corresponding carbinol accounted for a further 11–13%. Additionally, side chain oxidation took place at all three available sites, however oxidation at the 3-position, giving rise to C_6—C_2 metabolites, predominated. The metabolic pathways of zingerone in the rat, together with the amounts of the metabolites excreted in the urine, are shown in Fig. 4.2. The formation of several O-demethylated metabolites was shown to be due to the extensive biliary excretion of zingerone and some of its metabolites followed by their O-demethylation by intestinal micro-organisms.

III. Quinones

The information available on the metabolism of plant quinones is meagre. No data appear to be available on the fate of the benzoquinone deriva-tives, although the expected general route of metabolism is a reversible reduction to the corresponding dihydric phenol followed by conjugation and excretion (Fig. 4.3), as shown by Glock $et\ al.$ (1945) and Bray and Garrett (1961) with a few synthetic derivatives of p-benzoquinone.

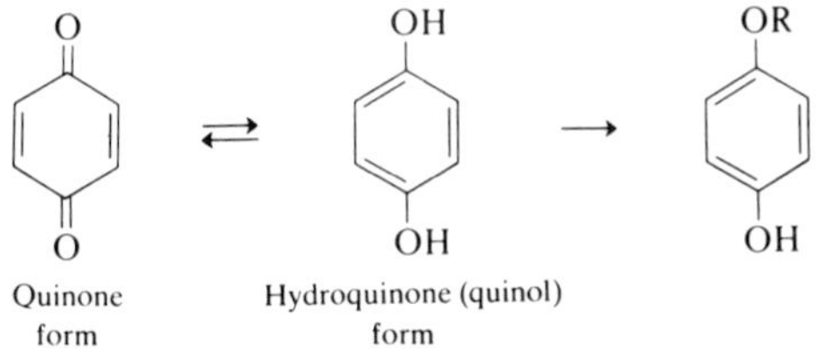

Quinone
form

Hydroquinone (quinol)
form

FIG. 4.3. General pathway of quinone metabolism. R = conjugating group.

A. NAPHTHOQUINONES

Interest in the metabolism of naphthoquinones is focused mainly on phyl-loquinone (Vitamin K_1) (83), although it seems justified, in view of the paucity of information dealing with quinone metabolism, to briefly sum-marize the metabolism of its synthetic homologue menadione (Vitamin K_3; 2-methyl-1,4-naphthoquinone) (84). Hoskin $et\ al.$ (1954) found that radioactive menadione (3–11 mg/kg, i.m.) undergoes little or no urinary excretion in the unchanged form in the rat and guinea pig. Instead, three metabolites were isolated. The major metabolite appeared by chemical analysis to be the diglucuronide of 2-methyl-1,4-naphthohydroquinone. Richert (1951) had noted previously that menadione given orally to rabbits caused an increased excretion of glucuronic acid. Further studies on the nature of the glucuronide formed by rat liver have been carried out by Thompson *et al.* (1972). Using mass spectral and g.l.c./mass spectral methods, they showed conclusively that menadione monoglucuronide was the metabolite excreted *in vivo* in bile or in the bile and perfusate in experiments with isolated perfused livers. It was not possible to determine if the conjugate was located at the 1- or 4-position but the results seemed to indicate the absence of a diglucuronide in the samples. The second metabolite identified by Hoskin *et al.* (1954) was shown to be the monosulphate, this confirming the report of Richert (1951) that menadione is metabolized to 4-hydroxy-2-methyl-1-naphthyl sulphate. The third metabolite was formed in lesser amounts and was shown to be a simple, easily hydrolysable conjugate of menadione. A subsequent investigation by Hart (1958) showed that this metabolite is the phosphate conjugate.

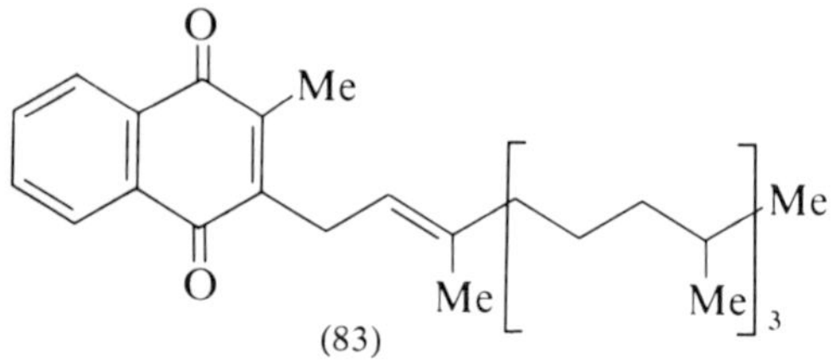

(83)
Phylloquinone

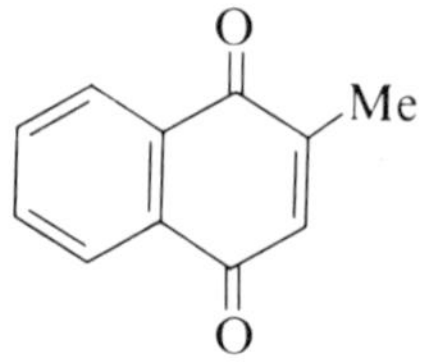

(84)
Menadione

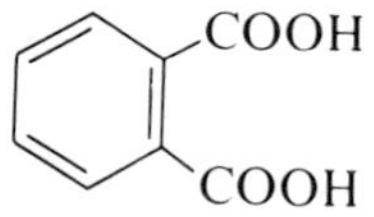

(85)

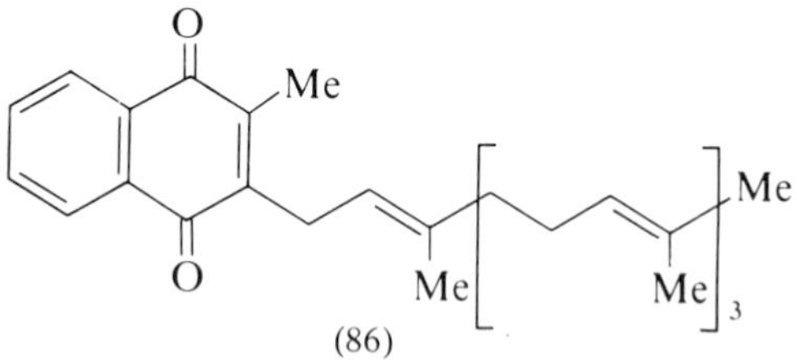

(86)

The formation of menadione metabolites noted above results from the reactions illustrated in the general pathway of quinone metabolism shown in Fig. 4.3. However, a report by Shemiakin and Schukina (1944) indicated that oxidation of the quinone ring also takes place. The s.c. administration of menadione to dogs resulted in the urinary excretion of phthalic acid (85). Neither unchanged compound nor the possible intermediate, 3-hydroxy-2-methyl-1,4-naphthoquinone, was excreted. Phthalic acid was also reported to be a metabolite of menadione in humans. A more notable menadione metabolite is the strongly lipophilic product menaquinone-4 (Vitamin K_2(20); 3-(geranyl-geranyl)-2-methyl-1,4-naphthoquinone) (86) shown by Martius and Esser (1958) to be formed in rats and chickens. This metabolic synthesis was confirmed in rats by Horth *et al.* (1966). Compound (86) is considered to be the active form of the K vitamins (see Martius, 1961; Taggart and Matschiner, 1969). A further aspect of menadione metabolism concerns the identity of the blue pigment first noted by Richert (1951) in rabbit urine. Later work by Losito *et al.* (1965) on the biliary metabolites of menadione in dogs showed the presence of a major metabolic fraction which had a greenish-blue colour and showed characteristics of the bile pigments. It is therefore possible that the metabolism and excretion of the quinone may also be associated with that of the bile pigments.

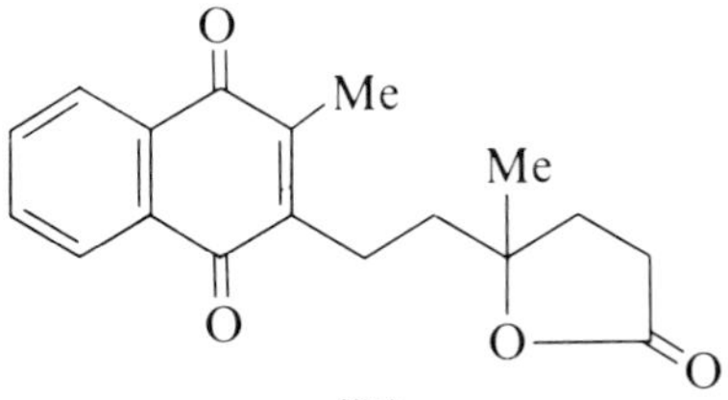

(87)

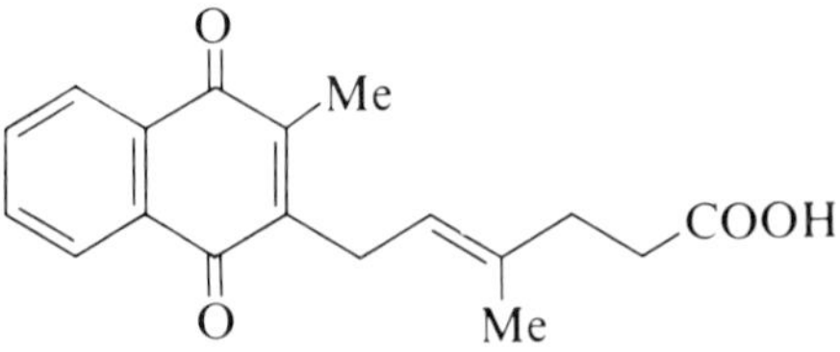

(88)

An early study of the metabolism of [^{14}C]-phylloquinone (83) in rats (Taylor *et al.*, 1956) showed that the radioactivity was only slowly excreted into the urine (10%, 17% and 19% in 24, 72 and 144 h, respectively, after i.v. administration of 26 mg/kg). It was also found that this material was present solely as metabolic products. Final faecal excretion was greater (43% after 144 h) and only 6·5% of this material was in the form of unchanged compound. However, the nature of the urinary and faecal metabolites was not determined. Later, Wiss and Gloor (1966) found that [^{14}C]-phylloquinone undergoes side chain oxidation in rats leaving a γ-lactone moiety containing seven carbon atoms. This metabolite, (3-(5′-carboxy-3′-hydroxy-3′-methylpentyl)-2-methyl-1,4-naphthoquinone lactone) (87), was detected both in urine and in extracts of liver and kidney in a conjugated form thought to be a glucuronide. However, for compound (87) to be able to undergo conjugation it would have to exist either in the reduced (hydroquinone) form or as the hydroxy acid. More recent studies on the metabolism of [^{14}C]-phylloquinone in man (Shearer and Barkhan, 1973) have shown that the lactone (87) is probably an artifact produced in acidic conditions and that the actual metabolite is likely the glucuronide conjugate of 3-(5'-carboxy-3'-methyl-2'-pentenyl)-2-methyl-1,4-naphthoquinone (88). In addition to this urinary metabolite containing a seven carbon side chain, evidence was also obtained which showed the excretion of another major terminal metabolite containing a five-carbon side chain, i.e. the 3-carboxybutyl group. It is worth noting that the metabolism of ubiquinone, a benzoquinone derivative not covered in this book, shows numerous points of similarity to that seen with phylloquinone (see Imada *et al.*, 1970). Phylloquinone is also metabolized by epoxidation (Caldwell *et al.*, 1974). Knauer *et al.* (1975) reported that the *cis* isomer of phylloquinone is a poor substrate for 2,3-epoxidation *in vitro* and *in vivo* in rats, in contrast to that seen with the *trans* isomer. The *cis* isomer lacked biological activity and it was suggested that the latter property is coupled with epoxidation of the quinone by the microsomal enzymes.

B. ANTHRAQUINONES

This group furnishes the largest number of quinones, many of which have been historically important because of their use as mordant dyes (alizarin is the principal colouring matter of madder) and as cathartics (cascara, rhubarb, senna). The parent compound, anthraquinone (89), has been detected in plant material although it most likely is an artifact. The naturally occurring anthraquinones are, with few exceptions, hydroxylated

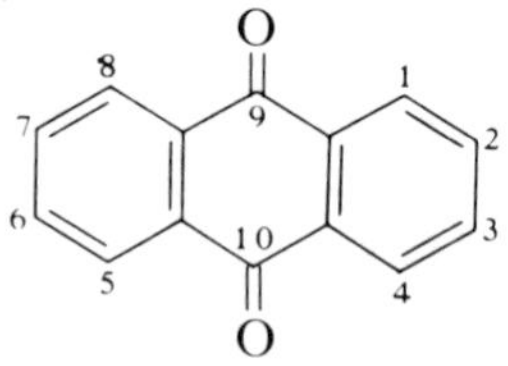

(89)

Anthraquinone

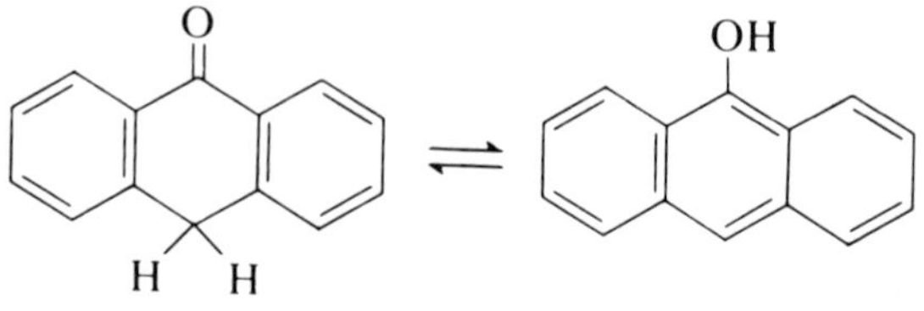

(90)   (91)

derivatives which may be present as methyl ethers or, especially, as glyco-sides. In addition, C-glycosides are found, e.g. barbaloin from aloe. Anthraquinones undergo reduction to anthrones (90), the latter existing also in the tautomeric anthranol forms (91).

The metabolism of **1-hydroxyanthraquinone** and **2-hydroxyanthra-quinone** given orally to rats (approx. 330 mg/kg) was studied by Fujita *et al.* (1961a). With the former compound, about 5% of the dose is found in the 48 h urine as unchanged compound and its conjugates. A further 2·5% of the dose was identified as alizarin (92) which was also excreted partly free and partly as conjugates. Faecal excretion of 1-hydroxyanthraquinone (13%) and alizarin (0·8%) was recorded during this period. Similar experiments with the 2-hydroxy isomer showed a urinary excretion of the unchanged compound and its conjugates of approx. 6% and a correspond-ing value of 0·7% for alizarin. Faecal excretion was 44% and 0·2%, respectively. These results indicate that hydroxylation is more extensive with 1-hydroxyanthraquinone than with the 2-hydroxy isomer. Only about a fifth of the dose of the former compound was accounted for whereas half of the latter was determined. However, it is not known if the low recoveries are due to metabolite excretion after 48 h or to formation of unidentified metabolites.

In experiments similar to those described above with the two mono-hydroxy anthraquinones, Fujita *et al.* (1961b) showed that the correspond-ing methyl ethers (approx. 165 mg/kg) were demethylated. Thus the urinary metabolites of the methoxy compounds were qualitatively similar to those detected when the hydroxyanthraquinones were given. In both cases the urinary excretion of the mono- and dihydroxy metabolites accounted for about 10% of the dose, however the proportion of alizarin (92) formed was higher in these experiments than in those with the hy-droxyanthraquinones. The unchanged compounds were not detected in the urine but were present in the faeces to the extent of about 13% (1-methoxyanthraquinone) and 27% (2-methoxyanthraquinone) of the dose. In the latter case, 2-hydroxyanthraquinone (6%) was also detected. Total metabolite recoveries were approx. 24% with 1-methoxyanthraquinone

(92)

Alizarin

and 44% with 2-methoxyanthraquinone. The results indicated that demethylation occurs more readily with the latter compound.

Mähner and Dulce (1968) studied the metabolism of **alizarin** (92) in rats (100 mg/kg, p.o.). They found alizarin monoglucuronide in the urine, the excretion of this metabolite being maximal 6–8 h after administration of the quinone. However, the fate of most of the compound is not known as only about 10% of the dose was recovered as the monoglucuronide in the 24 h urines. Similar experiments by Fujita *et al.* (1962) with alizarin (approx. 165 mg/kg) indicated that unchanged compound was the major urinary metabolite although unknown metabolites were detected by paper chromatography. **Alizarin-1-methyl ether** was shown in this study to be demethylated to alizarin which was then excreted in the urine. No data is available on the metabolism of the other monomethyl ether of alizarin, however Fujita *et al.* (1962) found that the synthetic dimethyl ether (approx. 165 mg/kg) was demethylated and excreted in the urine as alizarin-1-methyl ether (1·5%), alizarin (trace) and unchanged compound (trace). No 2-methyl ether was detected, however, and this preference for demethylation at the 2-position is similar to that described above with the isomeric methoxyanthraquinones.

Various derivatives of 1, 8-dihydroxyanthraquinone have been employed as cathartic agents. It has long been known that the naturally-occurring glycosides have little direct effect in this respect and that they must undergo metabolic change in order to be active. This was reported by Straub and Triendl (1937) and by Okada (1940) who suggested that intestinal bacteria convert the anthraquinone glycosides to hydroxy-anthraquinones. A similar conclusion was reached by Hardcastle and Wilkins (1970). They found no increase in colon moiety produced *in situ* by senna glycosides, however increased peristalsis occurred when the senna preparation was previously incubated with faeces or *E. coli*. This effect was similar to that produced by the rheinanthrone formed by chemical hydrolysis and reduction of senna. Breimer and Baars (1976) summarized the relationship between metabolism and laxative effects of the anthraquinone cathartics. They also studied the absorption, metabolism and excretion of 1,8-dihydroxyanthraquinone danthron in rats. This synthetic derivative has close structural similarity to the more complex naturally-occurring anthraquinones noted below. The results of Breimer and Baars indicated that significant absorption of danthron occurs in the small intestine, however only 30–40% or less of the dose (approx. 1–2 mg/kg, p.o.) could be accounted for in the urine and faeces. Urinary excretion of the metabolites took place during the first 24 h, the bulk of the material being present as glucuronide and sulphate conjugates. Metabolite excretion in the faeces was quantitatively less important, occurring mainly during the first 24 h period but sometimes extending beyond that. Additional pathways are therefore involved in the metabolism of danthron, however it was concluded that the unidentified material did not consist of the anthrone form which could be formed by bacterial reduction of the anthraquinone derivative in the large intestine.

(93)	(94)
Chrysophanic acid	Chrysarobin

The metabolism of **chrysophanic acid** (1,8-dihydroxy-3-methyl-anthraquinone) (93) which occurs in medicinal rhubarb and the antipsoriatic drug **chrysarobin** (3-methyl-1,8,9-anthracenetriol) (94) was investigated by Ippen (1959a) in a larger study concerning the interconversion of the anthraquinone and anthracenetriol (anthranol) forms (Ippen and Montag, 1958; Ippen, 1959a,b). This interconversion was shown to occur and it was found that the quinone is the form which predominates in the body and which is excreted in the urine. Reduction of the quinone takes place in the large intestine so that both forms are excreted in the faeces.

Chen *et al.* (1963) studied the absorption, distribution and excretion of anthraquinone derivatives of Chinese rhubarb, including **rhein** (1,8-dihydroxyanthraquinone-3-carboxylic acid) and **emodin** (1,3,8-trihydroxy-6-methylanthraquinone). The compounds were given to animals and humans by oral, i.m. or i.v. administration. The anthraquinone derivatives were readily absorbed (rhein more easily than emodin) and excreted, excretion occurring in the bile, urine and faeces. Excretion was complete within two days, however only slightly less than half of the administered dose was accounted for, being nearly equally divided in the urine and faeces. The nature of the metabolic products was not determined. Hsu *et al.* (1966) administered rhein, **aloe-emodin** (1,8-dihydroxy-3-(hydroxymethyl)-anthraquinone) and chrysophanic acid (93) orally to mice and found that 27%, 14% and 6%, respectively, was excreted in the 24 h urine as total anthraquinone derivatives.

References

Abou-Donia, M. B. (1976). *In* "Pesticide Reviews" (F. A. Gunther, Ed.), Vol. 61, pp. 125–160. Springer-Verlag, New York, Heidelberg, Berlin.

Abou-Donia, M. B. and Dieckert, J. W. (1974). *J. Nutr.* **104**, 754–760.

Abou-Donia, M. B. and Dieckert, J. W. (1975). *Toxic. appl. Pharmac.* **31**, 32–46.

Abou-Donia, M. B., Lyman, C. M. and Dieckert, J. W. (1970). *Lipids* **5**, 938–946.

Albrecht, J. E., Clawson, A. J. and Smith, F. H. (1972). *J. Anim. Sci.* **35**, 941–946.

Asahina, Y. and Ishidate, M. (1928). *Ber. dt. chem. Ges.* **61**, 533–536.

Asahina, Y. and Ishidate, M. (1933). *Ber. dt. chem. Ges.* **66**, 1673–1677.

Asahina, Y. and Ishidate, M. (1934). *Ber. dt. chem. Ges.* **67**, 71–77.

Asahina, Y. and Ishidate, M. (1935). *Ber. dt. chem. Ges.* **68**, 947–953.

Asano, M. and Yamakawa, T. (1950). *J. Biochem.* Tokyo **37**, 321–327.

Bielig, H.-J. and Hayasida, A. (1940). *Hoppe-Seyler's Z. physiol. Chem.* **266**, 99–111.

Bobik, A., Holder, G. M., Ryan, A. J. and Wiebe, L. I. (1975). *Xenobiotica* **5**, 65–72.

Boyland, E. (1940). *Biochem. J.* **34**, 1196–1201.

Boyland, E. and Chasseaud, L. F. (1967). *Biochem. J.* **104**, 95–102.

Boyland, E. and Chasseaud, L. F. (1968). *Biochem. J.* **109**, 651–661.

Boyland, E. and Chasseaud, L. F. (1970). *Biochem. Pharmac.* **19**, 1526–1528.

Bray, H. G. and Garrett, A. J. (1961). *Biochem. J.* **80**, 6P.

Bray, H. G., Thorpe, W. V. and White, K. (1951). *Biochem. J.* **48**, 88–96.

Bray, H. G., Thorpe, W. V. and White, K. (1952). *Biochem. J.* **52**, 423–430.

Breimer, D. D. and Baars, A. J. (1976). *Pharmacology* **14 Suppl. 1**, 30–47.

Bressani, R., Elias, L. G. and Braham, J. E. (1964). *J. Nutr.* **83**, 209–217.

Caldwell, P. T., Ren, P. and Bell, R. G. (1974). *Biochem. Pharmac.* **23**, 3353–3362.

Chen, C.-H., Kao, S.-M., Du, H.-F. and Yo, W.-H. (1963). *Acta pharm. sin.* **10**, 525–530.

Daly, J. W., Axelrod, J. and Witkop, B. (1960). *J. biol. Chem.* **235**, 1155–1159.

Dirscherl, W. and Brisse, B. (1966). *Hoppe-Seyler's Z. physiol. Chem.* **346**, 55–59.

Di Vincenzo, G. D., Kaplan, C. J. and Dedinas, J. (1976). *Toxic. appl. Pharmac.* **36**, 511–522.

Dodgson, K. S. (1950). *Biochem. J.* **47**, xi–xii.

Dodgson, K. S. and Williams, R. T. (1949). *Biochem. J.* **45**, 381–386.

Elliott, T. H., Parke, D. V. and Williams, R. T. (1959). *Biochem. J.* **72**, 193–200.

El Masry, A. M., Smith, J. N. and Williams, R. T. (1956). *Biochem. J.* **64**, 50–56.

Fischer, F. G. and Bielig, H.-J. (1940). *Hoppe-Seyler's Z. physiol. Chem.* **266**, 73–98.

Friedhoff, A. J., Schweitzer, J. W., Miller, J. and van Winkle, E. (1972). *Experientia* **28**, 517–519.

Friedmann, E. and Mai, H. (1931). *Biochem. Z.* **242**, 282–287.

Friedmann, E. and Türk, W. (1913). *Biochem. Z.* **55**, 425–431.

Fujita, M., Furuya, T. and Matsuo, M. (1961a). *Chem. pharm. Bull.* Tokyo **9**, 962–966.

Fujita, M., Furuya, T. and Matsuo, M. (1961b). *Chem. pharm. Bull.* Tokyo **9**, 967–970.

Fujita, M., Furuya, T. and Matsuo, M. (1962). *Chem. pharm. Bull.* Tokyo **10**, 909–911.

Glock, G. E., Thorp, R. H., Ungar, J. and Wien, R. (1945). *Biochem. J.* **39**, 308–313.

Grebennik, L. I., Gnevkovskaya, T. V. and Smirnov, G. A. (1963). *Vop. med. Khim. Akad. med. Nauk SSSR* **9**, 127–133. (*Chem. Abstr.* (1963) **59**, 13238b.)
Haggard, H. W., Miller, D. P. and Greenberg, L. A. (1945). *J. ind. Hyg. Toxicol.* **27**, 1–14.
Haley, T. J. and Bassin, M. (1952). *Proc. Soc. exp. Biol. Med.* **81**, 298–300.
Hämäläinen, J. (1909). *Skand. Arch. Physiol.* **23**, 297–301.
Hämäläinen, J. (1912). *Skand. Arch. Physiol.* **27**, 141–226.
Hardcastle, J. D. and Wilkins, J. L. (1970). *Gut* **11**, 1038–1042.
Hart, K. T. (1958). *Proc. Soc. exp. Biol. Med.* **97**, 848–851.
Hartles, R. L. and Williams, R. T. (1948). *Biochem. J.* **43**, 296–303.
Harvey, J. M. (1942). *Pap. Dep. Chem. Univ. Qd.* **1**, no. 23.
Hathway, D. E. (1975). *In* "Foreign Compound Metabolism in Mammals" (D. E. Hathway, Ed.), Vol. 3, pp. 201–365. The Chemical Society, London.
Heffter, A. (1895). *Arch. exp. Path. Pharmak.* **35**, 342–374.
Hildebrandt, H. (1901). *Arch. exp. Path. Pharmak.* **45**, 110–129.
Hildebrandt, H. (1902a). *Hoppe-Seyler's Z. physiol. Chem.* **36**, 441–451.
Hildebrandt, H. (1902b). *Hoppe-Seyler's Z. physiol. Chem.* **36**, 452–461.
Hinson, J. A. and Neal, R. A. (1972). *J. biol. Chem.* **247**, 7106–7107.
Hinson, J. A. and Neal, R. A. (1975). *Biochim. biophys. Acta* **384**, 1–11.
Horth, C. E., McHale, D., Jeffries, L. R., Price, S. A., Diplock, A. T. and Green J. (1966). *Biochem. J.* **100**, 424–429.
Hoskin, F. C. G., Spinks, J. W. T. and Jaques, L. B. (1954). *Can. J. Biochem. Physiol.* **32**, 240–250.
Hsu, G.-Y., Sun, C.-C., Chen, C.-H. and Wu, W. T. (1966). *Acta biochim. biophys. sin.* **6**, 110–117. (*Chem. Abstr.* (1966) **65**, 14287f.)
Ide, H. and Toki, S. (1970). *Biochem. J.* **119**, 281–287.
Imada, I., Watanabe, M., Matsumoto, N. and Morimoto, H. (1970). *Biochemistry N.Y.* **9**, 2870–2878.
Ippen, H. (1959a). *Dermatologica* **119**, 211–220.
Ippen, H. (1959b). *Planta med.* **7**, 423–426.
Ippen, H. and Montag, T. (1958). *Arzneimittel-Forsch.* **8**, 778–779.
Ishidate, M., Kawahata, H. and Nakazawa, K. (1941). *Ber. dt. chem. Ges.* **74**, 1707–1711.
Jaffé, M. and Cohn, R. (1887). *Ber. dt. chem. Ges.* **20**, 2311–2317.
James, S. P. and Waring, R. H. (1971). *Xenobiotica* **1**, 573–580.
Kamienski, F.-X. and Casida, J. E. (1970). *Biochem. Pharmac.* **19**, 91–112.
Kamil, I. A., Smith, J. N. and Williams, R. T. (1953). *Biochem. J.* **53**, 129–136.
Kawabata, H. (1943). *J. Pharm. Soc. Japan* **63**, 455–456.
Kiese, M. and Lenk, W. (1973). *Biochem. Pharmac.* **22**, 2575–2580.
Kiese, M. and Lenk, W. (1974). *Xenobiotica* **4**, 337–343.
Knauer, T. E., Siegfried, C., Willingham, A. K. and Matschiner, J. T. (1975). *J. Nutr.* **105**, 1519–1524.
Krieger, H. (1962). *Acta chim. fenn.* **35B**, 174–175. (*Chem. Abstr.* (1963). **58**, 3792d.)
Kuhn, R. and Livada, K. (1933). *Hoppe-Seyler's Z. physiol. Chem.* **220**, 235–246.
Kuhn, R. and Löw, I. (1938). *Hoppe-Seyler's Z. physiol. Chem.* **254**, 139–143.
Kuhn, R., Köhler, F. and Köhler, L. (1936). *Hoppe-Seyler's Z. physiol. Chem.* **242**, 171–197.
Leibman, K. C. and Ortiz, E. (1973). *Drug Metab. Disposit.* **1**, 543–551.
Losito, R., Millar, G. J. and Jaques, L. B. (1965). *Biochim. biophys. Acta* **107**, 123–125.

Magnus-Levy, A. (1907). *Biochem. Z.* **2**, 319–331.
Mähner, B. and Dulce, H.-J. (1968). *Z. klin. Chem. klin. Biochem.* **6**, 99–102.
Martius, C. (1961). *Am. J. clin. Nutr.* **9**, 97–103.
Martius, C. and Esser, H. O. (1958). *Biochem. Z.* **331**, 1–9.
Mayer, P. (1908). *Biochem. Z.* **9**, 439–441.
Meyer, T., Scheline, R. R. (1973). Unpublished observations.
Monge, P., Scheline, R. and Solheim, E. (1976). *Xenobiotica* **6**, 411–423.
Müller-Enoch, D., Thomas, H. and Holzmann, P. (1974). *Hoppe-Seyler's Z. physiol. Chem.* **355**, 1232–1233.
Neubauer, O. (1901). *Arch. exp. Path. Pharmak.* **46**, 133–154.
Okada, T. (1940). *Tohoku J. exp. Med.* **38**, 33–44. (*Chem. Abstr.* (1940). **34**, 5163^5.)
Opdyke, D. L. J. (1973a). *Fd Cosmet. Toxicol.* **11**, 95–115.
Opdyke, D. L. J. (1973b). *Fd Cosmet. Toxicol.* **11**, 477–495.
Parke, D. V. and Rahman, H. (1969). *Biochem. J.* **113**, 12P.
Paul, H. E., Austin, F. L., Paul, M. F. and Ellis, V. R. (1949). *J. biol. Chem.* **180**, 345–363.
Perry, T. L., Hansen, S., Hestrin, M. and MacIntyre, L. (1965). *Clinica chim. Acta* **11**, 24–34.
Prelog, V. and Meier, H. L. (1950). *Helv. chim. Acta* **33**, 1276–1284.
Prelog, V. and Würsch, J. (1951). *Helv. chim. Acta* **34**, 859–861.
Price, T. D. and Rittenberg, D. (1950). *J. biol. Chem.* **185**, 449–459.
Quick, A. J. (1932). *J. biol. Chem.* **97**, 403–419.
Reinarz, F. and Zanke, W. (1934a). *Ber. dt. chem. Ges.* **67**, 548–553.
Reinarz, F. and Zanke, W. (1934b). *Ber. dt. chem. Ges.* **67**, 589–593.
Reinarz, F. and Zanke, W. (1936). *Ber. dt. chem. Ges.* **69**, 2259–2262.
Reinarz, F., Zanke, W. and Faust, K. (1934). *Ber. dt. chem. Ges.* **67**, 1536–1542.
Richert, D. A. (1951). *J. biol. Chem.* **189**, 763–768.
Rimini, E. (1909). *Gazz. chim. ital.* **39 II**, 186–196. (*Chem. Abstr.* (1911) **5**, 689–690.)
Robertson, J. S. and Hussain, M. (1969). *Biochem. J.* **113**, 57–65.
Robertson, J. S. and Solomon, E. (1971). *Biochem. J.* **121**, 503–509.
Sakami, W. and Lafaye, J. M. (1951). *J. biol. Chem.* **193**, 199–203.
Sammons, H. G. and Williams, R. T. (1941). *Biochem. J.* **35**, 1175–1188.
Sammons, H. G. and Williams, R. T. (1946). *Biochem. J.* **40**, 223–227.
Sato, T., Suzuki, T., Fukuyama, T. and Yoshikawa, H. (1956a). *J. Biochem.* Tokyo **43**, 413–420.
Sato, T., Suzuki, T., Fukuyama, T. and Yoshikawa, H. (1956b). *J. Biochem.* Tokyo **43**, 421–429.
Scheline, R. R. (1970). Unpublished observations.
Scheline, R. R. (1972a). *Xenobiotica* **2**, 227–236.
Scheline, R. R. (1972b). Unpublished observations.
Schmiedeberg, O. and Meyer, H. (1879). *Hoppe-Seyler's Z. physiol. Chem.* **3**, 422–450.
Schwarz, L. (1898). *Arch. exp. Path. Pharmak.* **40**, 168–194.
Shearer, M. J. and Barkhan, P. (1973). *Biochim. biophys. Acta* **297**, 300–312.
Shemiakin, M. M. and Schukina, L. A. (1944). *Nature* Lond. **154**, 513.
Shimamoto, T. (1934a). *Sci. Pap. Inst. phys. chem. Res.* Tokyo **25**, 52–58.
Shimamoto, T. (1934b). *Sci. Pap. Inst. phys. chem. Res.* Tokyo **25**, 59–62.
Skutches, C. L. and Smith, F. H. (1974). *J. Am. Oil Chem. Soc.* **51**, 413–415.
Skutches, C. L., Herman, D. L. and Smith, F. H. (1973). *J. Nutr.* **103**, 851–855.

Smith, F. H. and Clawson, A. J. (1965). *J. Nutr.* **87**, 317–321.
Smith, J. N., Smithies, R. H. and Williams, R. T. (1954a). *Biochem. J.* **56**, 320–324.
Smith, J. N., Smithies, R. H. and Williams, R. T. (1954b). *Biochem. J.* **57**, 74–76.
Strand, L. P. and Scheline, R. R. (1975). *Xenobiotica* **5**, 49–63.
Straub, W. and Triendl, E. (1937). *Naunyn-Schmiedebergs Arch. exp. Path. Pharmak.* **185**, 1–19.
Taggart, W. V. and Matschiner, J. T. (1969). *Biochemistry* N.Y. **8**, 1141–1146.
Takenoshita, R. and Toki, S. (1974). *J. biol. Chem.* **249**, 5428–5429.
Tamura, Z. and Imanari, T. (1964). *Chem. pharm. Bull.* Tokyo **12**, 370–375.
Tamura, Z. and Imanari, T. (1970). *J. pharm. Soc. Japan* **90**, 506–508.
Taylor, J. D., Millar, G. J., Jaques, L. B. and Spinks, J. W. T. (1956). *Can. J. Biochem. Physiol.* **34**, 1143–1152.
Teppati, R. (1937). *Archs int. Pharmacodyn. Thér.* **57**, 440–449.
Teuchy, H., Quatacker, J., Wolf, G. and Van Sumere, C. F. (1971). *Archs int. Physiol. Biochim.* **79**, 573–587.
Thompson, R. M., Gerber, N., Seibert, R. A. and Desiderio, D. M. (1972). *Res. Commun. chem. Path. Pharmac.* **4**, 543–552.
Wiedemann, C. (1877). *Arch. exp. Path. Pharmak.* **6**, 216–232.
Williams, R. T. (1938). *Biochem. J.* **32**, 878–887.
Williams, R. T. (1940). *Biochem. J.* **34**, 690–697.
Williams, R. T. (1959). "Detoxication Mechanisms". Chapman and Hall, London.
Wiss, O. and Gloor, H. (1966). *Vitams Horm.* **24**, 575–586.
Wong, K. P. and Sourkes, T. L. (1966). *Can. J. Biochem.* **44**, 635–644.

5

METABOLISM OF ACIDS, LACTONES AND ESTERS

I. Acids

A. Aliphatic Acids

The aliphatic acids occurring in plants encompass a large number of compounds which include saturated and unsaturated volatile acids and non-volatile mono-, di- and tricarboxylic acids. All of the members of the homologous series of saturated, normal fatty acids from C_1 (formic acid) to C_{16} (palmitic acid) occur naturally as do many even-numbered members from C_{18} (stearic acid) upwards. The essential feature in the metabolism of these compounds is their oxidative degradation, a universal biochemical capacity among living organisms. This process involves β-oxidation which results in the formation of two-carbon fragments from the fatty acids by way of acylated coenzyme A derivatives. This subject falls outside the scope of this book, however, and will not be discussed further. The same applies to the metabolism of the di- and tricarboxylic acids which enter into the ubiquitous citric acid cycle. It is noteworthy, however, that some of the acids (e.g. citric and malic) produced by these mitochondrial reactions sometimes accumulate and are concentrated in the vacuolar sap of plants.

Among the unsaturated plant fatty acids, both sorbic acid, a C_6 acid, and a few acyclic terpenoid C_{10} acids have been studied metabolically. **Sorbic acid** (1), a useful fungistatic agent in foods, is known to enter into conventional fatty acid metabolism and Deuel *et al.* (1954) reported that its metabolism is identical with that of butanoic and hexanoic acid. Sorbic acid is thus converted extensively to CO_2 and water. Fingerhut *et al.* (1962), using 1-^{14}C-labelled compound, found that 85% of the radioactivity was excreted by rats as respiratory $^{14}CO_2$ following oral administration of 60–1200 mg/kg of the acid. It was absorbed nearly quantitatively from the intestine but no unchanged compound was excreted in the urine. Sorbic acid is not utilized in the formation of glycogen, however the detection of radioactivity in the lipids indicates that the acetyl coenzyme A formed is used for fatty acid synthesis. A minor route of metabolism is allylic hydroxylation and further oxidation of the terminal methyl group. Kuhn *et* *al.* (1937) reported that small amounts (<1% dose) of *trans-trans*-muconic acid (2) were excreted in the urine of rabbits fed large amounts of sorbic acid.

$$Me-CH=CH-CH=CH-COOH$$

(1)

Sorbic acid

$$HOOC-CH=CH-CH=CH-COOH$$

(2)

Oxidation similar to that noted above with sorbic acid was reported by Kuhn *et al.* (1936) to take place with **geranic acid** (3). About 16% of the total dose (55 g given orally to rabbits over a 6-day period) was isolated as the dicarboxylic acid known as Hildebrandt acid (4).

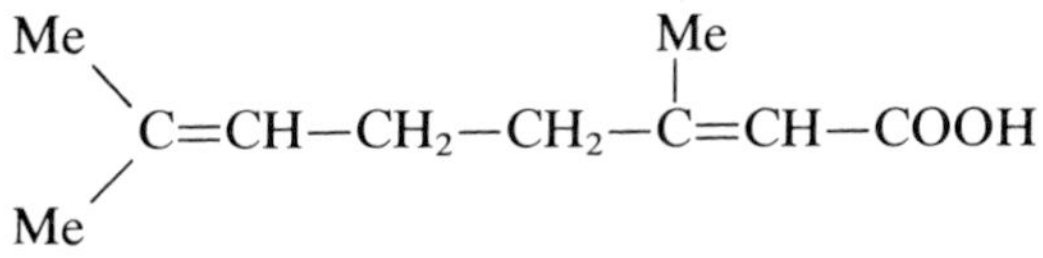

(3)

Geranic acid

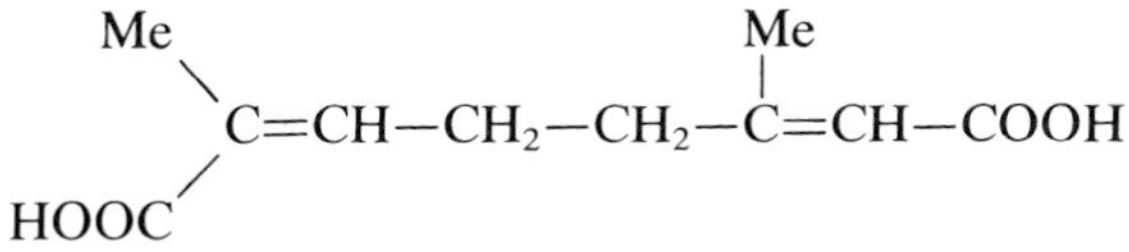

(4)

*d***-Citronellic acid** and *l***-rhodinic acid** are enantiomorphs of compound (5), which differs structurally from geranic acid only by the lack of the double bond at the 2,3-position. Asano and Yamakawa (1950) found that the metabolism of these acids in rabbits was analogous to that reported for geranic acid, the appropriate enantiomorphic dihydro Hildebrandt acids (6) being excreted in the urine.

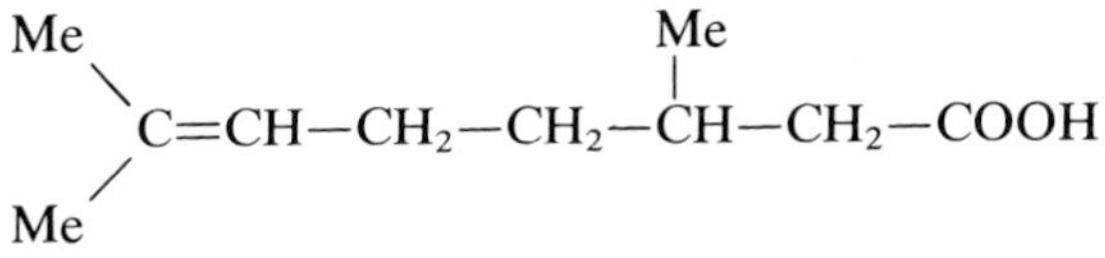

(5)

d-Citronellic acid
l-Rhodinic acid

$$\begin{matrix} \text{Me} & & & & \text{Me} \\ \diagdown & & & & | \\ & \text{C}{=}\text{CH}{-}\text{CH}_2{-}\text{CH}_2{-}\text{CH}{-}\text{CH}_2{-}\text{COOH} \\ \diagup & & & & \\ \text{HOOC} & & & & \end{matrix}$$

(6)

Oxalic acid (HOOC—COOH) is the simplest member of the series of unsubstituted aliphatic dicarboxylic acids and, except for succinic and fumaric acids which enter into the citric acid cycle, is the derivative which has received the most attention from a metabolic point of view. Oxalic acid is toxic by virtue of the formation of insoluble calcium oxalate in the body but is otherwise essentially metabolically inert. Both Weinhouse and Friedmann (1951) and Curtin and King (1955) reported that 1% or less was lost as $^{14}CO_2$ when rats were injected with a few mg of [^{14}C]-oxalic acid, however even this small amount of CO_2 may have arisen from impurities (e.g. formic acid) rather than from oxalic acid itself. No evidence was found for its conversion to other metabolites, however only roughly 20–40% of the dose (2–7 mg) was recovered unchanged in the urine (Weinhouse and Friedmann, 1951). This was believed to be due to the precipitation of calcium oxalate in the tissues. On the other hand, Elder and Wyngaarden (1960) reported that [^{14}C]-oxalic acid was excreted unchanged almost quantitatively (89–99%) in man. The metabolism of oxalic acid and also its formation from various precursors (e.g. ascorbic acid and glycine) have been reviewed by Hodgkinson and Zarembski (1968).

The higher dicarboxylic acids including the C_6, C_8 and C_{10} homologues (**adipic acid**, **suberic acid** and **sebacic acid**, respectively) are largely excreted unchanged when fed to dogs and humans (Bernhard and Andreae, 1937). Similar results showing the extensive excretion of unchanged C_9 compound (**azelaic acid**) were reported by Weitzel (1947) who gave large, daily doses of several of these higher dicarboxylic acids to humans.

(+)-**Tartaric acid** (7) is well absorbed following oral administration and subsequently excreted mainly unchanged in the urine. Underhill *et al.* (1931a) found that 90–100% of an oral dose (50 mg/kg) was excreted unchanged by rabbits. A similar picture was seen with rats and dogs and also in guinea pigs given the acid by injection. When administered orally to the latter species, however, little was detected in the urine and none could be found in the faeces. This suggests that tartaric acid undergoes alteration in the guinea pig gastrointestinal tract. Similar results were obtained in man by Underhill *et al.* (1931b) who recovered about 20% of the orally administered tartaric acid unchanged in the urine. The remainder was not detected in the faeces, however this loss was ascribed to bacterial metabolism in the large intestine. These conclusions were confirmed by Finkle (1933) who found an average of 17% excreted unchanged in the urine after oral doses of 200–400 mg. Similar findings were made using a 4 g dose, however nearly quantitative excretion of unchanged compound was recorded following the injection (i.m.) of the acid (1–2 g).

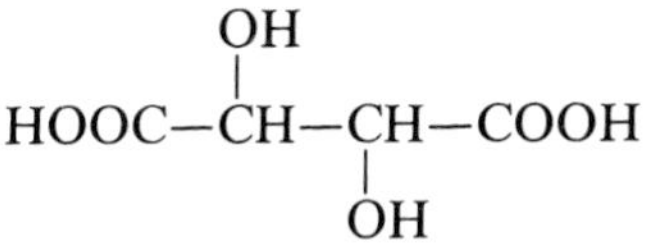

$$HOOC-\overset{\displaystyle OH}{\underset{\displaystyle OH}{CH}}-CH-COOH \qquad Me-\overset{\displaystyle Me}{\underset{\displaystyle Me}{N^+}}-CH_2-COO^-$$

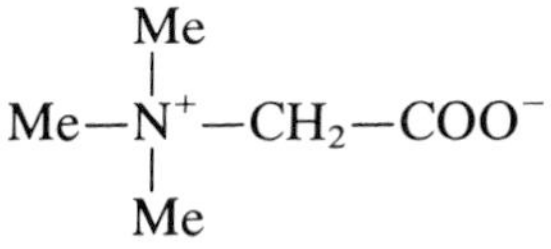

(7)   (8)

(+)-Tartaric acid   Betaine

Betaine (8) appears to be excreted mainly unchanged in the urine following its administration (p.o. or s.c.) to rabbits, cats and dogs (Kohlrausch, 1912). However, in rabbits it was stated that some conversion to trimethylamine also occurs.

B. ALICYCLIC ACIDS

This section deals mainly with (−)-**quinic acid** (1,3,4,5-tetrahydroxy-cyclohexanecarboxylic acid) (9), a commonly occurring plant constituent which was first shown by Lautemann (1863) to undergo metabolic aromatization. This transformation involves conversion to benzoic acid, this aromatic metabolite being excreted in the urine as its glycine conjugate (hippuric acid). The reaction sequence, again shown in man, was confirmed by Quick (1931). Vasiliu et al. (1938, 1940) reported that 50–60% of the quinic acid (10 g) given to sheep was excreted as hippuric acid but that little aromatization occurred in dogs. Values of 5–10% were found when dogs were maintained on a meat diet, however none was detected when a diet giving an acidic urine was fed. Bernhard (1937) also found that (−)-quinic acid was not converted to hippuric acid in dogs but instead excreted unchanged. Profound variations in aromatization ability among various

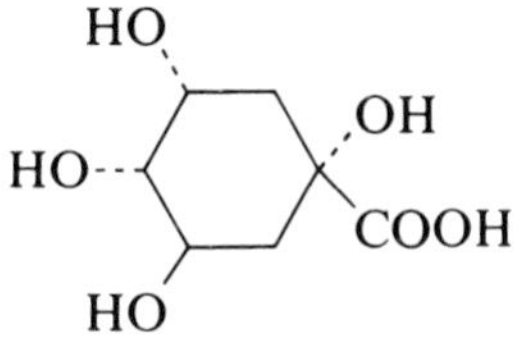

(9)

(−)-Quinic acid

animal species was also reported by Beer *et al.* (1951) who confirmed the high value (as much as 70%) in man but found very low values following oral dosage of ($-$)-quinic acid to rats and rabbits or after subcutaneous administration to guinea pigs and cats. Interestingly, Bernhard *et al.* (1955) found that as much as 50–80% of the dose (0·7–1 g/kg) of quinic acid was aromatized in guinea pigs when oral administration was employed. The significance of the route of administration used was clearly demonstrated by Cotran *et al.* (1960). They found that quinic acid was aromatized and excreted in the urine as hippuric acid in man and guinea pigs after oral administration. This did not occur when the guinea pigs were given the compound intraperitoneally. Furthermore, inhibition of the intestinal bacteria with neomycin suppressed quinic acid aromatization and it was concluded that this reaction is achieved by the intestinal bacteria. This effect of neomycin has been confirmed in rats (Asatoor, 1965) and in rhesus monkeys (Adamson *et al.*, 1970).

It therefore seems evident that the aromatization of quinic acid is dependent on the metabolic activity of the intestinal microorganisms and that the species differences in conversion may be related to differences in this activity of the microfloras. In addition to the species differences noted above, the investigation by Adamson *et al.* (1970) was carried out using a large number of animal species and indicated that extensive aromatization (20–60%) was confined to man and the Old World monkeys (baboon and rhesus and green monkeys). Low conversion levels (0–5%) were found in all of the other species studied which included New World monkeys (spider and squirrel monkeys, capuchin, bushbaby, slow lorris and three shrew), carnivores (ferret, cat and dog), rodents (lemming, mouse, rat, hamster and guinea pig) and the fruit bat, hedgehog and rabbit. Quinic acid doses of about 300 mg/kg (p.o.) were typical for most of these experiments. Martin (1975) confirmed the earlier reports of Vasiliu showing that sheep readily aromatize quinic acid. About a quarter of the quinic acid infused into the rumen was excreted in the urine as benzoic acid.

Fairly large variations in the extent of aromatization of ($-$)-quinic acid have been found in various experiments using the same animal species. This is evident from that noted above in guinea pigs and has also been recorded in rats. The data of Beer *et al.* (1951) indicate that only about 1% of the dose (2 g divided among six rats) was aromatized. The values of Adamson *et al.* (1970) averaged about 5% (dose = 600 mg/kg) whereas Teuchy *et al.* (1971), using a 50 mg dose, failed to find any conversion. Indahl and Scheline (1973) recorded the 48 h urinary hippuric acid excretion in 32 rats given 100 mg doses of quinic acid and found that this metabolite accounted for 12% of the dose. Interestingly, this value was not changed following the feeding of a purified diet containing 1% quinic acid for 24 or 48 days in the attempt to promote a metabolic adaptation of the microflora to this compound.

Aromatization of (−)-quinic acid in rats leads not only to the urinary excretion of hippuric acid but of catechol as well (Booth et al., 1960b). However, this pathway is of minor importance and Indahl and Scheline (1973) found that only 1% of the dose (100 mg, p.o.) was excreted as catechol. This metabolite has also been identified when quinic acid is incubated anaerobically with mixed cultures of rat faecal or caecal bacteria (Booth and Williams, 1963b; Scheline, 1968b). The in vitro formation of benzoic acid from quinic acid by rat caecal bacteria was also reported (Indahl and Scheline, 1973).

HO
|
HO---⟨ ⟩---COOH
|
HO
(10)

Shikimic acid

Shikimic acid (3,4,5-trihydroxy-1-cyclohexene-1-carboxylic acid) (10) is structurally closely related to quinic acid (9) and its metabolic fate in animals is similar to that described above for the latter compound. Asatoor (1965) reported that the oral administration of shikimic acid to rats led to the urinary excretion of larger quantities of hippuric acid than those formed from quinic acid. Aromatization of shikimic acid was likewise suppressed when the animals were treated with neomycin. The alternative pathway leading to the formation of catechol is also seen with shikimic acid (Booth et al., 1960b; Scheline, 1968b). Recently, Brewster et al. (1976, 1977a) administered generally-labelled [^{14}C]-shikimic acid (100 mg/kg, p.o.) to rats and found that 95–98% of the radioactivity was excreted within 24 h. The amounts recovered in the urine, faeces and expired air were 35–49%, 45–55% and 4–6% (as $^{14}CO_2$), respectively. The urinary hippuric acid accounted for 15–21% of the dose and 0·6–1·8% was found as urinary conjugated catechol. Significantly, it was shown that the intestinal bacteria effected the reduction of shikimic acid to cyclo-hexanecarboxylic acid. This reaction was demonstrated in vitro using mixed bacterial populations from caecal contents or faeces from the mouse, rat, guinea pig, rabbit and man but not the ferret. Balba and Evans (1977) confirmed this conversion using rat caecal contents or sheep rumen fluid. Additionally, Brewster et al. (1977a) showed that the perfused rat liver was able to extensively aromatize cyclohexanecarboxylic acid to benzoic acid. They proposed therefore that the bacteria are responsible for the reduction of shikimic acid to cyclohexanecarboxylic acid which, following absorption, is aromatized to benzoic acid and conjugated with glycine in the tissues. The aromatization of cyclohexanecarboxylic acid by liver enzymes by way of coenzyme A derivatives has been reported (Babior and Bloch, 1966). Nonetheless, several studies noted above identified aromatic metabolites when quinic acid or shikimic acid was incubated with intestinal microorganisms. These findings suggest that alternative sites and sequences for the aromatization of these compounds may exist.

Balba and Evans (1977) described an alternative metabolic pathway for the cyclohexanecarboxylic acid produced by the intestinal reduction of shikimic acid. Rather than undergoing aromatization, it is conjugated directly with glycine in herbivores (sheep, cattle, horse and elephant) and excreted in the urine as hexahydrohippuric acid. The latter compound was not detected in rat, pig or human urine. However, Brewster *et al.* (1977b) showed that hexahydrohippuric acid is a urinary metabolite of cyclohexanecarboxylic acid in rats. A partly aromatized conjugate, 3,4,5,6-tetrahydrohippuric acid, was also identified. The amounts of these two metabolites excreted varied from about a half of the hippuric acid excreted following a large dose (200 mg/kg) of cyclohexanecarboxylic acid to a sixth of that excreted after a small dose (0·5 mg/kg). Additional urinary metabolites were the ester glucuronides of benzoic acid and cyclohexanecarboxylic acid. The formation of the three glycine conjugates suggests that the main sequence of reactions in the metabolism of cyclohexanecarboxylic acid involves the initial formation of a coenzyme A ester which can then undergo dehydrogenation to the partially or fully aromatized CoA intermediates. All of these CoA esters are conjugated with glycine, however the bulk (about 50–80% of the dose) is excreted as the fully aromatized hippuric acid.

Bernhard and Müller (1938) fed small amounts of **chaulmoogric acid** (13-(2-cyclopenten-1-yl)tridecanoic acid) (11) to dogs and reported that it was well absorbed. However, no urinary metabolites were found.

$$\square\rangle-(CH_2)_{12}-COOH$$

(11)

Chaulmoogric acid

C. Aromatic Acids

Aromatic plant acids include aryl and aralkyl carboxylic acids, sometimes lacking ring substituents but generally containing hydroxyl and/or methoxyl groups. It is, in fact, these phenolic acids and their derivatives which we usually associate with the term aromatic plant acids. The general features of the metabolism of these compounds are dealt with in Chapter 1 which indicated that both tissue reactions and bacterial reactions may be involved. In the former case metabolic conjugation is the most important feature, the pathways involved utilizing a carbohydrate (usually glucuronic acid) or an amino acid (e.g. glycine, glutamine or taurine). Numerous other metabolic possibilities exist when other functional groups are present. Prominent among these are O-methylation of catechols and β-oxidation of some aralkyl acids. In the case of bacterial metabolism, decarboxylation reactions are sometimes observed and ring substituents may also be metabolized (e.g. dehydroxylation or O-demethylation). Examples of these various metabolic routes are given below in the summaries of the metabolic fate of particular aromatic plant acids.

(12)

Benzoic acid

(13)

Hippuric acid

The metabolism of **benzoic acid** (12), the simplest aromatic carboxylic acid, involves largely its conjugation with glycine to form hippuric acid (13). Conjugation with glucuronic acid to form an ester glucuronide is an additional metabolic pathway. In both cases a major point of interest has been the correlation of the extent of these conjugative pathways with the animal species used. Data on the conjugation of benzoic acid in numerous mammalian species are summarized in Table 5.1. Hippuric acid formation is the dominant pathway in most mammalian species and was found to be lacking only in the African fruit bat, although ferrets form, at higher dose levels, more benzoyl glucuronide than hippuric acid. The data shown in Table 5.1 are largely newer findings, often obtained with ^{14}C-labelled benzoic acid, and it is believed that no useful purpose would be served by including additional results from the early literature on benzoic acid conjugation. This latter material was collated by Williams (1959, pp. 349–353) who also reviewed the subject of glycine utilization in hippuric acid synthesis.

Other possible metabolic pathways of benzoic acid include decarboxylation and hydroxylation. Bernhard et al. (1955) reported that virtually no radioactivity was detected in the expired air of rats given the acid labelled with ^{14}C in the carboxyl group. This finding is in agreement with the results summarized in Chapter I, Section II.B.2 which show that decarboxylation

TABLE 5.1
Conjugation of benzoic acid in various species

Species	Dose mg/kg, p.o. unless otherwise indicated	% of dose excreted in 24 h urine	% excreted in 24 h as:			References
			Benzoic acid	Hippuric acid	Benzoyl glucuronide	
Rodents						
Mouse	56	55	trace	95	5	a
Rat	50	100	1	99	trace	a
Rat	50/rat			47		b
Rat	185			88		c
Gerbil	29	75	2	98	0	a
Hamster	52	99	1	97	1	a
Lemming	56	98	trace	100	0	a
Guinea pig	49	79	trace	98	3	a
Other						
Indian fruit bat	50	49, 54	12, 30	trace	88, 70	a
Indian fruit bat	50/bat, i.p.	98, 83	10, 19	$<0·01$	89, 81	d
Rabbit	49	60	0	100	0	a
Rabbit	200	86	trace	98	2	a
Rabbit	92			79[e]		f
Rabbit	500	94	1	84	15	g
European hedgehog	50	67, 78	5, 7	76, 86	19, 7	a
Pig	50	48, 51	15, 7	85, 93	trace	a
Pig	500–800	93	7	61	32	h
Sheep	200[i]			75		j
Elephant	100	64	9	90		k

	Dose (mg/kg)					Ref.
Carnivores						
Ferret	50	69	9	70	22	[a]
Ferret	198	78	9	47	44	[a]
Ferret	200	67	22	30	49	[a]
Ferret	100	86	1	40	59	[l]
Cat	51	29, 86	trace	100	0	[a]
Dog	51	94	0	82	18	[a]
Forest genet	75	54, 79	28, 23	67, 75	0	[m]
African civet	75	35, 44	17, 0	77, 95	0	[m]
Lion	75	92	15	84	0	
Primates						
Capuchin	50	57	0	100	trace	[a]
Squirrel monkey	50	46, 49	14, 18	81, 83	5, trace	[a]
Rhesus monkey	20	47	0	100	0	[a]
Man	1	[n]	0	100	0	[a]
Man	42					[o]
Man	43			50–85	5	[g]

[a] Bridges et al. (1970).
[b] Teuchy et al. (1971).
[c] Bernhard et al. (1955).
[d] Bababunmi et al. (1973).
[e] % of total dose.
[f] El Masri et al. (1956).
[g] Bray et al. (1951).
[h] Csonka (1924).
[i] 24 h infusion into abomasum.
[j] Martin (1966).
[k] Caldwell et al. (1975).
[l] Idle et al. (1976).
[m] French et al. (1974)
[n] 4 h excretion.
[o] Van Sumere et al. (1969).

requires the presence of a p-hydroxyl group in the aromatic acid. Bray *et al.* (1946) reported that benzoic acid did not form an ethereal sulphate, a finding which would exclude its metabolism by hydroxylation to a phenolic acid. However, Sato *et al.* (1956b) found that rat liver slices are capable of metabolizing small amounts of benzoic acid to the sulphate conjugate of 4-hydroxybenzoic acid. This subject was reinvestigated by Acheson and Gibbard (1962) who found that rats given [*carboxy*-^{14}C]-benzoic acid (500 mg/kg, i.p.) excreted about 0·25% of the dose as 2-, 3- and 4-hydroxybenzoic acids. At about one tenth of this dose level, only about 0·04% of the radioactivity was due to hydroxylated metabolites and the 2-hydroxy isomer was not present. This latter pattern was also observed in guinea pigs given a dose of about 200 mg/kg and the amount excreted was similarly about 0·04% of the dose. Thus, hydroxylation is a very minor metabolic route with benzoic acid.

$$\langle\text{C}_6\text{H}_5\rangle-\text{CH}_2-\text{COOR}$$

(14)

(a) Phenylacetic acid, R = H

(b) Phenacetylglutamine, R = $-\text{NH}-\overset{\overset{\displaystyle\text{COOH}}{|}}{\text{CH}}-(\text{CH}_2)_2-\overset{\overset{\displaystyle\text{O}}{\|}}{\text{C}}-\text{NH}_2$

(c) Phenaceturic acid,　　R = $-\text{NH}-\text{CH}_2-\text{COOH}$

(d) Phenacetyltaurine,　　R = $-\text{NH}-\text{CH}_2-\text{CH}_2-\text{SO}_3\text{H}$

Studies on the metabolism of **phenylacetic acid** (14a) show a general similarity to those dealing with benzoic acid insomuch as interest has been directed mainly towards conjugation reactions. Furthermore, investigations dealing with phenylacetic acid span a very long period of time. Little advantage is to be gained by summarizing the early material in this book since it is well covered in the review by Williams (1959, pp. 374–377). Instead, several recent reports including the extensive comparative study of James *et al.* (1972) will form the basis of the present discussion as they have considerably clarified the uncertainty formerly present about the conjugative pathways of phenylacetic acid metabolism. Nonetheless, the early reports by Thierfelder and Sherwin (1914, 1915) must be noted as they showed that phenylacetic acid was conjugated in man with glutamine rather than glycine as is seen in most other mammals. The species variations in phenylacetic acid conjugation are clearly shown in Table 5.2 which summarizes the major findings of James *et al.* (1972) as well as supplementary data on the ferret (Idle *et al.*, 1976) and values for the hyaena and elephant (Caldwell *et al.*, 1975). Several points of interest emerge from these data. Firstly, the presence of the pathway involving conjugation with glutamine which gives rise to phenacetylglutamine (14b) appears sharply in the evolutionary scale with the New World monkeys, being completely absent in the two prosimian species studied. This switch is concomitant with the reduction in the extent of glycine conjugation. The latter reaction forms phenaceturic acid (14c) and shows low to intermediate values in the New World monkeys, values of 1% or less in the Old World monkeys and is essentially absent in man. Also noteworthy is the detection of a new conjugate, phenylacetyltaurine (14d), in all the mammalian species except the vampire bat. The extent of this conjugative pathway is considerable in several species, especially the ferret and some monkeys. Conjugation of phenylacetic acid with glucuronic acid is not seen in most species including the rabbit, however Bray *et al.* (1946) reported that 5% of the dose was excreted in this form following large (0.75 g/kg) oral doses to rabbits.

$$\langle\!\!\!\!\!\!\;\bigcirc\!\!\!\!\!\!\;\rangle\!-\!CH\!\!=\!\!CH\!-\!COOH$$

(15)

Cinnamic acid

Investigations using several animal species make it clear that the major metabolic product of **cinnamic acid** (15) is hippuric acid. This was reported by Dakin (1909) who administered the acid (0.25–0.5 g/kg, s.c.) to cats and dogs. In addition, a small quantity of an intermediate oxidation product, 3-keto-3-phenylpropionic acid, was found as was some acetophenone which no doubt arose during the isolation procedure via decarboxylation of the keto acid. Dakin reported that cinnamoylglycine was a further minor metabolite, however no unchanged cinnamic acid was excreted in the urine. Rabbits extensively convert cinnamic acid to hippuric acid. El Masry *et al.* (1956) found a value of 74% following an oral dose of 300 mg/kg. Raper and Wayne (1928) reported the same percentage conversion to benzoic acid when cinnamic acid (115 mg/kg, s.c.) was given to dogs. El Masry *et al.* did not detect the cinnamoylglycine reported by Dakin in cat and dog urine in the urine of rabbits given cinnamic acid. This negative finding has recently been confirmed by Fahelbum and James (1977) who also reported that 60% of the dose (approx. 150 mg/kg) of cinnamic acid was excreted in the urine of rabbits as hippuric acid. Trace amounts of *p*-hydroxyhippuric acid were also detected. The urinary excretion of hippuric acid by rats receiving cinnamic acid was reported to be 44% of the dose (50 mg, i.p.) in 24 h (Teuchy *et al.*, 1971) and 67% of the dose (50 mg, p.o.) (Fahelbum and James, 1977). Similar experiments by Teuchy and Van Sumere (1971) with [3-^{14}C]-cinnamic acid resulted in about 48% of

TABLE 5.2
Conjugation of phenylacetic acid in various species[a]

Species	Dose mg/kg	% of dose excreted in 24 h urine	% excreted in 24 h as:				
			Phenylacetic acid	Glutamine conjugate	Glycine conjugate	Taurine conjugate	Glucuronic acid conjugate
Rodents							
Mouse	80, i.p.	54	32	—[b]	56	7	6
Rat	80, i.p.	95	—	—	99	1	—
Hamster	80, i.p.	85	52	—	47	1	—
Guinea pig	80, i.p.	63	5	—	94	1	—
Other							
Vampire bat	80, i.p.	75	—	—	100	—	—
Rabbit	80, i.p.	85	2	—	97	1	—
Elephant[c]	100, p.o.	31 (6h)	—	—	100 (6 h)		
Carnivores							
Ferret	80, i.p.	95	3	—	43	32	22
Ferret[d]	100, i.p.	68	3	—	63	21	—
Cat	80, i.p.	75	1	—	98	1	—
Dog	80, i.p.	81	—	—	94	4	2
Hyaena	25, p.o.	13	13	—	87	—	—
Prosimians							
Bushbaby	80, i.p.	69	—	—	87	13	—
Slow loris	80, i.p.	82	20	—	69	10	—

	Dose (mg/kg), route						
New World Monkeys							
Squirrel monkey	50, i.m.	67	4	75	2	18	—
Capuchin[e]	80, i.p.	28	4	64	10	20	—
Capuchin[f]	80, i.p.	34	4	29	23	44	—
Marmoset	80, i.m.	71	5	79	0·8	0·4	—
Old World Monkeys							
Rhesus monkey	80, i.p.	74	55	32	1	23, 1[g]	—
Cynomolgus monkey	80, i.p.	36, 79[g]	42, 5[g]	56, 90[g]	1	2, 4[g]	—
Green monkey	25, i.m.	87	12	79	0·5	4	—
Red bellied monkey	8, i.m.	45	8	87	1	3	—
Mona monkey	8, i.m.	15	45	32	1	21	—
Mangabey	8, i.m.	79	55	31	0·5	7	—
Drill	8, i.m.	52	65	28	0·4	7	—
Baboon	2, i.m.	100	5	85	0·1	10	—
Man	1, p.o.	98	—	93	<0·05	7	—

[a] Data taken from James *et al.* (1972) unless otherwise indicated

[b] — = not detected.

[c] Caldwell *et al.* (1975).

[d] Idle *et al.* (1976), unidentified metabolite amounted to 13% of the urinary radioactivity.

[e] *Cebus albifrons.*

[f] *Cebus nigrivittatus.*

[g] Separate values listed due to their large differences.

the radioactivity being recovered in the urine in 24 h and most of this (39% of the dose) was due to hippuric acid. Smaller amounts of benzoic acid and cinnamic acid were also found but no phenolic acid metabolites were detected. Not unexpectedly, no radioactivity was detected in the expired air. Bhatia *et al.* (1977) administered [1-^{14}C]-cinnamic acid to rats and recovered about 47% of the radioactivity in the urine in four days following an oral dose of 250 mg/kg. Interestingly, several phenolic metabolites were identified in this study. These included the 4-hydroxy-, 3,4-dihydroxy- and the isomeric monomethyl ethers of the 3,4-dihydroxy derivatives of cinnamic acid, i.e. *p*-coumaric, caffeic, ferulic and isoferulic acids. Several other urinary metabolites of cinnamic acid were detected and, of these, benzoic acid, 4-hydroxybenzoic acid and hippuric acid were identified. No additional glycine conjugates were found. The possibility that cinnamic acid or its metabolites may also be conjugated with glucuronic acid has not received much attention, however Quick (1928) reported that a larger proportion of the metabolically formed benzoic acid is conjugated with glucuronic acid than with glycine in the dog. New information on this point was obtained by Fahelbum and James in the studies noted above using rats and rabbits. They found that 3% and 10%, respectively, of the dose were excreted in the urine of these two species as material conjugated with glucuronic acid.

(16)

Indole-3-acetic acid

Another unsubstituted aromatic acid is **indole-3-acetic acid** (heteroauxin) (16). Erspamer (1955) found that rats excrete it partly unchanged and partly as the glycine conjugate, indoleaceturic acid. Recently, a comprehensive study of the metabolism of indole-3-acetic acid in 17 mammalian species was carried out by Bridges *et al.* (1974). This investigation, which employed most of the species used in the similar study with phenylacetic acid summarized above (James *et al.*, 1972), showed an appreciable urinary excretion of unchanged compound in most species. The glutamine conjugate was formed in man and monkeys but not in prosimians, carnivores, rodents or rabbits. In these latter species and also in New World monkeys, the glycine conjugate was found whereas this pathway was absent in man and Old World monkeys. Conjugation with taurine, which was noted above as a common metabolic route with phenylacetic acid, was prominent in ferrets and also in some of the monkeys. Interestingly, conjugation with glucuronic acid was observed only in man, in which case it amounted to 20–30% of the dose.

OH
|
〈benzene ring〉—CH—COOH

(17)

Mandelic acid

CH$_2$OH
|
〈benzene ring〉—CH—COOH

(18)

Tropic acid

Two simple aromatic hydroxy acids from plants are **mandelic acid** (17) and its higher homologue **tropic acid** (18), although the latter compound is more correctly a hydrolysis product obtained from some tropane alkaloids (e.g. atropine) rather than a true plant constituent. An important factor governing the metabolic fate of these acids is their acidity (pKa values of 3·4 and 4·1, respectively) which results in their extensive ionization at physiologic pH values. Accordingly, the most notable feature of their fate in the body is their extensive excretion in the unchanged state. Most of the investigations in this area are of an early date and a summary of these data was given by Williams (1959, pp. 380–381, 383–384). While some of this information is conflicting, it has been reported that mandelic acid may be dehydrogenated to the corresponding α-keto acid, phenylglyoxylic acid (benzoylformic acid), in dogs and man. This metabolite was recently reported by Ohtsuji and Ikeda (1971) to be a urinary metabolite of mandelic acid in rats. In addition to unchanged compound, some hippuric acid was also excreted which indicates that chain-shortening to a C_6—C_1 derivative takes place in the rat. This reaction was not detected in rabbits as El Masry *et al.* (1956) did not find increased urinary hippuric acid levels following the administration of racemic mandelic acid.

Gosselin *et al.* (1955) included [^{14}C]-tropic acid in their metabolic study of atropine in mice and rats. This preparation was labelled in the α-position and when injected (1 mg/kg, i.p.) it was excreted unchanged to the extent of 95–98% in 2 h, no radioactivity being found in the expired air. Ve and Scheline (1977), checking specifically for the ability of rats given tropic acid (100 or 400 mg/kg, p.o.) to oxidize the alcohol moiety, found only unchanged compound in the urine and no evidence for the dicarboxylic acid product, phenylmalonic acid.

The major part of this review of the metabolism of aromatic acids will be devoted to phenolic acids and their various ether derivatives, starting with the substituted benzoic acids (C_6—C_1 phenolic acids) and continuing with the higher homologues, of which the cinnamic acids (C_6—C_3 phenolic acids) are the most prominent. It is evident that due to the presence of both carboxyl and phenolic hydroxyl (or methoxyl) groups, numerous pathways of metabolism are available to the phenolic acids. Conjugation of the carboxyl group with an amino acid or glucuronic is possible and the latter reaction, as well as ethereal sulphate formation, may occur with the hydroxyl group. Ring hydroxylation and both O-methylation and O-demethylation reactions occur, the latter often carried out by intestinal bacteria. These microorganisms are also responsible for the dehydroxylation and decarboxylation of certain phenolic acids as well as double bond reduction in cinnamic acids. The reverse of the latter reaction, dehydrogenation of C_6-C_3 acids, is a tissue reaction. Our knowledge of this network of interrelated metabolic pathways has expanded considerably during the past decade or so and the following summaries of the metabolism of individual phenolic acids will give special consideration of these newer data.

The simplest plant phenolic acids are **salicylic acid** (2-hydroxybenzoic acid) (19) and 4-hydroxybenzoic acid. In view of the important medical uses of the former compound, it is hardly surprising that an abundant literature on its metabolic fate is available. A detailed discussion of this subject falls outside the scope of this book and the present summary aims primarily to illustrate the known metabolic pathways of salicylate in various animal species. As the pKa value of salicylic acid is approx. 3, it is understandable that an appreciable urinary excretion of unchanged compound occurs. This is seen in Table 5.3 which brings together quantitative data on the excretion of salicylic acid and its metabolites in several animal species. When considering the values for the excretion of free salicylate it is important to note that these are closely related to the urinary pH. In the investigation of Hollister and Levy (1965), the mean value of 10% shown in Table 5.3 includes values which range from 2–26%. It was noted that the individuals giving the lowest values had a consistently low urine pH (approx. 5) whereas the opposite was true in the individual giving the highest value. In the latter case the urinary pH values generally ranged between 6 and 7. The exceptionally high excretion of unchanged compound found by Schubert (1967) in the horse may be explained by the high urinary pH (range 6·9–8·1) of these animals. However, the data of Davis and Westfall (1972) indicate that urinary pH is not the sole factor influencing the extent of free salicylate excretion as the value found in horses (urine pH 7·6) was roughly 50% greater than that in goats (urine pH 8·2).

The main metabolic transformations of salicylic acid are illustrated in Fig. 5.1. and include conjugation with glycine to give salicyluric acid (20) or with glucuronic acid to give ether (21) or ester (22) glucuronides. Hydroxylation is a minor reaction but some gentisic acid (23) is usually

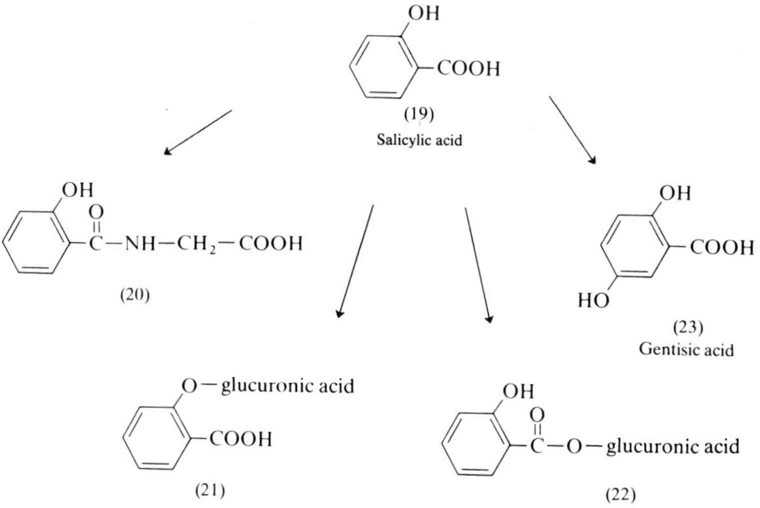

FIG. 5.1. Main metabolic pathways of salicylic acid. See text for description of minor or occasional metabolites.

detected and, rarely, a trace of the *o*-hydroxylation product, 2,3-dihydroxybenzoic acid. Not shown in Fig. 5.1 is the formation of an ethereal sulphate of salicylic acid which was reported by Haberland *et al.* (1957) to be excreted by rats. However, this metabolite was reported not to be formed in rabbits (Williams, 1938; Bray *et al.*, 1948) and in dogs and man (Alpen *et al.*, 1951b). Sato *et al.* (1956b), using rat liver supernatant, found no formation of the ethereal sulphate of salicylic acid. When liver slices were employed, three sulphate conjugates were detected but these were of gentisic acid and of other metabolites rather than of salicylic acid. Another possible metabolic pathway which is not encountered is decarboxylation. Schayer (1950) and Alpen *et al.* (1951a), using salicylate labelled with ^{14}C in the carboxyl group, did not find any $^{14}CO_2$ in the expired air of rats. Scheline (1966b) showed that rat caecal bacteria which are capable of decarboxylating numerous phenolic benzoic acids do not carry out this reaction with salicylic acid. A reported metabolite of salicylic acid about which some uncertainty exists is the uraminsalicylic acid first isolated from dog urine by Baldoni (1909). This compound was also reported by Kapp and Coburn (1942) to be excreted by man and their data indicated that it was a compound consisting of glycine, salicylic acid and gentisic acid. According to Haberland *et al.* (1957), uraminsalicylic acid is a molecular complex of gentisic acid and salicyluric acid. However, Alpen *et al.* (1951b) did not detect it in the urine of dogs or man given salicylic acid. Recently,

TABLE 5.3

Metabolism of salicylic acid in various species

Species	Dose	Urinary excretion period h	% of dose excreted as:					References
			Salicylic acid	Salicyluric acid	Salicyl ether glucuronide	Salicyl ester glucuronide	Gentisic acid	
Mouse	150–250 mg/kg i.p.	24					8	a
Rat	approx. 330 mg/kg p.o.	24	41–63	<1	1–3	0	18–34 + <1 as glucuronide	b
Rat	5 mg/kg i.p.	24	29	25	33 (total glucuronide)		8 + 3 as glucuronide	c
Rat	50–200 mg/kg i.p.	24	25–35	4–7	32–44 (total glucuronide)		10–17 + 5–15 as glucuronide	c
Rabbit	0.1–1.5 g/kg p.o.	24		approx. 5			<4	d
Rabbit	0.25–0.5 g/kg p.o.	24	85	0	5–14	3–4	4–5	e
Dog	38 mg/kg i.v.	24	38	29	33 (total glucuronide)		small amount	f
Dog	1 g i.v.	30–36	50	10	25	0	4–5	g

Goat	38 mg/kg i.v.	24	46	41	12	0	f
Pig	38 mg/kg i.v.	24	46	31	23	small amount	f
Horse	4–20 g i.v.	3–24	approx. 100	0		<0.5	h
Horse	38 mg/kg i.v.	24	65	25	10 (total glucuronide)	small amount in ♂	f
Rhesus monkey	9 mg/kg i.v.	16		72			i
Man	2–3 g daily	24–36	20	55	25 (total glucuronide)	4–8	j
Man	1 g i.v.		10–85	0–50	12–30 0–10	>1	g
Man	1 g p.o.		10[k]	69	21 (total glucuronide)		l
Man	approx. 1·5 g p.o.	24				3	a

[a] Roseman and Dorfman (1951).
[b] Quilley and Smith (1952).
[c] Haberland et al. (1957).
[d] Bray et al. (1950).
[e] Bray et al. (1948).
[f] Davis and Westfall (1972).
[g] Alpen et al. (1951b).
[h] Schubert (1967).
[i] Wan and Riegelman (1972).
[j] Kapp and Coburn (1942).
[k] See text.
[l] Hollister and Levy (1965).

Davis and Westfall (1972) and Davis *et al.* (1973) detected an unidentified metabolite of salicylic acid which they believed was uraminsalicylic acid. It was found only in the urine of adult cats and newborn dogs and pigs and was converted by acid hydrolysis to salicylic and gentisic acids and an amino acid. These investigators believed that the metabolite is formed in situations in which competing pathways are deficient. Their data showed that plasma salicylate levels decline much more slowly in cats than in the other species studied (dogs, goats, pigs and horses).

The data shown in Table 5.3 indicate that hydroxylation of salicylic acid generally occurs to a small extent. This reaction takes place in the position *para* to the hydroxyl group, forming gentisic acid (23), however the *ortho*-hydroxylated product, 2,3-dihydroxybenzoic acid, was reported by Bray *et al.* (1950) to be formed in trace amounts in rabbits. Dumazert and El Ouachi (1954) found that both gentisic acid and 2,3,5-trihydroxybenzoic acid were formed from salicylic acid in rabbits and man. The trihydroxy derivative was conjugated with glucuronic acid in rabbits and with glycine and glucuronic acid in man. Mitoma *et al.* (1956) reported the *in vitrℴ* formation of gentisic acid from salicylic acid by the rabbit liver microsomal system requiring TPNH and oxygen. Table 5.3 also shows that formation of the glucuronic acid conjugate of salicylic acid may take place at the hydroxyl or the carboxyl group, giving the ether glucuronide (21) or ester glucuronide (22), respectively. While many of the results indicated the total excretion of salicyl glucuronide, those which differentiated between the two forms reveal that the ether glucuronide predominates. Both glucuronides were isolated from urine by Robinson and Williams (1956).

Although falling outside the scope of this book, the pharmacokinetics of salicylate metabolism and elimination in animals and man has received a great deal of attention. Much useful information in this area was reviewed by Levy and Leonards (1966) and subsequently by Davison (1971). More recent relevant papers include those by Von Lehmann *et al.* (1973) on the renal contribution to salicyluric acid formation in man and by Davis *et al.* (1973) on salicylate pharmacokinetics in newborn animals.

4-Hydroxybenzoic acid contains the same functional groups as does salicylic acid and it is therefore not surprising that it is metabolized along the same conjugative pathways to form glycine and glucuronic acid derivatives and that it is also hydroxylated. In addition, the presence of the phenolic hydroxyl group in the 4- rather than the 2-position allows for some ethereal sulphate formation and also for metabolism by the bacteria of the gastrointestinal microflora. The quantitative data available are more limited with the 4-isomer and were obtained mainly using rabbits. The major findings from these studies are shown in Table 5.4. The isolation of

TABLE 5.4

Metabolism of 4-hydroxybenzoic acid in rabbits

Dose p.o.	Urinary excretion period h	% of dose excreted as:						References
		4-Hydroxybenzoic acid	Glycine conjugate	Ether glucuronide	Ester glucuronide	Ethereal sulphate	3,4-Dihydroxybenzoic acid	
0·365 g/kg	48					7		[a]
0·1–1·5 g/kg	24	30–60	10–30	0–19	2–16	4–8		[b]
0·25–0·35 g/kg	24			18	[c]	9		[d]
0·1–1·5 g/kg	24						<4	[e]
0·4 g/kg	24	52	23	8	7	5		[f]
0·8 g/kg	24	59	13	4	8	5		[f]

[a] Williams (1938).
[b] Bray et al. (1947).
[c] Little, if any, present.
[d] Hartles and Williams (1948).
[e] Bray et al. (1950).
[f] Tsukamoto and Terada (1964).

both the ether and ester glucuronides from the urine of rabbits given 4-hydroxybenzoic acid was reported by Tsukamoto and Terada (1962).

The metabolism of 4-hydroxybenzoic acid in rats appears to be qualitatively similar to that seen in rabbits. Booth *et al.* (1958) detected in the urine unchanged compound and relatively large amounts of its ethereal sulphate. Similar experiments in rats by Derache and Gourdon (1963) showed these two compounds as well as the glycine conjugate, the ether and ester glucuronides and an unidentified metabolite. In cats, oral doses (13, 26 and 260 mg/kg) of 4-hydroxybenzoic acid result in its metabolism to a single urinary metabolite, 4-hydroxyhippuric acid (Phillips *et al.*, 1977). Nearly the entire dose was recovered in the urine, mainly during the first 24 h. Quick (1932) isolated this glycine conjugate from the urine of dogs fed 4-hydroxybenzoic acid (approx. 300–500 mg/kg) but, interestingly, the major metabolite isolated was reported to be the diglucuronide conjugate. Quick found that the phenolic acid is excreted partly unchanged and partly conjugated with glycine in man. Tompsett (1961) also obtained evidence for this latter reaction in man. An *in vitro* study of the metabolism of 4-hydroxybenzoic acid using rat liver preparations was carried out by Sato *et al.* (1956a). The ethereal sulphate was formed by the supernatant fraction whereas liver slices produced this metabolite, the ethereal sulphate of 3,4-dihydroxybenzoic acid and a further product thought to be the ethereal sulphate of 4-hydroxyhippuric acid.

As noted above, the presence of the phenolic hydroxyl group *para* to the carboxyl group in 4-hydroxybenzoic acid furnishes a substrate which may undergo degradation by the gastrointestinal bacteria. Mixed bacterial populations from the rat caecum extensively decarboxylate this acid to phenol (Scheline, 1966b) and the same is true with human faecal bacteria (Curtius *et al.*, 1976). In a survey which included many common intestinal bacteria, Soleim and Scheline (1972) found that only *Aerobacter aerogenes* was capable of carrying out this reaction. Martin (1975) reported that 4-hydroxybenzoic acid, when infused into the rumen of sheep, was extensively decarboxylated and excreted in the urine as phenol.

Gentisic acid (2,5-dihydroxybenzoic acid) (23) was mentioned above as a metabolite of salicylic acid, however it has itself been the subject of a few studies, the first dating from the last century. Clarke and Mosher (1953) summarized the early work on gentisic acid metabolism and concluded that the conflicting and confusing results then available stemmed largely from questionable methodology. Stated very briefly, some of the earliest investigations indicated that part of the gentisic acid was excreted in the urine conjugated with sulphate whereas subsequent studies pointed to the excretion of gentisate largely in the unchanged form. Thus, Consden and Stanier (1951) reported that about 90% of the dose is excreted unchanged

METABOLISM OF ACIDS, LACTONES AND ESTERS 193

in the urine by man. Roseman and Dorfman (1951) recovered 61–77% of the dose (37 mg sodium gentisate/kg, p.o.) in the 24 h urine in man and found no increase in the amount of gentisate following acid hydrolysis of the samples. In the study of Clarke and Mosher, about 60% of the oral dose of either gentisic acid or its sodium salt was recovered unchanged in the 24 h urine in man and less than 1% was found in the faeces. However, acid hydrolysis of the urine samples liberated a small additional amount (roughly 10%) of gentisic acid and it was suggested that some conjugation, mainly forming the 5-O-sulphate derivative, took place. An interesting finding was the occasional collection of dark urines, sometimes when voided but usually as a result of exposure to air, which were thought to be due to the presence of the oxidized or quinone form (24) of gentisic acid. The evidence for the urinary excretion of conjugates of gentisic acid by man was more substantial in the study reported by Batterman and Sommer (1953). Following repeated administration of gentisate (3×300 mg daily, p.o.), they found that the average daily excretion of free gentisate accounted for 26% of the dose while a further 36% was excreted in a form liberated by acid hydrolysis. Not unexpectedly, alkalinization of the urine increased the amount of free gentisate excreted by 50%.

$$\text{(quinone)}\!-\!\text{COOH}$$

(24)

The results summarized above do not give a clear picture of the metabolism of gentisic acid in man, however a few additional studies, mainly using laboratory animals, provide some supplementary information. Haberland et al. (1957) administered [carboxy-^{14}C]-gentisic acid (100 mg/kg, i.p.) to rats and recovered more than 90% of the radioactivity in the 24 h urine. This material consisted of about two thirds unchanged compound and one third gentisic acid glucuronide. Gentisic acid is not among the phenolic acids which are decarboxylated by the rat intestinal bacteria (Scheline, 1966b). Both the most recent and the most detailed study of gentisic acid metabolism was carried out in dogs by Astill et al. (1964). They found that single oral doses of 190 or 310 mg/kg were excreted nearly quantitatively in the urine within three days. Most excretion occurred within 24 h and amounted to 84% and 71% of the lower and higher dose, respectively. The compounds excreted were shown to be free gentisic acid, gentisic acid 5-O-sulphate and gentisic acid 5-O-glucuronide, i.e. conjugation apparently only in the 5-position. Mention was made of unpublished results which indicated that these two conjugates are also formed in the rat. This indicates that the glucuronide previously reported by Haberland *et al.* (1957) was the 5-*O*-glucuronide. The values obtained for the excretion of unchanged gentisic acid, its *O*-sulphate and *O*-glucuronide were 63%, 28% and 7%, respectively (low dose), and 61%, 20% and 15%, respectively (high dose). The urine did not contain any salicylic acid or quinol, metabolites which could arise via dehydroxylation or decarboxylation, respectively, however some *O*-methylated product, 5-methoxysalicylic acid, was detected when the higher dose level was employed. This metabolite was previously reported by Sakamoto *et al.* (1959) to be formed from gentisic acid in man.

Although many of the metabolic pathways seen with the monohydric phenolic acids are also encountered with **protocatechuic acid** (3,4-dihydroxybenzoic acid), the presence of the catechol moiety in the 3,4-position affords two further possibilities, *O*-methylation and dehydroxylation. An early study using rabbits given protocatechuic acid (250 mg/kg, p.o.) indicated that 18% and 13% of the dose were excreted in the urine as glucuronic acid and ethereal sulphate conjugates, respectively (Dodgson and Williams, 1949). The remainder was thought to be excreted unchanged, however DeEds *et al.* (1955, 1957) reported that *O*-methylation of the 3-hydroxy group occurred, vanillic acid being excreted when rats or rabbits were given protocatechuic acid. Conjugation of the metabolically formed vanillic acid with glycine also occurs in rats (Masri *et al.*, 1959). Concurrent investigations by Scheline (1966a) and Wong and Sourkes (1966) confirmed that protocatechuic acid is metabolized to vanillic acid in rats and the former study showed that some of this material was in the form of acid-labile conjugated material. Additionally, both studies showed that some decarboxylation to catechol also occurred. The most detailed study of the metabolism of protocatechuic acid was carried out by Dacre and Williams (1962, 1968). Using material labelled with ^{14}C in the carboxyl group, they found that about 72% of the dose (100 mg/kg) was excreted in the urine (64%) and faeces (8%) in seven days by rats. Most of the urinary radioactivity was excreted within 24 h and quantitative analysis of this material indicated that it mainly consisted of free protocatechuic acid (21% of the dose) or its conjugates (15%, about half of which was due to the glycine conjugate). The second major metabolic pathway was *O*-methylation, giving free (4%) and conjugated (17%) vanillic acid. Dehydroxylation, a reaction of the intestinal microorganisms, occurred only to a limited extent, giving 3-hydroxy-, 4-hydroxy- and 3-methoxybenzoic acids to the extent of about 2%, 1% and 2%, respectively. Interestingly, when these experiments were repeated in rats treated orally with neomycin to inhibit the activity of the intestinal bacteria, the recovery of radioactivity excreted in the urine and faeces was quantitative. The main differences were an increase in total protocatechuic acid excretion from 36% in the normal group to 55% in the treated group and the very low values for dehydroxylated products. These results are best explained by the bacterial decarboxylation of the acid to catechol and $^{14}CO_2$ in the normal animals. Significantly, the latter metabolite was detected when the acid was incubated *in vitro* with rat gut contents. This decarboxylation reaction has been demonstrated in several investigations, first by Booth and Williams (1963b) using rat caecal or faecal extracts and subsequently by Scheline (1966a,b) using mixed cultures of rat caecal microorganisms. Soleim and Scheline (1972) studied a number of common intestinal bacteria for their ability to decarboxylate *p*-hydroxylated aromatic acids and found that *Aerobacter aerogenes* extensively decarboxylated protocatechuic acid.

The above summary of protocatechuic acid metabolism shows that besides excretion of the acid as such or as conjugates, the main pathways (in rats, at least) involve decarboxylation by the intestinal bacteria or *O*-methylation in the tissues. The latter reaction, effected by the enzyme catechol *O*-methyltransferase, has been extensively studied due to involvement of *O*-methylation in catecholamine metabolism and some of these data deal with the methylation of protocatechuic acid. Early examples of such studies include those of Pellerin and D'Iorio (1958) who used rat liver or kidney homogenates fortified with [*methyl*-^{14}C]-L-methionine and of Axelrod and Tomchick (1958) who used a rat liver preparation of the transferase and the methyl donor, *S*-adenosylmethionine. The latter compound was also employed by Dirscherl and Brisse (1966) who showed the conversion of protocatechuic acid to vanillic acid by rat and human liver homogenates. The general features of this reaction are dealt with in Chapter 1, Section I.D.4.b and will therefore not be repeated here. However, further studies which dealt specifically with protocatechuic acid were reported by Masri *et al.* (1962) who, using rat and rabbit liver slices, found that it was converted to vanillic acid and its glycine conjugate. Interestingly, isovanillic acid, the 4-*O*-methyl derivative of protocatechuic acid, was sometimes detected in small amounts. The likelihood that protocatechuic acid may undergo both 3- and 4-*O*-methylation in man had been previously suggested by Hill *et al.* (1959) and several more recent investigations using catechol-*O*-methyltransferase preparations clearly demonstrated that both 3- and 4-*O*-methylation take place (Creveling *et al.*, 1970, 1972). The ratio of 3- to 4-*O*-methylated products was found to be about 5·5:1, however the very small amounts of isovanillic acid actually excreted by animals suggests that the products ultimately excreted are also dependent upon *O*-demethylation of the methylated metabolites. Recently, Marzullo and Friedhoff (1975) described two forms of catechol

O-methyltransferase which differ in the ratio of methylated isomers produced from protocatechuic acid.

In contrast to that seen with most of the hydroxybenzoic acids, interest in the metabolism of **gallic acid** (3,4,5-trihydroxybenzoic acid) is of fairly recent date, mainly dealing with the reactions of O-methylation and decarboxylation. Furthermore, the investigations have used only rats and rabbits, as was the case with the initial study carried out by Booth *et al.* (1959b). When gallic acid was given to rats either in the diet (0·5% level) or in single 100 mg doses (p.o), the major urinary excretory products were unchanged compound and its 4-O-methyl ether. Also excreted, mainly as an acid-labile conjugate, was 2-O-methylpyrogallol. Similar results were obtained when rats were given gallic acid by i.p. injection except that some pyrogallol (1,2,3-trihydroxybenzene) was also detected. Rabbits fed a diet containing gallic acid also formed the 4-O-methylated acid, pyrogallol and 2-O-methylpyrogallol. The data indicated that the benzoic acid derivatives are largely excreted in the free forms although an acid-labile conjugate of 4-O-methylgallic acid was excreted by rabbits. Blumenberg and Dohrmann (1960) noted a marked increase in glucuronic acid excretion in rabbits fed gallic acid and identified its glucuronide in urinary extracts. Contrariwise, Watanabe and Oshima (1965) who carried out similar experiments in rabbits, claimed that little conjugated material was excreted. They found that unchanged gallic acid was the main excretory product, together with some 4-O-methylgallic acid and pyrogallol. Scheline (1966a) reported that rats given gallic acid (100 mg/kg, p.o.) excreted it and its 4-O-methyl ether both free and as acid-labile conjugates whereas the decarboxylated metabolites, pyrogallol and 2-O-methylpyrogallol, were excreted in conjugated form.

Both the O-methylation and decarboxylation reactions seen with gallic acid have been studied in further detail. Masri *et al.* (1962) found that rat or rabbit liver slices selectively methylated gallic acid in the 4-position, in agreement with the *in vivo* results. The same results were obtained when a catechol O-methyltransferase preparation from rat liver was used (Masri *et al.*, 1964). The decarboxylation reaction was investigated in rats by Scheline (1966a) who showed that it is carried out by the intestinal microorganisms, a finding which substantiates the proposal presented earlier by Tompsett (1958). Decarboxylation occurred readily *in vitro* when gallic acid was incubated anaerobically with rat intestinal contents or faeces. Interestingly, resorcinol which arises from the 2-dehydroxylation of pyrogallol, was also formed in these incubates and this compound was sometimes also detected in small amounts in the urine samples from rats given gallic acid. The structural requirement for decarboxylation, which was subsequently shown to be the presence of a free *para*-hydroxyl group

(Scheline, 1966b), indicates that the 2-*O*-methylpyrogallol excreted in the urine of animals given gallic acid is formed via the sequence gallic acid → pyrogallol → 2-*O*-methylpyrogallol rather than by the decarboxylation of 4-*O*-methylgallic acid.

The remaining benzoic acid derivatives to be covered in this section are methyl ethers of some of the phenolic acids covered above and also the methylenedioxy derivative, piperonylic acid. The acid in this group having the simplest structure is **anisic acid** (4-methoxybenzoic acid), which was found by Quick (1932) to be excreted by man conjugated with glycine or glucuronic acid. Following an oral dose of 3·7 g, the recovery of metabolites was nearly quantitative and about half was excreted in each form. The results obtained by Bray *et al.* (1955) in rabbits given anisic acid (400 mg/kg, p.o.) were similar, most of the dose being excreted as the ester glucuronide of anisic acid (57%), 38% as the glycine conjugate (4-methoxyhippuric acid) and only about 1% as unchanged compound. The metabolism of anisic acid in rats was studied by Cramer and Michael (1971) who found that about 85% of the dose (100 mg, i.p.) could be accounted for in the 24 h urines. Nearly 80% of the dose was due to anisic acid (6%) or its glycine (16%) and glucuronic acid (58%) conjugates, however a further 6% was found to consist of 4-hydroxybenzoic acid (1·4%) and its glycine and glucuronic acid conjugates (0·6% and 4·2%, respectively). Axelrod (1956) reported that rabbit liver microsomes can *O*-demethylate anisic acid.

The metabolism of **vanillic acid** (4-hydroxy-3-methoxybenzoic acid) shows similarity to that noted above with its demethylated derivative protocatechuic acid. Sammons and Williams (1941) fed vanillic acid (1 g/kg) to rabbits and accounted for 83% of the dose in the urine, 56% as free vanillic acid, 16% as glucuronide and 11% as ethereal sulphate. They also reported that a further 5% or so may undergo demethylation, probably to protocatechuic acid. When considering the metabolism of vanillic acid, it should be recalled that this compound is an important metabolite of vanillin, the metabolism of which is covered in Chapter 4, Section I.C. The metabolic pathways of vanillin in the rat are illustrated in Fig. 4.1 of Chapter 4 which indicates that some conjugation of vanillic acid with glycine also occurs. The pathway leading to protocatechuic acid is also shown and, interestingly, the acids may undergo decarboxylation to give guaiacol and catechol. Wong and Sourkes (1966) reported that vanillic acid is a source of urinary catechol in rats and it is now clear that this reaction is carried out by the intestinal bacteria. Scheline (1966b) incubated vanillic acid anaerobically with a mixed culture of rat caecal microorganisms and found that it underwent both decarboxylation and *O*-demethylation, giving guaiacol, protocatechuic acid and catechol.

The metabolism of **veratric acid** (3,4-dimethoxybenzoic acid) has not been studied in detail, however Sammons and Williams (1946) found that rabbits given oral doses (1 g/kg) excreted about 41% in the urine as unchanged compound and a further 28% as veratroylglucuronide. Veratroylglycine, vanillic acid derivatives and protocatechuic acid were not detected but a small amount of catechol was reported. Formation of the last metabolite indicates that O-demethylation and subsequent decarboxylation took place. Scheline (1966b) found that veratric acid undergoes O-demethylation at both the 3- and 4-positions when incubated anaerobically with rat caecal bacteria and, in view of that noted above with vanillic acid similarly treated, the initial formation of vanillic acid is sufficient to explain the subsequent degradation to catechol. However, a further site of O-demethylation is the liver and Thomas and Müller-Enoch (1974) reported that perfused rat liver O-demethylated veratric acid in both the 3- and 4-positions, with about 15 times as much of the latter metabolite (vanillic acid) being formed.

Piperonylic acid (3,4-methylenedioxybenzoic acid) is structurally similar to veratric acid and its metabolism not unexpectedly follows similar pathways. Heffter (1895) reported that about half of a 5 g dose was excreted in the urine by man as an unidentified acid. This was believed by Williams (1959, p. 371) to be the glycine conjugate which was reported to be excreted by rabbits together with the ester glucuronide of piperonylic acid. Acheson and Atkins (1961) found that the glycine conjugate is also a urinary metabolite of piperonylic acid in rats, some unchanged compound also being detected. Similar results were obtained with mice (Kamienski and Casida, 1970) and it was noted that the glycine conjugate was the major urinary metabolite detected. The latter investigation also looked into the question of cleavage of the methylenedioxy moiety, using material labelled with ^{14}C at this site. Only about 1% of the radioactivity was recovered in the expired air in 48 h and this indicates that this pathway is of minor importance. Furthermore, *in vitro* experiments with mouse liver microsomes failed to show any demethylenation of piperonylic acid to the corresponding catechol and CO_2. It seems reasonable to assume that the extensive ionization of this acid which takes place at physiological pH values will reduce its ability to effectively reach the oxidative enzymes in the lipoidal microsomal system.

The two relevant methyl ethers of gallic acid are **syringic acid** (3,5-dimethoxy-4-hydroxybenzoic acid) (25) and 3,4,5-trimethoxybenzoic acid. The metabolism of both of these acids given orally at doses of 800 mg/kg to rats was studied by Griffiths (1969). With syringic acid, large amounts of unchanged compound were excreted in the urine together with two minor metabolites identified as 3-O-methylgallic acid (26) and its 4-O-

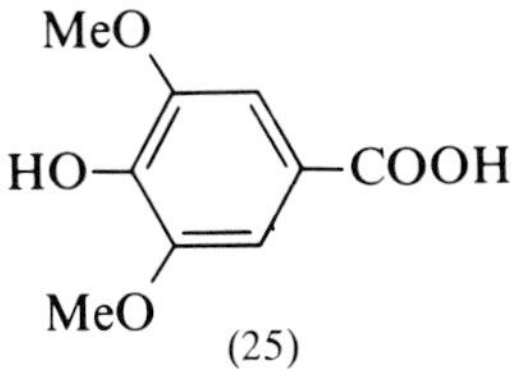

MeO

HO—⟨ ⟩—COOH

MeO

(25)

Syringic acid

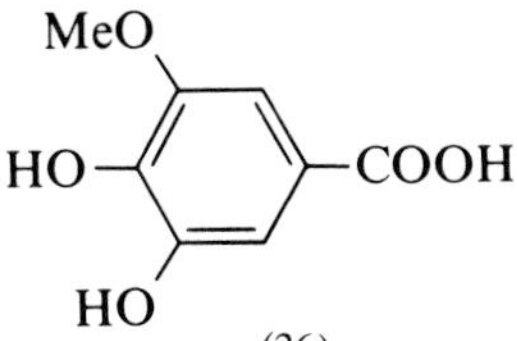

MeO

HO—⟨ ⟩—COOH

HO

(26)

MeO

MeO—⟨ ⟩—COOH

HO

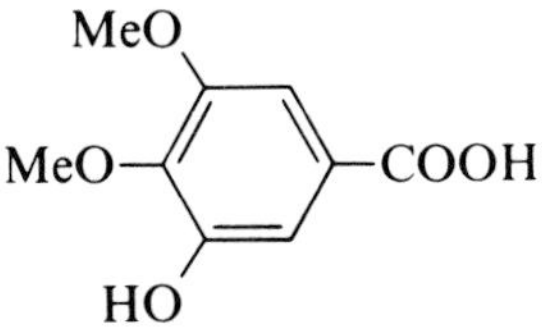

(27)

methylated product, 3,4-dimethoxy-5-hydroxybenzoic acid (27). Scheline (1966b) showed that the O-demethylation of syringic acid to compound (26) can be carried out by the intestinal microflora of rats.

Some uncertainty exists as to the metabolic fate of **3,4,5-tri-methoxybenzoic acid**. Numerof et al. (1955) reported that essentially the entire dose (1·75 mg/kg, p.o.) was rapidly excreted in the urine as unchanged material by mice. However, Griffiths (1969) reported that the major urinary metabolite of this acid (800 mg/kg, p.o.) in rats is the demethylated compound (27). A small amount of the other monode-methylated product, syringic acid (25), was also detected. The O-demethylation observed with many phenolic methoxybenzoic acids follow-ing anaerobic incubation with rat intestinal microorganisms was not seen with 3,4,5-trimethoxybenzoic acid (Scheline, 1966b).

Phenolic derivatives of phenylacetic acid are not common plant con-stituents and only a few examples are known. These include the 2- and 4-hydroxy derivatives and homogentisic acid (2,5-dihydroxyphenylacetic acid), however very little metabolic data are available on these compounds. **4-Hydroxyphenylacetic acid** appears to be mainly excreted unchanged and Ewins and Laidlaw (1910) found that man excreted 50% of a 0·5 g oral dose in this form in the urine within 36 h. They also referred to nineteenth century studies which reported corresponding values of 40–50% in dogs receiving 2 g and about 80% in humans receving 7·5 g. 4-Hydroxy-phenylacetic acid may undergo hydroxylation to a small extent. Wiseman-Distler et al. (1965) administered the acid (21 mg/kg) to rats and found that after intraperitoneal injection, about 4% of the dose was excreted in the 24 h urine as 4-hydroxy-3-methoxyphenylacetic acid. The probable

intermediate, 3,4-dihydroxyphenylacetic acid, was not detected and neither of these compounds was excreted when the 4-hydroxy acid was given
orally. The only other noteworthy point about this compound seems to be
its ability to undergo decarboxylation. This reaction results in the formation of p-cresol and has been shown to be carried out in vitro by faecal
bacteria from rats (Scheline, 1968a) and humans (Curtius et al., 1976).
Martin (1975) reported the same reaction when the acid was infused into
the rumen of sheep. The metabolic pathways of **homogentisic acid** have not
been elucidated except, of course, for that involved in the oxidative
metabolism of phenylalanine and tyrosine which involves transamination
to p-hydroxyphenylpyruvic acid and conversion of the latter to homogentisic acid which then undergoes ring scission and further oxidation. Homogentisic acid is not susceptible to decarboxylation or dehydroxylation when
incubated with mixed populations of rat caecal microorganisms (Scheline,
1968a).

The remaining phenolic phenylacetic acid derivative is **homoprotocatechuic acid** (3,4-dihydroxyphenylàcetic acid). Although this compound
appears to be primarily associated with lower rather than higher plants, it is
the only member of this group which has received significant attention
from a metabolic point of view. It is felt that a short summary of the main
findings is therefore desirable. Homoprotocatechuic acid metabolism is
interesting from a historical point of view since it was one of the
compounds included in the first reports on the biological O-methylation of
catecholic phenols in animals (DeEds et al., 1955; Booth et al., 1955). This
reaction has been a major point of interest in subsequent studies of the
metabolic fate of homoprotocatechuic acid and has been demonstrated
both in vitro and in vivo. Masri et al. (1962) found that rat or rabbit liver
slices methylated it mainly in the 3-position, giving rise to homovanillic
acid, but also noted that a small amount of 4-O-methylation (forming
homoisovanillic acid) occurs. Axelrod and Tomchick (1958) reported the
3-O-methylation of homoprotocatechuic acid by a catechol-O methyl-
transferase system from rat liver which requires S-adenosylmethionine.
The ability of this system to O-methylate at the 3- or 4-positions was
extensively studied by Creveling et al. (1970, 1972) who found a 7:1 ratio
between 3-O-methylated and 4-O-methylated product. The in vivo
formation and urinary excretion of O-methylated phenolic acids from
homoprotocatechuic acid has been investigated in rats, rabbits and man. In
rats, Wiseman-Distler et al. (1965) reported values of about 16–19%
following doses of roughly 10–50 mg/kg, i.p. and Dacre et al. (1968)
recorded a mean value of 19% with an oral dose of 100 mg/kg. At the
latter dose level, the urinary excretion of homovanillic acid in rabbits
accounted for only about 6% of the dose (Scheline et al., 1960; Dacre et

al., 1968) whereas, in man, 40% of a much smaller dose (2·5 mg, i.v.) was recovered in this form in the 24 h urine (Alton and Goodall, 1969). The amount of unchanged homoprotocatechuic acid excreted in these studies ranged from 18–44% (Wiseman-Distler *et al.*, 1965) and 55% (Dacre *et al.*, 1968) in rats, 63% in rabbits and 25–30% in man.

The second major point of interest in the metabolism of homoprotocatechuic acid concerns its dehydroxylation by the intestinal microflora. In their initial report, Booth *et al.* (1955) reported that rabbits given the catechol acid orally excreted in the urine a *p*-dehydroxylated product, 3-hydroxyphenylacetic acid. The dehydroxylation reaction in rabbits was quantitated by Scheline *et al.* (1960) who found that 14% of an oral dose (100 mg/kg) was excreted in the 44 h urine as 3-hydroxyphenylacetic acid. The other dehydroxylation product, 4-hydroxyphenylacetic acid, was also detected, but only to an extent 10% of this. The *p*-dehydroxylation of homoprotocatechuic acid *in vitro* by rat intestinal bacteria was demonstrated by Booth and Williams (1963b) and by Scheline (1967, 1968a). The last two studies also showed that a second bacterial reaction, decarboxylation, takes place with homoprotocatechuic acid. The product, 4-methylcatechol, was formed in the *in vitro* incubation experiments and was also excreted as an acid-labile conjugate in the urine of rats treated with oral doses (100 or 400 mg/kg) of the acid (Scheline, 1967).

The second major group of phenolic acids from plants includes the C_6—C_3-derivatives, mainly hydroxycinnamic acids. The most common representatives of this type are *p*-coumaric, caffeic, ferulic and sinapic acids. In some cases (e.g. melilotic and dihydrocaffeic acids), forms in which the side-chain double bond is reduced are also known. Some information is available on most of these cinnamic acid derivatives, however caffeic acid is the only compound studied extensively. Of the monohydroxy derivatives, both **o-coumaric acid** (28) and its reduced derivative, **melilotic acid** (29), occur naturally. As they are metabolically interconvertible, essentially the same picture is seen with both compounds. This was shown to be the case by Furuya (1958) and by Booth *et al.* (1959a) in rats and rabbits. The pathways of metabolism elucidated in these studies and in the investigation of Mead *et al.* (1958) who administered the two acids to rabbits are illustrated in Fig. 5.2.

Figure 5.2 shows that *o*-coumaric acid and melilotic acid are metabolized via several routes, however the details concerning some of the pathways are unclear. The pathways proposed in Fig. 5.2 should therefore not be considered to exclude other possible routes or intermediates which are not shown. The formation of 2-hydroxyphenylacetic acid (30) illustrates this lack of certainty about the actual mechanism involved. It is also known that metabolite (30) is formed from coumarin (31), however in that case an

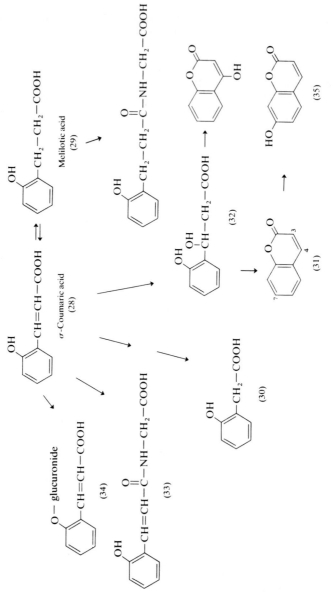

FIG. 5.2. Proposed metabolic pathways of O-coumaric acid and melilotic acid in rats and rabbits (see text).

accompanying metabolite of coumarin, 3-hydroxycoumarin, should also have been detected. The latter metabolite was not found but this need not preclude its formation as a transient intermediate. Booth *et al.* (1959a) suggested that the phenylacetic acid derivative (30) is formed from 2-hydroxyphenyllactic acid, a compound isomeric with 2-hydroxyphenylhydracrylic acid (32). Both of the last two compounds can be formed by hydration of the double bond of *o*-coumaric acid and, with the lactic acid derivative, oxidation would give an α-keto carboxylic acid derivative which could easily loose CO_2 to form metabolite (30). Some evidence for this mechanism was obtained by Booth *et al.* who detected only metabolite (30) and unchanged compound in the urine of rats fed 2-hydroxyphenyllactic acid. Furthermore, Flatow (1910) reported that guinea pigs converted the corresponding α-keto acid, 2-hydroxyphenylpyruvic acid, to metabolite (30).

Although most of the data on the metabolism of *o*-coumaric acid and melilotic acid is of a qualitative nature, Furuya (1958) also obtained some quantitative values in rabbits given *o*-coumaric acid (200 mg/kg, p.o.). Nearly half of the dose was accounted for in the 48 h urine and 20% was due to unchanged compound. The conjugates (33) and (34) accounted for 11% and 3·5%, respectively, and a further 8% was due to free melilotic acid (29). The other metabolites were excreted in fairly small amounts, however 4–5% was due to free and conjugated 7-hydroxycoumarin (35). It is noteworthy that the reduction of *o*-coumaric acid to melilotic acid can be carried out by the intestinal microorganisms (Scheline, 1968a). In view of the demonstrated formation of the β-hydroxy derivative (32), it might be expected that shortening of the side chain by β-oxidation would produce some salicylic acid. However, all three of the *in vivo* studies cited above failed to detect this compound as a urinary metabolite of the two C_6—C_3 acids.

The metabolism of **p-coumaric acid** (4-hydroxycinnamic acid) is straightforward and, in rats, the urinary metabolites are unchanged compound, its glycine conjugate, the reduced derivative 4-hydroxyphenylpropionic acid (phloretic acid) and the β-oxidation product 4-hydroxybenzoic acid (Booth *et al.*, 1960a; Griffiths and Smith, 1972). In the former study the ethereal sulphate of 4-hydroxybenzoic acid was also detected. Details of the β-oxidation of *p*-coumaric to 4-hydroxybenzoic acid by rat liver preparations were studied by Ranganathan and Ramasarma (1971, 1974). The enzyme carrying out this conversion is localized in the mitochondria and requires ATP. Mitochondrial enzymes were also reported to catalyse the interconversion of *p*-coumaric acid and phloretic acid (Ranganathan and Ramasarma, 1974). It must be added, however, that the reduction of the double bond can be carried out by rat intestinal microorganisms, as shown by Scheline (1968a) and by Griffiths and Smith (1972). The former investigation also showed that the microflora is able to decarboxylate p-coumaric acid but not phloretic acid. This reaction is carried out by a strain of *Bacillus* isolated from rat intestine (Indahl and Scheline, 1968) and by *Aerobacter aerogenes* (Soleim and Scheline, 1972). The significance of this decarboxylation reaction *in vivo* has not been assessed. Interestingly, the bacterial metabolism of both p-coumaric and phloretic acids in the sheep rumen leads to dehydroxylated rather than decarboxylated products (Martin, 1975).

Caffeic acid (36) is structurally a relatively simple compound, nonetheless it contains several metabolically active sites, a property which leads to the formation of a large number of metabolites. The reported metabolites of caffeic acid are shown in Fig. 5.3 which also illustrates their probable routes of formation. Of course, all of these metabolites have not been observed in a single experiment and the number of metabolites detected, as well as their amounts, will depend on many factors including the dose, route of administration and animal species. A subject of special interest in the study of caffeic acid metabolism has been the role of the intestinal bacteria which are now known to be solely responsible for several of the reactions shown in Fig. 5.3.

Several caffeic acid metabolites shown in Fig. 5.3 are 3-O-methyl derivatives and this reaction was first reported by DeEds *et al.* (1955) who identified ferulic acid (37) in the urine of rabbits fed caffeic acid. A subsequent detailed investigation of caffeic acid metabolism in rats and humans by Booth *et al.* (1957) revealed the formation of all of the hydroxymethoxy derivatives except 4-hydroxy-3-methoxyphenylhydracrylic acid (38), which was among the numerous metabolites of caffeic acid in man identified by Shaw and Trevarthen (1958). It is noteworthy that none of the metabolites shown in Fig. 5.3 possess the 3-hydroxy-4-methoxy-phenyl moiety. Neither the study of Booth *et al.* nor that of Shaw and Trevarthen, which are the most extensive studies available on the patterns of caffeic acid metabolism, reported any of these derivatives. However, Hill *et al.* (1959) suggested that the O-methylation of caffeic acid may occur at either hydroxyl group since several members of the isomeric catechol ethers corresponding to compounds (37), (38), (39), (40) and (41) were detected in human urine and were believed to arise from dietary chlorogenic acid, an ester of caffeic acid. This suggestion that both sets of O-methyl esters are formed is further supported by evidence obtained *in vitro* in studies using catechol-O-methyltransferase from liver. Pellerin and D'Iorio (1958) and Discherl and Brisse (1966) found that rat liver homogenates O-methylated caffeic acid to ferulic acid, however studies by Masri *et al.* (1962, 1964) and Creveling *et al.* (1972) showed that rat or rabbit

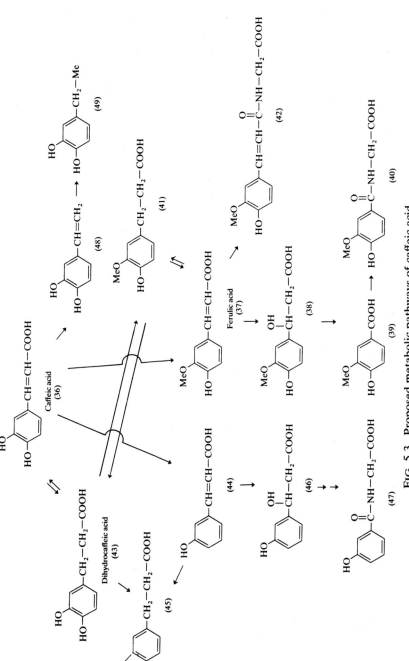

FIG. 5.3. Proposed metabolic pathways of caffeic acid.

liver slices or preparations of liver catechol-O-methyltransferase formed both isomers (ferulic and isoferulic acids). The latter study recorded a *meta*:*para* ratio of O-methylation of only $2·8:1$ and the usual lack of 4-methyl ether metabolites as urinary metabolites of caffeic acid seems therefore to be best explained by their greater susceptibility towards subsequent O-demethylation rather than due to their lack of formation.

Quantitative values for the excretion of caffeic acid and its metabolites have not been reported, however Booth *et al.* (1957) found that 3-hydroxyphenylpropionic acid (45) was the major metabolite in rats, relatively little chain-shortening to $C_6—C_1$ metabolites occurring. Similar findings in rats were reported by Scheline (1968a). On the other hand, Booth *et al.* found that humans produced larger amounts of a greater number of metabolites including the $C_6—C_1$ derivatives. The intermediate hydracrylic acids (38) and (46) have accordingly only been detected in human urine (Shaw and Trevarthen, 1958; Hill *et al.*, 1959). In addition to the glycine conjugates (40), (42) and (47) shown in Fig. 5.3, some of the other metabolites are excreted conjugated with glucuronic acid. Scheline (1968a) found that caffeic acid, ferulic acid and, to a minor extent, 3-hydroxyphenylpropionic acid (45) were partly excreted in rat urine as glucuronides and that both decarboxylated metabolites, 4-vinylcatechol (48) and 4-ethylcatechol (49) were detected only in conjugated form.

As noted above, the metabolism of caffeic acid by intestinal bacteria is an important feature which leads to the formation of several of the metabolites shown in Fig. 5.3. All of the 3-hydroxy derivatives are a result of this phenomenon and the role of the bacteria in caffeic acid dehydroxylation was first reported by Shaw *et al.* (1963). This reaction was subsequently shown to take place *in vitro* using incubates of mixed bacterial cultures from rat or rabbit caecal contents and sheep rumen fluid (Booth and Williams, 1963a). Scheline (1968a) confirmed this reaction using the rat caecal microflora. The ultimate hydroxylated product in these experiments is 3-hydroxyphenylpropionic acid (45), a finding which shows that reduction of the double bond is also a result of bacterial metabolism. Masri *et al.* (1962) reported that this reduction can also be carried out *in vitro* by rat or rabbit liver slices. This indicates that alternative sites of reduction are available, however that occurring in the intestine is probably quantitatively the most important. The dehydroxylation reaction takes place entirely by virtue of bacterial metabolism and can be prevented when the gut flora is suppressed by treatment with neomycin (Dayman and Jepson, 1969). Accordingly, 3-hydroxyphenylpropionic acid is not a urinary metabolite of caffeic acid in germ-free rats (Scheline and Midtvedt, 1970; Peppercorn and Goldman, 1972). These findings have stimulated interest in elucidating the nature of the microorganisms responsible for caffeic acid dehydroxylation. The first report on this subject was that of Perez-Silva *et al.* (1966) who isolated a strain of *Pseudomonas* from rat faeces which could convert the catechol to *m*-coumaric acid (44) and its reduction product (45). Perez Silva and Rodriguez Sanchez (1967) subsequently described five species of *Pseudomonas*, *P. fluorescens*, *P. viburni*, *P. insolita*, *P. myxogenes* and *P. chlororaphis*, that converted caffeic acid to 3-hydroxyphenylpropionic acid. Peppercorn and Goldman (1971) carried out a more extensive study of this problem and tested the ability of 12 microorganisms isolated from human faeces to metabolize caffeic acid. Interestingly, no single organism was able to degrade the acid to 3-hydroxyphenylpropionic acid, however the reduction to dihydrocaffeic acid (43) was demonstrated with *Clostridium perfringens* and a *Peptostreptococcus* sp. and this metabolite was then dehydroxylated by a mixed culture of *Escherichia coli* and *Streptococcus faecalis* var. *liquifaciens*. Peppercorn and Goldman (1972) reported that germ-free rats selectively infected with a mixture consisting of two *Lactobacillus* strains, a *Bacteroides* sp. and a *Streptococcus* group N gained the ability to convert caffeic acid to 3-hydroxyphenylpropionic acid.

A third reaction of caffeic acid known to take place by bacterial metabolism is that of decarboxylation. This reaction, leading to metabolites (48) and (49), was first shown to occur *in vivo* in rats fed caffeic acid by Scheline (1968a). These metabolites were also formed when caffeic acid was incubated anaerobically with rat caecal microorganisms. Several studies on the microbiology of the decarboxylation reaction have been made and it was shown to be carried out by a *Bacillus* sp. isolated from rat intestine (Indahl and Scheline, 1968), by *Streptococcus fecium* isolated from human faeces (Peppercorn and Goldman, 1971) and by *Aerobacter aerogenes* (Soleim and Scheline, 1972).

The metabolism of **dihydrocaffeic acid** (hydrocaffeic acid) (43) has been investigated in conjunction with several of the studies with caffeic acid noted above. According to the metabolic scheme shown in Fig. 5.3, the reduced compound, which may be dehydrogenated to caffeic acid, should give rise to the same metabolites as those formed from the latter compound. While this is probably true in a qualitative sense, the quantitative picture is undoubtedly different since the rate of the dehydrogenation reaction will influence the availability of compounds (36), (37) and (44) and thus many of the remaining reactions as well. In fact, Booth *et al.* (1957) found that dihydrocaffeic acid fed to rats produced a pattern of urinary metabolites identical to that seen with ferulic acid (37), rather than with caffeic acid. Other points of interest with dihydrocaffeic acid are that it is a substrate of catechol-*O*-methyltransferase (Creveling *et al.*, 1972) and that it undergoes dehydroxylation but not decarboxylation as a result of bacterial metabolism. The latter results have been obtained using both mixed cultures of intestinal bacteria (Scheline, 1968a) and several pure cultures (Peppercorn and Goldman, 1971; Soleim and Scheline, 1972).

Methoxylated cinnamic acid derivatives from plants include a few compounds derived from mono-, di- and trihydric phenolic acids. The sole example of the first type is **4-methoxycinnamic acid** (50) which Woo (1968) found to be oxidized to 4-methoxybenzoic acid (51) in rabbits. Solheim and Scheline (1973), who found that this cinnamic acid derivative was an intermediate in the metabolism of the alkenebenzenes estragole and anethole in rats, studied its metabolism following doses of 100 and 400 mg/kg (p.o. and i.p.). Excretion of unchanged compound was detected only at the higher dosage. The urinary metabolites excreted, including the postulated β-keto acid (52), are shown in Fig. 5.4 and indicate that, in rats, 4-methoxycinnamic acid is metabolized either by O-demethylation or by β-oxidation which leads mainly to benzoic acid derivatives.

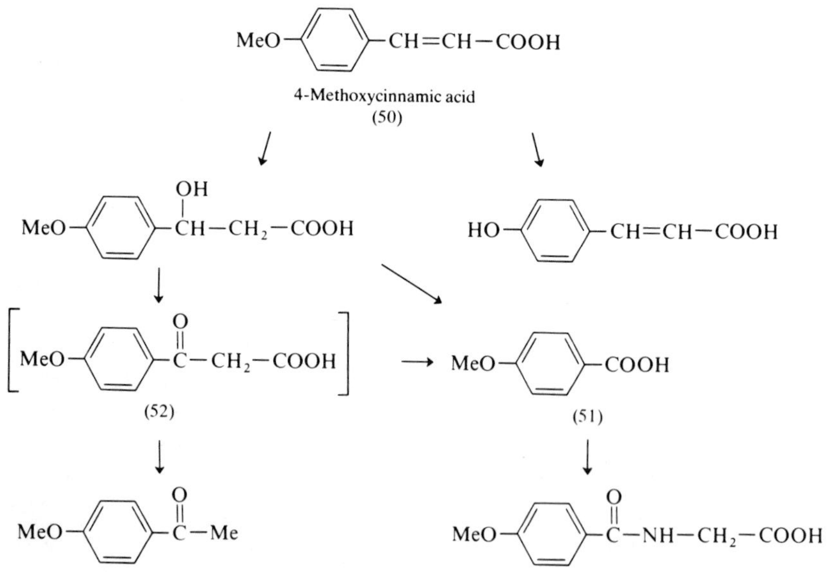

FIG. 5.4. Metabolic pathways of 4-methoxycinnamic acid in rats.

Both of the monomethyl ethers of caffeic acid, **ferulic acid** (37) and isoferulic acid (3-hydroxy-4-methoxycinnamic acid) occur naturally, but it is the former compound which is by far the most common and which has received attention from a metabolic point of view. As pointed out above in the review of caffeic acid metabolism, ferulic acid is formed from it by

3-O-methylation and in accordance with the pathways illustrated in Fig. 5.3, several of which are reversible, the metabolic fates of the two compounds should be qualitatively identical. The experimental data available indicate that this is largely true, especially with regard to the terminal metabolic products. Thus Shaw and Trerarthen (1958) found that the same 3-hydroxyphenyl and 3-methoxy-4-hydroxyphenyl acids were excreted in human urine following the ingestion of caffeic or ferulic acids. These metabolites are shown in Fig. 5.3 and include the 3-hydroxyphenyl derivatives (45), (46) and (47) and the hydroxymethoxy derivatives (37), (38), (39), (40), (41) and (42). Similarly, Booth et al. (1957) fed ferulic acid to rats and found that it was metabolized to the dehydroxylated compound (45) and the hydroxymethoxy derivatives (39), (40), (41) and (42). Teuchy and Van Sumere (1971) reported that 3-hydroxyphenylpropionic acid (45) was the main urinary metabolite of ferulic acid (150–190 mg/kg, i.p.) in rats and that vanillic acid (39) was also excreted. The conversion of ferulic acid to vanillic acid by rat liver homogenates was reported by Dirscherl and Brisse (1966).

The dehydroxylation reaction leading to 3-hydroxyphenylpropionic acid (45) is carried out by intestinal microorganisms (Scheline, 1968a). Indeed, this entire transformation is seen when ferulic acid is incubated anaerobically with a mixed culture of rat caecal microorganisms. The intermediate compounds dihydroferulic acid (41) and dihydrocaffeic acid (43) were also detected. The finding of Teuchy and Van Sumere noted above that 3-hydroxyphenylpropionic acid (45) is the major urinary metabolite of ferulic acid following its intraperitoneal dosage is explained by the fact that the conjugate, feruloyl glucuronide, is readily excreted in the bile of rats. Thus administration of ferulic acid to rats either orally or by injection will lead to a qualitatively similar pattern of metabolism since the biliary ferulic acid glucuronide is readily hydrolysed and further metabolized by the bacteria.

Another bacterial metabolite of ferulic acid is 4-vinylguaiacol (Scheline, 1968a). This decarboxylation product is formed in the same way that 4-vinylcatechol (48) arises from the degradation of caffeic acid (see Fig. 5.3). Metabolite (48) was shown to be a urinary metabolite of caffeic acid in rats (Scheline, 1968a) and it therefore seems reasonable to assume that 4-vinylguaiacol, in conjugated form, is a urinary metabolite of ferulic acid. This point has not been clarified however.

Little is known of the metabolism of **isoferulic acid** (3-hydroxy-4-methoxycinnamic acid), however the results of Scheline (1968a) obtained using anaerobic incubates of rat caecal microorganisms suggest that its fate will probably be fairly similar to that seen with its isomer, ferulic acid. These experiments showed that double-bond reduction as well as limited degradation to 3-hydroxyphenylpropionic acid (45) occurred. Direct decarboxylation of isoferulic acid, which lacks a free *p*-hydroxyl group, does not occur.

The metabolic pathways of **3,4-dimethoxycinnamic acid** (53) would be expected to be very similar to those shown above (Fig. 5.4) for the corresponding 4-methoxy derivative. Based upon data obtained by Solheim and Scheline (1976) in rats given the dimethoxy compound (100 and 400 mg/kg, p.o. and i.p.), this appears to be the case as both of the previously noted pathways of *O*-demethylation and β-oxidation leading to benzoic acid derivatives were found. However, a rather more extensive pattern of metabolism was seen with 3,4-dimethoxycinnamic acid, some of which was also excreted unchanged, and these pathways are therefore shown in Fig. 5.5. Smith and Griffiths (1974) reported that the rat caecal microflora can extensively *O*-demethylate 3,4-dimethoxycinnamic acid when incubated under anaerobic conditions.

The metabolic fate of **sinapic acid** (54) in animals has been investigated in rats and rabbits. Griffiths (1969) fed the compound (800 mg/kg) to rats and found that, in addition to unchanged compound, dihydrosinapic acid (55) and the two *O*-demethylated acids (56) and (57) were excreted in the urine. Interestingly, the maximal excretion of the demethylated acids occurred on the second and third days and the excretion of dihydrosinapic acid was also delayed. Some of the demethylated metabolites were excreted as acid-labile conjugates. Subsequently, Griffiths (1970) reported that the fully *O*-demethylated compound, 3,5-dihydroxyphenylpropionic acid (58) is also a urinary metabolite of sinapic acid in rats and also that it is the major metabolite of the acid (200 mg/animal) in rabbits. The latter investigation also showed that suppression of the intestinal microflora with an antibiotic resulted in the complete abolition of the excretion of *O*-demethylated metabolites in the urine. Also, metabolite (58) was formed when sinapic acid was incubated with rat intestinal bacteria. A detailed study of this bacterial degradation of sinapic acid was made by Meyer and Scheline (1972a) who found that it was reduced to dihydrosinapic acid (55) and then *O*-demethylated and dehydroxylated, forming finally 3,5-dihydroxyphenylpropionic acid (58). Additionally, Meyer and Scheline (1972b) reported that the latter metabolite was the only compound besides unchanged compound detected in the urine of rats following an oral dose (100 mg/kg) of sinapic acid. Both acids were excreted partly free and partly conjugated, sinapic acid only during the first 24 h but metabolite (58) for as long as two or three days. These results make it quite clear that the metabolic activities of the intestinal microflora play a key role in the metabolism of sinapic acid in rats and rabbits. The probable routes of metabolism are shown in Fig. 5.6, however the available data do not rule out alternate minor pathways. For example, the detection of the cinnamic

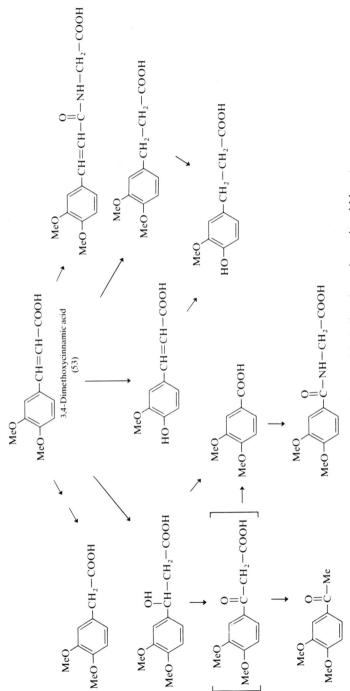

FIG. 5.5. Metabolic pathways of 3, 4-dimethoxycinnamic acid in rats.

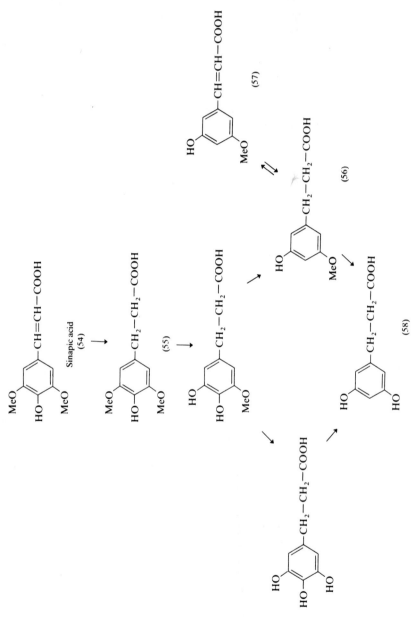

Fig. 5.6. Proposed metabolic pathways of sinapic acid in rats.

derivative (57) as a urinary metabolite (Griffiths, 1969) can best be explained by the dehydrogenation in the tissues of some of the 3-hydroxy-5-methoxyphenylpropionic acid (56) formed bacterially. However, its formation via O-demethylation and dehydroxylation without double bond reduction is also a possibility. It is noteworthy that benzoic acid derivatives have not been reported as sinapic acid metabolites. Attention is also drawn to the fact that the articles of Griffiths (1969) and Meyer and Scheline (1972a,b) also deal with the metabolism, both in rats and by rat caecal microorganisms, of several closely related derivatives of sinapic acid including **3,4,5-trimethoxycinnamic acid**, a seldomly encountered plant acid. The metabolism of the latter compound is similar to that of sinapic acid in rats insofar as the dihydric acid (58) is the ultimate metabolite in both cases. However, major urinary metabolites of the trimethoxy compound are unchanged compound and the mono-O-demethylated compound isomeric with sinapic acid, i.e. 3-hydroxy-4,5-dimethoxycinnamic acid. These metabolites are excreted both free and as conjugates. The phenylpropionic acid analogues of both of these compounds were also detected. The studies of Meyer and Scheline devoted special attention to the pathways of metabolism of 3,4,5-trimethoxycinnamic acid and indicated that tissue O-demethylation is limited to the initial m-demethylation. The alternative route, p-demethylation, was not seen and further demethylation and degradation to the final metabolites are reactions carried out by the intestinal bacteria.

Piperic acid

(59)

Piperic acid (59), which occurs as derivatives in pepper, was fed to rats at a dose level of 500–700 mg/kg and the urinary metabolites studied (Acheson and Atkins, 1961). Not unexpectedly, evidence for the β-oxidation of this unsaturated acid was obtained, the urine containing 3,4-methylenedioxycinnamoylglycine, 3,4-methylenedioxybenzoic acid (piperonylic acid) and greater amounts of the glycine conjugate of the latter acid. It was estimated that, within 72 h, about 0·3% of the dose of piperic acid was excreted as piperonylic acid and about 40% as piperonylglycine.

II. Lactones

Lactones are derived from hydroxy acids by the intramolecular loss of a molecule of water and may therefore be regarded as internal esters of hydroxy acids. The naturally occurring members of this group of *O*-heterocyclic compounds include those with ring systems which may be five-membered (γ-lactones derived from 4-hydroxy acids), six-membered (δ-lactones derived from 5-hydroxy acids) or many-membered (macrocyclic lactones). No metabolic information is available on the last group, the most well-known examples being C_{15} and C_{16} compounds exaltolide and ambrettolide which are noted for their musk-like odours. Diverse types of plant compounds contain the more common γ- and δ-lactone groups but, in many cases, it seems more reasonable to summarize their metabolism in other sections of this book. This is true with coumarin and its derivatives, lactones of *o*-hydroxycinnamic acid, which are covered in Chapter 7, Section III. Other lactones are more closely associated with their terpenoid properties and have therefore been included in the following chapter which deals with the metabolism of higher terpenoids. Examples include the sesquiterpenoid santonin (Chapter 6, Section I) and the cardiac glycosides (Chapter 6, Section III).

Parasorbic acid (60) is the lactone of 5-hydroxy-2-hexenoic acid. It is readily transformed, especially under alkaline conditions, to sorbic acid (2,4-hexadienoic acid). The latter compound is used as a fungistatic agent in foods and its metabolism is summarized above in Section I.A. However, parasorbic acid itself when administered to rats is known to give rise to an extensive reduction of liver glutathione levels (Boyland and Chasseaud, 1970). Furthermore, rat liver and kidney preparations contain glutathione *S*-transferase activity which catalyses the reaction of glutathione with parasorbic acid (Boyland and Chasseaud, 1967, 1968). These findings suggest that parasorbic acid may be metabolized to a mercapturic acid derivative.

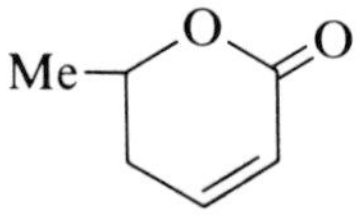

(60)
Parasorbic acid

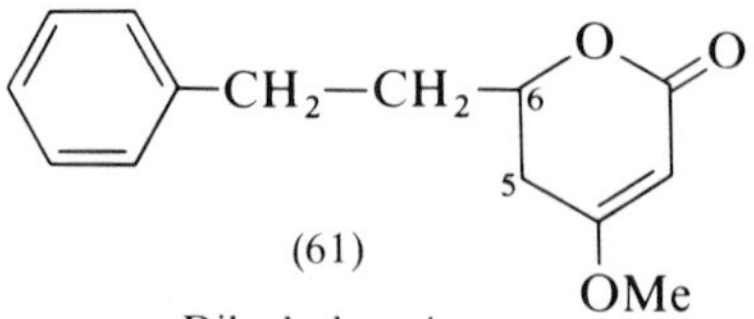

(61)
Dihydrokawain

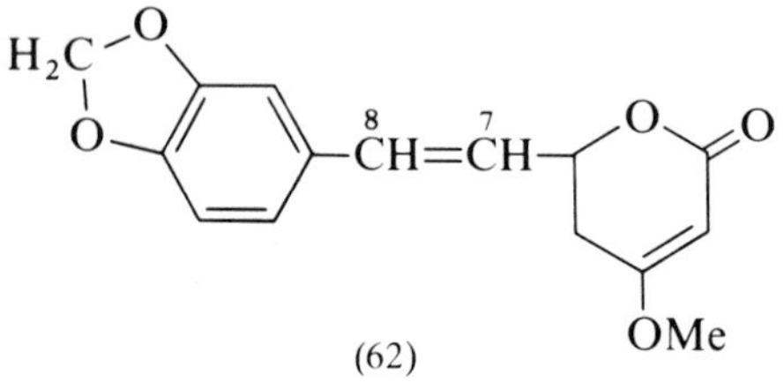

(62)
Methysticin

Some of the kava pyrones are 5,6-dihydro-α-pyrones and therefore, strictly speaking, lactones. Rasmussen *et al.* (1976) found that one of these, **dihydrokawain** (61), undergoes aromatic hydroxylation when given to rats (400 mg/kg, p.o.). Both the *p*-hydroxylated and a dihydroxylated compound were identified as urinary metabolites which were excreted in conjugated form. In addition, evidence was obtained which indicated that at least 5% of the dose was converted to hippuric acid. This metabolite is probably formed via the initial hydrolysis of the lactone which can be further metabolized by sequential β-oxidation to benzoic acid. In similar experiments with **methysticin** (62), two urinary metabolites were identified which indicated that the methylenedioxy group underwent demethylenation. These two compounds are the catechol analogues of methysticin and 7,8-dihydromyristicin. Attention is drawn to the fact that other closely related kawa pyrones have a double bond in the 5,6-position. Yangonin is such a compound and because it is a true derivative of α-pyrone (i.e. a coumalin derivative), it is placed among the pyrones which are covered in Chapter 7, Section I.A.

III. Esters

It is common knowledge that plants contain a large number of esters of diverse types. Besides the triglycerides, esters of glycerol with three fatty acid molecules which serve as food storage meterials, many simple esters of long-chain fatty acids which provide waxy protective coatings are also found. A discussion of the metabolism of these types falls outside the scope of this book. Other typical plant esters include those consisting of numerous lower alcohols combined with various aliphatic or aromatic acids. These are often volatile compounds and their aromaticity imparts pleasing or distinctive characteristics to many fruits. The general features of the hydrolysis of esters, a reaction which occurs readily in the tissues, were summarized in Chapter 1, Section I.C.1 and the further possibilities for this reaction as a result of the metabolic activities of the intestinal microflora were dealt with in Section II.A.2 of that chapter. These points will not be repeated here, however it is obvious that the subject of ester metabolism is to a large extent identical with that of the fate of the particular acid and alcohol components. This is conveniently illustrated by the study of Gallaher and Loomis (1975) who investigated the metabolism of **ethyl acetate** in rats. Blood levels of both ethyl acetate and ethanol were determined after intraperitoneal injections of the ester. Hydrolysis of the latter was very rapid (half-life = 5–10 min) and its disappearance was accompanied by a rapid increase in the blood ethanol levels which then fell relatively slowly. Partly because of this rapid hydrolysis of many esters to their component parts, but also because only a few of the myriad of plant esters have been the subject of specific metabolic investigations, the discussion which follows deals with a relatively limited number of compounds. In addition, a few terpenoid compounds containing ester groups are dealt with in Chapter 6. These include the diterpenoids cassaidine and cassaine and some carotenoid (tetraterpenoid) esters. Another complex ester covered elsewhere is the glucosinolate sinalbin. As described in Chapter 10, Section VI, the sinapine moiety in this compound is an ester consisting of sinapic acid and choline which undergoes hydrolysis when it is fed to rats.

(63)

(a) Pyrethrin I, R = Me,   R′ = −CH$_2$−CH=CH−CH=CH$_2$
(b) Pyrethrin II, R = −COOME, R′ = −CH$_2$−CH=CH−CH=CH$_2$
(c) Cinerin I,   R = Me,   R′ = −CH$_2$−CH=CH−Me
(d) Cinerin II,   R = −COOMe, R′ = −CH$_2$−CH=CH−Me
(e) Jasmolin I,   R = Me,   R′ = −CH$_2$−CH=CH−CH$_2$−Me
(f) Jasmolin II, R = −COOMe, R′ = −CH$_2$−CH=CH−CH$_2$−Me

The natural pyrethrins present in pyrethrum flowers are powerful insecticides which owe much of their usefulness to their low toxicity to mammals. These alicyclic esters include the well-known pyrethrins I and II (63a,b) and the closely related homologues cinerin I and II (63c,d) and jasmolin I and II (63e,f). Allethrin (63, R′ = −CH$_2$−CH=CH$_2$) is the synethetic allyl homologue of the I-series. Studies on the metabolism of the pyrethrins have been hampered by their complex chemistry, particularly their stereochemistry, however, the detailed investigation by Elliott *et al.* (1972) has provided an excellent foundation for the understanding of pyrethrin metabolism. Useful recent summaries in this field have appeared (Yamamoto *et al.*, 1971; Casida, 1973) and the present summary of the main features of the metabolism of the natural pyrethrins in mammals is taken from these three sources. Other equally important areas including the metabolism of pyrethrins in insects and the metabolism of numerous synthetic derivatives, including allethrin, are not covered in this book. However, the papers listed above provide convenient access to the literature devoted to these topics and Ueda *et al.* (1975) recently published a detailed investigation on the metabolism in rats of two synthetic insecticidal esters of chrysanthemummonocarboxylic acid (the acid moiety of pyrethrin I).

Inspection of the structures of **pyrethrin I** (63a) and **pyrethrin II** (63b) reveals that several potential sites of metabolic change are present. These include the ester linkage, the isobutenyl group of the acid moiety, the pentadienyl side-chain of the alcohol moiety and, in pyrethrin II, the methoxycarbonyl group. Based on results obtained using rats, it appears that hydrolysis of the central ester linkage in mammals is a reaction which occurs to only a very limited extent. This was made evident by the finding that the same radioactive urinary metabolites were excreted following administration of pyrethrin I labelled in the acid or alcohol moieties. This is in contrast to that seen with allethrin and several other related synthetic derivatives. The major identified metabolites of pyrethrin I and II are compounds resulting from oxidation of both the acid and alcohol moieties. These metabolites are shown in Fig. 5.7 which illustrates the probable pathways of metabolism of these pyrethrins in rats. Pyrethrin I was converted by the rat liver microsome-NADPH system to several neutral and acidic metabolites including the alcohol (64), aldehyde (65) and acid (66) intermediates. This scheme also includes the probable epoxide intermediate (67) which would be expected to undergo hydration to the 4',5'-diol (68a) or its conjugate (68b) (R = unidentified aromatic acid) or, by rearrangement, to the 2',5'-diol (69). Following oral administration of the two pyrethrins (3 mg/kg) to rats, these diol derivatives were excreted during a period of 100 h in the urine and faeces. Metabolites (68a), (68b) and (69) accounted for 14–21%, 4–6% and 3–4%, respectively, of the dose. The acidic precursor, metabolite (66) was not detected in the urine or faeces and appears therefore to be a transient product of metabolism. The unchanged compounds were not found in the urine but 18% and 4% of the original pyrethrin I and II, respectively, were recovered in the faeces. In addition, the excreta contained a few percent of the dose as less polar unidentified metabolites and 25–30%, about equally divided between the urinary and faecal routes, as more polar material. It was suggested that the latter may be conjugates, probably glucuronides, of the hydroxylated metabolites. It is noteworthy that none of the reported metabolites of pyrethrin II contain the original methyl ester group. The hydrolysis of this group was also shown by the excretion of $^{14}CO_2$ in the expired air of rats given the ^{14}C-methoxycarbonyl-labelled compound. A postulated pathway not shown in Fig. 5.7 is the reduction of the pentadienyl side-chain to an alkyl group which might then undergo aliphatic hydroxylation. This suggestion seems reasonable in view of the finding that some unabsorbed compound is excreted in the faeces. Reduction of double bonds is readily carried out by the intestinal bacteria and absorption of some of this reduced material might lead to its hydroxylation in the tissues.

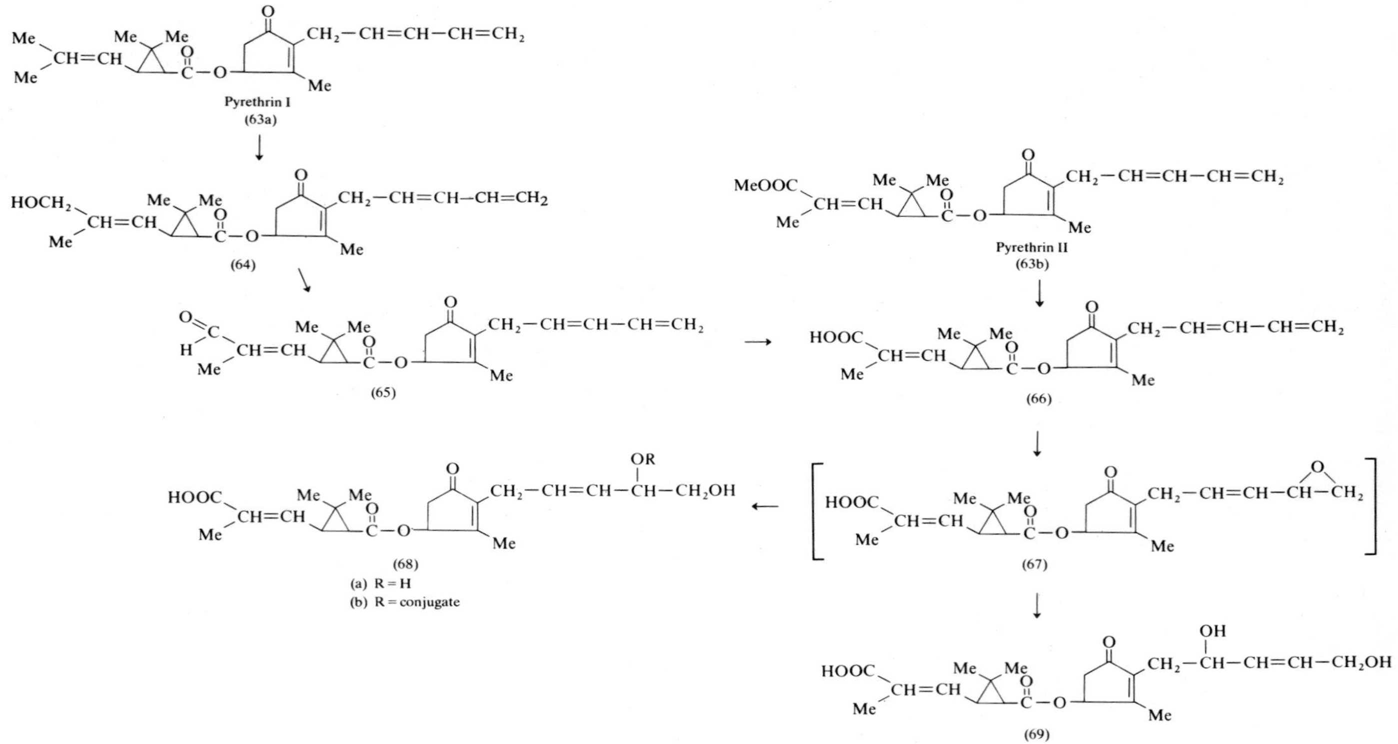

FIG. 5.7. Metabolic pathways of pyrethrin I and II in rats.

Casida (1973) proposed that the pathways of metabolism of **cinerin I** (63c) and **cinerin II** (63d) and of **jasmolin I** (63e) and **jasmolin II** (63f) are the same as those seen with the corresponding pyrethrins. Thus hydrolysis of the central ester linkage is probably very limited and metabolism proceeds instead via side-chain oxidation of the alcohol moiety. Two such pathways are likely, one involving allylic hydroxylation at the terminal methyl group (cinerins) or at the ω-1 position (jasmolins) and the other initiated with epoxidation of the double bond.

Simple aromatic esters are easily hydrolysed in the body and, as noted above, their metabolic fate is largely that of their acid and alcohol components. Excellent examples of this are seen with **benzyl acetate** (70), **benzyl benzoate** and **benzyl cinnamate**, naturally occurring esters which are rapidly and completely metabolized via benzyl alcohol to benzoic acid in man (Snapper *et al.*, 1925). More recently, Clapp and Young (1970) investigated the metabolism of benzyl acetate in rats. Following the administration of a large dose (1 g/kg, s.c.) hippuric acid was isolated from the 48 h urines. Another metabolite isolated, which accounted for only 0·2% of the dose, was benzylmercapturic acid (71). Fahelbum and James (1977) demonstrated the metabolic similarity of cinnamic acid and **methyl cinnamate** in rats and rabbits. Ester hydrolysis occurred very rapidly and no unchanged compound was detected in the peripheral blood. Only traces were found in the portal blood, however cinnamic acid and methanol were readily detected. The urinary metabolites of methyl cinnamate, being identical with those of cinnamic acid, are summarized above in Section I.C in the discussion of the metabolism of the latter compound.

Benzyl acetate
(70)

(71)

Methyl salicylate (wintergreen oil) was formerly commonly employed for its counterirritant properties and it is also useful as a flavouring agent. In an early study Baas (1890) recovered about 25% of the dose (5–10 g) in the urine of dogs as hydrolysis products, a value quite similar to those reported by Hanzlik and Wetzel (1920) in similar experiments using 1·4–2·9 g (200 mg/kg) doses. Small amounts (0·2–0·5%) of the unchanged ester were found in the urine following oral dosage, however nearly 15% was reported as urinary methyl salicylate when it was given intramuscularly. Urinary metabolites were detected for as long as six days.

However, newer experiments by Davison *et al.* (1961) clearly demonstrated that orally administered methyl salicylate is rapidly and completely hydrolysed in rats and dogs (dose, 300 mg/kg) and also, although to a somewhat lesser extent, in humans (dose, 7 mg/kg). Experiments in rats, rabbits, dogs and monkeys showed that the site of hydrolysis is mainly the liver. The further metabolism of methyl salicylate should then closely follow that shown for salicylic acid in Fig. 5.1 of this chapter. Williams (1959, p. 362) stated that glucuronide conjugates accounted for 12–55% of the dose (300 mg/kg) of the ester in rabbits. The ether glucuronide (21) was isolated from this material (Robinson and Williams, 1956). In addition, an interesting exception to that found with salicylic acid was reported in that as much as 10% of the dose was excreted as the ethereal sulphate. As this reaction is not observed with the acid, it appears that sulphate conjugation must precede hydrolysis.

Chlorogenic acid

(72)

Chlorogenic acid (3-caffeoylquinic acid) (72) is a commonly occurring plant ester which was first studied metabolically by Booth *et al.* (1957). The fate of the caffeic acid moiety was the point of interest in this study and the results showed a similar pattern of urinary metabolites derived from caffeic acid (see Fig. 5.3) following the administration of either caffeic acid or chlorogenic acid. Thus, rats given orally 100 mg of the latter compound excreted 3-hydroxyphenylpropionic acid (45) as the major urinary metabolite. Following ingestion of 1 g of chlorogenic acid, humans excreted mainly caffeic acid (36), dihydroferulic acid (41), the glucuronide of *m*-coumaric acid (44) and *m*-hydroxyhippuric acid (47). These results indicate that a larger number of urinary metabolites are detected following caffeic acid administration, however this might be expected in view of the slower appearance of caffeic acid in the organism when the ester is given. End products of metabolism would be more likely to predominate in the latter situation. It seems reasonable to assume that the most influential site of metabolism of chlorogenic acid is the intestine rather than the tissues as Scheline (1973) noted that its incubation with mixed cultures of rat caecal bacteria resulted in the formation of the ultimate metabolite, 3-hydroxy-phenylpropionic acid (45). This suggestion is supported by the findings of Czok *et al.* (1974) who detected chlorogenic acid in the serum and bile of rats following intravenous dosage but not when it was administered into the stomach or intestine. They also found that the amount of ester in the stomach decreased with time and that this occurred concomitantly with the appearance of the hydrolysis products, caffeic acid and quinic acid. However, tissue esterases are capable of hydrolysing chlorogenic acid following its intravenous administration as Michaud *et al.* (1971) detected both of the monomethyl ethers of caffeic acid in the bile of rats given the ester by this route. The same results were obtained when the closely related diester, 1,3-dicaffeoylquinic acid, was given.

Tannins are a complex group of plant polyphenols which are usually classified either as hydrolysable tannins or as condensed tannins. The former type is characterized by a central glucose unit attached by ester linkages to several gallic acid groups or groups built up of gallic acid. This product is known under several names including tannic acid, gallotannin and gallotannic acid and, upon hydrolysis, yields eight or nine gallic acid molecules for each glucose molecule. This indicates that tannic acid contains mainly residues of digallic acid (73) esterified with glucose. Condensed tannins are quite different compounds which are built up with much more stabile linkages. They probably arise by the oxidative poly-merization of flavonoid compounds of the catechin (3-hydroxyflavan) and 3,4-dihydroxyflavan types. Nearly all of the metabolic studies on tannins have been carried out with the former type, tannic acid, and Williams (1959, p. 306) summarized the older literature which indicated that it was excreted as gallic acid (3,4,5-trihydroxybenzoic acid) by rabbits, dogs and man. No unchanged tannic acid was detected in the urine but pyrogallol, the decarboxylation product of gallic acid was sometimes found. These results have been confirmed and extended in more recent investigations. Booth *et al.* (1959b) administered **tannic acid** (100 mg, p.o.) to rats and found that the urinary metabolites were the same as those excreted when gallic acid was given. As summarized above in Section I.C, these are mainly unchanged compound and its 4-*O*-methyl ether. Similarly, Watanabe and Oshima (1965) administered gallotannin from tea to rabbits and detected only gallic acid metabolites in the urine. These included pyrogallol in addition to the two metabolites mentioned above. Blumenberg *et al.* (1960) administered repeated daily doses (60 mg/kg, p.o.) of tannic acid, gallic acid and **digallic acid** (73) to rats and found that identical patterns of urinary metabolites were obtained chromatographically in all cases. Small amounts of gallic acid and increased amounts of glucuronides of it and a metabolite were detected. Tannic acid was not itself found in the urine.

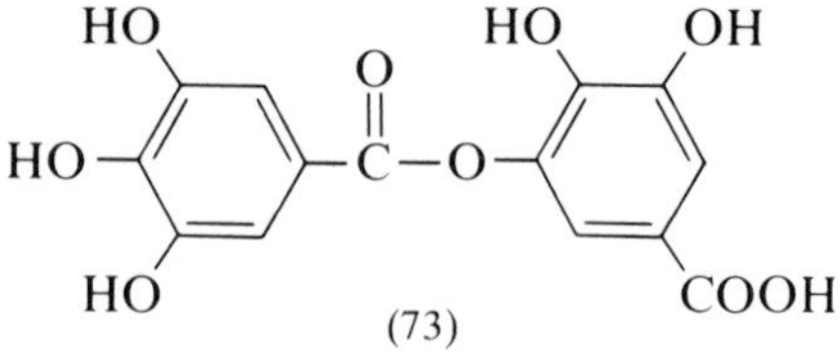

(73)

Digallic acid

Milić and Stojanović (1972) studied the metabolic fate of lucerne tannins in mice. This product is a complex mixture consisting of about 85% tannic substances. These consist of gallotannins, condensed tannins and free gallic acid in a ratio of about 16:3:1. Needless to say, the administration of this mixture of both hydrolysable and condensed tannins, which was fed in the diet at 0·5 or 4% levels, makes the results more difficult to interpret. This difficulty is compounded by the fact that only the faeces were analysed for metabolites. Information on the excretion of metabolites in the urine is likely to have given a clearer indication of the fate of the condensed tannins. The identified compounds in this fraction were (−)-gallocatechin gallate, (−)-epigallocatechin gallate, (+)-gallocatechin and (−)-epicatechin and their presence unchanged in the faeces was interpreted by Milić and Stojanović to indicate that they passed unaltered through the gastrointestinal tract. However, it seems more reasonable to assume that these faecal metabolites represent the unmetabolized portion of the material fed. As described in Chapter 7, Section IV.E, catechins are extensively degraded in the intestine by the microflora and give rise to C_6—C_3 acids containing catechol and m-hydroxyphenyl groups. Interestingly, several such compounds were detected in the faeces of mice fed the lucerne tannins. Another point of interest is the finding that O-methylated derivatives of protocatechuic and gallic acids were among the most prominent faecal metabolites. This is probably explained by the biliary excretion of various polyhydroxylated intermediates following their O-methylation in the tissues. In conclusion, it seems reasonable to assume that the faecal metabolites reported for lucerne tannins arise from a series of diverse but interrelated steps encompassing incomplete absorption, metabolism by intestinal microorganisms, metabolism by tissue enzymes and biliary excretion.

References

Acheson, R. M. and Atkins, G. L. (1961). *Biochem. J.* **79**, 268–270.
Acheson, R. M. and Gibbard, S. (1962). *Biochim. biophys. Acta* **59**, 320–325.

Adamson, R. H., Bridges, J. W., Evans, M. E. and Williams, R. T. (1970). *Biochem. J.* **116**, 437–443.
Alpen, E. L., Mandel, H. G. and Smith, P. K. (1951a). *J. Pharmac. exp. Ther.* **101**, 1.
Alpen, E. L., Mandel, H. G., Rodwell, V. W. and Smith, P. K. (1951b). *J. Pharmac. exp. Ther.* **102**, 150–155.
Alton, H. and Goodall McC. (1969). *Biochem. Pharmac.* **18**, 1373–1379.
Asano, M. and Yamakawa, T. (1950). *J. Biochem.* Tokyo **37**, 321–327.
Asatoor, A. M. (1965). *Biochim. biophys. Acta* **100**, 290–292.
Astill, B. D., Fassett, D. W. and Roudabush, R. L. (1964). *Biochem. J.* **90**, 194–201.
Axelrod, J. (1956). *Biochem. J.* **63**, 634–639.
Axelrod, J. and Tomchick, R. (1958). *J. biol. Chem.* **233**, 702–705.
Baas, H. K. L. (1890). *Hoppe-Seyler's Z. physiol. Chem.* **14**, 416–436.
Bababunmi, E. A., Smith, R. L. and Williams, R. T. (1973). *Life Sci.* **12 II**, 317–326.
Babior, B. M. and Bloch, K. (1966). *J. biol. Chem.* **241**, 3643–3651.
Balba, M. T. and Evans, W. C. (1977). *Biochem. Soc. Trans.* **5**, 300–302.
Baldoni, A. (1909). *Archo Farmac. sper.* **8**, 174–201.
Batterman, R. C. and Sommer, E. M. (1953). *Proc. Soc. exp. Biol. Med.* **82**, 376–379.
Beer, C. T., Dickens, F. and Pearson, J. (1951). *Biochem. J.* **48**, 222–237.
Bernhard, K. (1937). *Hoppe-Seyler's Z. physiol. Chem.* **248**, 256–276.
Bernhard, K. and Andreae, M. (1937). *Hoppe-Seyler's Z. physiol. Chem.* **245**, 103–106.
Bernhard, K. and Müller, L. (1938). *Hoppe-Seyler's Z. physiol. Chem.* **256**, 85–89.
Bernhard, K., Vuilleumier, J. P. and Brubacher, G. (1955). *Helv. chim. Acta* **38**, 1438–1444.
Bhatia, I. S., Bajaj, K. L. and Chakravarti, P. (1977). *Indian J. exp. Biol.* **15**, 118–120.
Blumenberg, F.-W. and Dohrmann, R. (1960). *Arzneimittel-Forsch.* **10**, 109–111.
Blumenberg, F.-W., Enneker, C. and Kessler, J.-J. (1960). *Arzneimittel-Forsch.* **10**, 223–226.
Booth, A. N. and Williams, R. T. (1963a). *Nature* Lond. **198**, 684–685.
Booth, A. N. and Williams, R. T. (1963b). *Biochem. J.* **88**, 66P–67P.
Booth, A. N., Murray, C. W., DeEds, F. and Jones, F. T. (1955). *Fed. Proc. Fedn. Am. Socs. exp. Biol.* **14**, 321.
Booth, A. N., Emerson, O. H., Jones, F. T. and DeEds, F. (1957). *J. biol. Chem.* **229**, 51–59.
Booth, A. N., Jones, F. T. and DeEds, F. (1958). *J. biol. Chem.* **223**, 280–282.
Booth, A. N., Masri, M. S., Robbins, D. J., Emerson, O. H., Jones, F. T. and DeEds, F. (1959a). *J. biol. Chem.* **234**, 946–948.
Booth, A. N., Masri, M. S., Robbins, D. J., Emerson, O. H., Jones, F. T. and DeEds, F. (1959b). *J. biol. Chem.* **234**, 3014–3016.
Booth, A. N., Masri, M. S., Robbins, D. J., Emerson, O. H., Jones, F. T. and DeEds, F. (1960a). *J. biol. Chem.* **235**, 2649–2652.
Booth, A. N., Robbins, D. J., Masri, M. S. and DeEds, F. (1960b). *Nature* Lond. **187**, 691.
Boyland, E. and Chasseaud, L. F. (1967). *Biochem. J.* **104**, 95–102.
Boyland, E. and Chasseaud, L. F. (1968). *Biochem. J.* **109**, 651–661.

Boyland, E. and Chasseaud, L. F. (1970). *Biochem. Pharmac.* **19**, 1526–1528.
Bray, H. G., Neale, F. C. and Thorpe, W. V. (1946). *Biochem. J.* **40**, 134–139.
Bray, H. G., Ryman, B. E. and Thorpe, W. V. (1947). *Biochem. J.* **41**, 212–218.
Bray, H. G., Ryman, B. E. and Thorpe, W. V. (1948). *Biochem. J.* **43**, 561–567.
Bray, H. G., Thorpe, W. V. and White, K. (1950). *Biochem. J.* **46**, 271–275.
Bray, H. G., Thorpe, W. V. and White, K. (1951). *Biochem. J.* **48**, 88–96.
Bray, H. G., Humphris, B. G., Thorpe, W. V., White, K. and Wood, P. B. (1955). *Biochem. J.* **59**, 162–167.
Brewster, D., Jones, R. S. and Parke, D. V. (1976). *Biochem. Soc. Trans.* **4**, 518–521.
Brewster, D., Jones, R. S. and Parke, D. V. (1977a). *Xenobiotica* **7**, 109.
Brewster, D., Jones, R. S. and Parke, D. V. (1977b). *Biochem. J.* **164**, 595–600.
Bridges, J. W., French, M. R., Smith, R. L. and Williams, R. T. (1970). *Biochem. J.* **118**, 47–51.
Bridges, J. W., Evans, M. E., Idle, J. R., Millburn, P., Osiyemi, F. O., Smith, R. L. and Williams, R. T. (1974). *Xenobiotica* **4**, 645–652.
Caldwell, J., French, M. R., Idle, J. R., Renwick, A. G., Bassir, O. and Williams, R. T. (1975). *FEBS Lett.* **60**, 391–395.
Casida, J. E. (1973). *In* "Pyrethrum. The Natural Insecticide" (J. E. Casida, Ed.), pp. 101–120. Academic Press, New York and London.
Clapp, J. J. and Young, L. (1970). *Biochem. J.* **118**, 765–771.
Clarke, N. E. and Mosher, R. E. (1953). *Circulation* **7**, 337–344.
Consden, R. and Stanier, W. M. (1951). *Biochem. J.* **48**, xiv.
Cotran, R., Kendrick, M. I. and Kass, E. H. (1960). *Proc. Soc. exp. Biol. Med.* **104**, 424–426.
Cramer, M. B. and Michael, W. R. (1971). *Life Sci.* **10 II**, 1255–1259.
Creveling, C. R., Dalgard, N., Shimizu, H. and Daly, J. W. (1970). *Molec. Pharmac.* **6**, 691–696.
Creveling, C. R., Morris, N., Shimizu, H., Ong, H. H. and Daly, J. (1972). *Molec. Pharmac.* **8**, 398–409.
Csonka, F. A. (1924). *J. biol. Chem.* **60**, 545–582.
Curtin, C. O'H. and King, C. G. (1955). *J. biol. Chem.* **216**, 539–548.
Curtius, H. C., Mettler, M. and Ettlinger, L. (1976). *J. Chromat.* **126**, 569–580.
Czok, G., Walter, W., Knoche, K. and Degener, H. (1974). *Z. ErnährWiss.* **13**, 108–112.
Dacre, J. C. and Williams, R. T. (1962). *Biochem. J.* **84**, 81P.
Dacre, J. C. and Williams, R. T. (1968). *J. Pharm. Pharmac.* **20**, 610–618.
Dacre, J. C., Scheline, R. R. and Williams, R. T. (1968). *J. Pharm. Pharmac.* **20**, 619–625.
Dakin, H. D. (1909). *J. biol. Chem.* **6**, 203–219.
Davis, L. E. and Westfall, B. A. (1972). *Am. J. vet. Res.* **33**, 1253–1262.
Davis, L. E., Westfall, B. A. and Short, C. R. (1973). *Am. J. vet. Res.* **34**, 1105–1108.
Davison, C. (1971). *Ann. N.Y. Acad. Sci.* **179**, 249–268.
Davison, C., Zimmerman, E. F. and Smith, P. K. (1961). *J. Pharmac. exp. Ther.* **132**, 207–211.
Dayman, J. and Jepson, J. B. (1969). *Biochem. J.* **113**, 11P.
DeEds, F., Booth, A. N. and Jones, F. T. (1955). *Fed. Proc. Fedn. Am. Socs. exp. Biol.* **14**, 332.
DeEds, F., Booth, A. N. and Jones, F. T. (1957). *J. biol. Chem.* **225**, 615–621.

Derache, R. and Gourdon, J. (1963). *Fd Cosmet. Toxicol.* **1**, 189–195.
Deuel, H. J., Calbert, C. E., Anisfeld, L., McKeehan, H. and Blunden, H. D. (1954). *Fd Res.* **19**, 13–19.
Dirscherl, W. and Brisse, B. (1966). *Hoppe-Seyler's Z. physiol. Chem.* **346**, 55–59.
Dodgson, K. S. and Williams, R. T. (1949). *Biochem. J.* **45**, 381–386.
Dumazert, C. and El Ouachi, M. (1954). *Annls Pharm. Fr.* **12**, 723–730.
Elder, T. D. and Wyngaarden, J. B. (1960). *J. clin. Invest.* **39**, 1337–1344.
Elliott, N., Janes, N. F., Kimmel, E. C. and Casida, J. E. (1972). *J. agric. Fd Chem.* **20**, 300–313.
El Masry, A. M., Smith, J. N. and Williams, R. T. (1956). *Biochem. J.* **64**, 50–56.
Erspamer, V. (1955). *J. Physiol. Lond.* **127**, 118–133.
Ewins, A. J. and Laidlaw, P. P. (1910). *J. Physiol. Lond.* **41**, 78–87.
Fahelbum, I. M. S. and James, S. P. (1977). *Toxicology* **7**, 123–132.
Fingerhut, M., Schmidt, B. and Lang, K. (1962). *Biochem. Z.* **336**, 118–125.
Finkle, P. (1933). *J. biol. Chem.* **100**, 349–355.
Flatow, L. (1910). *Hoppe-Seyler's Z. physiol. Chem.* **64**, 367–392.
French, M. R., Bababunmi, E. A., Golding, R. R., Bassir, O., Caldwell, J., Smith, R. L. and Williams, R. T. (1974). *FEBS Lett.* **46**, 134–137.
Furuya, T. (1958). *Chem. pharm. Bull.* Tokyo **6**, 706–710.
Gallaher, E. J. and Loomis, T. A. (1975). *Toxic. appl. Pharmac.* **34**, 309–313.
Gosselin, R. E., Gabourel, J. D., Kalser, S. C. and Wills, J. H. (1955). *J. Pharmac. exp. Ther.* **115**, 217–229.
Griffiths, L. A. (1969). *Biochem. J.* **113**, 603–609.
Griffiths, L. A. (1970). *Experientia* **26**, 723–724.
Griffiths, L. A. and Smith, G. E. (1972). *Biochem. J.* **128**, 901–911.
Haberland, G. L., Medenwald, H. and Köster, L. (1957). *Hoppe-Seyler's Z. physiol. Chem.* **306**, 235–246.
Hanzlik, P. J. and Wetzel, N. C. (1920). *J. Pharmac. exp. Ther.* **14**, 43–46.
Hartles, R. L. and Williams, R. T. (1948). *Biochem. J.* **43**, 296–303.
Heffter, A. (1895). *Arch. exp. Path. Pharmak.* **35**, 342–374.
Hill, G. A., Ratcliffe, J. and Smith, P. (1959). *Chemy Ind.*, 399.
Hodgkinson, A. and Zarembski, P. M. (1968). *Calcif. Tissue Res.* **2**, 115–132.
Hollister, L. and Levy, G. (1965). *J. pharm. Sci.* **54**, 1126–1129.
Idle, J. R., Millburn, P. and Williams, R. T. (1976). *Biochem. Soc. Trans.* **4**, 139–141.
Indahl, S. R. and Scheline, R. R. (1968). *Appl. Microbiol.* **16**, 667.
Indahl, S. R. and Scheline, R. R. (1973). *Xenobiotica* **3**, 549–556.
James, M. O., Smith, R. L., Williams, R. T. and Reidenberg, M. (1972). *Proc. R. Soc. B* **182**, 25–35.
Kamienski, F. X. and Casida, J. E. (1970). *Biochem. Pharmac.* **19**, 91–112.
Kapp, E. M. and Coburn, A. F. (1942). *J. biol. Chem.* **145**, 549–565.
Kohlrausch, A. (1912). *Z. Biol.* **57**, 273–308.
Kuhn, R., Köhler, F. and Köhler, L. (1936). *Hoppe-Seyler's Z. physiol. Chem.* **242**, 171–197.
Kuhn, R., Köhler, F. and Köhler, L. (1937). *Hoppe-Seyler's Z. physiol. Chem.* **247**, 197–220.
Lautemann, E. (1863). *Justus Liebigs Annln Chem.* **125**, 9–13.
Levy, G. and Leonards, J. R. (1966). In "The Salicylates" (M. J. H. Smith and P. K. Smith, Eds), pp. 5–48. Interscience Publishers, New York, London and Sydney.
Martin, A. K. (1966). *J. Sci. Fd Agric.* **17**, 496–500.

Martin, A. K. (1975). *Proc. Nutr. Soc.* **34**, 69A–70A.
Marzullo, G. and Friedhoff, A. J. (1975). *Life Sci.* **17**, 933–942.
Masri, M. S., Booth, A. N. and DeEds, F. (1959). *Archs Biochem. Biophys.* **85**, 284–286.
Masri, M. S., Booth, A. N. and DeEds, F. (1962). *Biochim. biophys. Acta* **65**, 495–505.
Masri, M. S., Robbins, D. J., Emerson, O. H and DeEds, F. (1964). *Nature Lond.* **202**, 878–879.
Mead, J. A. R., Smith, J. N. and Williams, R. T. (1958). *Biochem. J.* **68**, 67–74.
Meyer, T. and Scheline, R. R. (1972a). *Xenobiotica* **2**, 383–390.
Meyer, T. and Scheline, R. R. (1972b). *Xenobiotica* **2**, 391–398.
Michaud, J., Lesca, M. F. and Roudge, A. M. (1971). *Bull. Soc. pharm. Bordeaux* **110**, 65–72.
Milić, B. L. and Stojanović, S. (1972). *J. Sci. Fd Agric.* **23**, 1163–1167.
Mitoma, C., Posner, H. S., Reitz, H. C. and Udenfriend, S. (1956). *Archs Biochem. Biophys.* **61**, 431–441.
Numerof, P., Gordon, M. and Kelly, J. M. (1955). *J. Pharmac. exp. Ther.* **115**, 427–431.
Ohtsuji, H. and Ikeda, M. (1971). *Toxic. appl. Pharmac.* **18**, 321–328.
Pellerin, J. and D'Iorio, A. (1958). *Can. J. Biochem. Physiol.* **36**, 491–497.
Peppercorn, M. A. and Goldman, P. (1971). *J. Bact.* **108**, 996–1000.
Peppercorn, M. A. and Goldman, P. (1972). *Proc. natn Acad. Sci.* U.S.A. **69**, 1413–1415.
Perez-Silva, M. G. and Rodriguez Sánchez, D. (1967). *Boln R. Soc. esp. Hist. nat.* **65**, 401–405.
Perez-Silva, G., Rodriguez, D. and Perez-Silva, J. (1966). *Nature* Lond **212**, 303–304.
Phillips, J. C., Hardy, K., Richards, R., Cottrell, R. C. and Gangoli, S. D. (1977). *Toxicology* **7**, 257–263.
Quick, A. J. (1928). *J. biol. Chem.* **77**, 581–593.
Quick, A. J. (1931). *J. biol. Chem.* **92**, 65–85.
Quick, A. J. (1932). *J. biol. Chem.* **97**, 403–419.
Quilley, E. and Smith, M. J. H. (1952). *J. Pharm. Pharmac.* **4**, 624–630.
Ranganathan, S. and Ramasarma, T. (1971). *Biochem. J.* **122**, 487–493.
Ranganathan, S. and Ramasarma, T. (1974). *Biochem. J.* **140**, 517–522.
Raper, H. S. and Wayne, E. J. (1928). *Biochem. J.* **22**, 188–197.
Rasmussen, A., Scheline, R., Solheim, E. and Hänsel, R. (1976). Unpublished observations.
Robinson, D. and Williams, R. T. (1956). *Biochem. J.* **62**, 23P.
Roseman, S. and Dorfman, A. (1951). *J. biol. Chem.* **192**, 105–114.
Sakamoto, Y., Inamori, K. and Nasu, H. (1959). *J. Biochem.* Tokyo **46**, 1667–1669.
Sammons, H. G. and Williams, R. T. (1941). *Biochem. J.* **35**, 1175–1188.
Sammons, H. G. and Williams, R. T. (1946). *Biochem. J.* **40**, 223–227.
Sato, T., Suzuki, T., Fukuyama, T. and Yoshikawa, H. (1956a). *J. Biochem.* Tokyo **43**, 413–420.
Sato, T., Suzuki, T., Fukuyama, T. and Yoshikawa, H. (1956b). *J. Biochem.* Tokyo **43**, 421–429.
Schayer, R. W. (1950). *Archs Biochem.* **28**, 371–376.
Scheline, R. R. (1966a). *Acta pharmac. tox.* **24**, 275–285.

Scheline, R. R. (1966b). *J. Pharm. Pharmac.* **18**, 664–669.
Scheline, R. R. (1967). *Experientia* **23**, 493–494.
Scheline, R. R. (1968a). *Acta pharmac. tox.* **26**, 189–205.
Scheline, R. R. (1968b). *Acta pharmac. tox.* **26**, 332–342.
Scheline, R. R. (1973). *Pharmac. Rev.* **25**, 451–523.
Scheline, R. R. and Midtvedt, T. (1970). *Experientia* **26**, 1068–1069.
Scheline, R. R., Williams, R. T. and Wit, J. G. (1960). *Nature* Lond. **188**, 849–850.
Schubert, B. (1967). *Acta vet. scand.* Suppl. 21.
Shaw, K. N. F. and Trevarthen, J. (1958). *Nature* Lond. **182**, 797–798.
Shaw, K. N. F., Gutenstein, M. and Jepson, J. B. (1963). Proc. 5th int. Cong. Biochem., Moscow 1961 (N. M. Sissakian, Ed.), Vol. 9, p. 427. Pergamon Press, Oxford.
Smith, G. E. and Griffiths, L. A. (1974). *Xenobiotica* **4**, 477–487.
Snapper, J., Grünbaum, A. and Sturkop, S. (1925). *Biochem. Z.* **155**, 163–173.
Soleim, H. A. and Scheline, R. R. (1972). *Acta pharmac. tox.* **31**, 471–480.
Solheim, E. and Scheline, R. R. (1973). *Xenobiotica* **3**, 493–510.
Solheim, E. and Scheline, R. R. (1976). *Xenobiotica* **6**, 137–150.
Teuchy, H. and Van Sumere, C. F. (1971). *Archs int. Physiol. Biochim.* **79**, 589–618.
Teuchy, H., Quatacker, J., Wolf, G. and Van Sumere, C. F. (1971). *Archs int. Physiol. Biochim.* **79**, 573–587.
Thierfelder, H. and Sherwin, C. P. (1914). *Ber. dt. chem. Ges.* **47**, 2630–2634.
Thierfelder, H. and Sherwin, C. P. (1915). *Hoppe-Seyler's Z. physiol. Chem.* **94**, 1–9.
Thomas, H. and Müller-Enoch, D. (1974). *Naturwissenschaften* **61**, 222.
Tompsett, S. L. (1958). *J. Pharm. Pharmac.* **10**, 157–161.
Tompsett, S. L. (1961). *J. Pharm. Pharmac.* **13**, 115–120.
Tsukamoto, H. and Terada, S. (1962). *Chem. pharm. Bull.* Tokyo **10**, 91–95.
Tsukamoto, H. and Terada, S. (1964). *Chem. pharm. Bull.* Tokyo **12**, 765–769.
Ueda, K., Gaughan, L. C. and Casida, J. E. (1975). *J. agric. Fd Chem.* **23**, 106–115.
Underhill, F. P., Leonard, C. S., Gross, E. G. and Jaleski, T. C. (1931a). *J. Pharmac. exp. Ther.* **43**, 359–380.
Underhill, F. P., Peterman, F. I., Jaleski, T. C. and Leonard, C. S. (1931b). *J. Pharmac. exp. Ther.* **43**, 381–398.
Van Sumere, C. F., Teuchy, H., Verbeke, H. Pé. R. and Bekaert, J. (1969). *Clinica chim. Acta* **26**, 85–88.
Vasiliu, H., Timoşencu, A., Zaimov, C. and Coteleu, V. (1938). *Bul. Fac. Sti Agric. Chişinău (Comun. Lab. Chim. Agr.)* **2**, 56–62. (*Chem. Abstr.* (1938) **32**, 8514.)
Vasiliu, H., Timoşencu, A., Zaimov, C. and Coteleu, V. (1940). *Bul. Fac. Sti. Agric. Chişinău (Comun. Lab. Chim. Agr.)* **3**, 77–84. (*Chem. Abstr.* (1940) **34**, 4424.)
Ve, B. and Scheline, R. R. (1977). Unpublished observations.
Von Lehmann, B., Wan, S. H., Riegelman, S. and Becker, C. (1973). *J. pharm. Sci.* **62**, 1483–1486.
Wan, S. H. and Riegelman, S. (1972). *J. pharm. Sci.* **61**, 1284–1287.
Watanabe, A. and Oshima, Y. (1965). *Agric. biol. Chem.* Tokyo **29**, 90–93.
Weinhouse, S. and Friedmann, B. (1951). *J. biol. Chem.* **191**, 707–717.
Weitzel, G. (1947). *Hoppe-Seyler's Z. physiol. Chem.* **282**, 185–191.
Williams, R. T. (1938). *Biochem. J.* **32**, 878–887.
Williams, R. T. (1959). "Detoxication Mechanisms". Chapman and Hall, London.

Wiseman-Distler, M. H., Sourkes, T. L. and Carabin, S. (1965). *Clinica chim. Acta* **12**, 335–339.

Wong, K. P. and Sourkes, T. L. (1966). *Can J. Biochem.* **44**, 635–644.

Woo, W. S. (1968). *J. pharm. Sci.* **57**, 27–30.

Yamamoto, I., Elliott, M. and Casida, J. E. (1971). *Bull. Wld. Hlth Org.* **44**, 347–348.

6

METABOLISM OF HIGHER TERPENOIDS

Terpenoid compounds include a multitude of diverse plant constituents which are related by virtue of a common biosynthetic origin. Thus, their basic skeletons are derived from mevalonic acid and consist of C_5-units, i.e. the isoprene molecule $(CH_2=C(Me)-CH=CH_2)$. They are further classified according to the number of such units present, the simplest C_{10} derivatives containing two. These C_{10} compounds are known as monoterpenoids and in this book the choice was made to include them in other chapters rather than with their higher relatives. This decision was dictated by their close metabolic relationship to other classes of plant compounds, especially the alcohols, aldehydes and ketones, and, in contrast, the lack of similarity to the metabolism seen with many of the terpenoids included in this chapter. These latter compounds include the sesquiterpenoids (C_{15}), diterpenoids (C_{20}), triterpenoids (C_{30}) and tetraterpenoids (C_{40}).

I. Sesquiterpenoids

The sesquiterpenoids are C_{15} compounds which include both acyclic and cyclic derivatives. The most important member of the former group is **farnesol** (1), an alcohol found in many essential oils. Its metabolism in mammals is unclear, however Fischer and Bielig (1940) isolated an acid which they believed to be farnesenic acid, the acid analogue of farnesol, from the urine of a rabbit given the alcohol (6·7 g during a four-day period, i.p.).

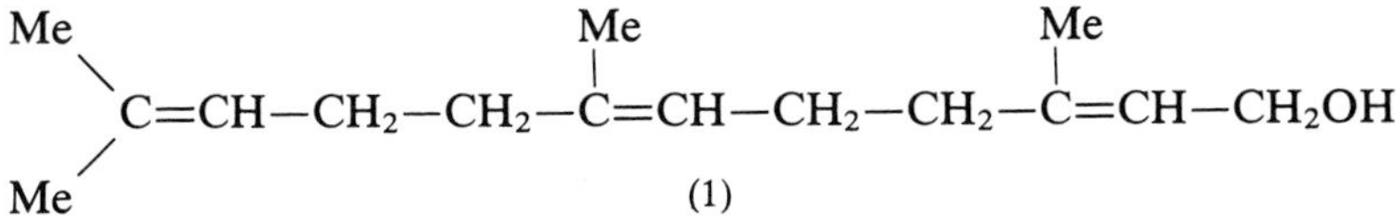

$$\text{Me}\diagdown\atop\text{Me}\diagup C=CH-CH_2-CH_2-\underset{\underset{\displaystyle Me}{|}}{C}=CH-CH_2-CH_2-\underset{\underset{\displaystyle Me}{|}}{C}=CH-CH_2OH$$

(1)

Farnesol

The hydroxylation of **cedrol** (2) in rabbits was studied by Bang and Ourisson (1975). Following an oral dose of about 330 mg/kg, the urine was collected for four days and hydrolysed enzymically with a β-glucuronidase

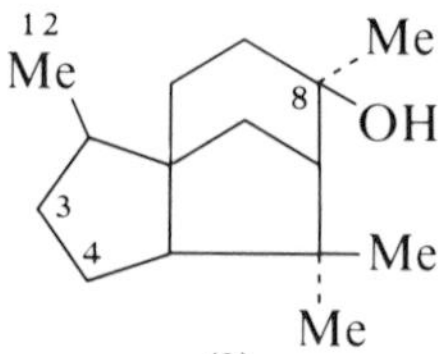

(2)   (3)

Cedrol

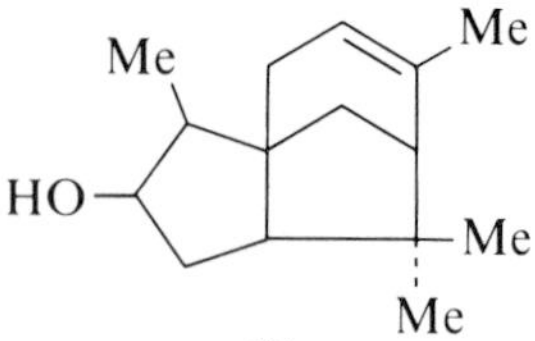

preparation. This resulted in the recovery of cedrol (5% of the dose), a mixture of alcohols (12%) and a mixture of diols (35%). Both of the alcohols had structure (3) and differed only in the orientation of the hydroxyl group at C3. The diols were shown to be 3-hydroxy-cedrol, the mixture likewise consisting of two stereoisomers differing in configuration at C3. Interestingly, 3-hydroxycedrol is also excreted in the urine when dogs are given cedrol, however only 3α-alcohol (---OH at C3) is present (Trifilieff *et al.*, 1975). The latter investigation employed an oral dose (2 g) of cedrol and also treated the urine samples with a β-glucuroni-dase preparation prior to metabolite analysis. Metabolites were present only in the initial 24 h urine sample and three oxidized metabolites in addition to 3α-hydroxycedrol were characterized. These were shown to be 4-hydroxycedrol, 12-hydroxycedrol and an acidic diol formed from cedrol by hydroxylation at C4 and replacement of the methyl group at C8 with a carboxyl group.

Bang *et al.* (1975) administered **patchouli alcohol** (patchouliol) (4a) to rabbits and dogs and found that it was oxidized to the primary alcohol metabolite (4b). The urine of rabbits fed patchouli alcohol (330 mg/kg) contained, after hydrolysis with β-glucuronidase, both the diol (4b) and the corresponding carboxylic acid (4c).

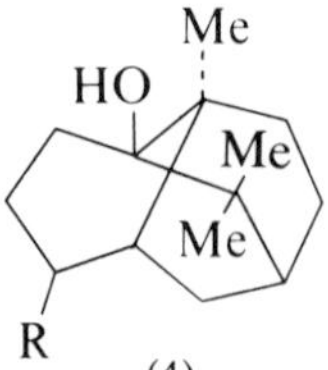

(4)

(a) Patchouli alcohol, R = Me
(b) R = CH₂OH
(c) R = COOH

Hildebrandt (1902) administered orally to rabbits the tricyclic alcohol **α-santalol** (5) and found evidence for a C₁₀-hydroxyacid in the urine. This metabolite corresponds to a compound having the groups at C3 replaced by —COOH and —CH₂OH. This suggested transformation is hardly convincing and it seems reasonable to suppose that the corresponding acid analogue of α-santalol would be readily formed.

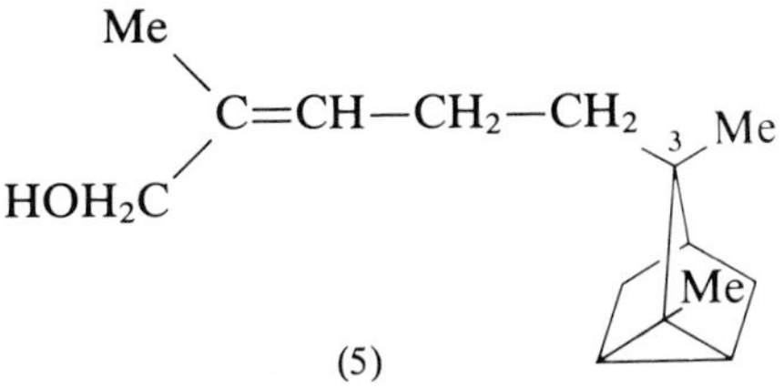

(5)

α-Santalol

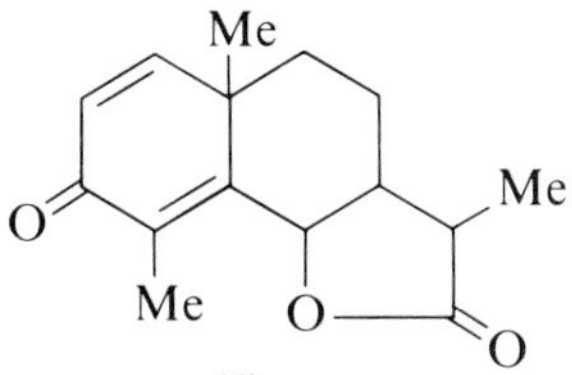

(6)

Santonin

The final example in this small group of sesquiterpenoids is the lactone **santonin** (6). Morishima (1961) found that it is excreted partly unchanged in the urine and faeces after oral or subcutaneous administration. In rabbits, a combined value of 3·5% of the dose was reported following oral dosage, whereas, by injection, 22% was found in the urine and much less in the faeces. Oral administration to man resulted in 19% of the dose being excreted in the 24 h urine. Kaya (1960) carried out similar experiments in dogs and found, following either route of administration, some unchanged santonin in the bile, urine and faeces. About 70% of the dose was excreted in the urine and very little in the faeces after injection whereas values of 4–6% and 7–9%, respectively, were found after oral dosage. A metabolite, α-hydroxysantonin, was detected in the serum, bile, intestinal juice, urine and faeces. Santonin contains an α,β-unsaturated group in common with many compounds of this type which are known to be conjugated with glutathione. However, Boyland and Chasseaud (1967) found that this glutathione S-transferase activity from rat liver did not utilize santonin as a substrate.

II. Diterpenoids

The diterpenoids comprise a chemically diverse group of C_{20}-compounds which, as with the lower terpenes, include hydrocarbons, alcohols, ethers and acids and their derivatives. However, the mammalian metabolism of these compounds has scarcely been studied and only a few reports are available.

Phytane (2,6,10,14-tetramethylhexadecane), while not a plant compound, is a saturated isoprenoid hydrocarbon showing close structural similarity to some unsaturated derivatives from plants. Its metabolism in rats was studied by Albro and Thomas (1974) who found that it was metabolized to a tertiary alcohol (probably the 2-ol) and a variety of

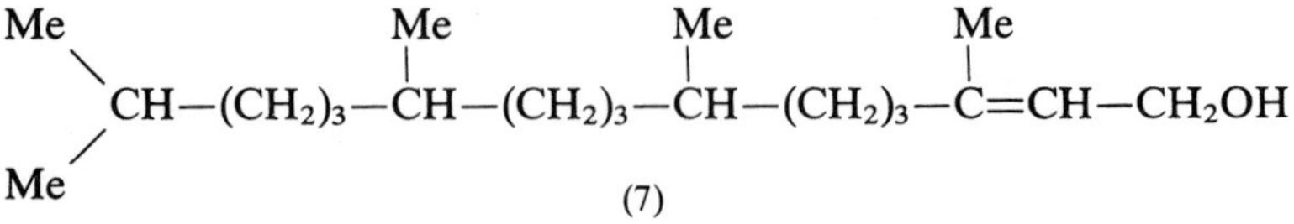

$$\underset{Me}{\overset{Me}{\diagdown}} CH-(CH_2)_3-\overset{\overset{Me}{|}}{CH}-(CH_2)_3-\overset{\overset{Me}{|}}{CH}-(CH_2)_3-\overset{\overset{Me}{|}}{C}=CH-CH_2OH$$

(7)

Phytol

short-chain carboxylic acids. The only important acyclic example is **phytol** (7) which is undoubtedly the most widely occurring of all diterpenes, being found as an esterified component of chlorophyll. Steinberg *et al.* (1965a) showed that phytol fed to rats is absorbed and partly converted to phytanic acid (8). Also, feeding the animals a diet containing 5% phytol for three

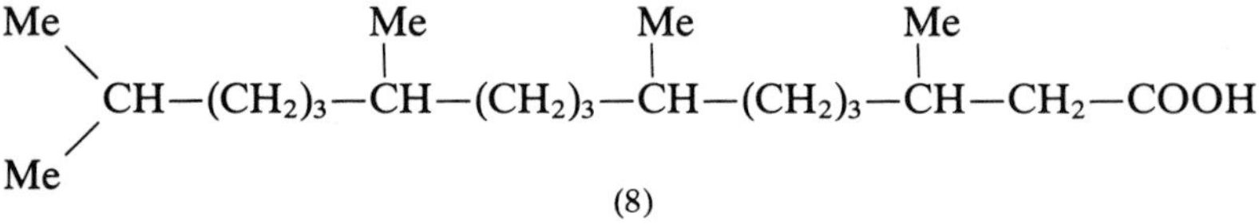

$$\underset{Me}{\overset{Me}{\diagdown}} CH-(CH_2)_3-\overset{\overset{Me}{|}}{CH}-(CH_2)_3-\overset{\overset{Me}{|}}{CH}-(CH_2)_3-\overset{\overset{Me}{|}}{CH}-CH_2-COOH$$

(8)

weeks resulted in an accumulation of the acid which accounted for roughly 20–30% of the fatty acids in the liver and plasma. A similar conversion in man was reported by Steinberg *et al.* (1965b) and, by using uniformly labelled $[^{14}C]$-phytol, complete oxidation to respiratory CO_2 was also shown. It appears that the normal pathway of metabolism of phytol is via phytanic acid to CO_2 but that the acid intermediate may accumulate when the dietary level of phytol is high. It is noteworthy that patients with Refsum's disease have an impaired ability to metabolize phytanic acid (Stoffel and Kahlke, 1965; Steinberg *et al.*, 1965b).

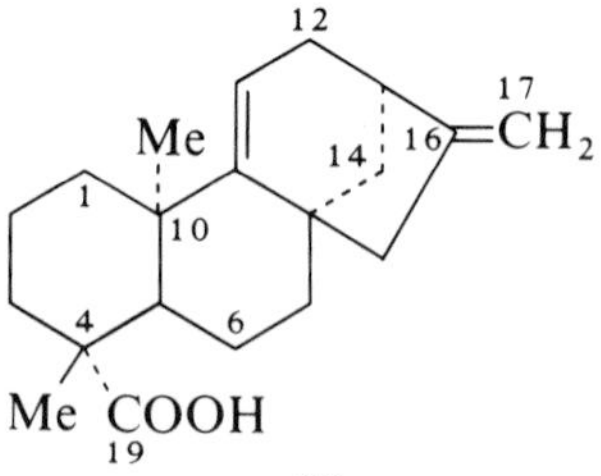

(9)

Grandiflorenic acid

Grandiflorenic acid (kaura-9(11),16-dien-19-oic acid) (9) is a tetracyclic diterpene. Neidlein and Stumpf (1977a) studied several aspects of its metabolism in rats, also using material labelled with ^{14}C in the 17-position. When male rats were given grandiflorenic acid (50 mg/kg, i.p.), only 5% of the dose was excreted in the urine whereas about 73% was lost in the faeces. The faecal material contained 11 metabolites, however unchanged compound accounted for over 40% of the faecal radioactivity. Several of the other metabolites were characterized. One was a hydroxylated derivative, probably formed by oxidation of a methyl group at either C4 or C10. This metabolite made up 6% of the faecal radioactivity. A further 9% consisted of the 12α-hydroxy derivative formed by the allylic hydroxylation of grandiflorenic acid. Further oxidation of this 12α-hydroxy metabolite occurred as 12-ketograndiflorenic acid was also identified. Another metabolite (6% of the faecal radioactivity) was hydroxylated at both the 6β and 14β positions and reduced at the exocyclic double bond at C16. The latter reaction was probably carried out by intestinal microorganisms as it was shown that grandiflorenic acid itself is converted to its 16,17-dihydro derivative when incubated with rat intestinal contents. Incubation of grandiflorenic acid with rat liver microsomes resulted in the formation of six metabolites, of which 12α-hydroxygrandiflorenic acid was the most prominent. Grandiflorenic acid was conjugated with glucuronic acid when incubated with rat liver preparations containing UDP-glucuronic acid. These findings suggest that grandiflorenic acid and its metabolites are extensively excreted in the bile of rats as glucuronide conjugates. This assumption was confirmed in a subsequent study of the absorption, metabolism and excretion of grandiflorenic acid in rats (Neidlein and Stumpf, 1977b). The biliary metabolites were subjected to an extensive enterohepatic circulation.

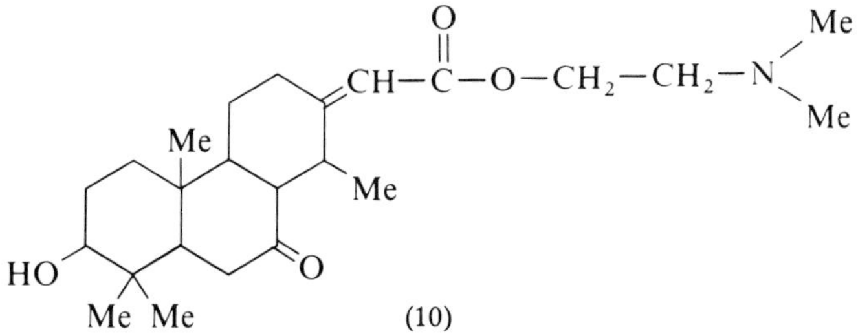

(10)

Cassaine

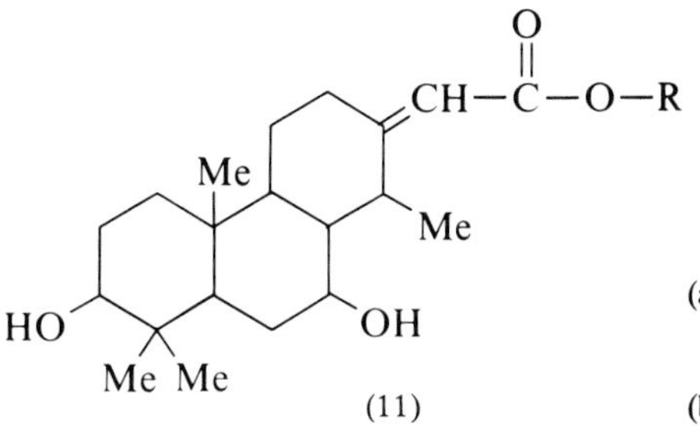

(a) Cassaidine, R = $-CH_2-CH_2-N \begin{smallmatrix} Me \\ Me \end{smallmatrix}$

(11)

(b) Cassaidic acid, R = H

Two cardiac diterpene alkaloids are **cassaine** (10) and **cassaidine** (11a). Both of these compounds contain an ester group and it is therefore not surprising that they are metabolized by hydrolysis. Zelck (1972) reported that cassaine was rapidly hydrolysed by liver microsomes. Cronlund (1976) found that cassaidic acid (11b) was a major urinary metabolite of cassaidine following intravenous administration of the latter to guinea pigs.

III. Triterpenoids

Triterpenoids include C_{30}-compounds or derivatives thereof and are often divided into the following classes: true triterpenes, steroids, saponins and cardiac glycosides. This classification is partly based on differences in use and properties rather than chemical differences as the last two classes are largely glycosides of triterpenes and steroids. The true triterpenes include the hydrocarbon squalene as the only important acyclic member. Among the cyclic triterpenoids lanosterol is a well-known tetracyclic derivative, however pentacyclic compounds are the most widely distributed triterpenoids. Very little information is available on the mammalian metabolism of most of these latter compounds and a discussion of the metabolism of some others including the aforementioned squalene and lanosterol which are important intermediates in the biosynthesis of cholesterol, falls outside the scope of this book. Plant sterols are similarly omitted, however discussion of the metabolism of the ubiquitous β-sitosterol was included in the review of dietary plant sterols by Subbiah (1973).

(12)

(a) Asiatic acid, R = R' = H
(b) Asiaticoside, R = H, R' = —glucose—glucose—rhamnose
(c) Madecassic acid, R = OH, R' = H

Several types of pentacyclic C_{30}-compounds are recognized according to the type of basic hydrocarbon skeleton present. One of these is the ursane type which is present in **asiatic acid** (12a), its glycoside **asiaticoside** (12b) and **madecassic acid** (12c). The metabolism of these three derivatives of ursolic acid was studied in rats by Chasseaud *et al.* (1971) and similar studies in dogs and humans were reported by Hathway (1973). The most striking finding in all species is the extensive faecal excretion of unchanged compound when asiatic acid or madecassic acid is administered orally. Nearly 90% of the dose (roughly 10–20 mg/kg) was recovered in the faeces of rats in four or six days when tritium-labelled asiatic acid or madecassic acid, respectively, was used. Average values of about 90% in four days were also obtained using dogs and, in man, about 85% of a single oral dose of the acids was recovered in the faeces after four days. However, the fact that the faecal material is due to the unchanged compounds is not attributable to lack of absorption but, in considerable degree, to biliary excretion of absorbed material. Absorption of either asiatic or madecassic acid in rats accounted for about half of the dose since the 36 h-bile contained 45% of the administered radioactivity, mainly as the glucuronides but partly as the sulphate conjugates of the acids. Steric considerations indicated that the 2α- or 3β-hydroxyl groups are the most likely sites of conjugation. The presence of unconjugated metabolites in the faeces indicates that the conjugate hydrolysis is carried out by the intestinal microflora. Likewise, the ester bond in asiaticoside (12b) is hydrolysed by the microflora and the metabolism of both the acid and its ester is therefore similar.

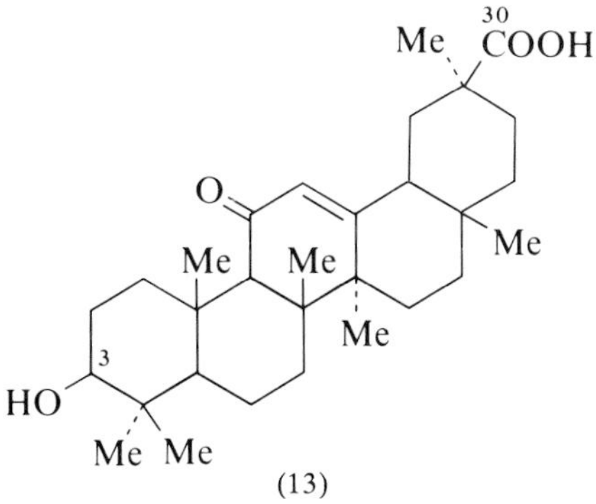

(13)

Glycyrrhetic acid

A triterpenoid closely related to those mentioned above is **glycyrrhetic acid** (glycyrrhetinic acid) (13), the aglycone of the diglucuronide **glycyrrhizic acid** (glycyrrhizin) which occurs in liquorice root. Carlat *et al.* (1959) administered 3–4 g oral doses of the aglycone or the ammonium salt of the glycoside, both labelled with ^{3}H, to man and found that about 98% of the radioactivity was excreted in the faeces within one day. Some glycyrrhetic acid was isolated from the small amount of material ($<1\%$ of the dose) excreted in the urine in both experiments. This indicates that the sugar residues are removed in the body. Helbing (1963) gave a single oral dose (4 g) or six daily doses of 2 g of glycyrrhetic acid to two patients and, using a chromatographic method of assay, recovered 53–61% of the dose in the faeces. This study showed prolonged excretion lasting over several days. The lower recoveries in these experiments are probably not due to bacterial degradation of glycyrrhetic acid in the gut since its concentration in a faecal suspension was not changed following incubation for four days. Findings of this sort suggest that glycyrrhetic acid may be subjected to an enterohepatic circulation, however Carlat *et al.* concluded that this does not occur in humans since $<0.5\%$ of the oral dose was found in the blood at 4 h or excreted in the bile within 4 h. A similar picture is not seen in rats and Parke *et al.* (1963) found that roughly 30–70% of an oral dose (25 mg/kg) and essentially all of that given intraperitoneally was excreted in the bile. Rapid and extensive excretion of β-glycyrrhetic acid in the bile was also found in mice given the compound intravenously (Miyake *et al.*, 1976). This led to the excretion of the material in the faeces. The results of Parke *et al.* using rats showed that biliary excretion of radioactive material is rapid following injection of the triterpenoid but that this occurred slowly during the first few hours following oral dosage. It is therefore possible that a longer collection period in humans would similarly indicate a high degree of biliary excretion of glycyrrhetic acid metabolites. Three metabolites were detected in rat bile by Parke *et al.* and none was identical with unchanged compound. Subsequently, Iveson *et al.* (1971) reported these to be 30-glucuronide, the 3-*O*-sulphate and, probably, the 3-*O*-glucuronide conjugates of glycyrrhetic acid. In intact rats, faecal excretion of radioactivity was very extensive whereas only about 1% of the dose (20–60 mg/kg, p.o.) was excreted in the urine. Iveson *et al.* reported that this faecal material consisted nearly entirely of unchanged glycyrrhetic acid following oral dosage whereas some conjugated material was also present after injection. This finding is probably related to the much faster biliary excretion of conjugates in the latter case with the result that some of this material could more easily escape hydrolysis by the intestinal bacteria. Synthetic derivatives of glycyrrhetic acid have been prepared which contain an ester group attached to the hydroxyl at C3. The metabolic fate of the succinate and acetate derivatives in rats was studied by Iveson *et al.* (1971) and Cameron *et al.* (1976), respectively.

Isojuripidine (14) was labelled with ^{3}H and administered orally to rats and dogs by Valzelli and Goldaniga (1973). A single dose (5 mg/kg) in dogs produced plasma levels which declined very slowly. Retention of the radioactivity was also seen in rats given doses of 1 or 10 mg/kg and very

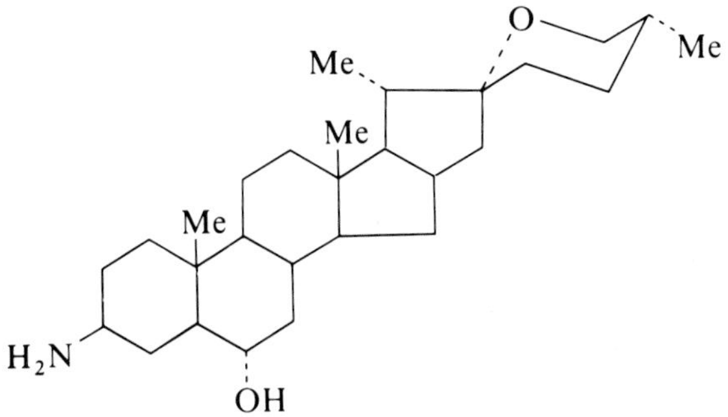

(14)

Isojuripidine

little excretion (1% in 32 h) occurred in the urine. However, faecal excretion accounted for 55% of the radioactivity during this period. Metabolites in the plasma and liver were investigated and it was found that N-acetyl isojuripidine was the major metabolite. Four additional metabolites were identified which show alterations at the C3- and/or C6-positions. Structures (15), (16), (17) and (18) illustrate these changes. The prolonged retention of the material in the body may be related to the fact that these metabolites are less polar than isojuripidine.

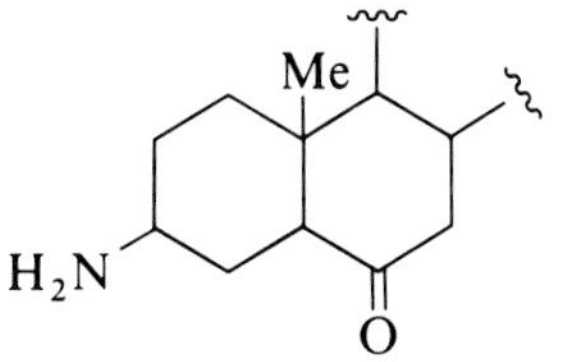

(15)

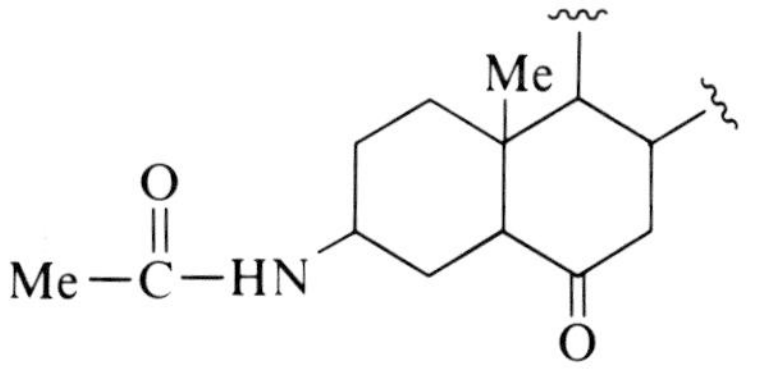

(16)

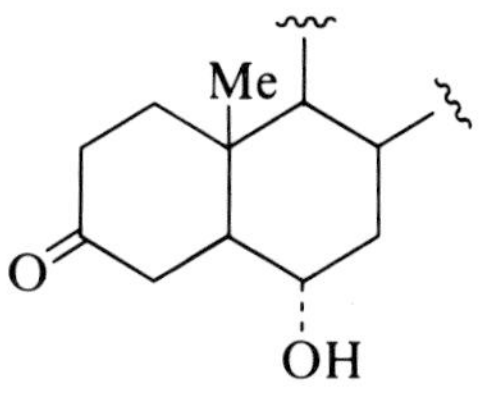

(17)

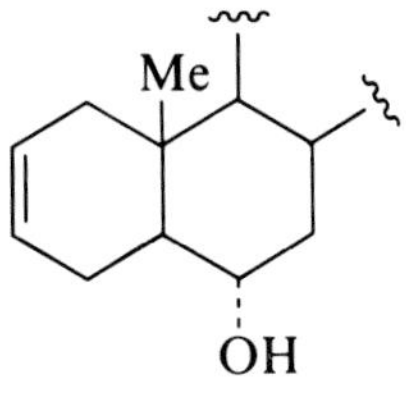

(18)

Saponins are triterpenoid glycosides which are powerful surface active and haemolytic agents. In view of these properties, it is hardly surprising that little is known of their metabolism, however their toxicity is related to the route of administration. Saponins are not usually considered to be readily absorbed after oral administration, however it is of interest to note that Han *et al.* (1976) found that Panax **saponin A** from Korean ginseng was readily absorbed in the gastrointestinal tract and excreted by renal and biliary routes in rabbits. They also reported that some of the compound appeared to be metabolized.

The remainder of this section on triterpenoids and related compounds is devoted to the cardiac glycosides or cardenolides. These drugs, formerly in the form of digitalis and more recently as pure glycosides, have held a central role in therapeutics for the past 200 years with the result that there is now a vast literature on this subject. Understandably, the factors of absorption, distribution, metabolism and excretion have figured prominently in this field because of their influence on the pharmacotherapeutic benefits which may be achieved. Useful results in these areas have largely appeared during the past 20 or so years following the development of sensitive and selective detection methods and, most recently, the understanding of the pharmacokinetics of the cardiac glycosides has increased rapidly. A thorough discussion of the biological disposition of these compounds is clearly beyond the scope of this book and, indeed, even an exhaustive survey of the pathways of metabolism of the cardiac glycosides would result in a significant increase in the length of the text. The following summary, while giving a comprehensive account of the metabolic changes known to occur with the cardiac glycosides, will therefore not attempt to be exhaustive. The review by Doherty (1973) and the book edited by Storstein (1973) are convenient sources of information on the biological disposition, particularly in relation to clinical implications, of cardiac glycosides. These publications deal primarily with digitoxin and digoxin, the two most commonly used compounds.

Both the nomenclature and the chemistry of the cardiac glycosides are complex and it is therefore essential that the main points of these subjects are summarized prior to any discussion of their metabolic fate. The central structural feature of these compounds is a steroid nucleus (see Structure (19)) which is similar to that found in the saponins but which contains the distinguishing feature of a *cis*-configuration of the substituents attached at C13 and C14 and the presence of a 14β-hydroxyl group. Further characteristic features are the unsaturated lactone group at C17 and the presence of unusual sugar residues attached at the C3-hydroxyl group. Structural variations of this basic pattern involve the addition of hydroxyl groups to the ring system, the replacement of the C19-methyl with an aldehyde group, the presence of a six-membered lactone ring at C17 and, of course, variations in the types and numbers of sugar moieties attached at C3. Table 6.1 summarizes the nomenclature and chemical relationships of most of the cardiac glycosides subsequently discussed in this section and also lists their botanical sources. The latter are informative because cardiac glycoside terminology is mainly based on botanical origins rather than chemical structures. From Table 6.1 it can be seen that these compounds consist of an aglycone, also termed genin, and from one to four sugar residues. These residues can vary with respect to their stability towards hydrolysis and the natural glycosides (also called native or genuine glycosides) which occur in the plant easily undergo hydrolysis of the terminal glucose unit and, when present, of an acetyl group to form the glycosides commonly employed in therapeutics (e.g. digitoxin, digoxin). Further hydrolysis of the rare sugar moieties, commonly digitoxose, requires stronger conditions chemically (e.g. acid hydrolysis) and gives rise to the aglycones.

Digitoxin is one of the two most widely used cardiac glycosides and it is therefore understandable that our knowledge of its metabolic fate is fairly extensive. Nevertheless, the data presently available do not give a complete picture of the metabolism of digitoxin as some reported metabolites remain unidentified. However, it is clear that the glycoside undergoes stepwise hydrolysis of the sugar moieties to give the bis-digi-toxoside, then the mono-digitoxoside and finally the aglycone, digitoxi-genin (19). Hydroxylation of the steroid nucleus at several sites occurs but that at C12 has received the greatest amount of interest as this pathway forms the cardioactive derivative digoxin or its cleavage products. Epi-merization of the hydroxyl group at C3 occurs with digitoxigenin, this reaction changing the orientation of the group from β to α via the 3-keto intermediate. Conjugation of the metabolites with glucuronate or sulphate also takes place, this being especially the case with 3-epidigitoxigenin.

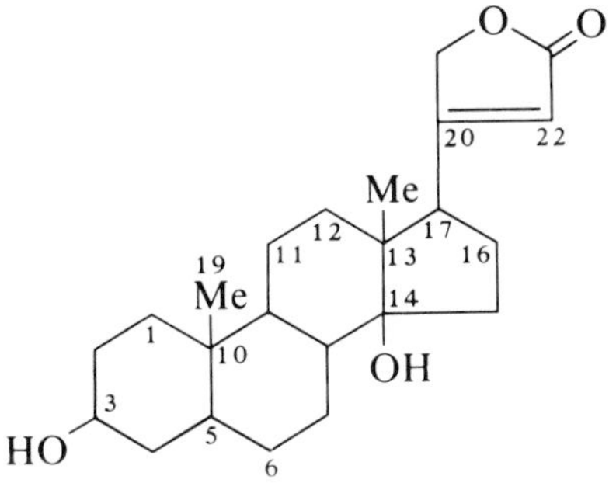

(19)

Digitoxigenin

The cleavage of the sugar residues of digitoxin was shown to occur in rat liver slices (Lauterbach and Repke, 1960a) and in similar preparations from guinea pig, rabbit, cat, dog and man (Herrmann and Repke, 1964a) with the result that the aglycone digitoxigenin was formed. Sequential

TABLE 6.1
Chemical components of some cardiac glycosides

Type and plant source	Aglycone	Sugar moieties at C3
Digitalis *Digitalis purpurea*	Digitoxigenin	—digitoxose—digitoxose—digitoxose—glucose
		Digitoxin
		Purpurea glycoside A (Desacetyldigilanide A)
Digitalis *Digitalis lanata*	Digitoxigenin	—digitoxose—digitoxose—digitoxose-3-acetyl—glucose
		Digitoxin
		Lanatoside A (Digilanide A)
Digitalis *Digitalis lanata*	Digoxigenin	—digitoxose—digitoxose—digitoxose-3-acetyl—glucose
		Digoxin
		Lanatoside C (Digilanide C, Cedilanid)
Thevetin *Thevetia neriifolia*	Digitoxigenin	—thevetose—glucose—glucose
		Neriifolin
		Thevebioside
		Cerberoside (Thevetin B)

Thevetin	*Thevetia neriifolia*	Cannogenin —thevetose—glucose—glucose Peruvoside Thevetin A
Strophanthus	*Strophanthus kombé*	Strophanthidin —cymarose—glucose—glucose Cymarin **Strophanthin (K-Strophanthin-β)** **K-Strophanthoside (K-Strophanthin-γ)**
Strophanthus	*Strophanthus gratus*	Ouabagenin —rhamnose—glucose Ouabain (G-Strophanthin) G-Strophanthoside
Convallaria	*Convallaria majalis*	Strophanthidin —rhamnose Convallatoxin
Convallaria	*Convallaria majalis*	Strophanthidol —rhamnose Convallatoxol
Squill	*Urginea maritima*	Scillarenin —rhamnose—glucose Proscillaridin A Scillaren A

hydrolysis of the sugar moieties was also reported by Kolenda *et al.* (1971b) when digitoxin was perfused through the isolated guinea pig liver. The aglycone is often not detected *in vivo* (Repke, 1959) and, in a review of cardiac glycoside metabolism, Repke (1963) concluded that this finding is due to slow hydrolysis of the glycosidic linkages compared with the rate of subsequent metabolism of the released aglycone (e.g. 3-epimerization and conjugation). Some recent investigations of digitoxin metabolism have likewise failed to detect the aglycone as a urinary metabolite. This result was obtained by Rietbrock and Vöhringer (1974) in the rat and in man by Wirth *et al.* (1976). On the other hand, Okita (1964) reported digitoxigenin in the urine of patients receiving [^{3}H]-digitoxin and it was also detected in most of the tissues analysed 18 h following administration of digitoxin to rats (Castle and Lage, 1973).

The hydroxylation of digitoxin is of interest because this metabolic pathway can furnish metabolites showing cardioactive properties. An excellent example of this pathway is the 12β-hydroxylation reaction which forms digoxin. Brown *et al.* (1957) and Ashley *et al.* (1958) identified digoxin in the urine of both rats and humans given digitoxin and Repke (1959) showed that the partially cleaved glycosides containing the 12β-hydroxyl group were also formed. Species differences in the hydroxylation of digitoxin at C12 were noted by Herrmann and Repke (1964a) who found it to occur in liver slices from rats and humans but not from guinea pigs, cats, rabbits and dogs. Hydroxylation at some unidentified position occurred with these latter species and Herrmann and Repke (1964b) suggested that this might be at C6. Recently, a series of investigations has appeared on this subject which studied the metabolism of the aglycone digitoxigenin *in vitro* by rat or rabbit liver preparations. The formation of hydroxylated metabolites was suggested by the data using rat liver homogenates (Stohs *et al.*, 1971) or rat liver microsomes (Spratt, 1973) and Talcott and Stohs (1973) identified several monohydroxylated metabolites using the latter system. In addition to 12β-hydroxydigitoxigenin, slightly larger amounts of the 5β- and smaller amounts of the 1β- and 16β-hydroxylated derivatives were also found. Interestingly, only the 5-hydroxylase activity was increased when the animals were pretreated with inducing agents. The formation of 5β-hydroxydigitoxigenin from digitoxigenin by rabbit liver homogenates was reported by Bulger and Stohs (1973), however Bulger *et al.* (1974) subsequently found that most of the hydroxylation under these conditions takes place at C6, the product being 6β-hydroxy-3-epidigitoxigenin. This finding thus confirms the earlier suggestion of Herrmann and Repke noted above. The knowledge that the biliary conjugates of digitoxin in rats are those of digitoxigenin monodigitoxoside rather than of digitoxigenin or epidigitoxigenin prompted Schmoldt *et al.* (1975) to investigate the microsomal metabolism of digitoxin. This was found to include both hydroxylation at C12 giving digoxigenin glycosides and the cleavage of the sugar residues giving bis- and mono-digitoxosides of digitoxigenin and digoxigenin. The free genins or their epi derivatives were not detected.

The subject of epimerization (inversion from β to α configuration) of digitoxigenin shown to occur by Repke and coworkers was reviewed by Repke (1963) who noted that the reaction occurs mainly in the liver. The reaction sequence involves the formation of the 3-dehydro (i.e. 3-keto) derivative and subsequent hydrogenation to the epigenin. Repke and Samuels (1964) carried out an enzymic study of the $3\alpha,\beta$-hydroxysteroid dehydrogenases in rat liver and noted that this sequence is a process of detoxification since both the keto and epi derivatives possess very weak biological activities. Enzyme activity was present both in the supernatant fraction and the microsomal fraction. Subsequent investigations have shown the formation of the 3-keto and 3-epi derivatives from digitoxigenin using rat liver homogenates (Stohs *et al.*, 1971) or microsomes (Spratt, 1973). 3-Epidigitoxigenin is also the major metabolite of digitoxigenin when homogenates of guinea pig liver are used (Bulger, 1975). As noted above, rat liver microsomes did not cleave digitoxin past the monodigitoxoside and the keto- and epigenins are therefore not formed *in vitro* from digitoxin (Schmoldt *et al.*, 1975).

Many studies of the metabolism of digitoxin have detected very polar metabolites and the identities of some of these are known. Repke (1959) noted that the cleavage of the sugar moieties leads finally to a conjugate rather than the aglycone and Herrmann and Repke (1964c) found that incubation of digitoxigenin with liver slices from rat, guinea pig, rabbit, dog or man resulted in the formation of a total of six highly polar metabolites. The sulphate conjugates of the genin and the epigenin were formed in all species except the dog which formed a conjugate which was probably a diglucuronide of a hydroxylated digitoxigenin derivative. The glucuronide of 3-epidigitoxigenin was formed by liver slices of rabbit, dog and man. The quantitative aspects of metabolite formation are summarized below and indicate that complete loss of the sugar moieties is not a major pathway in the metabolism of digitoxin. It is therefore of interest to note that Vöhringer and Rietbrock (1974) found the glucuronide of digitoxigenin mono-digitoxoside as a urinary metabolite of digitoxin in man. Recently, Storstein (1977) reported a comprehensive study of the metabolism of digitoxin in man following maintenance or single doses. The methodology employed allowed for the quantitation of unchanged compound, hydrolysed and hydroxylated derivatives and conjugates of these. Inactive metabolites formed by epimerization were not measured.

All known cardioactive metabolites of digitoxin were detected and the serum and urine patterns of these were found to be quite similar. In the steady-state group, digitoxin was the main cardioactive substance present and little ($< 1\%$) digoxin was formed. About a third of the material was present in the serum or urine as glucuronide and/or sulphate conjugates. In contrast to these results which indicate that hydrolysis and conjugation are more important than hydroxylation in the steady-state group, subjects given a single dose formed appreciable amounts of hydroxylated product. Thus, digoxin accounted for a fourth of the cardioactive material in the urine. It was therefore emphasized that differences in dosage regimens give different patterns of metabolism, the three pathways noted above appearing to be of about equal importance following a single dose of digitoxin.

The summary given above on the metabolic reactions of digitoxin and its degradation products included relatively little about either the quantitative importance of the various pathways or the excretory routes of the metabolites. These points will be considered now. It is important to note that digitoxin is completely absorbed from the intestine following oral dosage, a fact clearly related to the relatively low degree of hydroxylation of the steroid moiety and concomitant non-polar properties. Most of the absorbed digitoxin is excreted in the urine, in part as unchanged compound but also as metabolites. Various values for the urinary excretion of unchanged digitoxin have been reported depending on the dosage regimen. In patients receiving the drug, Jelliffe et al. (1970) found that about 30% of the daily losses were attributable to urinary digitoxin. Marcus (1975) reported a value of only 8% under similar conditions and this corresponds well with the 6–10% value reported in an early study by Okita et al. (1953). It is therefore noteworthy that Vöhringer and Rietbrock (1974) found that 60% of the material excreted in the urine following a single oral dose (1 mg) of [^{3}H]-digitoxin was unchanged compound. Smaller amounts of digoxin and the bis-digitoxoside of both of the genins were also detected. Storstein (1977) reported that about 57% of the total cardioactive material and conjugates in the urine was excreted unchanged after a single dose (0·6 mg) of digitoxin in man. The corresponding figure in subjects receiving maintenance doses was 87%. A similar high value was found by Wirth et al. (1976) who reported that 79% of the urinary radioactivity following the intravenous administration of [^{3}H]-digitoxin to cardiac patients receiving the drug daily was unchanged compound. None of the remaining urinary material was identified, however about 5% and 12% were non-polar and polar metabolites, respectively. Jelliffe et al. (1970) stated that roughly 8% of the daily loss of digitoxin from the body is in the form of digoxin, the 12-hydroxylated metabolite. As noted above, Storstein (1977) reported that only about 1% of the total cardioactive and conjugated material excreted in the urine was represented as digoxin following maintenance doses of digitoxin. However, values of about 25% were noted following a single dose. In any case, extensive conversion of digitoxin to digoxin seems unlikely in view of the relatively short biological half-life of digoxin and the well-known persistence of digitoxin in the body. Of course, these findings in man need not apply to other species and, in fact, Rietbrock and Völringer (1974) found that metabolite excretion in rats occurred largely by the faecal route and that over twice as much digoxin as digitoxin was excreted in the urine.

An important feature in the biological disposition of digitoxin is its biliary excretion. This allows for both its enterohepatic circulation and the faecal excretion of metabolites in spite of complete initial absorption. The enterohepatic circulation of digitoxin in man seems to have been originally proposed by Okita et al. (1955) and this phenomenon has subsequently been shown to occur in other species as well. Katzung and Meyers (1965) found that the half-life of digitoxin was markedly shortened in animals with biliary fistulas. The bile contained both digitoxin and digoxin in the non-polar fraction as well as two polar metabolites which released digitoxigenin on hydrolysis (Katzung and Meyers, 1966). Russell and Klaassen (1973) found 46% of the dose in the bile (12 h) in dogs. The corresponding value in rabbits was 53%. Experiments on rats have given various results. Cox and Wright (1959) found that about 10% of the administered digitoxin was excreted in the bile (5 h) and Lauterbach (1964b) stated that its entero-hepatic circulation was small in rats. However, Russell and Klaassen (1973) reported a value of 71% (12 h). Züllich et al. (1975) found values between 31% and 39% and also determined the nature of the biliary metabolites. The most abundant metabolite was a conjugate of digitoxi-genin mono-digitoxoside. Moderate amounts of the bis-digitoxosides of both digitoxigenin and digoxigenin were present as well as a few percent of both of these aglycones. Similar results in guinea pigs were reported by Ingwersen (1974). Marzo and Ghirardi (1977) found that nearly 70% of the administered radioactivity was excreted in the bile in 5 h when guinea pigs were given [^{3}H]-digitoxin intravenously. Unchanged compound accounted for only a small part of this material, most of which was due to more polar metabolites. Beermann et al. (1971) stated that less than 10% of the dose of [^{3}H]-digitoxin was excreted in the bile in man in 24 h. Storstein (1975) reported a value of only 1·5% for this period in man, however this was a measurement of digitoxin and cardioactive metabolites and it seems likely that considerable amounts of inactive products were also present.

Digoxin differs from digitoxin by the presence of a hydroxyl group at C12, the aglycone being called digoxigenin (20). As noted above, digoxin is

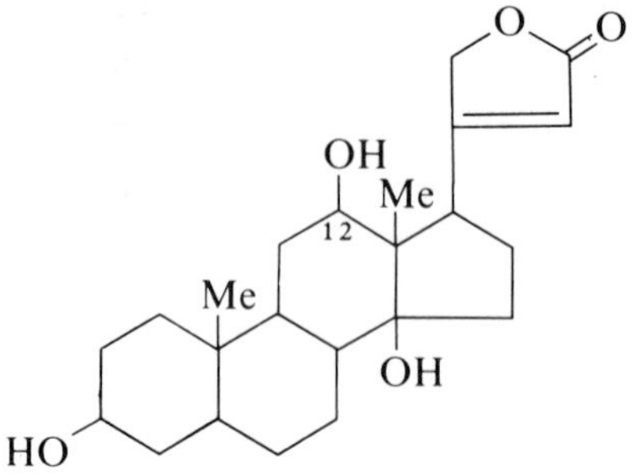

(20)

Digoxigenin

formed metabolically from digitoxin. However, interest in the biological disposition of the former compound is considerable quite apart from this fact due to its expanding use as a cardioactive glycoside. The relatively small chemical difference between the two glycosides is nonetheless sufficient to give rise to considerable differences in their fate in the body. As summarized by Doherty *et al.* (1971), the absorption following oral dosage of digoxin in man is only 80–90% complete. Its persistence in the body is also much shorter, the half-life being about 1·5 days compared with five to seven days for digitoxin. Metabolism is much less extensive with digoxin which is largely excreted unchanged in the urine (Marcus *et al.*, 1964) than with digitoxin which is appreciably excreted in the form of metabolites. The high proportion of urinary excretion of unchanged compound is also seen in rats (Bergmann *et al.*, 1972) and dogs (Marcus *et al.*, 1967) but not in guinea pigs which excrete mainly metabolites which are either less or more polar than digoxin.

Some of the metabolites of digoxin have been identified. The transformation product detected by Ashley *et al.* (1958) in the urine of rats and humans was found to be the same as one of the metabolites excreted following digitoxin administration. Wright (1962) subsequently identified it as the bis-digitoxoside of digoxin. This picture of extensive urinary excretion of unchanged compound and some excretion of the bis-digitoxoside derivative was also give by Okita (1964) who noted that small or trace amounts of several other metabolites including the mono-digitoxoside and the aglycone were also excreted by man. A recent study in man by Gault *et al.* (1976) showed that metabolic products were mainly excreted during the first day but that, even then, values of 90–95% for unchanged digoxin were typical. Reduction of the unsaturated lactone moiety to give dihydrodigoxin was reported by Abel *et al.* (1965) to occur in dogs. This transformation has recently been investigated in man using more selective methods and dihydrodigoxin was found to be the major dihydro derivative of digoxin (Watson *et al.*, 1973). Subsequently, Clark and Kalman (1974) reported that the urinary excretion of dihydrodigoxin by 50 patients given maintenance doses of digoxin averaged 13% of the material found in the methylene chloride fraction (contains also digoxin, mono- and bis-digitoxosides and genins). Greenwood and Snedden (1976) found an average excretion value for the dihydro derivative of 16% of the total oral dose for patients maintained on digoxin. Several studies have pointed towards the formation of highly polar metabolites of digoxin and Dwenger and Haberland (1971) obtained results from three subjects which indicated that 9–34% of the material excreted during the first 24 h was in this form. The identity of these metabolites was not determined, however they are not identical with the mono- and bis-digitoxosides of digoxin or its aglycone. It is perhaps pertinent that Kolenda *et al.* (1971b) found that the isolated perfused guinea pig liver converted digoxin to conjugates with glucuronic and/or sulphuric acid.

From the above it can be seen that cleavage of the sugar residues is usually not a metabolic pathway which is extensively utilized in the case of digoxin. If the mono-digitoxoside itself is given to man it is rapidly metabolized and excreted in the urine mainly as conjugates of itself and of 3-epidigoxin (Kuhlmann *et al.*, 1974). *In vivo* experiments in rats using the aglycone, digoxigenin, also showed rapid metabolism with 3-epimerization a major metabolic pathway (Thomas and Wright, 1965). This subject of the metabolism of sugar cleavage products of digoxin was studied in detail in rats by Abshagen and Rietbrock (1973). The general resistance of digoxin to loss of the sugar moieties found in the *in vivo* experiments noted above is also seen when *in vitro* techniques are used. Wong and Spratt (1964) were able to demonstrate the formation of only trace amounts of the genin when digoxin was incubated with rat liver slices or homogenates. Subsequently, Lage and Spratt (1965) were able to demonstrate cleavage of the sugar residues using rat liver slices but not homogenates. The *in vitro* hepatic metabolism of digoxin using preparations from mice, rats, guinea pigs, rabbits, cats and dogs was investigated by Lage and Spratt (1968). *In vitro* experiments have also been carried out with digoxigenin. Talcott *et al.* (1972) found that this aglycone was not hydroxylated by rat liver homogenates. Experiments using rat liver microsomes indicated that the hydroxylation of digoxigenin does not occur, in contrast to that observed with digitoxigenin (Talcott and Stohs, 1973).

The role of bile as an excretory route for digoxin has been the subject of a number of investigations which clearly indicate that this is a factor of importance. The picture in man is perhaps least clear and Doherty *et al.* (1970) concluded that the enterohepatic circulation of digoxin, which was found to be the major product in the bile, is not extensive and amounts to roughly 7% of the dose. However, the data of Caldwell and Cline (1976) indicated that it may be higher. They found that about 30% of an intravenous dose of digoxin was excreted in the bile in 24 h and that most of this material is presumably reabsorbable, biologically active drug. In other species it is apparent that a considerable portion of the administered digoxin is excreted in the bile. This was shown by Cox and Wright (1959) who found that 40% of the dose was excreted by rats in 5 h. This material consisted of unchanged compound and a metabolite, which as noted above was later identified as the bis-digitoxoside, in about a 1:3 ratio. Several more recent studies using rats indicated values of 32% in 2 h and 59% in 12 h (Russell and Klaassen, 1973), 45% in 12 h (Bergmann et al., 1972) and 61% in 11 h (Abshagen et al., 1972). The last investigation showed clearly that enterohepatic circulation is an important feature in the biological disposition of the compound in rats. The bulk of the material excreted in the bile is unchanged compound and, in contrast to the findings of Cox and Wright noted above, the ratio between digoxin and its bis-digitoxoside derivative was about 4:1. About 10% of the total was due to water-soluble metabolites. Biliary excretion of digoxin in dogs was shown by Russell and Klaassen (1973) to account for about 14% of the dose in 12 h. This value is close to that of 15% reported by Harrison et al. (1966) who collected bile for about 4·5 days. Only about 10% of this material was due to unchanged compound. According to Marcus et al. (1966, 1967), much of the material excreted in dog bile is due to water-soluble metabolites. The extent of biliary excretion of digoxin is large in both rabbits and guinea pigs. Russell and Klaassen (1973) reported values of 39% (2 h) and 60% (12 h) of the administered radioactivity following the intravenous dosage of [³H]-digoxin in rabbits. Using the isolated perfused guinea pig liver Kolenda et al. (1971a) found that nearly 70% of the added digoxin was excreted in the bile in 4 h and that this material was mainly in the form of polar metabolites. Haass et al. (1972) also demonstrated the extensive and rapid biliary excretion of digoxin in the guinea pig. In a recent study using [³H]-digoxin, Marzo and Ghirardi (1977) found that about 72% of the radioactivity was excreted in the bile in 5 h. However, only 5–8% of this material was due to unchanged compound, the bulk being due to more polar metabolites.

The above summaries have dealt with the metabolism of the two most commonly used cardiac glycosides, digitoxin and digoxin. As shown in Table 6.1, the natural glycosides occurring in the plant contain an added sugar residue, a terminal glucose moiety. Some of these compounds obtained from *Digitalis lanata* and termed lanatosides are employed therapeutically. This is true of **lanatoside A** and **lanatoside C** and some metabolic information is available on these compounds. It is useful to note that the presence of an extra sugar residue furnishes compounds of greater polarity than seen with the desglucose glycosides. This difference in physicochemical properties has a bearing on the biological disposition of these natural glycosides. The increase in polarity might be expected to result in reduced absorption following oral administration. However, the data of Beermann (1972) indicate that the total recovery of radioactivity in the urine of man following administration of [^{3}H]-lanatoside C is only slightly less than that found after giving digoxin. Dengler *et al.* (1973) also reported that the excretion of urinary radioactivity (4–6 days) following the administration of lanatoside C was nearly as great as that shown with digoxin (about 80%). In fact, similarity between these two compounds in this respect may be due to the fact that the terminal glucose unit of the natural glycosides appears to be removed not by the action of tissue enzymes but by those of the intestinal bacteria (see Repke, 1963). Lauterbach and Repke (1960b) reported that species of *Clostridium* are mainly involved. Hawksworth *et al.* (1971) found β-glucosidase activity in all of the six major groups of intestinal bacteria tested but noted that the enterococci were most active in converting lanatoside C to acetyldigoxin and, in some cases, to digoxin. These two metabolites are rapidly formed when lanatoside C is incubated with human faeces (Beermann, 1972). Aldous and Thomas (1977) studied the absorption and metabolism of lanatoside C in humans after oral administration and found that its conversion to digitoxin and perhaps further breakdown products takes place in the intestine, partly by acid hydrolysis and partly by bacterial action. Little or no unchanged compound was detected in most of the plasma samples. Interestingly, Beermann (1972) found that, although the total uptakes of digoxin and lanatoside C were similar, the sites of absorption were different. Lanatoside C was absorbed to a far greater extent in the distal parts of the intestine.

Another consequence of the presence of an additional sugar residue in lanatosides A and C compared with digitoxin and digoxin is their increased biliary excretion. Cox and Wright (1959) found that 70–75% of the dose was recovered in the bile in 5 h when rats were given the lanatosides intravenously. Corresponding values for digitoxin and digoxin were 10% and 40%, respectively. Furthermore, the latter compounds were partly excreted as metabolites whereas the lanatosides were excreted in the bile as such. The results of Beermann (1972) in man indicate that the biliary excretion of lanatoside C is less extensive than seen in the rat and also that the nature of the biliary metabolites is dependent on the route of administration. Unchanged compound is most abundant following intravenous dosage whereas various hydrolysis products were mostly detected after oral dosage. In regard to urinary metabolites of lanatoside C, Ashley *et al.* (1958) found that these were identical in both rats and man. Brown and

Wright (1956) showed the similarity between the excretory products of lanatoside C and digoxin in rats and Beermann (1972) found that most of the urinary material following the administration of lanatoside C to man is digoxin.

A further point of interest with the cardiac glycosides based on digoxin is the occasional use of various derivatives, often semi-synthetic compounds containing methyl or acetyl groups. A summary of the metabolism of these derivatives is not included in this section, however relevant data on β-methyldigoxin (4'''-methyldigoxin) are found in the papers of Voigtländer *et al.* (1972), Rietbrock *et al.* (1972, 1975) and Abshagen *et al.* (1974). Similar information on acetylated derivatives of digoxin, formed by the enzymic hydrolysis of lanatoside C, was given by Buchtela *et al.* (1968), Förster and Schulzeck (1968), Ruiz-Torres and Burmeister (1972), Bodem *et al.* (1974, 1975) and Flasch *et al.* (1977).

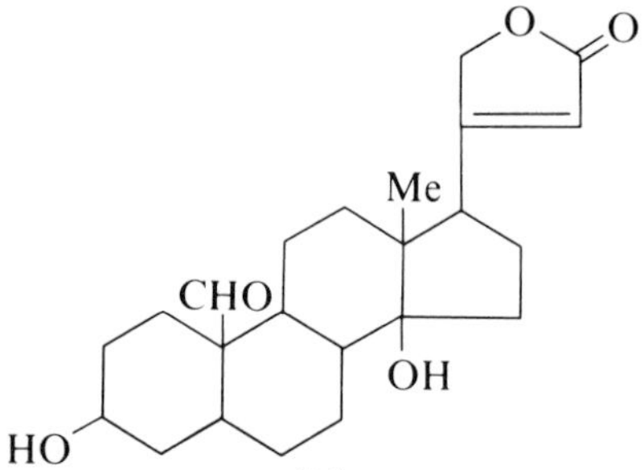

(21)

Cannogenin

Several cardiac glycosides including **cerberoside** (thevetin B) and **thevetin A** have been isolated from the yellow oleander, a tropical bush. Their chemical relationships are shown in Table 6.1 and it is seen that loss of the two terminal glucose units gives the glycosides **neriifolin** and **peruvoside**, respectively. Digitoxigenin (19) is the aglycone of neriifolin and cannogenin (21) of peruvoside. Raudonat and Engler (1957) gave cerberoside orally to rats and identified neriifolin (0·9% of dose) in the urine together with another metabolite which, however, was not the aglycone digitoxigenin. No unchanged compound was detected. In the faeces, about 2·6% of the dose was recovered as neriifolin. Engler *et al.* (1958) noted that neriifolin is found in both the urine and faeces of rats following oral or subcutaneous administration of cerberoside and proposed the existence of an enterohepatic circulation of these glycosides. Of interest in this regard is the finding of Lauterbach and Repke (1960b) that cerberoside is converted to neriifolin when incubated with rat faeces. A more recent study of Frölich *et al.* (1972) showed that similar amounts of radioactivity were excreted in the urine following the oral or intravenous administration of peruvoside to man. About 33–37% of the dose was recovered in 48 h and none of this was due to unchanged drug. However, two unidentified metabolites were detected in the urine and faeces. In view of the fact that peruvoside is intermediate in polarity between the highly-metabolized digitoxin and the less well-metabolized digoxin, it is reasonable that a fairly high rate of metabolism was found. It is noteworthy that considerable material was excreted in the bile, however this appears to be lost by the faecal route rather than entering an enterohepatic circulation.

Another main group of cardioactive glycosides includes the strophanthus glycosides, however the therapeutic use of these compounds has been much more limited than that of the digitalis glycosides and ouabain is the only member of the group which has had much importance. The nomenclature of the strophanthus glycosides and their chemical relationships are shown in Table 6.1. Engler *et al.* (1958) gave the natural glycoside **K-strophanthoside** to rats both orally and subcutaneously and found that cymarin, the glycoside formed by loss of both glucose residues, was excreted in the urine and faeces. Two unidentified metabolites were also detected, however these were not identical with unchanged K-strophanthoside, strophanthin or the aglycone strophanthidin (22) (Raudonat and Engler, 1957). The material found in the bile was K-strophanthoside, not cymarin, and the participation of an enterohepatic circulation was concluded to be of importance in determining the metabolic fate of this glycoside in rats. This belief is strengthened by the finding that cymarin formation was not seen when several tissue preparations were used but was extensive when K-strophanthoside was incubated with rat faeces. The latter finding was confirmed by Lauterbach and Repke (1960b). The biological disposition of [^{3}H]-K-strophanthoside in guinea pigs has been studied

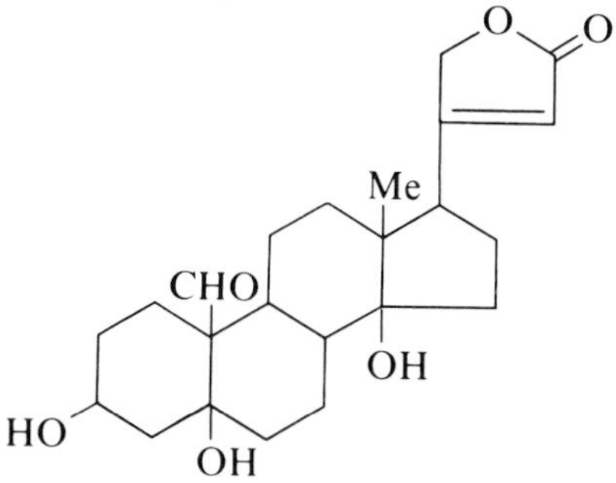

(22)

Strophanthidin

following rectal administration (Marzo *et al.*, 1973), intravenous administration (Marzo *et al.*, 1974) and intravenous and intraduodenal administration (Marzo and Ghirardi, 1977). In the first case the results indicated that the compound was rapidly absorbed, mainly as unchanged K-strophanthoside. About 55% of the dose was absorbed in 15 h. The urinary excretion at this time was 17–19%, this material consisting of unchanged compound and cymarin in a ratio of 9 : 1. When given intravenously to bile duct cannulated guinea pigs, the glycoside was excreted unchanged in the urine and bile in approximately equal amounts. These values were about 20% in 5 h and 30–40% in 24 h. Only about 2% of the dose was excreted in the urine and bile in 5 h when K-strophanthoside was given intraduodenally.

A few investigations have studied the further metabolism of **cymarin** in rats. Lauterbach (1964a) reported that the aldehyde group was reduced by rat liver slices or homogenates to give cymarol, the corresponding alcohol, and that the aglycones of these glycosides, strophanthidin and strophanthidol, were also formed. Moerman (1965) found that cymarin is partly metabolized *in vivo* in rats and that uncnanged compound, cymarol and strophanthidin are excreted in the bile and urine. The metabolism of the semi-synthetic diacetyl derivative of cymarol in rats was studied by Boutagy and Thomas (1977). Reduction of the aldehyde group can also take place with the aglycone strophanthidin (Lauterbach, 1964a). The same occurs with a closely related glycoside, **helveticoside**. This compound differs from cymarin only by the presence of a digitoxose moiety in place of the cymarose unit.

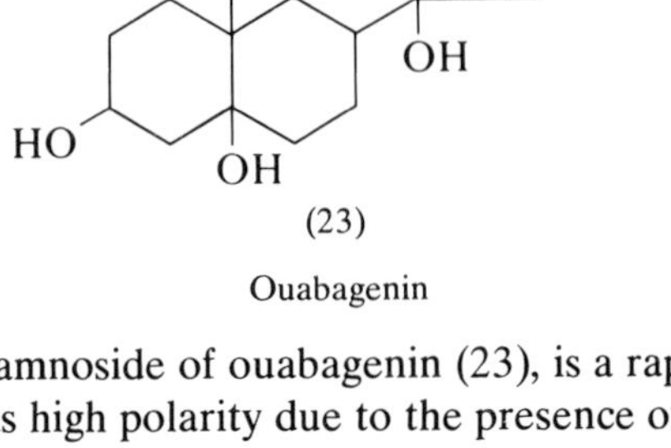

(23)

Ouabagenin

Ouabain, the rhamnoside of ouabagenin (23), is a rapidly acting cardiac glycoside which has high polarity due to the presence of five free hydroxyl groups attached to the steroid nucleus. An early study by Farah (1946) indicated that it was extensively excreted in the bile in rats and that little metabolic alteration occurred. Cox *et al.* (1959) reinvestigated this subject and confirmed the earlier results. Following intravenous administration, about 90% of the dose was excreted in the bile in 5 h whereas only 4% was recovered in the urine. Similarly, Kupferberg and Schanker (1968) recovered 85% of the unchanged ouabain in the bile of rats only 90 min after its intravenous dosage. Values of 42% in 2 h and 55% in 12 h, again as unchanged compound, were recorded by Russell and Klaassen (1972, 1973). The biliary excretion of ouabain in several other species has been studied and it appears that appreciable species variations occur, both with regard to the extent of excretion and the material excreted. The investigations of Russell and Klaassen showed that the 12 h values for rabbits and dogs were only 4·4% and 1·3%, respectively. In the latter case this material appears to be unchanged compound but about one-third of that excreted in rabbit bile is in a more water-soluble form. Seldon *et al.* (1974) recovered only about 5% of the radioactivity in the bile collected for four days when dogs were given [³H]-ouabain intravenously. Biliary excretion of ouabain in the guinea pig is also very low. This was shown by Kolenda *et al.* (1971a) using the isolated, perfused guinea pig liver and it was also shown that the glycoside was not degraded by the liver under these conditions (Kolenda *et al.*,1971b). Marzo *et al.* (1974) and Marzo and Ghirardi (1977) recovered less than 4% of the dose in the 5 h bile after intravenous administration of ouabain to guinea pigs. The corresponding urinary excretion was about 25%. The material excreted by both routes was shown to be unchanged compound. Ouabain is also excreted preferentially in the urine of sheep (Dutta *et al.*, 1963). Marks *et al.* (1964) found that the same is true in man and that the urinary material is primarily unchanged glycoside. Seldon *et al.* (1974) recovered only about 5% of the radioactivity in the bile of man following intravenous administration of [³H]-ouabain whereas 46–48% was excreted in the urine in five days. Interestingly, the results indicated that the faecal excretion of ouabain may arise from intestinal excretion at sites in addition to the biliary tract. The gastrointestinal absorption of ouabain is very poor. Marchetti *et al.* (1971) recorded values of less than 10% of the dose in both guinea pigs and man.

The convallaria (lily of the valley) glycosides are closely related to the foregoing strophanthus glycosides in that they are derivatives of the same genin, strophanthidin (22) or its corresponding C19-alcohol, strophanthidol. As shown in Table 6.1, **convallatoxin** is the rhamnoside of strophanthidin whereas **convallatoxol** is the rhamnoside of strophanthidol. The latter glycoside is fairly polar due to the presence of three hydroxyl groups and Greenberger *et al.* (1969) found that its intestinal absorption was lower than that of digitoxin and other nonpolar glycosides. Similar properties also account for the fact that about 80% of the absorbed convallatoxin is excreted in the bile of rats (Lauterbach, 1964b). Lauterbach (1964a) also showed that the aldehyde group of convallatoxin can undergo metabolic reduction to the alcohol. Convallatoxol was formed when incubates using rat liver slices or homogenates were used, however no reduction was seen with guinea pig liver and only traces of the alcohol were formed by cat liver.

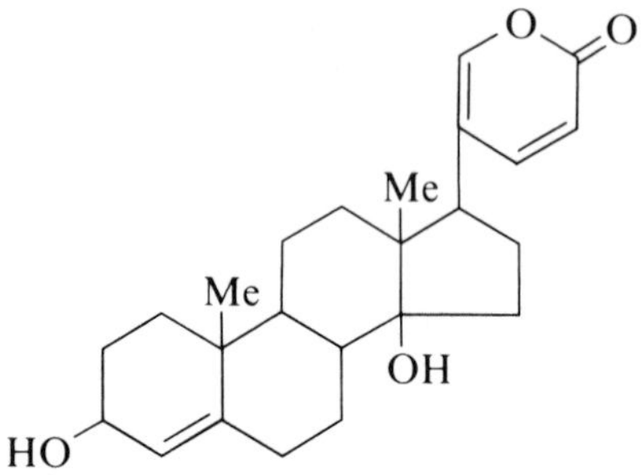

(24)

Scillarenin

The squill glycosides, termed scilladienolides, differ chemically from the compounds mentioned above by the presence at C17 of a six-membered lactone ring containing two double bonds. The chemical components of the natural glycoside **scillaren A** are listed in Table 6.1 which indicates that the aglycone scillarenin (24) is attached via the hydroxyl group at C3 to rhamnose, giving the glycoside **proscillaridin A** and then to the terminal glucose moiety. Scillaren A is extensively excreted in the bile following intravenous administration to rats (Simon and Wright, 1960). They found roughly 80–90% of the dose excreted as unchanged compound by this route in 5 h. No glycosides or metabolites were detected in the 12 h urine. However, Lauterbach and Repke (1960a) found that scillaren A was extensively hydrolysed to proscillaridin A when incubated with a rat liver preparation. The latter desglucose glycoside was not detected when incubations with rat faeces were carried out, however this appears to be due to rapid further metabolism rather than lack of ability to remove the terminal glucose moiety. A few reports on the biological disposition of proscillaridin A have appeared and Greenberger *et al.* (1969) found that it, like other nonpolar glycosides (e.g. digitoxin and digoxin), was more rapidly absorbed from guinea pig intestine than were several more polar glycosides. Davis *et al.* (1969) suggested that it is eliminated rapidly. However, the metabolic fate of proscillaridin A is unclear. It is perhaps noteworthy that clinical trials with this glycoside have shown a relatively poor bioavailability after oral administration. Andersson *et al.* (1977b) calculated that the absorption of active proscillaridin A amounted to only about 7% of the dose during the first 4 h in man. The absorbed material is then excreted extensively in the bile in conjugated form (Andersson *et al.*, 1977a). Accordingly, attention has been turned to a semi-synthetic derivative, methylproscillaridin. Rietbrock and Staud (1975) and Staud *et al.* (1975) recently reported on its metabolism in man.

A few studies on the biological disposition of cardiac glycosides have included some of the less common derivatives. Angarskaya *et al.* (1967) found that the desglucose derivative of **cheirotoxin** (strophanthidin-lyxose-glucose) was excreted unchanged in the bile following intravenous administration to rats. **Adonitoxin**, the rhamnoside of adonitoxigenin which is a structural isomer of strophanthidin, was however excreted partly unchanged and partly as an oxidation product in similar experiments. The investigations of Herrmann and Repke (1964b) and Repke and Samuels (1964) dealt with the epimerization and hydroxylation of cardiac glycoside aglycones, mainly digitoxigenin. In addition, several other aglycones (structural differences from digitoxigenin (19) shown in parentheses) were studied including **uzarigenin** (*trans* junction of A/B rings), **sarmentogenin** (hydroxyl group at C11), **gitoxigenin** (hydroxyl group at C16) and **diginatigenin** (hydroxyl group at C12 and C16). The studies of Lauterbach and Repke (1960a,b) on the cleavage of the sugar residues of cardiac glycosides by tissue enzymes and by the intestinal bacteria also included **evomonoside** (digitoxigenin α-rhamnoside) and **somalin** (digitoxigenin cymaroside).

IV. Tetraterpenoids

The most familiar tetraterpenoids are the carotenoids which are yellow to red, lipid-soluble plant pigments. More than 300 of these compounds are now known. Some derivatives are esters and Booth (1947) studied the metabolism in rats of the ester of citraurin. This compound, obtained from orange peel, was hydrolysed during passage through the gastrointestinal tract. Similar findings were reported with **physalien**, the dipalmitate of zeaxanthol, and **taraxien**, a diester of taraxanthin (Booth, 1964). A third of the esters was recovered in the faeces, mainly as free carotenol formed by ester hydrolysis, whereas the remainder was not accounted for. Some of the taraxien was also excreted as the monoester. β-Carotene is both the most widespread and widely studied carotenoid. The latter fact is related to the important role of β-carotene as a source of vitamin A. This topic, which deals with the metabolism of not only β-carotene but a number of other carotenoid vitamin A precursors, falls outside the scope of this book. Reviews on β-carotene metabolism and the formation of vitamin A include those of Thommen (1971) and Pitt (1971). It is worthy of note that Thommen also dealt with the subject of the metabolism of numerous carotenoids in birds.

References

Abel, R. M., Luchi, R. J., Peskin, G. W., Conn, H. L. and Miller, L. D. (1965). *J. Pharmac. exp. Ther.* **150**, 463–468.

Abshagen, U. and Rietbrock, N. (1973). *Naunyn-Schmiedeberg's Arch. Pharmac.* **276**, 157–166.

Abshagen, U., Bergmann, K. v. and Rietbrock, N. (1972). *Naunyn-Schmiedeberg's Arch. Pharmac.* **275**, 1–10.

Abshagen, U., Rennekamp, H., Küchler, R. and Rietbrock, N. (1974). *Eur. J. clin. Pharmac.* **7**, 177–181.

Albro, P. W. and Thomas, R. O. (1974). *Biochim. biophys. Acta* **372**, 1–14.

Aldous, S. and Thomas, R. (1977). *Clin. Pharmac. Ther.* **21**, 647–658.

Andersson, K.-E., Bergdahl, B. and Wettrell, G. (1977a). *Eur. J. clin. Pharmac.* **11**, 273–276.

Andersson, K.-E., Bergdahl, B., Dencker, H. and Wettrell, G. (1977b). *Eur. J. clin. Pharmac.* **11**, 277–281.

Angarskaya, M. A., Sokolova, V. E., Lyubartseva, L. A. and Lutokhin, S. I. (1967). *Farmak. Toks.* **30**, 438–440. (*Chem. Abstr.* (1967) **67**, 80862d.)

Ashley, J. J., Brown, B. T., Okita, G. T. and Wright, S. E. (1958). *J. biol. Chem.* **232**, 315–322.

Bang, L. and Ourisson, G. (1975). *Tetrahedron Lett.* 1881–1884.

Bang, L., Ourisson, G. and Teisseire, P. (1975). *Tetrahedron Lett.* 2211–2214.

Beermann, B. (1972). *Eur. J. clin. Pharmac.* **5**, 11–18.

Beermann, B., Hellström, K. and Rosen, A. (1971). *Circulation* **43**, 852–861.

Bergmann, K. von, Abshagen, U. and Rietbrock, N. (1972). *Naunyn-Schmiedeberg's Arch. Pharmac.* **273**, 154–167.

Bodem, G., Wirth, K., Gernand, E. and Dengler, H. J. (1974). *Archs int. Pharmacodyn. Thér.* **208**, 102–116.

Bodem, G., Wirth, K. and Dengler, H. J. (1975). *Arzneimittel-Forsch.* **25**, 1448–1452.

Booth, V. H. (1947). *Biokhimiya* **12**, 21–25.

Booth, V. H. (1964). *Biochim. biophys. Acta* **84**, 188–191.

Boutagy, J. and Thomas, R. (1977). *Xenobiotica* **7**, 267–278.

Boyland, E. and Chasseaud, L. F. (1967). *Biochem. J.* **104**, 95–102.

Brown, B. T. and Wright, S. E. (1956). *J. biol. Chem.* **220**, 431–437.

Brown, B. T., Wright, S. E. and Okita, G. T. (1957). *Nature* Lond. **180**, 607–608.

Buchtela, K., Drexler, K., Hackl, H., Königstein, M. and Schläger, J. (1968). *Arzneimittel-Forsch.* **18**, 295–303.

Bulger, W. H. (1975). *Diss. Abstr. Int.* **35B**, 4075.

Bulger, W. H. and Stohs, S. J. (1973). *Biochem. Pharmac.* **22**, 1745–1750.

Bulger, W. H., Stohs, S. J. and Wheeler, D. M. S. (1974). *Biochem. Pharmac.* **23**, 921–929.

Caldwell, J. H. and Cline, C. T. (1976). *Clin. Pharmac. Ther.* **19**, 410–415.

Cameron, B. D., Chasseaud, L. F., Hawkins, D. R. and McCormick, D. J. (1976). *Arzneimittel-Forsch.* **26**, 1680–1683.
Carlat, L. E., Margraf, H. W., Weathers, H. H. and Weichselbaum, T. E. (1959). *Proc. Soc. exp. Biol. Med.* **102**, 245–248.
Castle, M. C. and Lage, G. L. (1973). *Archs int. Pharmacodyn. Thér.* **203**, 323–334.
Chasseaud, L. F., Fry, B. J., Hawkins, D. R., Lewis, J. D., Sword, I. P., Taylor, T. and Hathway, D. E. (1971). *Arzneimittel-Forsch.* **21**, 1379–1384.
Clark, D. R. and Kalman, S. M. (1974). *Drug Metab. Disposit.* **2**, 148–150.
Cox, E. and Wright, S. E. (1959). *J. Pharmac. exp. Ther.* **126**, 117–122.
Cox, E., Roxburgh, G. and Wright, S. E. (1959). *J. Pharm. Pharmac.* **11**, 535–539.
Cronlund, A. (1976). *Acta pharm. suec.* **13**, 43–50.
Davis, S. H., Van Dyke, K. and Robinson, R. L. (1969). *Archs int. Pharmacodyn. Thér.* **177**, 231–237.
Dengler, H. J., Bodem, G. and Wirth, K. (1973). *Arzneimittel-Forsch.* **23**, 64–74.
Doherty, J. E. (1973). *Ann. intern. Med.* **79**, 229–238.
Doherty, J. E., Flanigan, W. J., Murphy, M. L., Bulloch, R. T., Dalrymple, G. L., Beard, O. W. and Perkins, W. H. (1970). *Circulation* **42**, 867–873.
Doherty, J. E., Hall, W. H., Murphy, M. L. and Beard, O. W. (1971). *Chest* **59**, 433–437.
Dutta, S., Marks, B. H. and Smith, C. R. (1963). *J. Pharmac. exp. Ther.* **142**, 223–230.
Dwenger, A. and Haberland, G. (1971). *Naunyn-Schmiedeberg's Arch. Pharmak.* **270**, 102–104.
Engler, R., Holtz, P. and Raudonat, H. W. (1958). *Naunyn-Schmiedeberg's Arch. exp. Path. Pharmak.* **233**, 393–408.
Farah, A. (1946). *J. Pharmac. exp. Ther.* **86**, 248–257.
Fischer, F. G. and Bielig, H.-J. (1940). *Hoppe-Seyler's Z. physiol. Chem.* **266**, 73–98.
Flasch, H., Schumpelick, V. and Koch, G. (1977). *Arzneimittel-Forsch.* **27**, 656–659.
Förster, W. and Schulzeck, S. (1968). *Biochem. Pharmac.* **17**, 489–496.
Fröhlich, J. C., Falkner, F. C., Watson, J. T. and Scheler, F. (1972). *Eur. J. clin. Pharmac.* **5**, 65–71.
Gault, M. H., Ahmed, M., Symes, A. L. and Vance, J. (1976). *Clin. Biochem.* **9**, 46–52.
Greenberger, N. J., MacDermott, R. P., Martin, J. F. and Dutta, S. (1969). *J. Pharmac. exp. Ther.* **167**, 265–273.
Greenwood, H. and Snedden, W. (1976). Int. Symp. Mass Spectrometry Drug Metab., Milan, Italy. Summaries, p. 12.
Haass, A., Lüllmann, H. and Peters, T. (1972). *Eur. J. Pharmac.* **19**, 366–370.
Han, B. H., Lee, E. B., Yoon, U. C. and Woo, L. K. (1976). *Hanguk Saenghwa Hakhol Chi* **9**, 21–27. (*Chem. Abstr.* (1976) **85**, 186438g.)
Harrison, C. E., Brandenburg, R. O., Ongley, P. A., Orvis, A. L. and Owen, C. A. (1966). *J. Lab. clin. Med.* **67**, 764–777.
Hathway, D. E. (1973). *In* "International Aspects of Drug Evaluation and Usage" (A. J. J. Jouhar and M. F. Grayson, Eds), pp. 19–31. Churchill Livingstone, Edinburgh and London.
Hawksworth, G., Drasar, B. S. and Hill, M. J. (1971). *J. med. Microbiol.* **4**, 451–459.
Helbing, A. R. (1963). *Clinica chim. Acta* **8**, 756–762.

Herrmann, I. and Repke, K. (1964a). *Naunyn-Schmiedeberg's Arch. exp. Path. Pharmak.* **247**, 35–48.

Herrmann, I. and Repke, K. (1964b). *Naunyn-Schmiedeberg's Arch. exp. Path. Pharmak.* **248**, 351–369.

Hermann, I. and Repke, K. (1964c). *Naunyn-Schmiedeberg's Arch. exp. Path. Pharmak.* **248**, 370–386.

Hildebrandt, H. (1902). *Hoppe-Seyler's Z. physiol. Chem.* **36**, 441–451.

Ingwersen, F. (1974). *Naunyn-Schmiedeberg's Arch. Pharmac.* **282**, R38.

Iveson, P., Lindup, W. E., Parke, D. V. and Williams, R. T. (1971). *Xenobiotica* **1**, 79–95.

Jelliffe, R. W., Buell, J., Kalaba, R., Sridhar, R. and Rockwell, R. (1970). *Math. Biosci.* **6**, 387–403.

Katzung, B. G. and Meyers, F. H. (1965). *J. Pharmac. exp. Ther.* **149**, 257–262.

Katzung, B. G. and Meyers, F. H. (1966). *J. Pharmac. exp. Ther.* **154**, 575–580.

Kaya, K. (1960). *Nippon Yakurigaku Zasshi* **56**, 368–376. (*Chem. Abstr.* (1961) **55**, 21372h.)

Kolenda, K.-D., Lüllmann, H., Peters, T. and Seiler, K.-U. (1971a). *Br. J. Pharmac.* **41**, 648–660.

Kolenda, K.-D., Lüllmann, H. and Peters, T. (1971b). *Br. J. Pharmac.* **41**, 661–673.

Kuhlmann, J., Abshagen, U. and Rietbrock, N. (1974). *Eur. J. clin. Pharmac.* **7**, 87–94.

Kupferberg, H. J. and Schanker, L. S. (1968). *Am. J. Physiol.* **214**, 1048–1053.

Lage, G. L. and Spratt, J. L. (1965). *J. Pharmac. exp. Ther.* **149**, 248–256.

Lage, G. L. and Spratt, J. L. (1968). *J. Pharmac. exp. Ther.* **159**, 182–193.

Lauterbach, F. (1964a). *Naunyn-Schmiedeberg's Arch. exp. Path. Pharmak.* **247**, 71–86.

Lauterbach, F. (1964b). *Naunyn-Schmiedeberg's Arch. exp. Path. Pharmak.* **247**, 391–411.

Lauterbach, F. and Repke, K. (1960a). *Naunyn-Schmiedeberg's Arch. exp. Path. Pharmak.* **239**, 196–218.

Lauterbach, F. and Repke, K. (1960b). *Naunyn-Schmiedeberg's Arch. exp. Path. Pharmak.* **240**, 45–71.

Marchetti, G. V., Marzo, A., de Ponti, C., Scalvini, A., Merlo, L. and Noseda, V. (1971). *Arzneimittel-Forsch.* **21**, 1399–1403.

Marcus, F. I. (1975). *Am. J. Med.* **58**, 452–459.

Marcus, F. I., Kapadia, G. J. and Kapadia, G. G. (1964). *J. Pharmac. exp. Ther.* **145**, 203–209.

Marcus, F. I., Petterson, A., Salel, A., Scully, J. and Kapadia, G. G. (1966). *J. Pharmac. exp. Ther.* **152**, 372–382.

Marcus, F. I., Pavlovich, J., Burkhalter, L. and Cuccia, C. (1967). *J. Pharmac. exp. Ther.* **156**, 548–556.

Marks, B. H., Dutta, S., Gauther, J. and Elliott, D. (1964). *J. Pharmac. exp. Ther.* **145**, 351–356.

Marzo, A. and Ghirardi, P. (1977). *Naunyn-Schmiedeberg's Arch. Pharmac.* **298**, 51–56.

Marzo, A., Ghirardi, P., Croce, G. and Marchetti, G. (1973). *Naunyn-Schmiedeberg's Arch. Pharmac.* **279**, 19–29.

Marzo, A., Ghirardi, P. and Marchetti, G. (1974). *J. Pharmac. exp. Ther.* **189**, 185–193.

Miyake, T., Asano, K., Saito, M., Yoshida, M. and Shimura, K. (1976). *Kaku Igaku* **13**, 451–458. (*Chem. Abstr.* (1976) **85**, 153696m.)
Moerman, E. (1965). *Archs int. Pharmacodyn. Thér.* **156**, 489–493.
Morishima, E. (1961). *Nippon Yakurigaku Zasshi* **57**, 353–362. (*Chem. Abstr.* (1963) **58**, 839c.)
Neidlein, R. and Stumpf, U. (1977a). *Arzneimittel-Forsch.* **27**, 1162–1166.
Neidlein, R. and Stumpf, U. (1977b). *Arzneimittel-Forsch.* **27**, 1384–1390.
Okita, G. T. (1964). *Pharmacologist* **6**, 45.
Okita, G. T., Kelsey, F. E., Talso, P. J., Smith, L. B. and Geiling, E. M. K. (1953). *Circulation* **7**, 161–168.
Okita, G. T., Talso, P. J., Curry, J. H., Smith, F. D. and Geiling, E. M. K. (1955). *J. Pharmac. exp. Ther.* **115**, 371–379.
Parke, D. V., Pollock, S. and Williams, R. T. (1963). *J. Pharm. Pharmac.* **15**, 500–506.
Pitt, G. A. J. (1971). *In* "Carotenoids" (O. Isler, Ed.), pp. 717–742. Birkhäuser Verlag, Basel and Stuttgart.
Raudonat, H. W. and Engler, R. (1957). *Naunyn-Schmiedeberg's Arch. exp. Path. Pharmak.* **232**, 295–297.
Repke, K. (1959). *Naunyn-Schmiedeberg's Arch. exp. Path. Pharmak.* **237**, 34–48.
Repke, K. (1963). *In* "New Aspects of Cardiac Glycosides" (W. Wilbrandt and P. Lindgren, Eds), Vol. 3, pp. 47–73, Proc. First Int. Pharmac. Meeting. Pergamon Press, Oxford, London, New York and Paris.
Repke, K. and Samuels, L. T. (1964). *Biochemistry* N.Y. **3**, 689–695.
Rietbrock, N. and Staud, R. (1975). *Eur. J. clin. Pharmac.* **8**, 427–432.
Rietbrock, N. and Vöhringer, H.-F. (1974). *Biochem. Pharmac.* **23**, 2567–2575.
Rietbrock, N., Rennekamp, C., Rennekamp, H., Bergmann, K. von and Abshagen, U. (1972). *Naunyn-Schmiedeberg's Arch. Pharmac.* **272**, 450–453.
Rietbrock, N., Abshagen, U., Bergmann, K. von and Rennekamp, H. (1975). *Eur. J. clin. Pharmac.* **9**, 105–114.
Ruiz-Torres, A. and Burmeister, H. (1972). *Klin. Wschr.* **50**, 191–195.
Russell, J. Q. and Klaassen, C. D. (1972). *J. Pharmac. exp. Ther.* **183**, 513–519.
Russell, J. Q. and Klaassen, C. D. (1973). *J. Pharmac. exp. Ther.* **186**, 455–462.
Schmoldt, A., Benthe, H. F. and Haberland, G. (1975). *Biochem. Pharmac.* **24**, 1639–1641.
Seldon, R., Margolies, M. N. and Smith, T. W. (1974). *J. Pharmac. exp. Ther.* **188**, 615–623.
Simon, M. and Wright, S. E. (1960). *J. Pharm. Pharmac.* **12**, 767–768.
Spratt, J. L. (1973). *Biochem. Pharmac.* **22**, 1669–1671.
Staud, R., Rietbrock, N. and Fassbender, H. P. (1975). *Eur. J. clin. Pharmac.* **9**, 99–103.
Steinberg, D., Avigan, J., Mize, C. and Baxter, J. (1965a). *Biochem. biophys. Res. Commun.* **19**, 412–416.
Steinberg, D., Avigan, J., Mize, C., Eldjarn, J., Try, K. and Refsum, S. (1965b). *Biochem. biophys. Res. Commun.* **19**, 783–789.
Stoffel, W. and Kahlke, W. (1965). *Biochem. biophys. Res. Commun.* **19**, 33–36.
Stohs, S. J., Reinke, L. A. and El-Olemy, M. M. (1971). *Biochem. Pharmac.* **20**, 437–446.
Storstein, L. (1975). *Clin. Pharmac. Ther.* **17**, 313–320.
Storstein, L. (1977). *Clin. Pharmac. Ther.* **21**, 125–140.
Storstein, O. (1973). Symposium on Digitalis. Gyldendal Norsk Forlag, Oslo.

Subbiah, M. T. R. (1973). *Am. J. clin. Nutr.* **26**, 219–225.

Talcott, R. E. and Stohs, S. J. (1973). *Res. Commun. chem. Path. Pharmac.* **5**, 663–672.

Talcott, R. E., Stohs, S. J. and El-Olemy, M. M. (1972). *Biochem. Pharmac* **21**, 2001–2006.

Thomas, R. E. and Wright, S. E. (1965). *J. Pharm. Pharmac.* **17**, 459.

Thommen, H. (1971). *In* "Carotenoids" (O. Isler, Ed.), pp. 637–668. Birkhäuser Verlag, Basel and Stuttgart.

Trifilieff, E., Bang, L. and Ourisson, G. (1975). *Tetrahedron Lett.* 4307–4310.

Valzelli, G. and Goldaniga, G. (1973). *Biochem. Pharmac.* **22**, 911–918.

Vöhringer, H. F. and Rietbrock, N. (1974). *Clin. Pharmac. Ther.* **16**, 796–806.

Voigtländer, W., Schaumann, W., Koch, K. and Zielske, F. (1972). *Naunyn-Schmiedeberg's Arch. Pharmac.* **272**, 46–64.

Watson, E., Clark, D. R. and Kalman, S. M. (1973). *J. Pharmac. exp. Ther.* **184**, 424–431.

Wirth, K. E., Frölich, J. C., Hollifield, J. W., Falkner, F. C., Sweetman, B. S. and Oates, J. A. (1976). *Eur. J. clin. Pharmac.* **9**, 345–354.

Wong, K. C. and Spratt, J. L. (1964). *Biochem. Pharmac.* **13**, 489–495.

Wright, S. E. (1962). *J. Pharm. Pharmac.* **14**, 613–614.

Zelck, U. (1972). *In* "Biochemie und Physiologie der Alkaloide", Int. Symp. 4th 1969 (K. Mothes, Ed.), pp. 141–144. Akademi-Verlag, Berlin, East Germany.

Züllich, G., Damm, K. H., Braun, W. and Lisboa, B. P. (1975). *Archs int. Pharmacodyn. Thér.* **215**, 160–167.

7

METABOLISM OF OXYGEN HETEROCYCLIC COMPOUNDS

I. Pyrones

The pyrone structure is found in several classes of natural compounds, including chromones, coumarins, flavones, isoflavones and xanthones. The metabolism of these compounds is discussed subsequently in this chapter and the present section will deal only with the simpler derivatives of α-pyrone (1) and γ-pyrone (2).

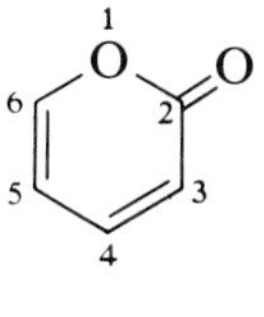

(1) 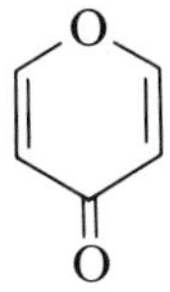   (2)

A. α-PYRONES

The simpler derivatives of α-pyrone furnish a very restricted group of plant compounds and metabolic data has been obtained on only one of these, **yangonin** (3). This compound, a kava pyrone, was found by Rasmussen *et al.* (1976) to undergo *O*-demethylation when administered orally to rats. Evidence for this included the identification of the *p*-hydroxy analogue, in conjugated form, as a urinary metabolite and also the recovery of $^{14}CO_2$ in the expired air when material labelled with ^{14}C in the *p*-methoxy group of phenyl moiety was employed.

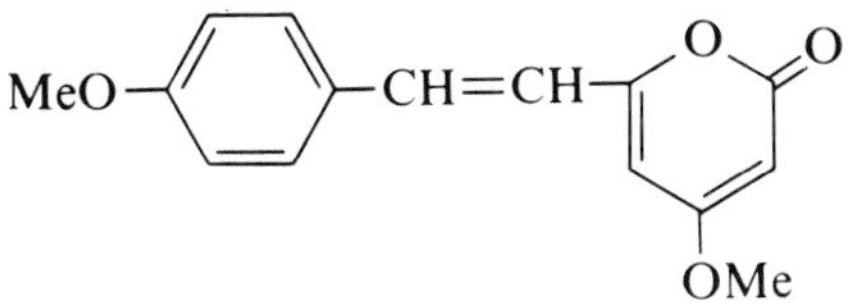

Yangonin

(3)

B. γ-Pyrones

Maltol (4), the sole naturally occurring γ-pyrone studied metabolically, is employed by the food industry as a flavour enhancer. Rennhard (1971), in an investigation which also included the synthetic homologue ethyl maltol,

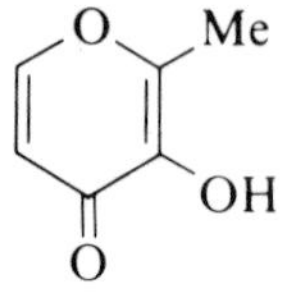

Maltol
(4)

found that maltol (10 mg/kg) was excreted in the urine as glucuronide and sulphate conjugates following i.v. administration to dogs. Conjugate excretion was rapid, occurring mainly within 6 h, and about 60% of the dose was accounted for after 48 h. It is not known if maltol is metabolized along other pathways and the fate of the remaining 40% of the dose is unknown. However, faecal excretion of maltol is probably very low as virtually none of the ethyl homologue or its conjugates was detected in the faeces following a large (200 mg/kg) oral dose.

II. Chroman, Chromene, Chromanone and Chromone Derivatives

A. Chroman Derivatives

The derivatives of chroman (5) which have been studied metabolically are the dye haematoxylin, some tocopherol derivatives and several cannabinoids. Gautrelet and Gravellat (1906) reported that **haematoxylin** was excreted partly unchanged and partly in a reduced form. However, nothing is known of the actual metabolic change occurring with this phenolic derivative of chroman.

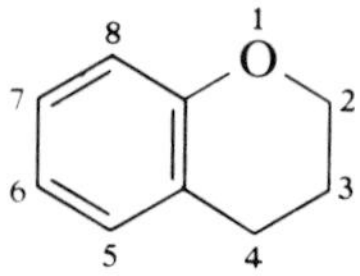

(5)

The metabolism of tocopherols is highly interesting, partly because the chroman ring undergoes hydrolytic scission at the 1,2-bond and partly because the isoprenoid side chain is extensively degraded to acidic metabolites, including a hydroxy acid which readily forms a γ-lactone. This pattern of metabolism is also seen with phylloquinone (Chapter 4, Section III.A). Simon *et al.* (1956a,b) showed that **α-tocopherol** (6), administered as the succinate ester, was metabolized and excreted in the urine by rabbits and man mainly as conjugates of tocopheronic acid (2-(3-hydroxy-3-methyl-5-carboxypentyl)-3,5,6-trimethylbenzoquinone) (8) and its γ-lactone (tocopheronolactone) (9). The conjugates are most likely glucuronides, probably not of the benzoquinones themselves but of their reduced hydroquinone derivatives. The latter, upon liberation, would be expected to be fairly easily oxidized by air. The initial step in the metabolism of α-tocopherol is considered to be hydrolysis of the chroman ring to give an α-tocopherylquinone (7) (Simon *et al.*, 1956b), a metabolite which is known to be formed in the animal body (Csallany *et al.*, 1962).

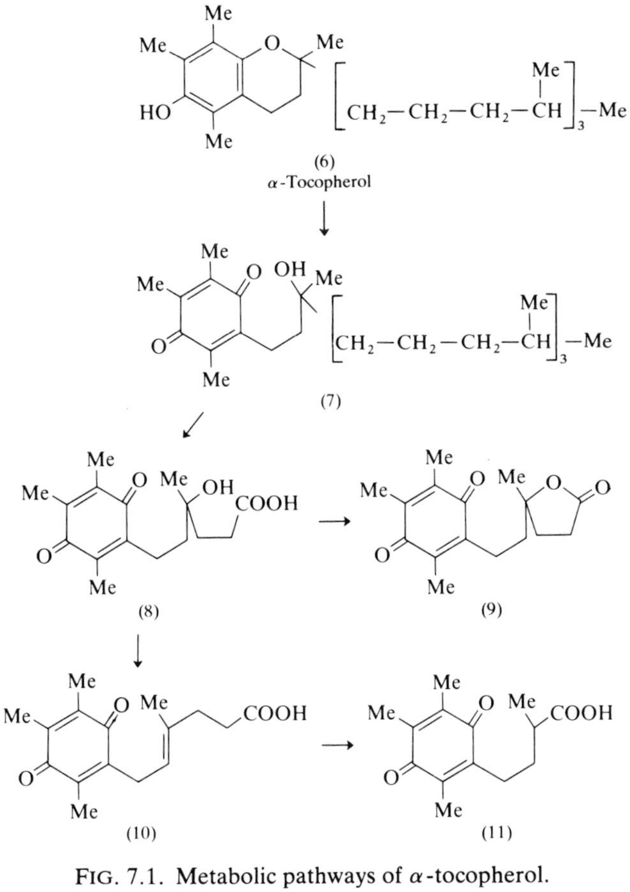

FIG. 7.1. Metabolic pathways of α-tocopherol.

Furthermore, this latter metabolite, when itself given orally to rats, is converted to the lactone (9) in higher amounts than that observed with α-tocopherol (Wiss and Gloor, 1966). The metabolic pathways illustrated in Fig. 7.1 show that tocopherylquinone (7) undergoes extensive side chain oxidation to form tocopheronic acid (8) which is excreted in the urine together with its γ-lactone (9). Further details of the metabolism of α-tocopherol are available in the review by Draper and Csallany (1969). In addition, Watanabe *et al.* (1974) found that the acidic metabolites (10) and (11) were also present in the urine of rabbits given α-tocopherol acetate. These compounds probably arise by dehydration of (8) followed by reduction and β-oxidation.

The term cannabinoid is used to describe a group of C_{21} compounds typical of and present in *Cannabis sativa*, their carboxylic acids, analogues and transformation products (Mechoulam, 1970). This group is included among the chroman derivatives although some representatives belong strictly to other chemical classes. For example, cannabinol (31) is a substituted chromene and cannabidiol (33), which lacks the heterocyclic ring, might be placed among the phenols. However, the biologically most important cannabinoids are the tetrahydrocannabinols which are derivatives of chroman. A prerequisite in any discussion of cannabinoid metabolism is an explanation of their chemical numbering. Two numbering systems are presently used with these compounds. The system used in this book has a biogenic basis, the cannabinoids being regarded as substituted monoterpenoids. With this system the major psychoactive cannabinoid is designated *trans*-Δ^1-tetrahydrocannabinol or merely Δ^1-tetrahydrocannabinol (12). The second system uses dibenzopyran numbering and is

(12)

Δ^1-Tetrahydrocannabinol
(monoterpenoid numbering)

based on the formal chemical rules used for numbering pyran-type compounds. This system is employed with the tetrahydrocannabinols and, with it, the compound noted above is designated (−)-*trans*-Δ^9-tetrahydrocannabinol (13). However, some cannabinoids are not pyran deriva-

$$\text{Me} \quad CH_2-CH_2-CH_2-CH_2-Me$$

(13)

Δ^9-Tetrahydrocannabinol
(formal numbering)

tives and this system is then no longer applicable. An important chemical feature of Δ^1-tetrahydrocannabinol, which is an oily, water-insoluble liquid, is its instability to air, light, high temperatures and acidic conditions. In the last case it isomerizes to Δ^6-tetrahydrocannabinol (24). Both the Δ^1- and Δ^6-isomers undergo oxidation in air to cannabinol (31) which is not psychoactive. Δ^6-Tetrahydrocannabinol is psychoactive but is usually a minor component of marihuana. Besides Δ^1-tetrahydrocannabinol, the other main cannabinoids found in *Cannabis sativa* are cannabinol (31) and cannabidiol (33). Additional neutral and acidic cannabinoids which are not psychoactive have also been isolated from the plant. Further information on cannabinoid chemistry is found in the articles by Mechoulam (1970, 1973), Neumeyer and Shagoury (1971) and Mechoulam *et al.* (1976).

In keeping with the general theme of this book, the present summary of the metabolic fate of cannabinoids deals nearly exclusively with their metabolic alterations. Consequently, several closely related areas including their absorption and distribution have been largely omitted. Information on these topics is available in the reviews by Lemberger (1972), Agurell *et al.* (1972), Burstein (1973), Paton and Pertwee (1973), Nahas (1973), Paton (1975) and Lemberger and Rubin (1976).

Δ^1-Tetrahydrocannabinol (12) is a highly lipophilic compound and little or no unchanged compound would therefore be expected to be excreted in the urine. This has been borne out in several investigations in which Δ^1-tetrahydrocannabinol was not detected in the urine after administering tetrahydrocannabinol to rats (Joachimoglu *et al.*, 1967), a cannabis preparation to man (Christiansen and Rafaelsen, 1969) or Δ^1-tetrahydrocannabinol to man (Lemberger *et al.*, 1971a; Hollister *et al.*, 1972) or rhesus and squirrel monkeys (Würsch *et al.*, 1972). In those studies in which some unchanged Δ^1-tetrahydrocannabinol has been found in the urine, the amounts present were very small, often about 0·01% of the administered dose. This has been shown in rats (Agurell *et al.*, 1969; Mikes *et al.*, 1971; Klausner and Dingell, 1971), rabbits (Agurell *et al.*,

1970), dogs (Hunt, 1977) and man (Lemberger *et al.*, 1970; Hollister *et al.*, 1974). On the other hand, it might seem reasonable to assume that the phenolic 3'-hydroxyl group could undergo direct conjugation to give a urinary glucuronide or sulphate of Δ^1-tetrahydrocannabinol. However, these conjugates have not been reported as metabolites and several studies have shown that enzymic hydrolysis of the urine from rats (Agurell *et al.*, 1969; Mikes *et al.*, 1971), rabbits (Agurell *et al.*, 1970) and man (Lemberger *et al.*, 1970) does not increase the amount of detectable Δ^1-tetrahydrocannabinol. Although unchanged Δ^1-tetrahydrocannabinol may be found in the faeces of rats (Mikes *et al.*, 1971; Turk *et al.*, 1973) the amounts present are very small. Unchanged compound was not found in the faeces of humans given Δ^1-tetrahydrocannabinol by i.v. injection (Lemberger *et al.*, 1970) although the opposite finding has been reported (Wall, 1975). Furthermore, the biliary route is not important in the excretion of the unchanged compound in rats (Joachimoglu *et al.*, 1967; Turk *et al.*, 1973; Widman *et al.*, 1974; Siemans and Kalant, 1975).

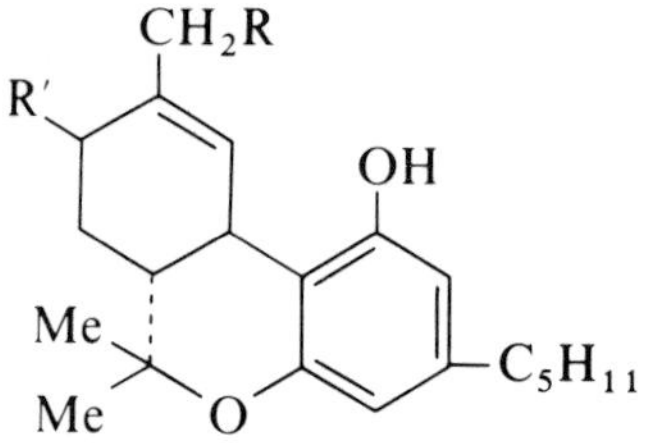

(14)
(a) R = OH, R' = H
(b) R = H, R' = ···OH
(c) R = H, R' = —OH
(d) R = R' = OH

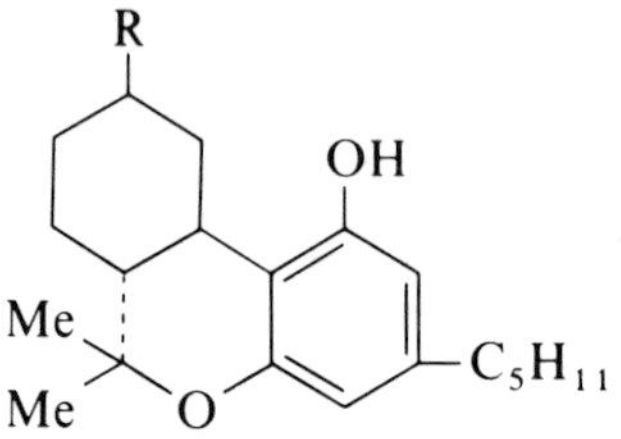

(15)
(a) R = CH$_2$OH
(b) R = ···COOH
(c) R = —COOH

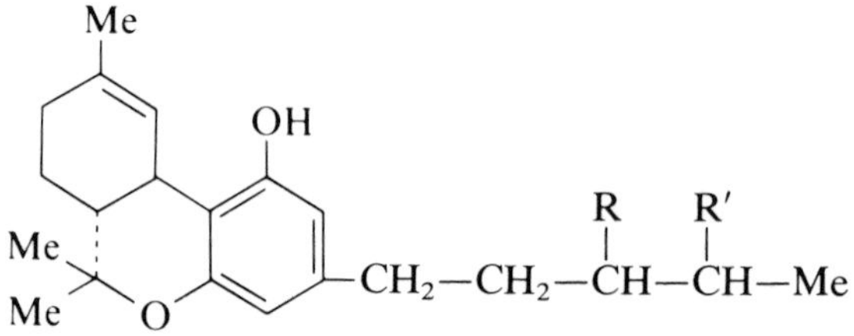

(16)
(a) R = OH, R' = H
(b) R = H, R' = OH

The identified metabolites of Δ^1-tetrahydrocannabinol are listed in Table 7.1. Many of the investigations cited refer also to unidentified metabolites, but it is not possible to ascertain to what extent these may be

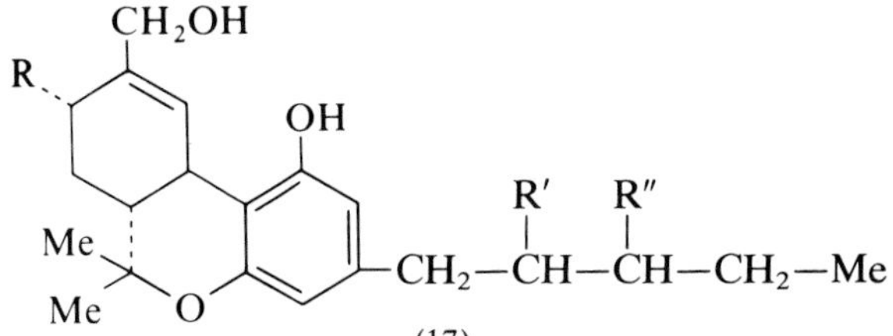

(17)

(a) R = R″ = H, R′ = OH
(b) R = R′ = H, R″ = OH
(c) R = R′ = OH, R″ = H
(d) R = R″ = OH, R′ = H

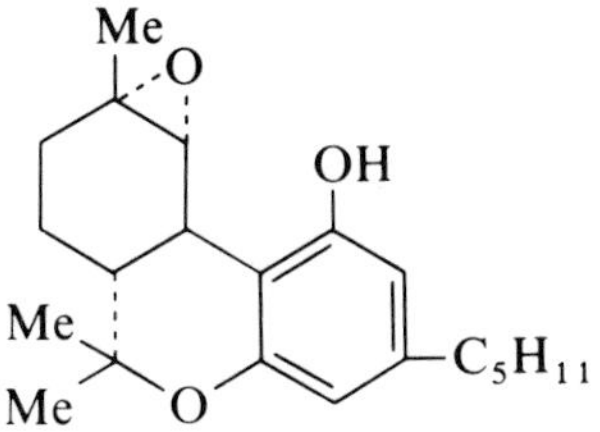

(18)

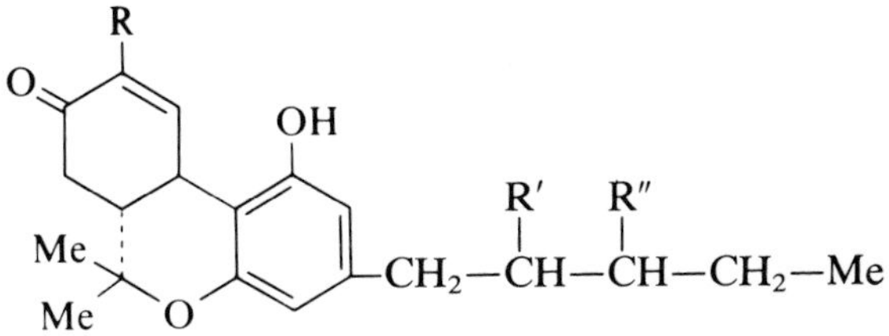

(19)

(a) R = Me, R′ = R″ = H
(b) R = CH_2OH, R′ = R″ = H
(c) R = Me, R′ = OH, R″ = H
(d) R = Me, R′ = H, R″ = OH
(e) R = CH_2OH, R′ = OH, R″ = H
(f) R = CH_2OH, R′ = H, R″ = OH

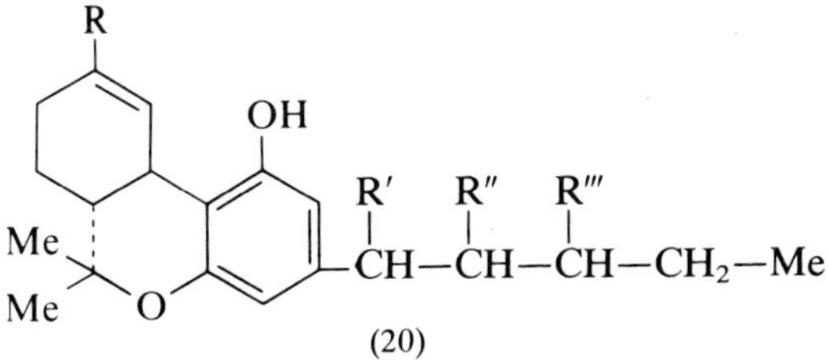

(20)

(a) R = CHO, R′ = R″ = R‴ = H
(b) R = COOH, R′ = R″ = R‴ = H
(c) R = COOH, R′ = OH, R″ = R‴ = H
(d) R = COOH, R′ = R‴ = H, R″ = OH
(e) R = COOH, R′ = R″ = H, R‴ = OH

identical to metabolites identified in other investigations and therefore
listed in Table 7.1. There is now abundant information which indicates that
the liver is the principal site of metabolism and that the major pathway
involves hydroxylation at the 7-position, i.e. allylic hydroxylation. This
activity is associated with the liver microsomes and the mono-oxygenase

TABLE 7.1

Metabolites of Δ^1-tetrahydrocannabinol (12)

Metabolite	Structure	Species and conditions	Reference
7-Hydroxy-Δ^1-tetrahydrocannabinol	(14a)	Mouse, various tissues in vivo	Christensen et al. (1971)
		Mouse, brain in vivo	Gill and Jones (1971)
		Mouse, liver and brain in vivo	Jones et al. (1974b)
		Mouse, liver in vivo	Harvey and Paton (1976)
			Harvey et al. (1977b)
		Rat, liver in vitro[a]	Wall et al. (1970)
			Wall (1971)
			Burstein and Kupfer (1971)
			Gill et al. (1973)
		Rat, liver and blood in vivo	Wall (1971)
		Rat, brain in vivo	Ho et al. (1973)
		Rat, various tissues in vivo	Leighty (1973)
			Willinsky et al. (1974)
			Nilsson et al. (1970)
			Wall (1971)
		Rabbit, liver in vitro[a]	Ben-Zvi and Burstein (1975)
		Dog, liver in vitro[a]	Widman et al. (1975b)
		Dog, perfused lung	Widman et al. (1975b)
		Squirrel monkey, brain in vivo	Ho et al. (1972)
		Marmoset monkey, various organs in vivo	Just et al. (1975)
		Marmoset monkey, brain in vivo	Erdmann et al. (1976)
		Man, jejunum in vitro	Greene and Saunders (1974)
		Man, blood in vivo	Lemberger et al. (1971b)
			Wall et al. (1972)
			Perez-Reyes et al. (1973)
		Man, urine and faeces	Lemberger et al. (1970)
		Man, faeces	Wall (1975)
		Man, plasma, urine and faeces	Wall et al. (1976)
7β-Hydroxyhexahydrocannabinol	(15a)[b]	Mouse, liver in vivo	Harvey et al. (1977c)
Fatty acid conjugates of 7-hydroxy-Δ^1-tetrahydrocannabinol		Rat, liver in vitro[a] and in vivo, various organs in vivo	Leighty et al. (1976)

6α-Hydroxy-Δ¹-tetrahydrocannabinol	(14b)	Mouse, liver *in vitro*[a]	Jones *et al.* (1974b) Ben-Zvi *et al.* (1974b)
			Harvey and Paton (1976) Harvey *et al.* (1977b)
		Mouse, brain *in vivo* Rabbit, liver *in vitro*[a] Dog, liver *in vitro*[a] Dog, perfused lung Man, blood *in vivo*	Jones *et al.* (1974b) Ben-Zvi and Burstein (1975) Widman *et al.* (1975b) Widman *et al.* (1975b) Wall *et al.* (1972)
6β-Hydroxy-Δ¹-tetrahydrocannabinol	(14c)	Mouse, liver *in vivo* Rabbit, liver *in vitro*[a] Dog, liver *in vitro*[a] Dog, perfused lung Man, blood *in vivo* Man, faeces	Harvey *et al.* (1977b) Wall (1971) Widman *et al.* (1975b) Widman *et al.* (1975b) Wall *et al.* (1972) Wall *et al.* (1976)
3″-Hydroxy-Δ¹-tetrahydrocannabinol	(16a)	Mouse, liver *in vivo* Dog, liver *in vitro*[a] Dog, perfused lung	Harvey *et al.* (1977b) Widman *et al.* (1975b) Widman *et al.* (1975b)
4″-Hydroxy-Δ¹-tetrahydrocannabinol	(16b)	Dog, liver *in vitro*[a] Dog, perfused lung	Widman *et al.* (1975b); Widman *et al.* (1975b)
6,7-Dihydroxy-Δ¹-tetrahydrocannabinol	(14d)	Mouse, various organs *in vivo* Mouse, liver *in vivo*	Christensen *et al.* (1971) Harvey and Paton (1976) Harvey *et al.* (1977b)
		Rat, liver *in vitro*[a]	Wall *et al.* (1970) Wall (1971) Wall (1971)
		Rat, liver and blood *in vivo* Rat, urine and faeces Rabbit, liver *in vitro*[a] Rabbit, liver *in vitro*[a] Man, blood *in vivo*	Mikes *et al.* (1971) Ben-Zvi and Burstein (1975) Wall (1971) Wall *et al.* (1972) Perez-Reyes *et al.* (1973)
		Man, urine Man, faeces[c]	Perez-Reyes *et al.* (1973) Wall (1975)

TABLE 7.1—*continued*

Metabolite	Structure	Species and conditions	Reference
2″,7-Dihydroxy-Δ^1-tetrahydrocannabinol	(17a)	Mouse, liver *in vivo*	Harvey *et al.* (1977b)
3″,7-Dihydroxy-Δ^1-tetrahydrocannabinol	(17b)	Mouse, liver *in vivo*	Harvey *et al.* (1977b)
2″,6α,7-Trihydroxy-Δ^1-tetrahydrocannabinol	(17c)	Mouse, liver *in vivo*	Harvey *et al.* (1977b)
3″,6α,7-Trihydroxy-Δ^1-tetrahydrocannabinol	(17d)	Mouse liver *in vivo*	Harvey *et al.* (1977b)
1,2-Epoxyhexahydrocannabinol	(18)	Rabbit, liver *in vitro* [a]	Ben-Zvi and Burstein (1975)
		Dog, liver *in vitro* [a]	Widman *et al.* (1975b)
		Squirrel monkey, liver *in vitro* [a]	Gurny *et al.* (1972)
Δ^1-Tetrahydrocannabinol-6-one	(19a)	Mouse, liver *in vitro* [a]	Jones *et al.* (1974b)
		Squirrel monkey, liver *in vitro* [a]	Gurny *et al.* (1972)
7-Hydroxy-Δ^1-tetrahydrocannabinol-6-one	(19b)	Mouse, liver *in vivo*	Harvey *et al.* (1977b)
2″-Hydroxy-Δ^1-tetrahydrocannabinol-6-one	(19c)	Mouse, liver *in vivo*	Harvey *et al.* (1977b)
3″-Hydroxy-Δ^1-tetrahydrocannabinol-6-one	(19d)	Mouse, liver *in vivo*	Harvey *et al.* (1977b)
2″,7-Dihydroxy-Δ^1-tetrahydrocannabinol-6-one	(19e)	Mouse, liver *in vivo*	Harvey *et al.* (1977b)
3″,7-Dihydroxy-Δ^1-tetrahydro-cannabinol-6-one	(19f)	Mouse, liver *in vivo*	Harvey *et al.* (1977b)
7-Oxo-Δ^1-tetrahydrocannabinol	(20a)	Rat, liver *in vitro* [a]	Ben-Zvi and Burstein (1974)
Δ^1-Tetrahydrocannabinol-7-oic acid	(20b)	Mouse, liver *in vivo*	Harvey and Paton (1976)
			Harvey *et al.* (1977b)
		Man, faeces [d]	Wall (1975)
		Man, plasma, urine, faeces [d]	Wall *et al.* (1976)
Hexahydrocannabinol-7α-oic acid	(15b)	Mouse, liver *in vivo*	Harvey *et al.* (1977c)
Hexahydrocannabinol-7β-oic acid	(15c)	Mouse, liver *in vivo*	Harvey *et al.* (1977c)
6α-Hydroxy-Δ^1-tetrahydrocannabinol-7-oic acid	(21a)	Mouse, liver *in vivo*	Harvey and Paton (1976)
			Harvey *et al.* (1977b)
1″-Hydroxy-Δ^1-tetrahydrocannabinol-7-oic acid	(20c)	Rabbit, urine	Burstein *et al.* (1972)
2″-Hydroxy-Δ^1-tetrahydrocannabinol-7-oic acid	(20d)	Mouse, liver *in vivo*	Harvey and Paton (1976)
			Harvey *et al.* (1977b)
3″-Hydroxy-Δ^1-tetrahydrocannabinol-7-oic acid	(20e)	Rabbit, urine	Burstein *et al.* (1972)
		Mouse, liver *in vivo*	Harvey and Paton (1976)
			Harvey *et al.* (1977b)

Compound		Source	Reference
5″-Nor-Δ^1-tetrahydrocannabinol-4″-oic acid	(22a)	Guinea pig, liver *in vivo*	Martin *et al.* (1976a)
4″,5″-Bisnor-Δ^1-tetrahydrocannabinol-3″-oic acid	(22b)	Mouse, liver *in vivo* Guinea pig, liver *in vivo* Rabbit, liver *in vivo*	Martin *et al.* (1976a) Martin *et al.* (1976a) Martin *et al.* (1976a)
3″,4″,5″-Trisnor-Δ^1-tetrahydrocannabinol-2″-oic acid	(22c)	Guinea pig, liver *in vivo*	Martin *et al.* (1976a)
2″,3″,4″,5″-Tetranor-Δ^1-tetrahydro-cannabinol-1″-oic acid[e]	(22d)	Rabbit, liver *in vivo*	Martin *et al.* (1976a)
2″,6α-Dihydroxy-Δ^1-tetrahydrocannabinol-7-oic acid	(21b)	Mouse, liver *in vivo*	Harvey and Paton (1976) Harvey *et al.* (1977b)
3″,6α-Dihydroxy-Δ^1-tetrahydrocannabinol-7-oic acid	(21c)	Mouse, liver *in vivo*	Harvey and Paton (1976) Harvey *et al.* (1977b)
4″,5″-Bisnor-Δ^1-tetrahydrocannabinol-7,3″-dioic acid	(23)	Rabbit, urine	Nordqvist *et al.* (1974)
Cannabinol	(31)	Rat, blood *in vivo* Rat, bile Rat, faeces	McCallum *et al.* (1975) Widman *et al.* (1974) Mikes *et al.* (1971)
Cannabinol-7-oic acid	(32b)	Rhesus monkey, urine	Ben-Zvi *et al.* (1974a, 1976)

[a] Employed microsomal fraction or microsomal + soluble fractions.
[b] See text.
[c] After administering compound (14a).
[d] Also after administering compound (14a).
[e] Tentative identification.

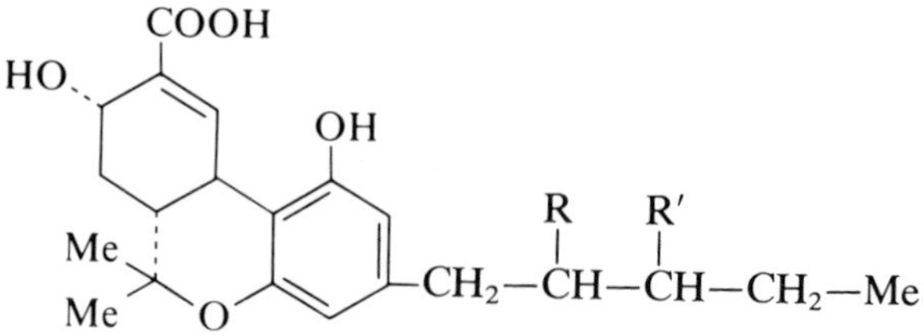

(21)

(a) R = R' = H
(b) R = OH, R' = H
(c) R = H, R' = OH

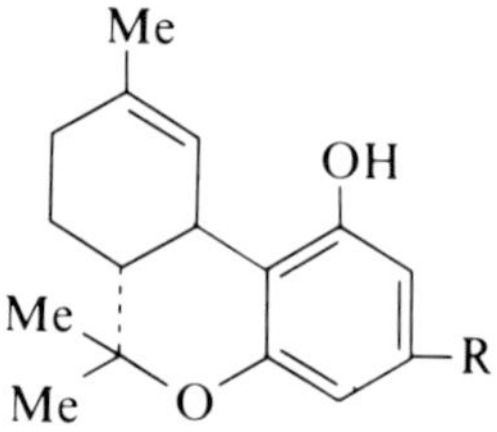

(22)

(a) R = CH$_2$—CH$_2$—CH$_2$—COOH
(b) R = CH$_2$—CH$_2$—COOH
(c) R = CH$_2$—COOH
(d) R = COOH

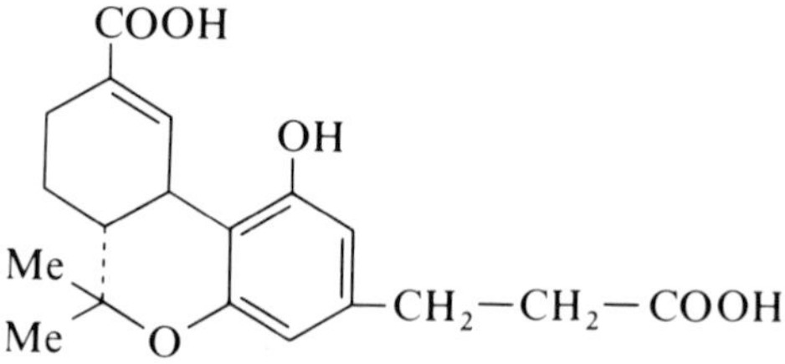

(23)

system (Burstein and Kupfer, 1971). Allylic hydroxylation can also take place at the 6-position so that 6- or 7-monohydroxy- and 6,7-dihydroxy-tetrahydrocannabinols are formed. Harvey *et al.* (1977b) identified both the 6α,7- and 6β,7-dihydroxy metabolites in the liver of mice given Δ^1-tetrahydrocannabinol. Hydroxylation of the pentyl moiety also occurs and 1″-, 2″-, 3″- and 4″-hydroxy metabolites have been detected (Table 7.1). It is likely that ω-hydroxylation also takes place as this reaction can partly account for the formation of several metabolites in which the pentyl side-chain has been shortened. Recent evidence indicates that reduced metabolites (hexahydrocannabinols) may be formed in small amounts (Harvey *et al.*, 1977c). In addition to the identified metabolites (15a), (15b) and (15c), three derivatives of compounds (15b) or (15c) were detected which contained a hydroxyl group, located probably in the pentyl side-chain.

Although 7-hydroxylation occurs rapidly and extensively, it is noteworthy that this metabolite is not prominent among those actually excreted in the urine. It sometimes is seen as a major faecal metabolite, as with man (Wall, 1975), but this is not the case in rats (Turk *et al.*, 1973). In the latter species extensive biliary and faecal excretion of Δ^1-tetrahydrocannabinol metabolites occurs, but these are mostly of an acidic and highly polar nature (Klausner and Dingell, 1971; Turk et al., 1973; Widman et al., 1974). Wall et al. (1976) studied especially the formation of highly polar acidic metabolites in man. These compounds made up the major fraction of metabolites in both plasma and urine and, due to their highly polar nature, were excreted largely unconjugated. Some of the data shown in Table 7.1 indicate that several quite polar metabolites are formed in other species as well. Harvey et al. (1977a) studied the conjugated metabolites of Δ^1-tetrahydrocannabinol produced in vivo in mouse liver. They found that only trace amounts of glucuronide conjugates were formed. This result contrasts markedly with that observed with cannabinol and cannabidiol (see below).

Leighty (1973) reported that rats showed prolonged tissue retention of a non-polar metabolite of Δ^1-tetrahydrocannabinol. Further information on this subject was obtained by incubating 7-hydroxy-Δ^1-tetrahydrocannabinol with a rat liver microsomal enzyme system (Leighty et al., 1976). The non-polar material formed was shown to consist of several fatty acid conjugates of the 7-hydroxy metabolite. The most abundant esters contained palmitic (C_{16}) and stearic (C_{18}) acids, however a small amount of material was esterified with unsaturated C_{18}-acids (presumably oleic and linoleic). A similar metabolic reaction was discovered by Mikes et al. (1971) who identified the diacetate of 7-hydroxy-Δ^1-tetrahydrocannabinol in the bile of rats given Δ^1-tetrahydrocannabinol.

Several references in Table 7.1 deal with the extra-hepatic metabolism of Δ^1-tetrahydrocannabinol to identified metabolites. Also, Nakazawa and Costa (1971) found that rat lung preparations produced in vitro a different pattern of metabolites including some compounds not observed with liver preparations. Several metabolites formed by the isolated perfused dog lung have been identified by Widman et al. (1975b) and are listed in Table 7.1.

Studies in the mouse indicated that the metabolism of the unnatural enantiomorph $(+)$-Δ^1-tetrahydrocannabinol is similar to that shown by the $(-)$-isomer (Jones et al., 1974a).

Δ^6-Tetrahydrocannabinol (24), which is designated Δ^8-tetrahydrocannabinol under the dibenzopyran numbering system, is a psychoactive cannabinoid usually present in varying but small amounts in cannabis. Although the data available on the metabolism of this compound are considerably less than that summarized above for the Δ^1-isomer, it can nevertheless be concluded that the general features of metabolism are similar with both isomers. Especially noteworthy is the well-documented fact that allylic hydroxylation (at the 5- or 7-positions) is also a prominent initial feature of the metabolism of Δ^6-tetrahydrocannabinol. The data

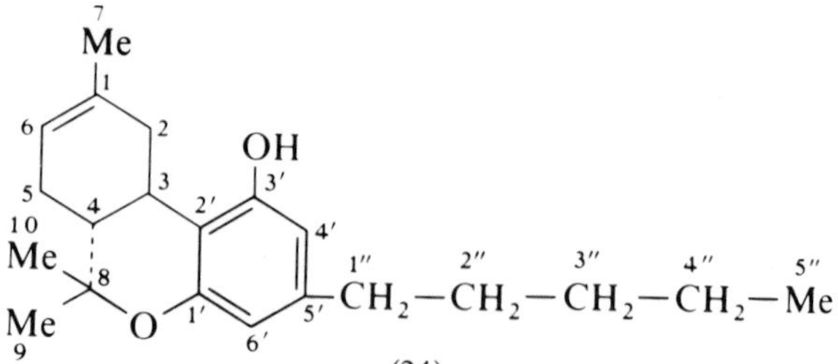

(24)

Δ^6-Tetrahydrocannabinol

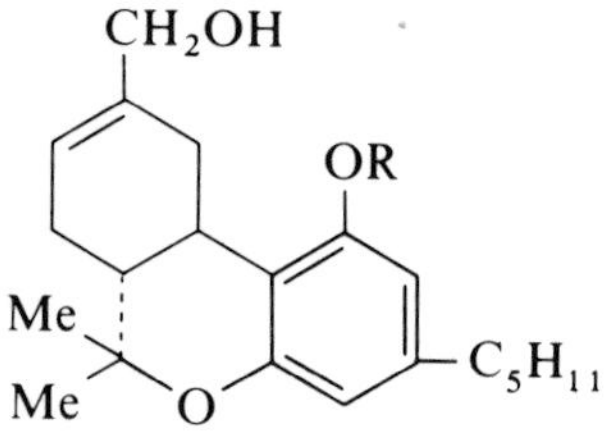

(25)

(a) R = H
(b) R = Me

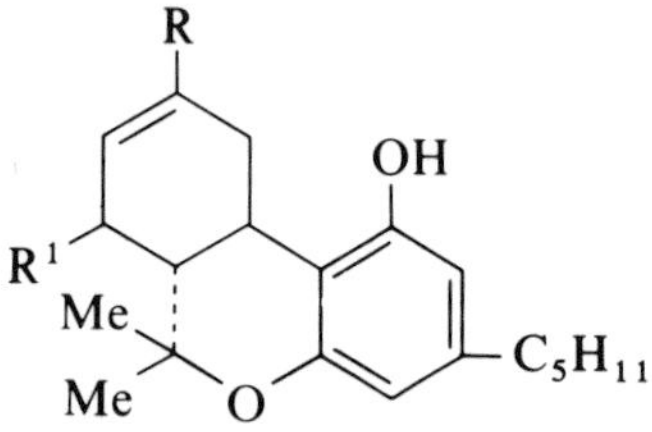

(26)

(a) R = Me, R′ = ···OH
(b) R = Me, R′ = —OH
(c) R = Me, R′ = =O
(d) R = CH₂OH, R′ = ···OH
(e) R = CH₂OH, R′ = —OH

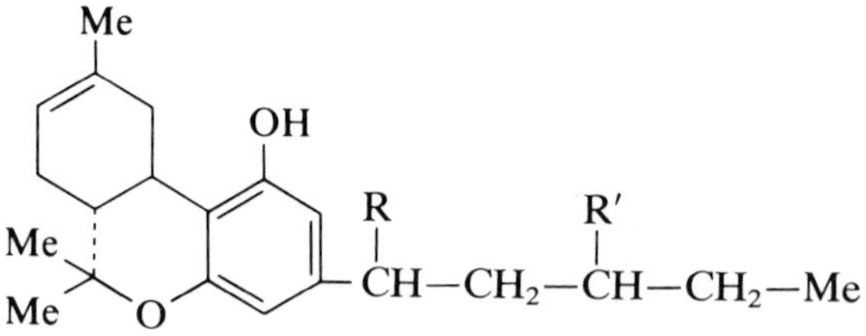

(27)

(a) R = OH, R′ = H
(b) R = H, R′ = OH

summarized in Table 7.2 show may other similarities to that in Table 7.1 including the identification of the epoxide intermediate (28) analogous to metabolite (18) included in Table 7.1. Interestingly, the corresponding diol (29a) and the two further oxidized metabolites (29b) and (29c) were also demonstrated. The latter metabolites and several others listed in Table 7.2 demonstrate that multiple sites of oxidation in the pentyl side chain are also present in Δ^6-tetrahydrocannabinol. Several reduced metabolites (hexahydrocannabinols) have been shown to be metabolites of both Δ^1- and Δ^6-tetrahydrocannabinol (Harvey et al., 1977c). These, produced in

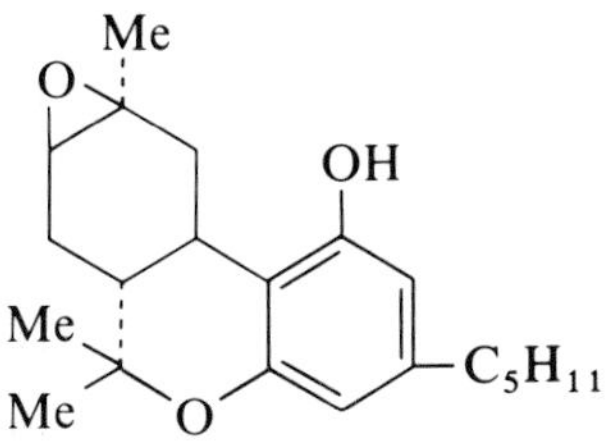

(28)

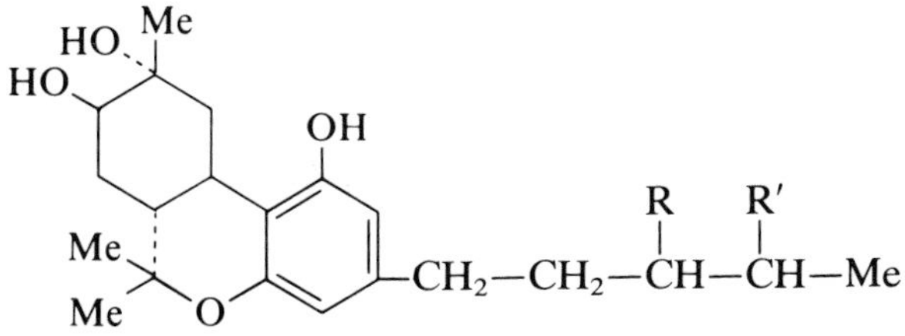

(29)

(a) R = R' = H
(b) R = OH, R' = H
(c) R = H, R' = OH

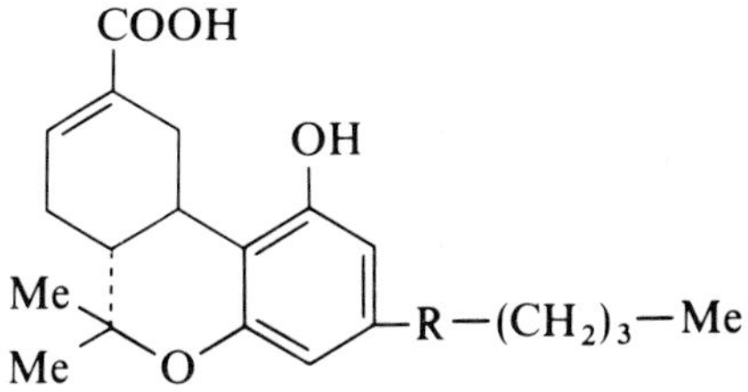

(30)

(a) R = CH$_2$
(b) R = CHOH
(c) R = C=O

small amounts in mouse liver, are compounds (15a), (15b) and (15c) and the three hydroxy derivatives of (15b) or (15c) noted above in the summary of Δ^1-tetrahydrocannabinol metabolism. In identical experiments to those described above using Δ^1-tetrahydrocannabinol and its 7-hydroxy derivative, Leighty et al. (1976) showed that the Δ^6-isomer also forms fatty acid conjugates. These also are esters formed from the 7-hydroxy metabolite and palmitic (C_{16}) and stearic (C_{18}) acids and lesser amounts of unsaturated C_{18}-acids. The in vivo formation of glucuronide conjugates of Δ^6-tetrahydrocannabinol metabolites in mouse liver is, as with the Δ^1-isomer, a minor metabolic pathway (Harvey et al., 1977a). Interestingly, Mechoulam et al. (1977) found that rabbit UDP-glucuronyl transferase formed a C-glucuronide from Δ^6-tetrahydrocannabinol and UDP-glucuronic acid. In addition to the identified metabolites of Δ^6-tetrahydrocannabinol, some unidentified major metabolites have been detected in vitro with liver preparations (Maynard et al., 1971), in various organs of animals treated with the cannabinoid (Estevez et al., 1973; Just et al., 1975) and in the urine of treated rabbits (Nilsson et al., 1973a). While some of these compounds may correspond to metabolites identified in other investigations (Table 7.2), it seems reasonable to assume that the overall metabolic pattern is fairly complex and that several of the ultimate metabolites are

TABLE 7.2

Metabolites of Δ^6-tetrahydrocannabinol (24)

Metabolite	Structure	Species and conditions	Reference
7-Hydroxy-Δ^6-tetrahydrocannabinol	(25a)	Rat, liver in vitro[a]	Foltz et al. (1970) Wall (1971)
		Rat, liver in vivo	Foltz et al. (1970)
		Rat, brain in vivo	Ho et al. (1973)
		Rabbit, urine	Burstein et al. (1970) Ben-Zvi et al. (1970) Just et al. (1975)
		Marmoset monkey, various organs in vivo	
		Marmoset monkey, brain in vivo	Erdmann et al. (1976)
		Rhesus monkey, urine and faeces	Gau et al. (1974)
		Man, faeces[b]	Wall (1975)
7β-Hydroxyhexahydrocannabinol	(15a)	Mouse, liver in vivo	Harvey et al. (1977c)
Fatty acid conjugates of 7-hydroxy-Δ^6-tetrahydrocannabinol	[c]	Rat, liver in vitro and in vivo, various organs in vivo	Leighty et al. (1976)
7-Hydroxy-3'-O-methyl-Δ^6-tetrahydrocannabinol[b]	(25b)	Rat, liver and brain in vivo	Estevez et al. (1973)
5α-Hydroxy-Δ^6-tetrahydrocannabinol	(26a)	Squirrel monkey, liver in vitro[a]	Gurny et al. (1972)
5β-Hydroxy-Δ^6-tetrahydrocannabinol	(26b)	Squirrel monkey, liver in vitro[a]	Gurny et al. (1972)

Compound		Species/conditions	Reference
1″-Hydroxy-Δ^6-tetrahydrocannabinol	(27a)	Dog, liver *in vitro*[a]	Maynard *et al.* (1971)
3″-Hydroxy-Δ^6-tetrahydrocannabinol	(27b)	Dog, liver *in vitro*[a]	Maynard *et al.* (1971)
5α,7-Dihydroxy-Δ^6-tetrahydrocannabinol	(26d)	Rat, liver *in vitro*[a]	Wall (1971)
5β,7-Dihydroxy-Δ^6-tetrahydrocannabinol	(26d)	Rat, liver *in vitro*[a]	Wall (1971)
5,7-Dihydroxy-Δ^6-tetrahydrocannabinol[b]	(26d,e)	Rat, liver and brain *in vivo*	Estevez *et al.* (1973)
1,6-Epoxyhexahydrocannabinol	(28)	Mouse, liver *in vivo*	Harvey and Paton (1977)
1α,6β-Dihydroxyhexahydrocannabinol	(29a)	Mouse, liver *in vivo*	Harvey and Paton (1977)
1α,3″,6β-Trihydroxyhexahydrocannabinol	(29b)	Mouse, liver *in vivo*	Harvey and Paton (1977)
1α,4″,6β-Trihydroxyhexahydrocannabinol	(29c)	Mouse, liver *in vivo*	Harvey and Paton (1977)
Δ^6-Tetrahydrocannabinol-5-one	(26c)	Squirrel monkey, liver *in vitro*[a]	Gurny *et al.* (1972)
Δ^6-Tetrahydrocannabinol-7-oic acid[b]	(30a)	Rabbit, urine	Nilsson *et al.* (1973a)
		Man, faeces	Wall (1975)
Hexahydrocannabinol-7α-oic acid	(15b)	Mouse, liver *in vivo*	Harvey *et al.* (1977c)
Hexahydrocannabinol-7β-oic acid	(15c)	Mouse, liver *in vivo*	Harvey *et al.* (1977c)
1″-Hydroxy-Δ^6-tetrahydrocannabinol-7-oic acid[d]	(30b)	Rhesus monkey, urine and faeces	Gau *et al.* (1974)
1″-Keto-Δ^6-tetrahydrocannabinol-7-oic acid	(30c)	Rhesus monkey, urine and faeces	Gau *et al.* (1974)
Cannabinol	(31)	Rat, blood *in vivo*	McCallum *et al.* (1975)

[a] Employed microsomal fraction or microsomal+ soluble fractions.
[b] After administering compound (25a).
[c] See text.
[d] Hydroxyl group may, however, be at the 2″-position.

highly polar compounds (Nilsson *et al.*, 1973a), similar to the hydroxylated acids formed from Δ^1-tetrahydrocannabinol (see above).

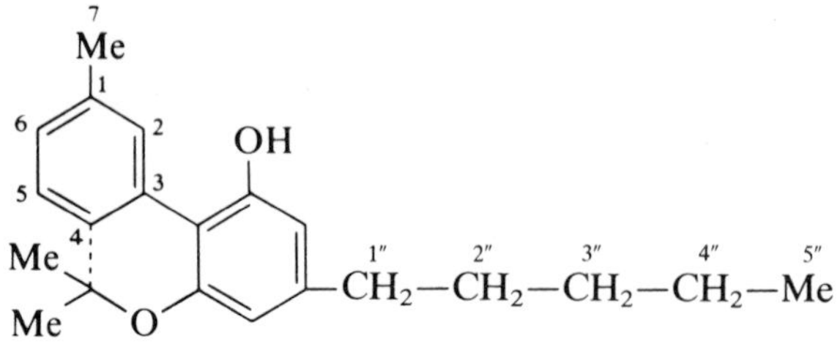

(31)

Cannabinol

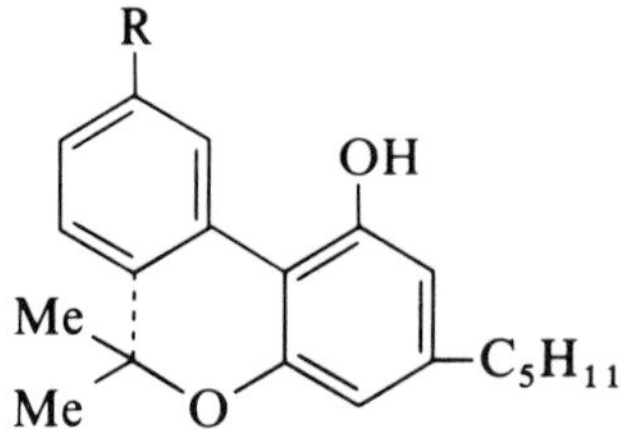

(32)

(a) R = CH$_2$OH
(b) R = COOH

As noted initially in this summary of cannabinoid metabolism, the most important naturally occurring constituents other than Δ^1-tetrahydrocannabinol itself are cannabinol and cannabidiol. However, these compounds are not psychoactive and interest in their metabolism has accordingly been much less than that shown towards the tetrahydrocannabinols. **Cannabinol** (31) appears to be a transient or minor metabolite of Δ^1-tetrahydrocannabinol (Table 7.1). Several investigations have shown that cannabinol is metabolized *in vitro* by rat or rabbit liver preparations mainly to 7-hydroxycannabinol (32a) (Agurell *et al.*, 1971; Widman *et al.*, 1971; Wall, 1971; Widman *et al.*, 1975a). These results indicate that the liver microsomal enzyme system can hydroxylate the methyl group at the 1-position of the aromatic ring of cannabinol as well as the allylic methyl group at this position in the tetrahydrocannabinols. Wall (1971) also detected two other hydroxylated metabolites which were tentatively identified as the 2″-hydroxy and 2″,7-dihydroxy derivatives of cannabinol. The study by Widman *et al.* (1975a) showed that liver preparations are able to hydroxylate most of the positions in the pentyl side chain of cannabinol. With rat liver, positions 2″, 3″, 4″ and 5″ were hydroxylated, but these metabolites together accounted for only about 2% of the converted material. On the other hand, rabbit liver showed a much greater ability to carry out hydroxylation of the pentyl group, forming 1%, 30% and 5% of the 3"-, 4"- and 5"-hydroxy metabolites, respectively. Fonseka and Widman (1977) found that rat liver microsome preparations formed four dihydroxylated derivatives of cannabinol. These were the 1",7- 2",7- 3",7- and 4",7-dihydroxy compounds, however side chain oxidation occurred predominantly at the 4"- and 3"-positions.

Findings similar to those noted above have been made when cannabinol is given to animals and man. Following the i.v. administration of cannabinol to mice, both unchanged compound and 7-hydroxycannabinol (32a) were detected in the brain 15, 30 and 60 min after injection (Ho et al., 1973). Burstein and Varanelli (1975) studied the urinary and faecal metabolism of [14C]-cannabinol given by s.c. injection to mice. They found that it was hydroxylated at the 7-position and in the side-chain and that further oxidation to acidic metabolites occurred. The presence of cannabinol-7-oic acid (32b) in the faeces was established and the inter- mediate 7-hydroxy derivative was tentatively identified. The acidic deriva- tive (32b) was also a major metabolite in the urine, as was a compound in which the pentyl side-chain was replaced by a carboxyl group. The metabolites definitely or tentatively identified represented only a portion of the large number of metabolites of cannabinol in the mouse and it was concluded that the metabolic pathways probably are of similar complexity to those seen with other cannabinoids. Harvey et al. (1977a) found that mouse liver contained three glucuronide conjugates following cannabinol administration to the animals. These were identified as the phenolic glucuronides of cannabinol, 7-hydroxycannabinol and cannabinol-7-oic acid. Yisak et al. (1977) found that cannabinol (100 mg/kg, i.v.) was slowly eliminated by rats, mainly in the faeces, and they identified nine oxy- genated metabolites in the neutral fraction from faeces. These included 7-hydroxycannabinol (32a) and all of the monohydroxylated side-chain derivatives. The relative proportions of these, in decreasing order, were: 7-hydroxy-, 4"-hydroxy-, 1"-hydroxy-, 2"-hydroxy-, 3"-hydroxy- and 5"- hydroxycannabinol. Evidence was also obtained for the presence of the 7-aldehyde derivative of cannabinol. Only two dihydroxylated metabolites were detected, 1",7-dihydroxy- and smaller amounts of 4",7-dihydroxy- cannabinol. The metabolism of cannabinol in man following the i.v. administration of ³H-labelled material (18 mg) was studied by Wall et al. (1976). They found that polar acidic metabolites were the most abundant plasma metabolites. The 7-carboxy derivative (32b) accounted for about a fifth of this amount, however the intermediate 7-hydroxycannabinol was present only in very small quantities. The urine is a minor excretory route and only about 8% of the radioactivity was excreted in the urine in 72 h. Most of this material consisted of unconjugated acidic metabolites, especially polar acids. Faecal excretion accounted for 35% of the radioactivity in 72 h, the major fractions consisting of monohydroxy derivatives and cannabinol-7-oic acid. The monohydroxy compounds were mainly the 7-hydroxy and, tentatively, the 4″-hydroxy derivatives.

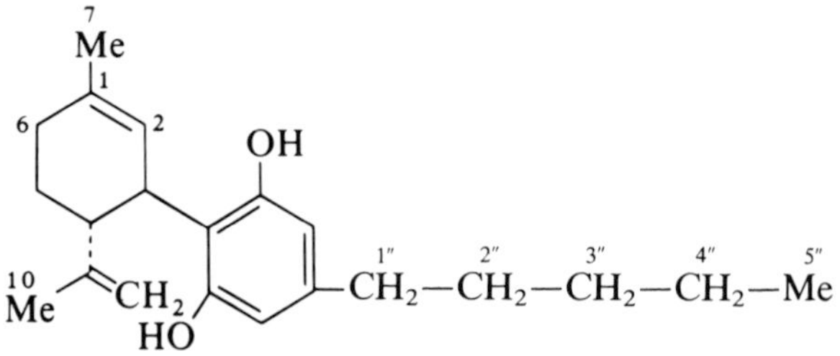

(33)

Cannabidiol

The metabolism of **cannabidiol** (33) closedly resembles that of cannabinol (31). Several investigations using rat liver microsomal preparations showed that hydroxylation at multiple sites occurred. Nilsson *et al.* (1971) reported that the 7-, 10- and 1″-hydroxy derivatives were formed. Subsequently, Nilsson *et al.* (1973b) added 3″-hydroxycannabidiol to this list and also found that the 7-hydroxy derivative (34b) was the main metabolite detected. The latter finding was confirmed by Martin *et al.* (1976b) who showed that the 6α-hydroxy derivative was the second most abundant metabolite. Only traces of 6β-hydroxycannabidiol were found. Hydroxylation also occurred at all positions of the pentyl side chain, the ratio of yields of the 4″-, 3″-, 1″-, 2″- and 5″-hydroxylated compounds being about 4:2:1:1:1. Further experiments of this type indicated that cannabidiol is also converted by rat liver preparations to a large number of dioxygenated metabolites (Martin *et al.*, 1976c). The dihydroxy compounds identified were the 6,7-, 1″,7-, 3″,7-, 4″,7-, 5″,7-, 2″,6-, 3″,6β-and, tentatively, 4″,6β-isomers. Additionally, 3″-hydroxy-6-oxo- and 4″-hydroxy-6-oxocannabidiol were found. The relative abundances of these metabolites reflected the amounts of the monohydroxylated compounds produced and 7-hydroxy derivatives were again the most abundant. Nilsson *et al.* (1973b) found no evidence for the cyclization of cannabidiol to Δ^1-tetrahydrocannabinol.

The nature of the *in vivo* conjugates of cannabidiol in mouse liver was studied by Harvey *et al.* (1976, 1977a). A subsequent report (Martin *et al.*, 1977) also dealt with this subject as well as the oxidized derivatives of cannabidiol formed in mouse liver. The glucuronide conjugates of

(34)

(a) R = $CH_2CH_2CH_2CH_2Me$, R' = Me, R'' = OH
(b) R = $CH_2CH_2CH_2CH_2Me$, R' = CH_2OH, R'' = H
(c) R = $CH(OH)CH_2CH_2CH_2Me$, R' = CH_2OH, R'' = H
(d) R = $CH_2CH_2CH_2CH_2Me$, R' = CH_2OH, R'' = OH
(e) R = $CH_2CH_2CH_2CH(OH)Me$, R' = CH_2OH, R'' = H
(f) R = $CH_2CH_2CH_2CH_2COOH$, R' = Me, R'' = H
(g) R = CH_2CH_2COOH, R' = Me, R'' = H
(h) R = COOH, R' = Me, R'' = H
(i) R = $CH_2CH_2CH_2CH_2COOH$, R' = CH_2OH, R'' = H
(j) R = $CH_2CH_2CH_2COOH$, R' = CH_2OH, R'' = H
(k) R = CH_2CH_2COOH, R' = CH_2OH, R'' = H
(l) R = COOH, R' = CH_2OH, R'' = H
(m) R = $CH_2CH_2CH_2COOH$, R' = Me, R'' = =O
(n) R = CH_2CH_2COOH, R' = Me, R'' = =O
(o) R = CH_2COOH, R' = Me, R'' = =O
(p) R = $CH_2CH_2CH_2CH_2Me$, R' = COOH, R'' = H
(q) R = $CH_2CH(OH)CH_2CH_2Me$, R' = COOH, R'' = H
(r) R = $CH_2CH_2CH(OH)CH_2Me$, R' = COOH, R'' = H
(s) R = $CH_2CH_2CH_2CH(OH)Me$, R' = COOH, R'' = H

cannabidiol, 7-hydroxycannabidiol and 6-hydroxycannabidiol (probably the 6-α isomer) were identified, conjugation taking place with a phenolic hydroxyl group in all cases. Most of the cannabidiol was metabolized by conjugation and cannabidiol glucuronide was the most prominent product. The identities of the oxidized metabolites were determined by gas chromatographic-mass spectrometric methods which allowed a particularly extensive list of compounds to be identified. Nineteen of these metabolites were characterized. These include compounds (34a) to (34s) and the structures of these cannabidiol derivatives indicate the formation of several types of metabolites including hydroxylated derivatives ((34a)–(34e)), side-chain acidic derivatives ((34f)–(34h)), 7-hydroxy side-chain acidic derivatives ((34i)–(34l)), 6-oxo side-chain acidic derivatives ((34m)–(34o)) and cannabidiol-7-oic acid (34p) and its hydroxylated derivatives ((34q)–(34s)). The predominance of hydroxylation at C7 is similar to that noted above by rat liver, however dihydroxylated cannabidiol derivatives are much less abundant in mouse liver than in rat liver. Such intermediates appear to be readily oxidized to carboxylic acid derivatives in the mouse. Interestingly, further oxidation to dicarboxylic acids, as observed with Δ^1-tetrahydro-cannabinol in rabbits, was not detected. Wall *et al.* (1976) administered [^{3}H]-cannabidiol by i.v. injection to humans and measured the amounts of unchanged compound and various types of oxidized metabolites in plasma, urine and faeces after various intervals. The 7-carboxy derivative (34p) was an abundant plasma metabolite, however even larger quantities of more polar acids were present. About a third of the radioactivity was excreted in the faeces, mainly as unconjugated metabolites. Interestingly, unchanged cannabidiol was the most abundant of these.

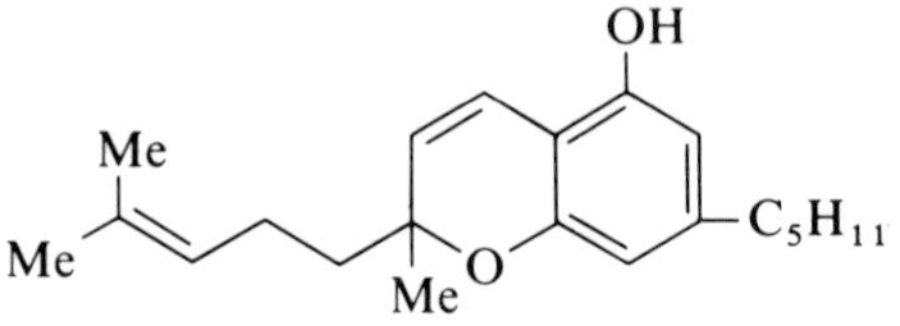

(35)

Cannabichromene

In their study of the glucuronides formed *in vivo* in mouse liver from various cannabinoids, Harvey *et al.* (1977a) also administered cannabis tincture. This material contained **cannabichromene** (35) as well as the propyl analogues of cannabinol and cannabidiol in addition to the more common cannabis constituents and the results indicated that glucuronide conjugates of these minor compounds were also formed.

The widespread interest in cannabinoid metabolism has, since about 1970, generated an abundant literature which has already been the subject of numerous reviews including those by Lemberger (1972), Agurell *et al.* (1972), Burstein (1973), Nahas (1973), Hanuš and Krejčí (1974) and Lemberger and Rubin (1976).

B. CHROMENE DERIVATIVES

Nearly all naturally occurring derivatives of chromene (36) are 2,2-dimethylchromenes, but no metabolic information is available on the simpler types. As noted in the previous section, cannabinol (31) is a substituted chromene but its metabolism is more conveniently summarized

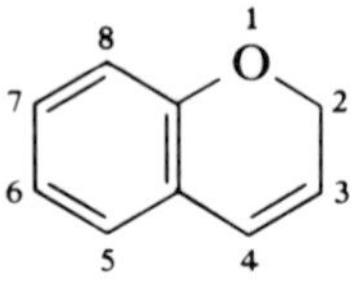

(36)

together with the other cannabinoids. An example in which the chromene group is incorporated into a more complex structure is seen with calophyllolide (70). However, this compound is also a coumarin derivative and its metabolism is therefore covered below in Section III.C.

C. Chromanone Derivatives

Metabolic information on the derivatives of chromanone (37) is limited to a short report by Krishnaswamy *et al.* (1971) on the fate of 2,2-dimethyl-7-hydroxychromanone (38). Although this compound does not appear to be

(37)   (38)   (39)

naturally occurring, it is included here because it is the sole representative of the chromanone group for which data are available and because its metabolism shows an interesting similarity to that observed with flavanone (76) (see Section IV.C). After oral administration (200 mg/kg) to rabbits, about 23% of the original compound was recovered in the 48 h urine free (19·5%) or conjugated (3·5%). In addition, about 1% of the dose was excreted as a highly fluorescent metabolite which was believed to be 2,2-dimethyl-7-hydroxychromene (39). This metabolite was probably formed by dehydration of the corresponding 4-ol intermediate.

D. Chromone Derivatives

The chromone (40) group is found in several classes of natural compounds including the flavones (2-phenylchromones) and isoflavones (3-phenylchromones). The metabolism of these groups is discussed below in Sections

(40)

IV.A and G, respectively. However, other chromones are not widely encountered in Nature. The present section includes a single compound, 3-(hydroxymethyl)-8-methoxychromone (41a). This derivative is not naturally occurring but contains both methoxyl and hydroxymethyl substituents which are typical of plant chromones. A discussion of its metabolism may therefore have relevance to other compounds in this group.

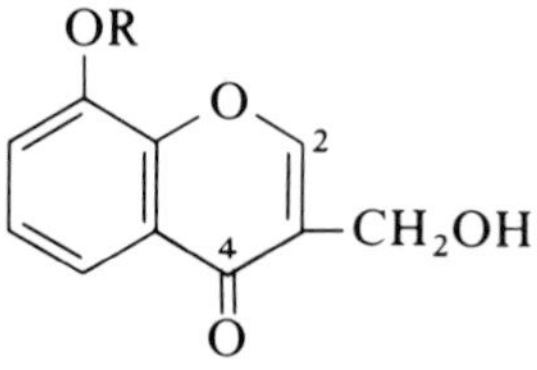

OR

$-CH_2OH$

(41)

(a) R = Me
(b) R = H

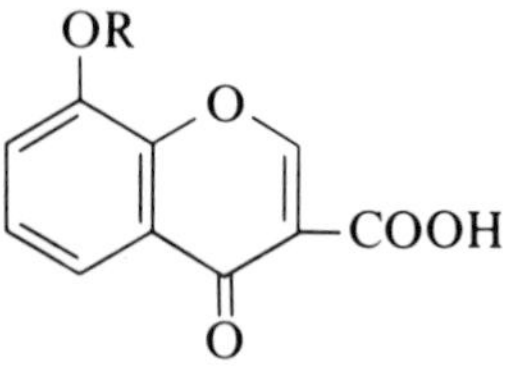

OR

—COOH

(42)

(a) R = Me
(b) R = H

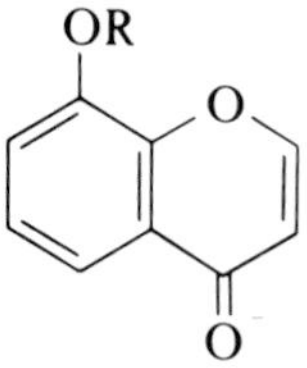

OR

(43)

(a) R = Me
(b) R = H

RO   OH

$-\overset{O}{\overset{\|}{C}}-Me$

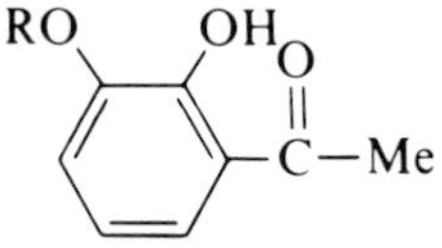

(44)

(a) R = Me
(b) R = H

Crew *et al.* (1976a,b) administered ^{14}C-labelled 3-(hydroxymethyl)-8-methoxychromone (41a) orally to rats and dogs. In the former species about 75–80% of the radioactivity was excreted within 24 h, with 40–50% appearing in the urine. Two labelled compounds were used, either with the ^{14}C in the 4-position or in both the 2-position and the hydroxymethyl group at C3. Many of the urinary metabolites identified are those to be expected from a compound containing a methoxyl and a hydroxymethyl group. Thus, demethylated compound (41b) was detected as was the oxidation product (42a). In addition, some evidence was also obtained for the presence of the demethylated acid (42b). These conventional metabolites were excreted both free and as their glucuronide and/or sulphate conjugates. Also excreted in the urine were small quantities of the decarboxylated metabolite 8-methoxychromone (43a) and relatively large amounts of 8-hydroxychromone (43b), which was the major urinary metabolite in rats. Interestingly, the two acetophenone derivatives (44a) and (44b) were also excreted in the urine. Di Carlo *et al.* (1976) found that plasma from rats given 3-(hydroxymethyl)-8-methoxychromone orally contained unchanged compound and metabolites (42a), (43a) and (44a) in unconjugated form. The plasma levels of these compounds were determined over a period of 48 h. The formation of metabolites (44a) and (44b) demonstrates that scission of the heterocyclic ring at the 1,2-position must have taken place. Also pertinent in this regard is the finding that administration of compound (41a) in which the ^{14}C was located in the 4-position did not lead to the formation of respiratory $^{14}CO_2$. However, roughly 30–40% of the administered radioactivity was recovered in the CO_2 when the other labelled compound was given. The finding that 8-methoxy- and 8-hydroxychromone were excreted in the urine indicates that much of the $^{14}CO_2$ must derive from the hydroxymethyl group. Crew *et al.* (1976b) noted that the 3-carboxy compounds are fairly readily decarboxylated. It is therefore possible that an undetermined extent of the decarboxylation may occur chemically rather than enzymically. In fact, the presence of acetophenone derivatives as urinary metabolites also suggests that other purely chemical transformations may be involved in the overall metabolism of 3-hydroxymethyl-8-methoxychromone. Although the findings of Crew *et al.* (1976a,b) do not reveal the mechanism involved in scission of the γ-pyrone ring or the point in the sequence of metabolism at which this occurs, it is possible that a β-keto acid intermediate may be produced. If so, this highly unstable compound would be expected to be decarboxylated to an acetophenone derivative as has been found with several closely related aromatic compounds metabolized via β-keto acids (Solheim and Scheline, 1973, 1976).

III. Coumarins

A. COUMARIN

Coumarin (5,6-benz-2-pyrone) (45) was formerly used in foods and drugs for its flavouring properties but this has been discontinued because of the liver toxicity produced by the compound (see Feuer, 1974). Coumarin is not a typical representative of this class of compounds insofar as it lacks an oxygen atom at the 7-position, a characteristic of nearly all of the naturally occurring coumarins.

The metabolism of coumarin, while being neither overly complex nor unexpected, is nonetheless extensive and leads to a large number of metabolites. These fall into two main groups: (1) hydroxylated products, and (2) products in which the α-pyrone ring has been cleaved although, as noted below, formation of the scission products is sometimes dependent upon initial ring hydroxylation. The metabolic pathways shown in Fig. 7.2 illustrate both the hitherto detected metabolites of coumarin and the probable pathways by which the metabolites are formed. It must be

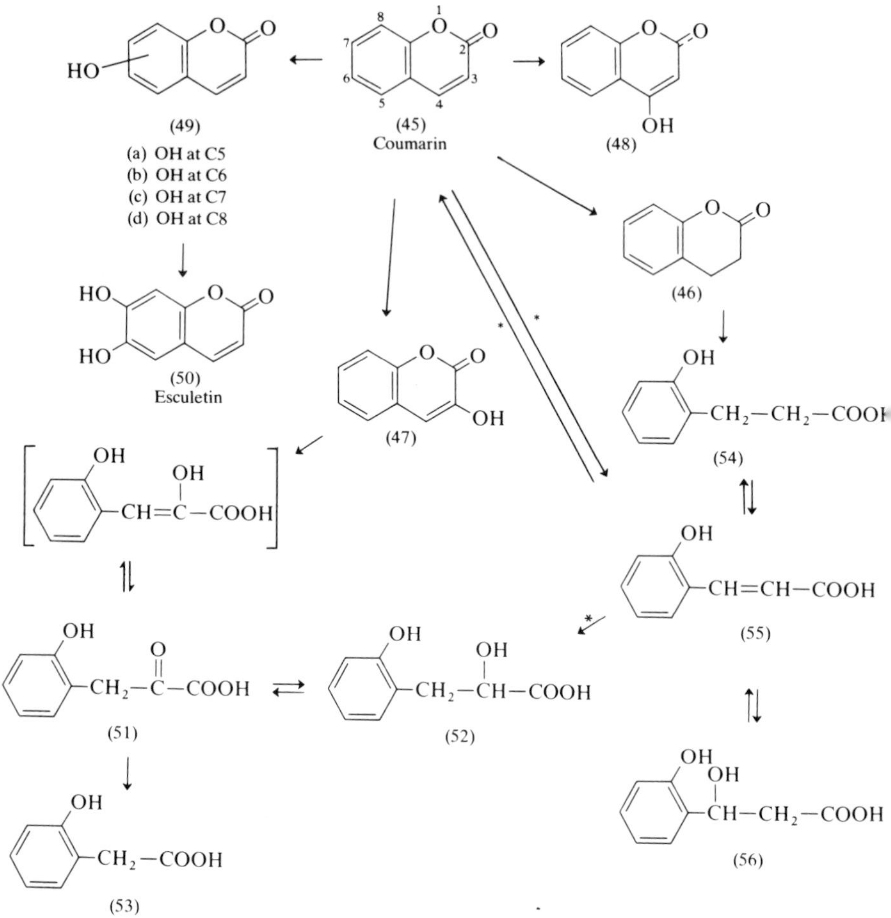

FIG. 7.2. Metabolic pathways of coumarin (*see text).

stressed that certain metabolites have not been consistently detected, even in experiments using the same animal species. A good example of the variability in metabolite formation is that seen with melilotic acid (54) in rats (Booth *et al.*, 1959; Feuer, 1974; Kaighen and Williams, 1961; Mead *et al.*, 1958b; Scheline, 1968a). Furthermore, considerable species variations in metabolism have been noted and these will be discussed below.

The proposed scheme of coumarin metabolism shown in Fig. 7.2 is based upon the identification of urinary metabolites in *in vivo* studies reported by Furuya (1958b), Mead *et al.* (1958b), Booth *et al.* (1959), Kaighen and Williams (1961), Feuer *et al.* (1966), Scheline (1968a), Pekker and Schäfer

(1969), Shilling *et al.* (1969), Van Sumere and Teuchy (1971) and Feuer (1974). However, several studies using microsomal preparations have duplicated many of these findings (Creaven *et al.*, 1965; Fink and Kerekjarto, 1966; Kratz and Staudinger, 1965). Figure 7.2 shows that hydroxylation occurs at all possible positions in the coumarin molecule. This was shown in the study of Kaighen and Williams (1961) using rabbits. However, several of the hydroxycoumarins are formed in small amounts and have not been detected in all of the studies. This is especially true of the 5- and 6-hydroxy derivatives which are never found in very large amounts in the urine of coumarin-treated animals. Also, polyhydroxylated compounds are seldom encountered although Pekker and Schäfer (1969) detected 6,7-dihydroxycoumarin (esculetin) (50) as a minor coumarin metabolite in rabbits. The 4- and 8-hydroxycoumarins are somewhat more abundant and may account for a few percent of the dose. The major hydroxylated metabolites are the 3- and 7-derivatives. The relative proportions of these can differ greatly and Shilling *et al.* (1969) found that 7-hydroxylation is the major pathway in humans, amounting to about 80% of the dose. On the other hand, probably no more than 6% undergoes 3-hydroxylation.

As shown in Fig. 7.2, 3-hydroxylation gives rise not only to the hydroxylated compound itself but also to several phenolic acids resulting from ring scission. The most characteristic of these is *o*-hydroxyphenylacetic acid (53) which is invariably a prominent coumarin metabolite. It is now generally agreed that this C_6–C_2 acid is formed via 3-hydroxycoumarin (47) and Kaighen and Williams (1961) showed that when the latter compound is fed to rabbits it is mainly excreted as its glucuronide but is also metabolized to *o*-hydroxyphenylpyruvic (51), -lactic (52) and -acetic (53) acids. They found that this pathway was more prominent in rats in which case the main urinary metabolite of (47) was (53).

The mechanisms involved in the formation of *o*-coumaric acid (55) and *o*-hydroxyphenylhydracrylic acid (56) are not entirely clear. Also, it must be noted that not all investigations have detected compound (55) as a coumarin metabolite. Mead *et al.* (1958b) and Kaighen and Williams (1961) reported that it was absent from the urine of coumarin-treated rabbits although Booth *et al.* (1959), Furuya (1958b) and Scheline (1968a) reported that it was excreted by both rabbits and rats. It is perhaps significant that in those experiments in which *o*-coumaric acid was detected, its reduction product melilotic acid (54) was also reported. If metabolites (54) and (55) do appear concomitantly, the formation of the latter may be solely via the intermediate 3,4-dihydrocoumarin (46) and no direct formation from coumarin itself need be proposed. A further point of interest regarding the pathway via metabolite (46) is that this compound appears to be formed by the intestinal bacteria and not the tissue enzymes (Scheline, 1968a). Thus, the absence of (54) and (55) as urinary metabolites of coumarin in some cases may be explained by variations in the metabolic activities of the intestinal microflora of the animals involved. Some further support for this proposal is available in the results of Fink and Kerekjarto (1966) who found that liver microsomes formed all of the possible hydroxy derivatives from coumarin as well as the acidic metabolites (52) and (53). However, compounds (54) and (55) and (56) were not detected.

Another point of interest regarding the pathways shown in Fig. 7.2 is that the transformation between metabolites (52) and (55) does not appear to be reversible. Whereas administration of (54) or (55) to animals resulted in the urinary excretion of the phenylacetic acid derivative (53), only the latter metabolite was excreted when compound (52) was given (Booth *et al.*, 1959). The phenolic acids formed from coumarin are excreted partly free, partly as glucuronide conjugates and, in the case of coumaric (55) and melilotic (54) acids, also as glycine conjugates (Booth *et al.*, 1959; Furuya, 1958b; Mead *et al.*, 1958b).

Coumarin contains an α,β-unsaturated group in common with many compounds of this type which are known to be conjugated with glutathione. However, Boyland and Chasseaud (1967) found that this glutathione S-transferase activity from rat liver did not utilize coumarin as a substrate.

Considerable species variations in the metabolism of coumarin have been observed. Quantitative data by Kaighen and Williams (1961) and by Feuer (1974) show that the excretion of hydroxycoumarins in rats is relatively low (approx. 5–10% of the dose) compared with 55–60% in rabbits (Kaighen and Williams, 1961), 80% in man (Shilling *et al.*, 1969) and 60% in the baboon (Gangolli *et al.*, 1974). In the last two species ring hydroxylation takes place at the 7-position. The pathway via 3-hydroxy-coumarin leading to acidic metabolites is of minor importance in man and accounts for only about 6% of the dose (Shilling *et al.*, 1969). In rats, differing results have been reported for these metabolites, ranging from approx. 20% (Kaighen and Williams, 1961) to nearly 70% (Feuer, 1974). In rabbits, the corresponding value for these metabolites ((52) and (53)) is 20–25% (Kaighen and Williams, 1961). Another species difference shown by coumarin is the high degree of faecal excretion of metabolites seen in the rat only (Kaighen and Williams, 1961). These metabolites include compound (53) and unidentified compounds but little unchanged compound. Their presence in the faeces is no doubt explained by the greater facility of the rat in excreting compounds in the bile.

B. Coumarins Containing Hydroxyl and/or Methoxyl Groups

As noted above, the true parent of the great majority of naturally occurring coumarins is 7-hydroxycoumarin (umbelliferone) (57). This section will therefore deal mainly with coumarin derivatives containing an oxygen function in the 7-position although, for comparative purposes, the metabolism of the other isomeric monohydroxylated compounds will be briefly reviewed.

The most comprehensive study of the metabolism of the hydroxy-coumarins was carried out by Mead *et al.* (1958a) who administered the compounds orally (200 mg/kg) to rabbits. After giving the 3-, 4-, 5-, 6- and 8-hydroxycoumarins the 24 h urines were found to contain from 40 to 75% of the dose as the corresponding conjugates, although in several cases the figures obtained may have been low due to incomplete hydrolysis of the glucuronide conjugates. The compounds administered were excreted mainly as their glucuronides but about 10–30% of the dose was accounted for as ethereal sulphates, except with 4-hydroxycoumarin which did not form a sulphate. This was considered likely to be due to the higher acidity of the 4-hydroxyl group compared with that of the other isomers. An early study of the metabolism of 3-hydroxycoumarin in man (Flatow, 1910) showed that it is excreted nearly entirely as the corresponding glucuronide. More recently, Kaighen and Williams (1961) found that 3-hydroxy-coumarin (approx. 250 mg/kg, p.o.) is excreted in rabbits not only as the corresponding glucuronide and sulphate but also as metabolites formed by the scission of the heterocyclic ring. These products were shown to be *o*-hydroxyphenylpyruvic (51), *o*-hydroxyphenyllactic (52) and *o*-hydroxy-phenylacetic (53) acids (see Fig. 7.2).

4-Hydroxycoumarin (84 mg/kg) given i.v. to dogs was excreted in the urine within 24 h as unchanged compound (approx. 50%) and as the corresponding glucuronide (25%) (Roseman *et al.*, 1954). The fate of the remaining 25% was not determined although it was shown that the urine did not contain increased amounts of steam-distillable phenols or ethereal sulphates and that no salicylic acid was formed. Hydroxylation of the aforementioned monohydroxy coumarins does not appear to take place readily and this reaction has been reported only with 6-hydroxycoumarin which is partly converted to 6,7-dihydroxycoumarin (esculetin) (50) in rabbits (Mead *et al.*, 1958a).

Umbelliferone (57) is of widespread occurrence and has been studied in a number of metabolic investigations. Sieburg (1921), using rabbits, found that doses of 300–500 mg/kg, i.p. were excreted in the urine as conjugated material which formed the original compound upon hydrolysis. Similar

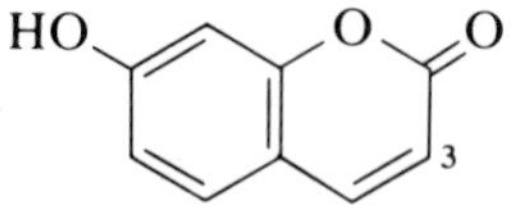

(57)

Umbelliferone

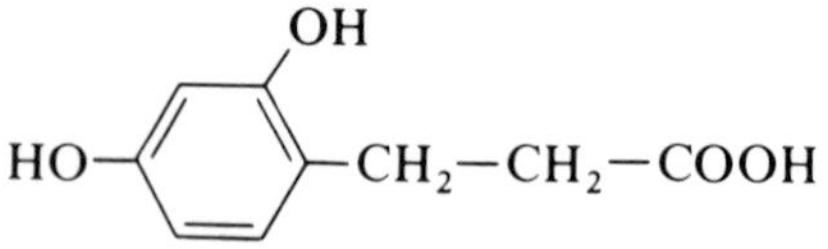

(58)

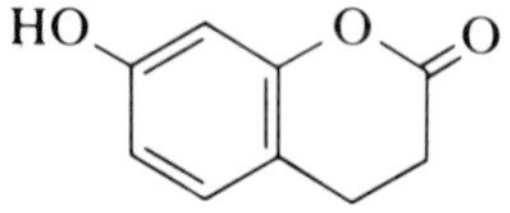

(59)

experiments by Mead *et al.* (1955) employing a dose of 200 mg/kg confirmed this and showed that the conjugates consisted of ethereal sulphate (approx. 20%) and twice as much or more of umbelliferone glucuronide. These findings were further substantiated by Fujita and Furuya (1958a) in similar experiments. They found that excretion was complete within 48 h and also that about 20% of the dose was excreted unchanged. None of these investigations using rabbits has demonstrated the occurrence of biological hydroxylation or ring cleavage of umbelliferone. However, a later report shows that both of these reactions take place when the compound (100 mg/kg) is administered to rats (Indahl and Scheline, 1971). While confirming the excretion of both free and conjugated umbelliferone, this study showed that ring scission giving 2,4-dihydroxyphenylpropionic acid (58) occurs and that, after i.p. but not oral dosage, a small amount of 3,7-dihydroxycoumarin was formed. Further evidence indicating the hydroxylation of the umbelliferone structure in rats was obtained with its homologue, 4-methylumbelliferone (Tatematsu *et al.*, 1972). The major urinary metabolites, both as free compounds and glucuronides, were found to be the 3,5,7-trihydroxy-, 6-hydroxy-7-methoxy- and 7-hydroxy-6-methoxy-derivatives of 4-methylcoumarin.

Ring cleavage of umbelliferone to the phenylpropionic acid derivative (58) was found to be carried out by the intestinal microflora (Indahl and Scheline, 1971), a result which supported the earlier conclusion mentioned above with coumarin itself (Scheline, 1968a). Again, formation of the C_6—C_3 acid appears to be due to reduction of the coumarin to the corresponding 3,4-dihydro compound (59) which is easily hydrolysed to compound (58). However, there was no indication of ring cleavage of umbelliferone leading to C_6—C_2 acids as described above with coumarin. Another investigation of umbelliferone metabolism in the rat (Van Sumere and Teuchy, 1971) showed that it is excreted in the urine mainly free and in conjugated form. These metabolites accounted for about half of the dose

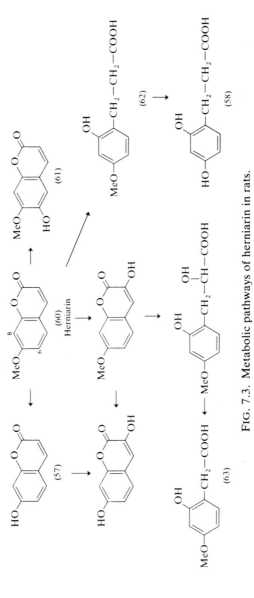

FIG. 7.3. Metabolic pathways of herniarin in rats.

(approx. 45 mg/rat, i.p.) and were excreted within 24 h. The only other urinary metabolite detected was 2,4-dihydroxybenzoic acid (β-resorcylic acid) which may arise via β-oxidation of 2,4-dihydroxycoumaric acid formed as a result of the opening of the lactone ring. The reaction of ring opening takes place with coumarin itself (Fig. 7.2), however the formation of a C_6—C_1 phenolic acid has not been reported previously in studies of the metabolism of coumarin and its derivatives.

The metabolism of the methyl ether of umbelliferone, **herniarin** (60), was investigated in rabbits (Fujita and Furuya, 1958b) and rats (Indahl and Scheline, 1971). In the former study only about 30% of the dose (200 mg/kg) could be accounted for, chiefly as umbelliferone (57) which was excreted free and conjugated in a ratio of about 2:1. A small amount of herniarin itself was excreted. As with umbelliferone, no hydroxylation or ring cleavage of herniarin was noted in the rabbit. In contrast, the pathways of its metabolism in the rat are more complex and are shown in Fig. 7.3. The major metabolite of herniarin is 2-hydroxy-4-methoxy-phenylacetic acid (63) which, in a manner analogous to that described above for the formation of o-hydroxyphenylacetic acid from coumarin, most likely arises via 3-hydroxylation of herniarin as shown in Fig. 7.3. A lesser amount of hydroxylation occurred at the 6-position to give metabolite (61) and the results also suggested that trace amounts of the 8-hydroxy derivative (not shown) were also excreted. As with umbelliferone, herniarin was metabolized by the intestinal bacteria to C_6—C_3 phenolic acids (compounds (62) and (58)). These metabolites were not excreted in the urine of germ-free rats given herniarin. Both (62) and (58) were readily formed when herniarin was incubated with rat caecal microorganisms *in vitro*.

(64)   (65)   (66)

Daphnetin   Scopoletin   Esculetin dimethyl ether

The remaining plant coumarins in this group that have been studied metabolically are dihydroxy compounds and their derivatives. **Esculetin** (6,7-dihydroxycoumarin) (50) (see Fig. 7.2) and the isomeric 7,8-dihydroxy compound **daphnetin** (64) were found by Sieburg (1921) to be excreted in the urine of rabbits as conjugation products which would liberate the original compounds upon hydrolysis. In rats, the administration of esculetin results in the urinary excretion of some unchanged compound (Yang *et al.*, 1958) as well as its sulphate and glucuronide conjugates (Braymer, 1960). In addition, the latter investigation showed that *O*-methylation occurred to give both the 6- and 7-methyl ethers. Similar results were obtained when **esculin** (esculetin-6-glucoside) was given, a finding undoubtedly related to the hydrolysis of the glycoside in the intestine by the microflora. Hawksworth *et al.* (1971) and Drasar and Hill (1974) reported that this reaction was carried out by nearly all of the common groups of intestinal bacteria.

Studies have been carried out on the metabolism of **scopoletin** (7-hydroxy-6-methoxycoumarin) (65) in rats (Braymer, 1960; Braymer *et al.*, 1960) and rabbits (Furuya, 1958a) and of its methylated derivative esculetin dimethyl ether (66) in rabbits (Furuya, 1958a). Scopoletin is excreted in the urine unchanged and as glucuronide and sulphate conjugates in both animal species. In addition, it is demethylated to a small extent to esculetin (50). **Esculetin dimethyl ether** (66), lacking a free hydroxyl group, is of necessity subjected to a greater degree of demethylation prior to excretion. Both mono-demethylated isomers were formed and excreted in the urine of rabbits given a dose of 200 mg/kg orally (Furuya, 1958a). However, demethylation at the 6-position was found to be about five times as extensive as that occurring at the 7-position. Most of the recovered dose was accounted for as the mono-demethylated metabolites, both free and conjugated, but about 3% was shown to be excreted as a conjugate of the fully demethylated esculetin (50).

C. COUMARINS CONTAINING MORE COMPLEX SUBSTITUENTS

This group is represented at present by coumarins containing substituents in the pyrone ring. 3-Phenylcoumarins are uncommon natural products and metabolic data are available only on some of the simpler synthetic analogues (Krishnaswamy *et al.*, 1971). 3-Phenyl-7-hydroxycoumarin (3-phenylumbelliferone) (67) was completely excreted within 48 h following oral dosage (200 mg/kg) to rabbits. About 20% was excreted unchanged while most of the dose (approx. 70%) was found as the corresponding glucuronide and sulphate conjugates. With 3-phenylherniarin, the methyl ether of (67), only about 30% of the dose was accounted for, nearly

(67)

(68)

entirely as 3-phenylumbelliferone. In this case the ratio of free to conjugated compound was 2:1. Furthermore, a trace (0·2%) of the administered 3-phenylherniarin was detected in the urine.

The metabolism of 4,7-dihydroxy-3-phenylcoumarin (68) in rabbits was found to differ from that observed with the mono-hydroxy compound (67) (Krishnaswamy *et al.*, 1971). The main urinary metabolite detected was the unchanged compound (22%) and no conjugates were found in the 48 h urines. A small amount (0·2%) of the 4-methyl ether of (68) was found in two of five 24 h urine samples.

Coumestrol (69), an oestrogenic substance found in alfalfa and several species of clover, shows a close structural resemblance to compound (68). Its metabolism in mammals has not been studied but Cayen and Common (1965) found that it is extensively converted to both less and more polar metabolites in the hen. While the nature of these products was not ascertained, it was found that coumestrol was not converted to equol. The latter compound is a metabolite of the structurally related isoflavone genistein (see Section IV.H).

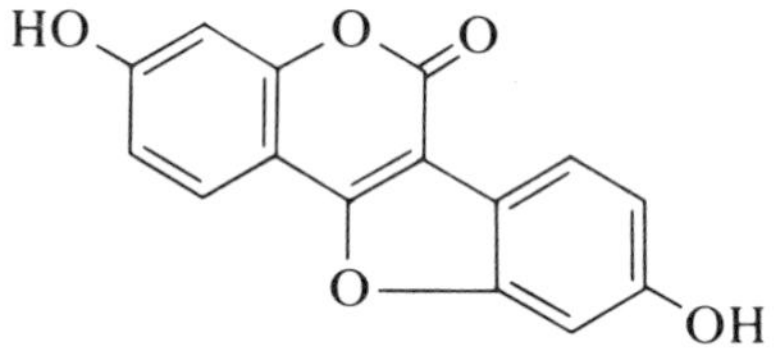

(69)

Coumestrol

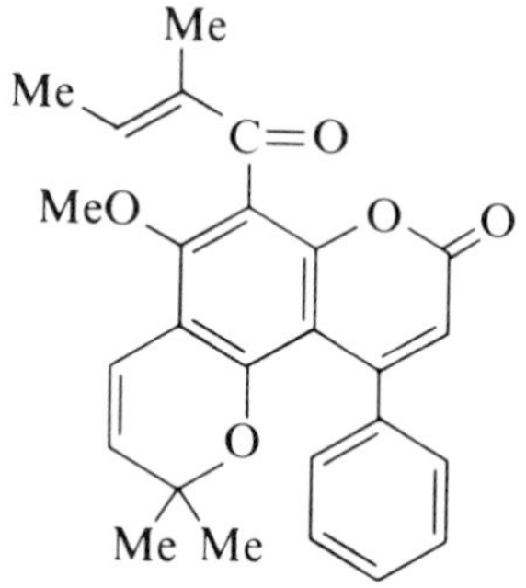

(70)

Calophyllolide

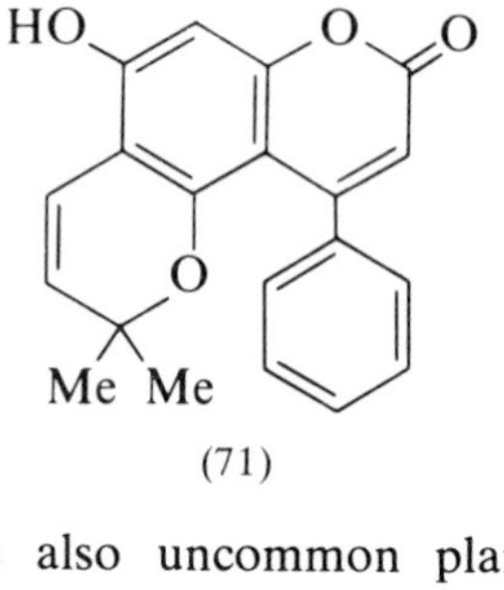

(71)

4-Phenylcoumarins are also uncommon plant compounds and the metabolism of only a single representative, calophyllolide (70), has been reported together with data on two simpler synthetic analogues, 4-phenylumbelliferone and its methyl ether, 4-phenylherniarin (Arora *et al.*,

1966; Krishnaswamy *et al.*, 1971). 4-Phenylumbelliferone (200 mg/kg, p.o.) is metabolized in rabbits similarly to that described above for both umbelliferone and its 3-phenyl derivative. Thus 75% of the dose was accounted for in the 48 h urines, partly free (25%) and partly as glucuronide and sulphate conjugates (50%). When the methoxy compound was given, only about 30% of the dose was recovered in the urine, nearly exclusively as the demethylated compound. The ratio of free to conjugated metabolite was 2:1, similar to that noted above with the isomeric 3-phenylherniarin. No information on the fate of the remainder of these compounds was obtained, but it was stated that oxidation or cleavage of the lactone ring was not observed.

When **calophyllolide** (70) (50 mg/kg) was given orally to rabbits only about 2% of the dose could be accounted for (Krishnaswamy *et al.*, 1971). This was divided fairly evenly between a free metabolite and its conjugate.

The metabolite (71) is a simpler derivative in which demethylation has occurred at the 7-position and the acyl group lost at the 8-position. The latter reaction is unusual as the 8-acyl moiety forms, in fact, an aromatic ketone. This provides a metabolically active site and it is possible that part of the great majority of the calophyllolide still unaccounted for may be the carbinol derivative. Additionally, allylic hydroxylation at the two available sites in the 8-acyl group may occur.

IV. Flavonoids

A. FLAVONES

Nearly 130 naturally occurring flavones are known, excluding flavonols and the glycosides of these two groups (Venkataraman, 1975). Flavonols are 3-hydroxyflavones and are considered separately below in Section B. **Flavone** (72) itself occurs naturally, in contrast to that found with the

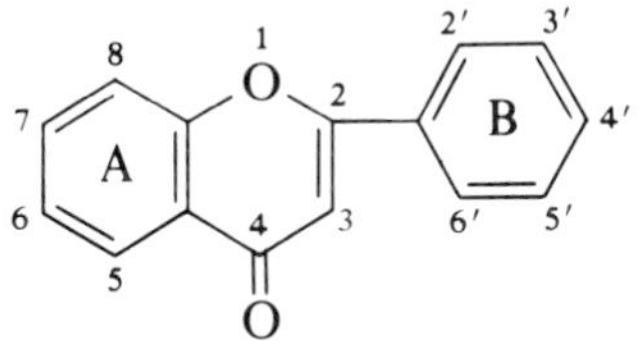

(72)

Flavone

unsubstituted parent compound of most of the other flavonoid groups. Demole (1962) found that flavone was not excreted in the urine unchanged following its administration to mice, rats, guinea pigs and rabbits.

However, two unidentified urinary metabolites were detected when flavone (200 mg/kg) was given to rabbits. Das and Griffiths (1966) reported that 4′-hydroxyflavone was excreted in the urine of guinea pigs given flavone orally or by i.p. injection. However, only about 3% of the dose (100 mg/kg) was accounted for as the 4′-hydroxylated metabolite. Smaller amounts of 3′,4′-dihydroxyflavone were detected in the urine after oral administration of flavone and, in contrast to the results of Demole (1962), a small amount of unchanged compound was also found. With the possible exception of salicylic acid which was detected in trace quantities, no phenolic acids arising from ring fission of the flavone molecule were found in the urine.

(73)

(a) Chrysin, R = R′ = R″ = H
(b) Apigenin, R = R″ = H, R′ = OH
(c) Acacetin, R = R″ = H, R′ = OMe
(d) Diosmetin, R = OH, R′ = OMe, R″ = H
(e) Tricetin, R = R′ = R″ = OH
(f) Tricin, R = R″ = OMe, R′ = OH
(g) R = R′ = R″ = OMe

The first study to reveal the general pattern of metabolism of substituted flavones was that of Booth *et al.* (1958a) who found that the major urinary metabolite in rats of **diosmetin** (73d) and its 7-rhamnoglucoside (**diosmin**) is *m*-hydroxyphenylpropionic acid. Also detected were traces of the corresponding cinnamic acid derivative *m*-coumaric acid, and of diosmetin. When the latter compound was administered, some diosmetin glucuronide was also excreted. Thus, the metabolism of this flavone derivative proceeds, in effect, by way of fission at the 1,2-bond and at the carbonyl group in the 4,5-region leading to a C_6—C_3 phenolic acid.

Subsequently, the metabolism of a group of flavones was studied (Griffiths and Smith, 1972a,b) and the results have greatly clarified the structural requirements for ring fission in this group of flavonoids. The simplest flavone studied was **chrysin** (73a), which is unsubstituted in the B-ring. After its oral administration to rats, some unchanged chrysin was excreted in the urine together with its 4′-hydroxy derivative, apigenin (73b). No urinary phenolic acids resulting from ring fission were observed and no detectable metabolites were found following the anaerobic incubation of chrysin with rat caecal microorganisms. **Tectochrysin** is the 7-O-methyl ether of chrysin (73a) and similar experiments produced the same general results in that no phenolic acids were excreted and no metabolism was found in the incubates. The main urinary metabolite appeared to be apigenin (73b) but small amounts of its 7-O-methyl ether (genkwanin) and unchanged tectochrysin were also excreted. Lack of ring fission to acidic metabolites and lack of microbial degradation were also noted with 7,4'-dihydroxyflavone. Oral administration in rats led to the urinary excretion of unchanged compound and an unidentified metabolite. DeEds (1968) reported that **nobiletin** (5,6,7,8,3',4'-hexamethoxyflavone) is resistant to metabolic degradation.

In contrast to the general metabolic picture summarized above showing that the dihydroxyflavones are resistant to fission of the heterocyclic ring system, administration of the commonly occurring trihydroxyflavone **apigenin** (73b) orally to rats led to quite different results (Griffiths and Smith, 1972a). The major urinary metabolites were p-hydroxyphenylpropionic acid, p-coumaric acid and p-hydroxybenzoic acid together with unchanged apigenin and two conjugates of apigenin, one of which was a glucuronide. Incubation experiments with apigenin showed that it was degraded to p-hydroxyphenylpropionic acid by the intestinal microorganisms. Apigenin is not converted to ring fission products in germ-free rats (Griffiths and Barrow, 1972a). When **acacetin** (73c), the 4'-O-methyl ether of apigenin, was given to normal rats, the extent of ring fission was greatly reduced and only traces of p-methoxyphenylpropionic acid were excreted (Griffiths and Smith, 1972a). This degradation was also observed in incubates. The results summarized above suggest that a free hydroxyl group at the 4'-position favours ring fission in the flavones and this point was investigated further in studies using the pentahydroxyflavone **tricetin** (73e) and its methyl ethers **tricin** (73f) and compound (73g) (Griffiths and Smith, 1972b). The 3',4',5'-trihydroxy compound tricetin was excreted in the urine partly unchanged after its oral administration to rats but some 3,5-dihydroxyphenylpropionic acid was also detected. This phenolic acid and some m-hydroxyphenylpropionic acid were formed *in vitro* in incubates using rat caecal microorganisms. In experiments using the 4'-hydroxy-3',5'-dimethoxy compound tricin, similar results were obtained. Both unchanged tricin and 3,5-dihydroxyphenylpropionic acid were excreted in the urine and small amounts of the latter compound were formed in the incubates. This report is in variance with those of Bickoff *et al.* (1964) and Stelzig and Ribeiro (1972) who were unable to detect degradation products in the urine of tricin treated rats. With the trimethoxy compound (73g), Griffiths and Smith (1972b) found that fission of the heterocyclic ring system was less extensive. No phenylpropionic acid derivative was formed in incubation experiments and only traces of 3,5-dihydroxyphenyl propionic acid were detected in the urine of rats given compound (73g). Interestingly, large amounts of both unchanged compound and its O-demethylated derivative (73f) were excreted in the faeces. The extensive faecal excretion of tricin (73f) by rats was also reported by Stelzig and Ribeiro (1972).

The above findings clearly indicate that the presence of a 4′-hydroxyl group in the flavones is important with regard to the degree of ring fission and production of phenolic acids that can be expected. The results mentioned above with diosmetin (73d), a 4′-methoxy compound, need not contradict this as its extensive metabolism to m-hydroxyphenylpropionic acid most likely occurs following initial O-demethylation of the flavone. Based upon the finding noted above that 7,4′-dihydroxyflavone is not degraded, it appears that the susceptibility to ring fission may be related to the presence of a free hydroxyl group at position 5, as is observed with the flavonols (see Section B). Nothing definite is presently known of the intermediates formed in the degradation of flavones to phenolic acids. However, the general similarities in end-products of metabolism between this group of flavonoids and flavanones (Section C) and dihydrochalcones (Section F) suggest that these three groups may be degraded in a similar fashion. It is of interest to note that intermediates have been identified only in those flavonoid groups containing a 3-hydroxyl group (flavonols and catechins).

B. FLAVONOLS

Derivatives of flavonol (74), which are 3-hydroxyflavones, are of widespread occurrence in higher plants. While more than a hundred of these

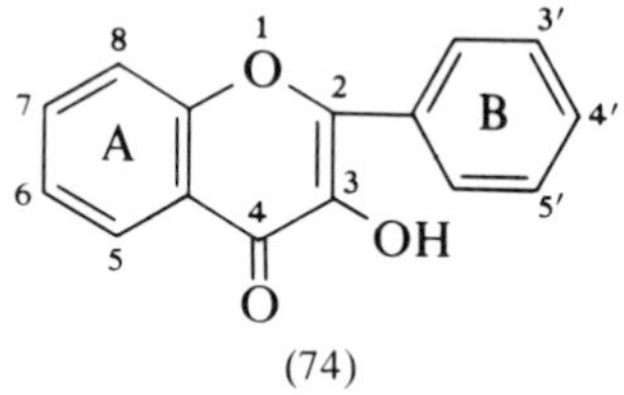

(74)

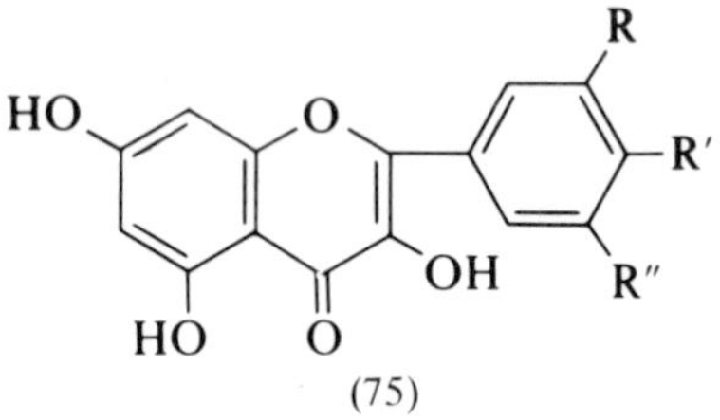

(75)

(a) Kaempferol, R = R″ = H, R′ = OH
(b) Quercetin, R = R′ = OH, R″ = H
(c) Myricetin, R = R′ = R″ = OH

compounds are known, only kaempferol (75a), quercetin (75b) and myricetin (75c) are common. Information on the metabolic fate of flavonols is limited nearly entirely to these three compounds and especially to quercetin or its glycoside rutin (quercetin 3-rutinoside). The latter compound is the most common glycoside of quercetin, of which many are known. In fact, flavonols generally occur in glycosidic combination.

Flavonol itself does not occur naturally and nothing is known of its metabolism in animals. The simplest flavonol studied metabolically is 3',4'-dihydroxyflavonol which is also not known to occur naturally. An early report by Ozawa (1951) indicated that only about a tenth of the administered dose was excreted unchanged in the urine and faeces of rabbits. However, three unidentified metabolites were excreted in the urine.

As noted above, most of the data on flavonol metabolism deal with **quercetin** (75b) and **rutin** (quercetin 3-rutinoside). The often conflicting results of a number of early studies in this field have been reviewed by DeEds (1968). In brief, these results pointed to little or no urinary excretion of rutin taking place following its oral administration. On the other hand, injection of rutin led to the appearance of some of the compound in the urine. Clark and MacKay (1950) found that large oral doses of rutin (50 mg/kg) in humans resulted neither in its urinary excretion nor its recovery in the faeces. Recently, Gugler *et al.* (1975) carried out a pharmacokinetic study of the disposition of quercetin in man. They found that an oral dose (4 g) was absorbed to an extent of 1% or less and that no unchanged material was excreted in the urine. However, about half the dose was recovered in the faeces. These results suggest the extensive bacterial degradation of quercetin in the gut. Interestingly, Clark and MacKay reported that rutin was destroyed upon incubation with faecal samples, however no breakdown products were identified. The first investigation to clearly identify metabolites of quercetin and rutin was carried out by DeEds' group (Murray *et al.*, 1954; Booth *et al.*, 1955, 1956b). Oral administration of quercetin to rats, guinea pigs, rabbits or humans resulted in the urinary excretion of *m*-hydroxyphenylacetic acid, 3,4-dihydroxyphenylacetic acid (homoprotocatechuic acid) and 4-hydroxy-3-methoxyphenylacetic acid (homovanillic acid). These metabolites were believed to arise as a result of fission of the heterocyclic ring at the 1,2- and 3,4-bonds. Thus, these metabolites are derived from the B-ring and not from the phloroglucinol part of the molecule (A-ring). The formation of these $C_6—C_2$ phenolic acids from quercetin in rats has been confirmed in studies using randomly labelled [^{14}C]-quercetin (Petrakis *et al.*, 1959; Masri *et al.*, 1959) and by others using non-labelled quercetin (Braymer, 1960; Nakagawa *et al.*, 1965). However, the overall metabolic picture is more complex as derivatives of benzoic, phenylpropionic and cinnamic acids as well as a neutral compound characterized as an *o*-dihydroxyphenyl lactone (Braymer, 1960) have been detected in the urine of rats given quercetin orally. These metabolites include m-hydroxy-benzoic acid, 4-hydroxy-3-methoxybenzoic acid (vanillic acid), m-hydroxyphenylpropionic acid and m-coumaric acid (Petrakis et al., 1959; Braymer, 1960; Nakagawa et al., 1965). The finding of Petrakis et al. (1959) that about 15% of the radioactivity was recovered in the respiratory CO_2 following oral administration of randomly labelled $[^{14}C]$-quercetin also indicates that at least a portion of the molecule undergoes extensive metabolism.

Following the initial reports describing some metabolic products of quercetin and rutin, interest was directed towards the site or sites of their formation. Lang and Weyland (1955) described a mitochondrial enzyme system that utilizes quercetin and rutin anaerobically. Highest activities were found in liver and kidney with lower amounts present in heart, brain and muscle. Douglass and Hogan (1958) reported that rat kidney homo-genates were able to metabolize quercetin to 3,4-dihydroxybenzoic acid (protocatechuic acid) under aerobic conditions. However, the phenylacetic acid derivatives mentioned above were not formed by these preparations and it was suggested that they may arise from bacterial or digestive action in the gastrointestinal tract. This possibility was partially tested by Braymer (1960) who found that quercetin was not degraded in vivo in the ligated rat stomach. Westlake et al. (1959) reported that various moulds, streptomy-cetes and bacteria were able to degrade rutin and Booth and Williams (1963) made the significant finding that anaerobic incubates of rat caecal or faecal microorganisms converted rutin to m-hydroxyphenylpropionic acid. As noted above, this C_6-C_3 phenolic acid is a urinary metabolite of quercetin in rats. Further studies of the anaerobic degradation of rutin and also quercetin by rat caecal microorganisms were carried out by Scheline (1968b) who found that both compounds were degraded to m-hydroxy-phenylacetic acid and m-hydroxyphenylpropionic acid. From this it is clear that several types of the main urinary metabolites of these flavonols can be formed entirely by the intestinal microflora. Variations in ring substitution due to dehydroxylation also arise in this way whereas subsequent tissue reactions including dehydrogenation or β-oxidation of the side chain and O-methylation explain the formation of other urinary metabolites. Further evidence for the key role played by the intestinal microflora in the degradation of quercetin or rutin was obtained by Nakagawa et al. (1965) who reported that the C_6-C_2 phenolic acids normally appearing in the urine of quercetin-treated rats are absent when the animals are also given neomycin sulphate orally. Also, phenolic acid metabolites of rutin are not excreted by germ-free rats (Griffiths and Barrow, 1972a).

Interest in the degradation of flavonols by gastrointestinal bacteria has also been directed towards the activity of rumen microorganisms. Incuba-tion of rutin with bovine rumen fluid resulted in its degradation to phenolic compounds which appeared to be similar to those formed in the rat intestinal tract (Simpson *et al.*, 1969). Numerous strains of *Butyrivibrio sp.* were subsequently isolated from bovine rumen contents and shown to degrade rutin anaerobically (Cheng *et al.*, 1969). The metabolic products formed included phloroglucinol, 3,4-dihydroxybenzaldehyde and 3,4-dihydroxyphenylacetic acid (Krishnamurty *et al.*, 1970). These latter results indicate that different patterns of degradation are produced by bovine rumen and rat intestinal microorganisms.

The finding that rumen microorganisms degrade rutin to phloroglucinol (Simpson *et al.*, 1969; Krishnamurty *et al.*, 1970) is of interest in regard to the fate of the A-ring. While Kallianos *et al.* (1959) reported that quercetin is converted to phloroglucinol, phloroglucinol carboxylic acid and 3,4-dihydroxybenzoic acid in the stomach of the rat, there is good reason to believe that these compounds may also be formed as a result of chemical degradation during sample preparation (Masri *et al.*, 1959). Nonetheless, the latter group found subsequently (DeEds, 1968) that small amounts of phloroglucinol may be formed from quercetin when incubated for short times with rat faecal microorganisms. The transient nature of this metabolite has been noted in experiments with bovine rumen microorganisms (Simpson *et al.*, 1969). At the present time, the detailed pathways involved in the degradation of quercetin are not clear although it seems reasonable to assume that multiple pathways exist. One of these allows the A-ring to remain intact and gives rise to phloroglucinol carboxylic acid and/or phloroglucinol. Another pathway which has been prominent in most studies of flavonol metabolism to date involves destruction of the A-ring leading to CO_2 and excretion of the remainder of the quercetin molecule as $C_6—C_2$ and sometimes $C_6—C_3$ phenolic acids. It is further possible that phenolic lactone intermediates (Braymer, 1960) similar to those encountered in catechin degradation (see Section E) may be involved in this pathway. It is likely that the relative significance of these routes may be a reflection of the types and relative numbers of microorganisms in the particular intestinal microflora.

Experiments with normal and antibiotic-treated animals and with cultures of intestinal microorganisms have shown that the general metabolic picture seen with other flavonols is similar to that summarized above with quercetin and rutin. **Kaempferol** (75a) and its 7-glycoside **robinin** are degraded to the expected *p*-hydroxyphenylacetic acid when incubated with rat caecal microorganisms and this phenolic acid is excreted in the urine of rats given these flavonols (Griffiths and Smith, 1972a). **Myricetin** (75c) and its 3-rhamnoside **myricitrin** are metabolized in rats and by rat caecal microorganisms to phenylacetic acid derivatives, mainly 3,5-dihydroxyphenylacetic acid (Griffiths and Smith, 1972b). The conversion in rats was prevented when the animals were treated with neomycin before and during administration of the flavonols. The degradation of myricetin is also absent in germ-free rats (Griffiths and Barrow, 1972a).

Some information is available on the structural requirements for flavonol degradation. The conclusions are similar to those noted in Section A on flavone metabolism. The presence of a 5-hydroxyl group appears to be essential as **robinetin** (3,7,3',4',5'-pentahydroxyflavone), which differs from myricetin only by the absence of this group, is not degraded when given orally to rats or incubated with caecal microorganisms (Griffiths and Smith, 1972b). DeEds (1968) reported that **rhamnetin** (quercetin-7-methyl ether), **azaleatin** (quercetin-5-methyl ether) and quercetin-3-methyl ether are not degraded in the rat. Likewise, no urinary metabolites of **tangeretin** (3,5,6,7,4'-pentamethoxyflavone) were detected. Lack of degradation to phenolic metabolites was also noted by Braymer (1960) with the 5,7,3',4'-tetramethyl ether of quercetin in the rat.

C. FLAVANONES

Flavanone (76) is chemically closely related to flavone (72), differing only in the lack of a double bond between C2 and C3. Furthermore, flavanones are isomeric with chalcones which are formed by ring opening at the 1,2-position. The relationship of 3-hydroxyflavanones (flavanonols, dihydroflavonols) (77) to flavanones is similar to that found between flavonols

(76)   (77)

(Section B) and flavones (Section A). However, in contrast to the relatively large number of investigations dealing with the metabolism of flavonols, the literature on the 3-hydroxyflavanones is restricted to a few reports. Therefore, both the flavanones and their 3-hydroxy derivatives are treated together in this section.

Flavanone (76) itself does not occur naturally but its metabolism in animals is described here in order to allow a comparison with that observed with flavone (72) described above. Das *et al.* (1973) administered flavanone to rats and found that the urine contained unchanged compound, flavone (72) and flav-3-ene (78). These three compounds accounted for approx. 25% of the dose and, although several unidentified metabolites were detected, the fate of the remainder is unknown. However, no evidence was obtained for the formation of aromatic acids arising from ring fission. Interestingly, formation of a 4'-hydroxylated metabolite, as seen with flavone (Section A), was not noted.

(78)

(79)

(a) Naringenin, R = H, R' = OH
(b) Eriodictyol, R = R' = OH
(c) Hesperetin, R = OH, R' = OMe
(d) Homoeriodictyol, R = OMe, R' = OH

Metabolic studies of flavanones are limited to four closely related compounds: naringenin (79a), eriodictyol (79b), hesperetin (79c) and homoeriodictyol (79d) as well as the 7-rhamnoglucosides of (79a) and (79c). The latter compounds are known as naringin and hesperidin, respectively. **Naringenin** (79a) was first shown by Booth *et al.* (1956a) to be degraded to *p*-hydroxyphenylpropionic acid, which was excreted in the urine of rabbits given the flavanone orally. A more extensive report appeared later (Booth *et al.*, 1958b) which showed that, in rats, smaller amounts of *p*-coumaric acid and the ethereal sulphate of *p*-hydroxybenzoic acid were also excreted. Both the unchanged flavanone and its glucuronide conjugate were also detected in the urine. The same pattern of urinary metabolites was seen when **naringin** (naringenin-7-rhamnoglucoside) was given. The latter results were largely confirmed by Griffiths and Smith (1972a) who, with large oral doses (approx. 600 mg/kg) in rats, detected *p*-hydroxyphenylpropionic, *p*-hydroxycinnamic (*p*-coumaric) and *p*-hydroxybenzoic acids and the aglycone naringenin in the urines. Interestingly, a species difference has been noted with naringin which, when given orally to a human volunteer, was excreted only as naringenin and its glucuronide in the urine (Booth *et al.*, 1958b). In addition to the urinary route, excretion of naringin and unidentified conjugates in the bile of rats has been reported (Barrow and Griffiths, 1971; Griffiths and Barrow, 1972b).

In vivo metabolic studies with the remaining flavanones, **hesperetin** (79c) and its glycoside **hesperidin, eriodictyol** (79b) and **homoeriodictyol** (79d), were carried out by Booth *et al.* (1956a, 1958a). The main finding in rats was that the major urinary metabolite in all cases was *m*-hydroxyphenylpropionic acid. Thus, all of the flavanones studied are metabolized to phenylpropionic acid derivatives, i.e. C_6—C_3 phenolic acids. Noteworthy in this regard is the fact that flavone derivatives, which differ only in having a double bond at the 2,3-position, are also degraded in this fashion (Section A). In addition to the major metabolite noted above, other urinary metabolites in rats were: from hesperidin and hesperetin, m-hydroxycinnamic acid and a conjugate of hesperetin; from eriodictyol, m-hydroxycinnamic acid, eriodictyol glucuronide and homoeriodictyol; from homoeriodictyol, m-hydroxycinnamic acid, 3-methoxy-4-hydroxyphenylpropionic acid (dihydroferulic acid), unchanged compound and homoeriodictyol glucuronide. Species variations in metabolism have been noted with hesperidin in rabbits and man (Booth *et al.*, 1958a). In the case of rabbits, the nature and extent of the urinary metabolites varied with the type of diet given. When maintained on a purified diet and given hesperidin orally, the following compounds were detected in the urine: hesperetin, hesperetin glucuronide, 3,4-dihydroxyphenylpropionic acid, 3-methoxy-4-hydroxyphenylpropionic acid, m-hydroxycinnamic acid, m-hydroxyphenylpropionic acid, m-hydroxyhippuric acid, m-hydroxybenzoic acid and vanillic acid. In man, the major urinary metabolite of hesperidin and hesperetin was shown to be 3-hydroxy-4-methoxyphenylhydracrylic acid (80).

Meo—⟨ring⟩—CH—CH$_2$—COOH, with HO and OH substituents

(80)

Honohan *et al.* (1976) prepared hesperetin labelled with ^{14}C in the 3-position and studied its metabolism in rats. This preparation allowed the dose given (approx. 1·6 mg/kg, p.o. and 0·8 mg/kg, i.p.) to be reduced to 1% or less of that employed in the earlier studies. Excretion of radioactivity in the urine, faeces, bile and respiratory air was maximal during the first 24 h and essentially complete after 48 h. A third of the radioactivity was excreted in the urine following oral dosage and most of this was shown to be due to the methoxy-hydroxy-, dihydroxy- and m-hydroxy-derivatives of phenylpropionic acid. A further 40% of the dose was lost as ^{14}CO$_2$ and this indicates that these C_6—C_3 metabolites also undergo extensive β-oxidation to benzoic acid derivatives. Biliary excretion of radioactivity was studied after both oral and intraperitoneal administration. Recoveries were 57% and 100%, respectively, in bile duct cannulated rats. The latter figure is noteworthy as it indicates conclusively that if any tissue metabolism of hesperetin had occurred, no C_6—C_1 metabolites could have been formed as this would have entailed loss of radioactivity as CO$_2$. Therefore, no conversion of hesperetin to C_6—C_3 derivatives could have occurred in this experiment either. The *in vivo* experiments of Honohan *et al.* indicated that this metabolic pathway is mediated by the intestinal microflora. This conclusion was further substantiated by *in vitro* experiments using caecal microorganisms as noted below.

The investigations summarized above in which flavanones were administered to animals have been supplemented by studies which show that these compounds and their glycosides can be fully degraded to phenylpropionic acid derivatives by intestinal microorganisms. Thus, the anaerobic incubation of hesperidin or hesperetin with rat caecal micro-organisms resulted in their degradation to *m*-hydroxyphenylpropionic acid (Scheline, 1968b). Similar experiments with naringin resulted in the formation of *p*-hydroxyphenylpropionic acid (Griffiths and Smith, 1972a). Honohan *et al.* (1976) incubated [3-^{14}C]-hesperetin anaerobically with mixed cultures of rat caecal microorganisms and found extensive degrada-tion to varying proportions of the three phenylpropionic acid derivatives noted above. Very little radioactivity was lost as $^{14}CO_2$, indicating that further bacterial metabolism to C_6—C_1 metabolites is insignificant. Both hesperidin and naringin are degraded to water-soluble products by bovine rumen microorganisms (Simpson *et al.*, 1969). When these two flavanones were given orally to germ-free rats, none of the phenolic acid metabolites was detected in the urine (Griffiths and Barrow, 1972a). It is therefore evident that these intestinal reactions are able to fully account for the major urinary metabolites of the flavanones. Subsequent dehydrogenation and β-oxidation of the phenylpropionic acids in the tissues give rise to the cinnamic and benzoic acid derivatives that are also usually detected.

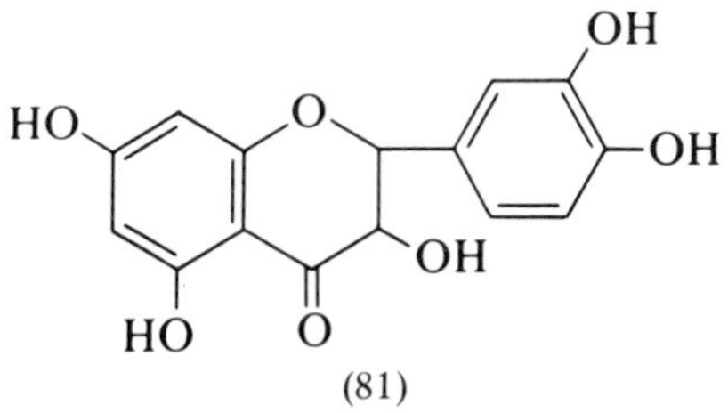

(81)

Taxifolin

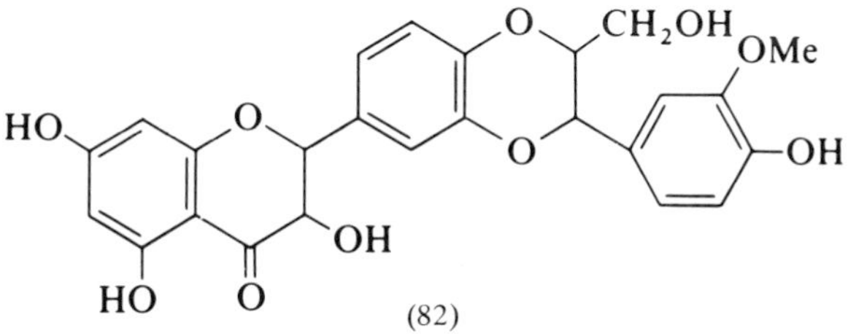

(82)

Silybin

Metabolic studies of the flavanonols (77) are limited, dealing with only two compounds. Booth and DeEds (1958) and DeEds (1968) reported that **taxifolin** (dihydroquercetin) (81) is metabolized in rats and humans to m-hydroxyphenylacetic acid, 3,4-dihydroxyphenylacetic acid (homoprotocatechuic acid) and 4-hydroxy-3-methoxyphenylacetic acid (homovanillic acid). In addition to these C_6—C_2 phenolic acids, a small amount of m-hydroxyphenylpropionic acid was also detected in rat urine. Thus, the metabolism of this 3-hydroxyflavanone resembles that observed with the flavonols (Section B) rather than that of the flavanones. **Silybin** (82) is a more complex flavanonol which, after oral or i.v. administration to rats, is excreted partly in the urine but mainly in the bile (Bülles *et al.*, 1975). The urinary material is mainly unchanged silybin whereas glucuronide and sulphate conjugates of silybin and, perhaps, 2,3-dehydrosilybin are excreted in the bile.

D. ANTHOCYANINS

The anthocyanins are intensely coloured, water-soluble pigments which are to a large extent responsible for the attractive scarlet, pink, red, mauve, violet and blue colours in flowers, leaves and fruits of higher plants.

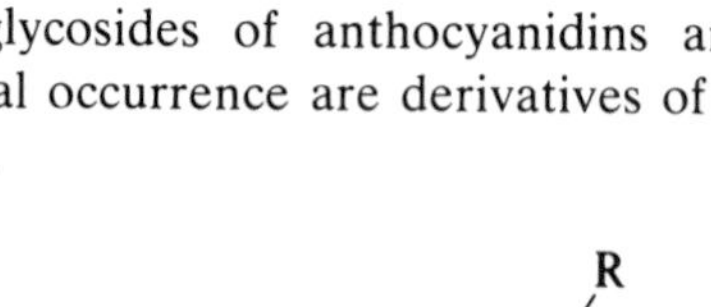

(83)

Anthocyanins are glycosides of anthocyanidins and all of the latter compounds of natural occurrence are derivatives of the parent flavylium cation structure (83).

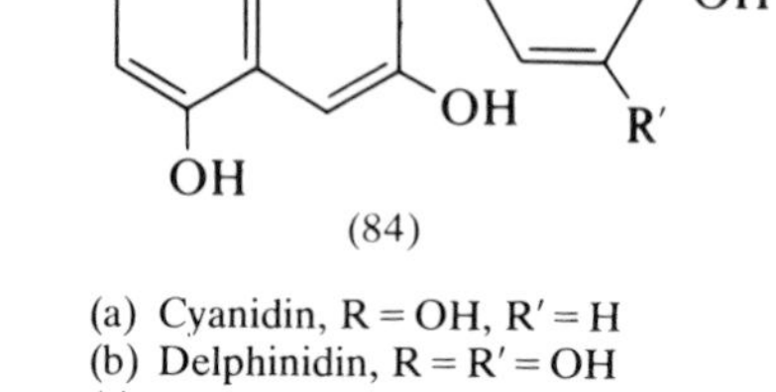

(84)

(a) Cyanidin, R = OH, R′ = H
(b) Delphinidin, R = R′ = OH
(c) Petunidin, R = OMe, R′ = OH
(d) Malvidin, R = R′ = OMe
(e) Pelargonidin, R = R′ = H

An early investigation of the metabolism of the anthocyanin pigment from Concord grapes was reported by Horwitt (1933). The precise nature of these anthocyanins was unknown at that time but a subsequent phytochemical study showed that the fruit of the Concord grape contains several glycosides of cyanidin (84a), delphinidin (84b), petunidin (84c) and malvidin (84d) (Timberlake and Bridle, 1975). Horwitt found that s.c. injection of the pigment (approx. 100 mg/kg) in rats resulted in considerable excretion in the urine, apparently in unchanged form. However, oral administration at about twice this dose level did not lead to detectable urinary excretion of the pigment. This is also the case in dogs but, on the other hand, a few percent of the dose is excreted in the urine of rabbits fed the pigment. Some pigment is excreted in the faeces in both rats and rabbits following oral administration of the anthocyanins. In the dog, biliary excretion was found to be an alternative route of excretion of injected pigment. The possibility that the pigment was metabolized by intestinal bacteria was studied using incubates containing extracts of human faeces. However, no loss of colour was observed.

The degradation of **cyanidin** chloride (84a) by rat caecal microorganisms was studied by Scheline (1968b) and Griffiths and Smith (1972a) who found that it was not metabolized to phenolic compounds. The aforementioned results suggest that the flavylium cation is metabolically stable, not an unreasonable situation in view of its ionic and hydrophilic nature. Nonetheless, studies by Griffiths and Smith (1972a,b) have shown that some of these compounds can be metabolized when given orally to rats or incubated with rat intestinal microorgansisms. Thus **pelargonin,** the 3,5-diglucoside of pelargonidin (84e), was converted in incubation experiments to a phenolic compound tentatively identified as *p*-hydroxyphenyllactic acid (85). Similar experiments with **delphinidin** (84b) resulted in the formation of two unidentified metabolites having, respectively, neutral and acidic properties. The neutral metabolite was also detected in the urine of a rat given delphinidin (100 mg) orally. **Malvin,** the 3,5-diglucoside of malvidin (84d), gave rise *in vivo* to three unidentified neutral urinary metabolites but none of these was detected in the incubation experiments.

$$HO-\!\!\!\bigcirc\!\!\!-CH_2-\overset{\displaystyle OH}{\overset{|}{CH}}-COOH$$

(85)

The limited results presently available indicate that flavylium compounds may undergo metabolic alteration but that this occurs to a much more limited extent than is the case with related flavonoids, e.g. catechins, lacking the cationic group.

E. Catechins

Catechins are derivatives of flavan-3-ol (86) and differ structurally from many other types of flavonoids by their lack of a 4-keto group. The flavan-3,4-diols (leucoanthocyanidins) (87) are close relatives of the catechins but are also closely related to the anthocyanidins (Section D) as they

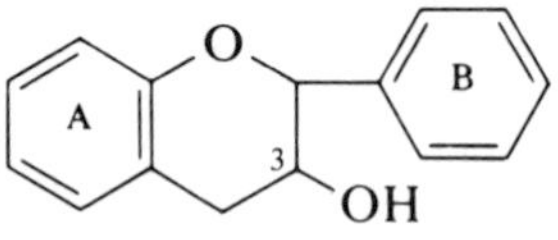

(86)

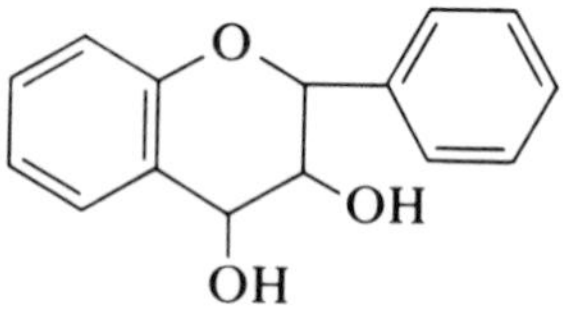

(87)

are readily converted to the latter in the presence of acid. However, knowledge of the metabolism of leucoanthocyanidins in animals is limited to a single compound, leucocyanidin (95), which will be dealt with at the end of this section. Both the catechins and leucoanthocyanidins are mainly found in the woody parts of plants, but it should be noted that they also occur in tea leaves and cacao beans. Also noteworthy is the presence of these flavonoids in condensed tannins.

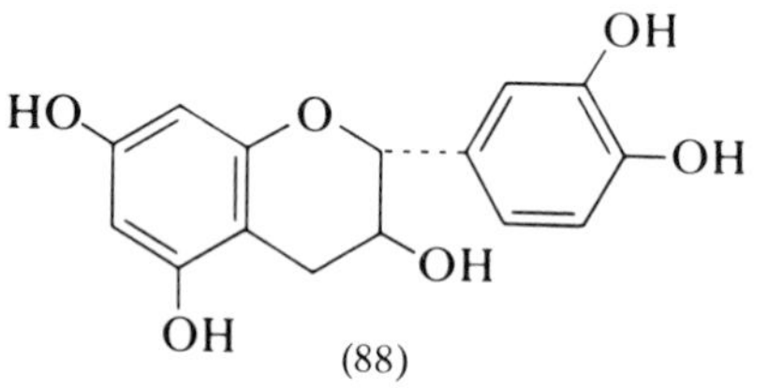

(88)
(+)-Catechin

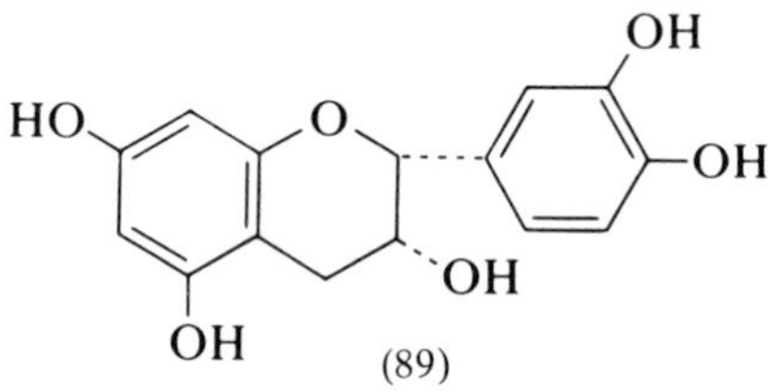

(89)
(−)-Epicatechin

The most common flavan-3-ols are the diastereoisomeric pair (+)-catechin (88) and (−)-epicatechin (89). The metabolism of both has been studied but most of the investigations have dealt with the former compound. In fact, our knowledge of the metabolism of (+)-catechin is more extensive than that of any other single flavonoid compound. The metabolism of (+)-**catechin** (88) was first studied in rabbits by Oshima *et al.* (1958) and Oshima and Watanabe (1958) who found that it was degraded to several simple phenolic acids and to neutral compounds. The major acidic metabolites were *m*-hydroxybenzoic acid, 3,4-dihydroxybenzoic acid (protocatechuic acid) and 4-hydroxy-3-methoxybenzoic acid (vanillic acid) and the neutral metabolites were postulated to be phenolic derivatives of phenyl-γ-valerolactone. A detailed investigation of the latter compounds showed that they were δ-(3-hydroxyphenyl-γ-valerolactone (90a), δ-(3,4-dihydroxyphenyl)-γ-valerolactone (90b) and δ-(4-hydroxy-3-methoxyphenyl)-γ-valerolactone (90c) (Watanabe, 1959a,b,c,d). All three of these valerolactones have also been identified as urinary metabolites when (+)-catechin is administered to rats and guinea pigs (Das and Griffiths, 1968, 1969), monkeys (*Macaca iris sp.*) (Das, 1974) and man (Das, 1971).

(90)

(a) R = OH, R′ = H
(b) R = R′ = OH
(c) R = OMe, R′ = OH

While the initial investigation of (+)-catechin metabolism indicated that C_6—C_1 phenolic acids were excreted in the urine of rabbits, a subsequent study showed that a C_6—C_3 acid, m-hydroxyphenylpropionic acid, was a major urinary metabolite of (+)-catechin in rats (Griffiths, 1962). As the metabolite showed a somewhat delayed excretion profile, it was suggested that it may be formed by the metabolic action of intestinal microorganisms. This prediction was confirmed by Booth and Williams (1963) who detected m-hydroxyphenylpropionic acid following the anaerobic incubation of (+)-catechin with rat faecal or caecal microorganisms. This reaction sequence was confirmed using rat intestinal microorganisms by Scheline (1968b) and by Das (1969). Interestingly, two of the phenylvalerolactones (compounds (90a) and (90b)) were also detected in the incubates in the latter investigation. These compounds were shown to be intermediates in the degradation of (+)-catechin to phenolic acids as the oral adminis- tration of compound (90a) to guinea pigs results in its partial metabolism to m-hydroxyphenylpropionic acid, m-hydroxybenzoic acid and m-hy- droxyhippuric acid which are excreted in the urine (Das and Griffiths, 1968). Similar experiments in rats resulted in the excretion of the first two of these phenolic acids (Das and Griffiths, 1969).

From the above results it appears that a similar pattern of degradation of (+)-catechin exists in the different animal species studied. Thus, the flavonoid is metabolized to phenylvalerolactones and partly further to C_6—C_3 phenolic acids by the intestinal microflora. Following the absorp- tion of these metabolites, some of the catecholic compounds may be O-methylated and the acids may be metabolized further by β-oxidation to benzoic acid derivatives which may then be conjugated with glycine giving hippuric acid derivatives. Not unexpectedly, phenolic acid metabolites were not detected in the urine of germ-free rats given (+)-catechin (Griffiths and Barrow, 1972a). Excretion of C_6—C_1 phenolic acids and/or their conjugates is seen in rabbits (Oshima et al., 1958) and also predominates in guinea pigs (Das and Griffiths, 1968). On the other hand, m-hydroxyphenylpropionic acid is excreted in increased amounts in rats (Griffiths, 1964; Das and Griffiths, 1969) and is the major urinary phenolic acid in man following the oral administration of (+)-catechin (Das, 1971). m-Hydroxyphenylhydracrylic acid (91), the β-oxidation product of m-hydroxyphenylpropionic acid, is also excreted by man and this compound is the major urinary phenolic acid in the monkey (Macaca iris sp.) (Das, 1974). These findings may indicate a species difference in metabolism in higher animals and man as a phenylhydracrylic acid derivative has also been identified as a urinary metabolite of a flavanone in man but not in laboratory animals (see Section C). Some unchanged (+)-catechin and usually also its conjugates are found in the urine of (+)-catechin-treated rats (Griffiths, 1964), guinea pigs (Das and Griffiths, 1968), monkeys (Das, 1974) and man (Das, 1971).

$$\text{HO} \quad \text{OH}$$
$$\underset{(91)}{\overbrace{\qquad\qquad}}\text{—CH—CH}_2\text{—COOH}$$

(91)

Although the details of the degradation of (+)-catechin to phenylvalerolactones and of the latter compounds to phenolic acids are not presently clear, some data on these points are available. When ^{14}C-labelled (+)-catechin is given to rats or guinea pigs, [14]CO_2 is found in the expired air (Das and Griffiths, 1969). The production of [14]CO_2 is greater with ring A-labelled compound than with randomly-labelled compound. These results show that part of the A-ring can be completely oxidized. While the degradation of (+)-catechin and the phenylvalerolactones to phenolic phenylpropionic acid derivatives is well established, it is also known that the flavonoid, when incubated anaerobically with rat or especially rabbit intestinal microorganisms, is converted to C_6—C_5 phenolic acids (Scheline, 1970). These compounds were shown to be 5-(3-hydroxyphenyl)-valeric acid (92a) and 5-(3,4-dihydroxyphenyl)-valeric acid (92b). However, it is not known if these metabolites are intermediates leading to the C_6—C_3 phenolic acids or if they represent terminal metabolites formed under special conditions. They have not been reported to be excreted in the urine of animals given (+)-catechin.

R
R′—⟨benzene ring⟩—(CH$_2$)$_4$—COOH

(92)

(a) R = OH, R′ = H
(b) R = R′ = OH

While the results summarized above deal mainly with the urinary excretion of (+)-catechin and its metabolites, excretion in both the faeces and bile has also been studied. Das and Griffiths (1969) found that only about 1% of the radioactivity was excreted in the 48 h faeces following the oral administration of [^{14}C]-(+)-catechin to rats or guinea pigs. Similar experiments in monkeys indicated about 2% excretion in 5 days (Das, 1974). (+)-Catechin and m-hydroxyphenylpropionic acid were identified as faecal metabolites in monkeys, a finding similar to that reported earlier in man (Das, 1971). However, nearly 20% of the orally administered (+)-catechin (83 mg/kg) was recovered unchanged in the faeces in man. In rats, m-hydroxyphenylpropionic acid and the phenylvalerolactones (90a) and (90b) were detected in the faeces (Griffiths, 1964; Das and Griffiths, 1969). The biliary excretion of (+)-catechin and its metabolites has been studied in rats. Griffiths and Barrow (1972b) found that unchanged compound, two conjugates of (+)-catechin and an unknown compound were excreted in the bile following oral or i.p. administration. Two further conjugates were also detected in the latter case. Das and Sothy (1971) reported that 33–44% of the radioactivity was excreted in the bile of rats within 24 h after i.v. injection of [^{14}C]-(+)-catechin. The biliary metabolites were of the same types as those reported by Griffiths and Barrow (1972b) and these results indicate that the biliary metabolites of (+)-catechin are largely different from those excreted in the urine.

The metabolism of (−)-**epicatechin** (89), a stereoisomer of (+)-catechin, has been studied in rabbits (Oshima et al., 1960). Following its oral administration, the same acidic metabolites (m-hydroxybenzoic acid, protocatechuic acid and vanillic acid) and the same neutral lactones (compounds (90a), (90b) and (90c)) were excreted in the urine as were found earlier with (+)-catechin (Oshima et al., 1958; Oshima and Watanabe, 1958). (+)-Catechin and (−)-epicatechin have different optical properties and this difference was also confirmed in metabolite (90b). When (−)-epicatechin was given this metabolite was levorotatory whereas the dextrorotatory form was produced from (+)-catechin.

On the basis of incubation experiments using rat intestinal micro-organisms, it appears that the degradation of (−)-**epiafzelechin** (93)

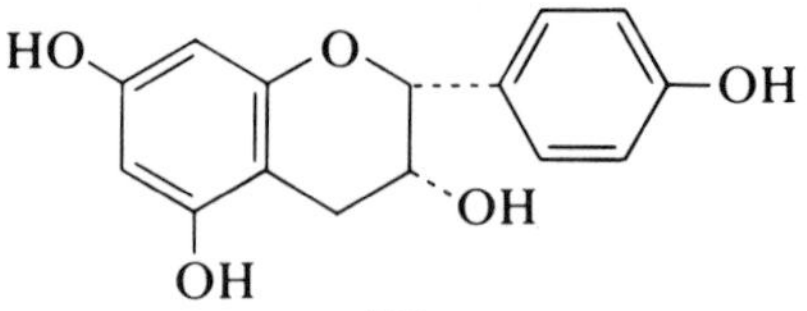

(93)

(−)-Epiafzelechin

follows the same general pattern as that described above with the isomers of catechin. Thus, Griffiths and Smith (1972a) detected p-hydroxy-phenylpropionic acid and two neutral metabolites in the incubates. Subsequently, Griffiths and Barrow (1972b) identified a neutral metabolite of (−)-epiafzelechin as δ-(p-hydroxyphenyl)-γ-valerolactone, i.e. isomeric with compound (90a).

In addition to the above studies using pure catechins, the metabolism of mixtures of catechins in animals has also been reported. The fate of tea catechins was studied in rabbits (Watanabe and Oshima, 1965) and in guinea pigs (Fedurov, 1966) and some metabolic results of feeding lucerne tannins to mice have been reported (Milić and Stojanović, 1972). However, these products contain a mixture of catechins as well as other related compounds and it is difficult to determine the origin of particular metabolites. Both products contain epicatechins and gallocatechins, the general formula of which is shown in structure (94). In addition, the lucerne preparation contains gallic acid (Milić, 1972). It is therefore not surprising that a wide variety of $C_6—C_1$ and $C_6—C_3$ phenolic acids and catechin derivatives were detected in the faeces of mice given the lucerne tannin. The urinary metabolites of the tea catechin preparation in both rabbits and guinea pigs were generally the same as those derived from gallic acid (see Chapter 5, Section I.C).

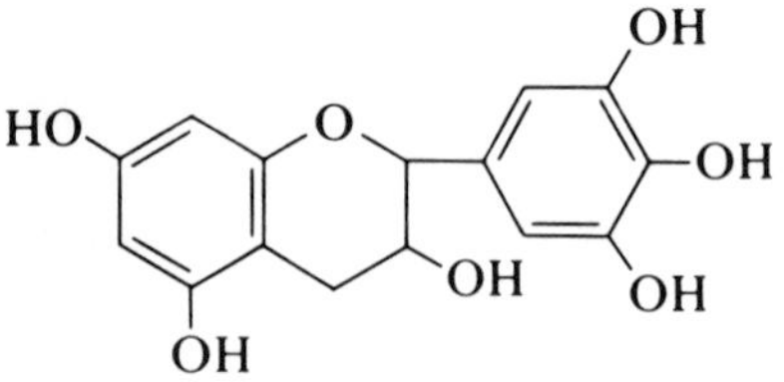

(94) 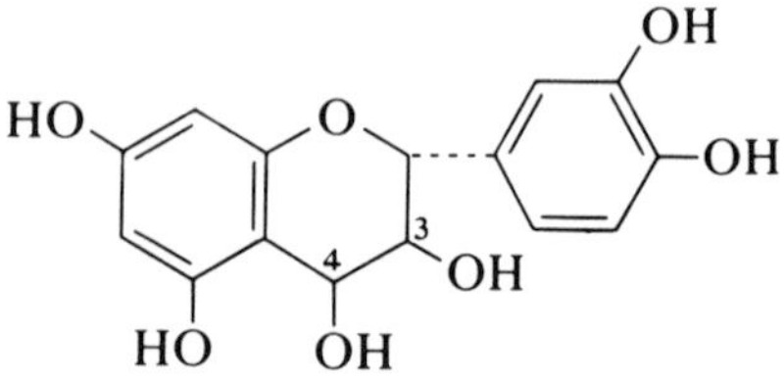   (95)

Leucocyanidin

The sole report on the metabolism of leucoanthocyanidins is that of Masquelier et al. (1965) who administered **leucocyanidin** (95) orally or i.p. to rats. They detected 3,4-dihydroxyphenylacetic acid, homovanillic acid and phloroglucinol glucuronide in the urine. This indicates that the 3,4-bond has been broken and that the metabolic degradation leading to phenolic acids is similar to that observed with the flavonols (Section B). However, the excretion of phloroglucinol is not typical of that found after administration of the latter group of compounds to laboratory animals. This is an interesting finding which should be confirmed as it may shed light on the mechanisms involved in flavonoid metabolism.

F. DIHYDROCHALCONES AND CHALCONES

Dihydrochalcone (96) is the parent of a small group of plant compounds which are structurally closely related to flavanones. This similarity is more easily seen when the structural formula is drawn as in (97). While dihydrochalcone itself occurs naturally, most of these compounds contain

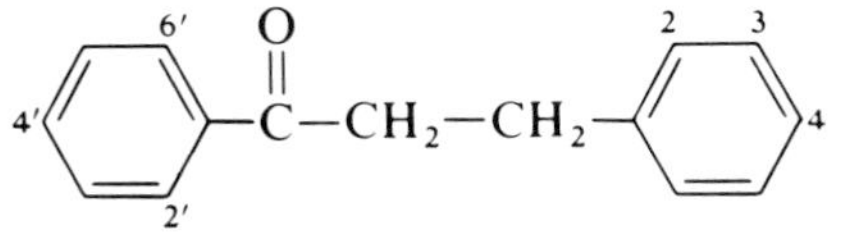

(96)

Dihydrochalcone

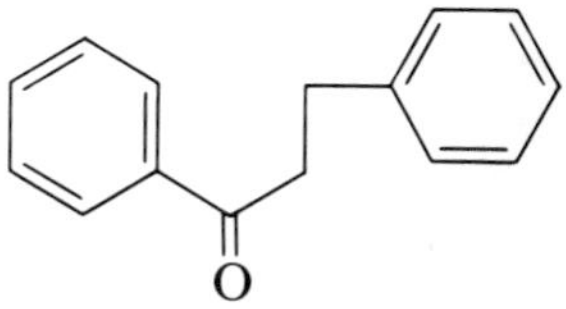

(97)

hydroxyl groups or their derivatives in one or usually both of the aromatic rings. Metabolic data on this group are not extensive, being confined mainly to phloretin (98) and its 2'-O-glucoside, phloridzin.

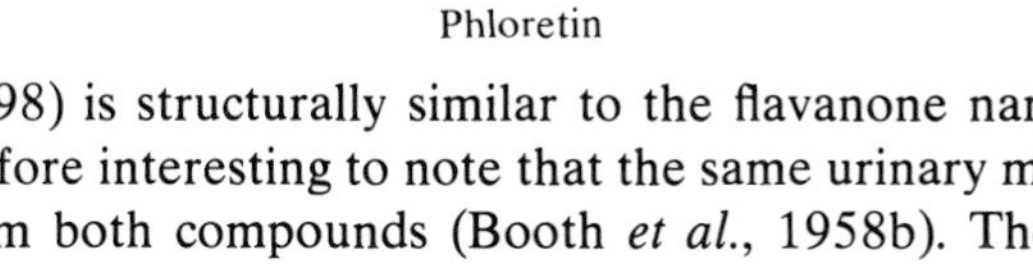

(98)

Phloretin

Phloretin (98) is structurally similar to the flavanone naringenin (79a) and it is therefore interesting to note that the same urinary metabolites are produced from both compounds (Booth et al., 1958b). They found that rats given large doses of phloretin (100 mg/rat, p.o. or s.c.) excreted p-hydroxyphenylpropionic acid (phloretic acid) as the major urinary metabolite together with some p-coumaric acid and ethereal sulphate of p-hydroxybenzoic acid. The same metabolites were detected when **phloridzin** (phloretin-2'-glucoside) was administered. The latter results were confirmed by Griffiths and Smith (1972a). In addition to these degradation products, some phloretin and its glucuronide were excreted in the urine following phloretin or phloridzin administration (Booth *et al.*, 1958b). The metabolism of the glucoside phloridzin in rabbits was the subject of an early investigation by Schüller (1911) who reported that it was excreted in the urine mainly as phloridzin glucuronide. This reaction also occurs in the dog (Braun *et al.*, 1957), although somewhat less of the glucuronide than of phloridzin itself is excreted. The location of attachment of the glucuronic acid moiety is not known although Schüller believed that this was at the 4-position rather than at one of the hydroxyl groups of the phloroglucinol ring.

The dihydrochalcone of naringin is closely related to phloridzin, being phloretin-4-rhamnoglucoside. Not surprisingly, the urinary metabolites of this compound in the rat were reported to be phloretic acid and *p*-coumaric acid (DeEds, 1968). The aglycone phloretin was not detected. A difference in metabolic behaviour has been reported with neohesperidin dihydrochalcone (99) which, like the dihydrochalcone of naringin, has not been reported to occur naturally. This compound did not undergo conversion to phenolic acids but was excreted in the urine of rats partly unchanged but mainly as the aglycone (DeEds, 1968).

(99)

In addition to the urinary route of excretion of dihydrochalcones and their metabolites, biliary excretion was shown by Barrow and Griffiths (1971) and Griffiths and and Barrow (1972b) to be involved in phloridzin elimination in rats. Following i.p. administration of this glucoside, unchanged compound as well as three acid-labile conjugates of increased polarity were detected in the bile. Oral administration of phloridzin led to two different biliary metabolites, of which one was an acid-labile conjugate.

It has been stressed in the preceding sections on flavonoid metabolism that the intestinal microflora plays an essential part in the degradative metabolism of these compounds. Based on the rather limited data available on the bacterial metabolism of the dihydrochalcones, this situation appears also to be the case with these compounds. DeEds (1968) described experiments in which the dihydrochalcones of naringin and neohesperidin were incubated anaerobically with caecal microorganisms. The former compound was degraded to phloretic acid, which is noted above to be a major urinary metabolite of this dihydrochalcone in rats. Similar experiments with the dihydrochalcone of neohesperidin led to the formation of 3-hydroxy-4-methoxyphenylpropionic and 3-hydroxyphenylpropionic acids. A subsequent report by Griffiths and Smith (1972a) described the metabolism of phloridzin in anaerobic cultures of rat caecal microorganisms. Chromatographic analysis of the incubates showed the presence of the aglycone phloretin (98), the expected phenolic acid phloretic acid and also a third metabolite which was identified as phloroglucinol. This finding is of considerable interest because it suggests that the degradation of dihydrochalcones may depend upon the action of a hydrolase which splits the molecule between the ketone group and the phloroglucinol ring. Furthermore, this metabolic pathway differs from the major route of metabolism occurring with other types of flavonoids described above in which the breakdown does not lead to the formation of phloroglucinol. Of relevance in this regard is the fact that the flavanone naringin and its aglycone naringenin (79a) when given to rats or incubated with caecal microorganisms both give rise to phloretic acid but not to phloretin or phloroglucinol.

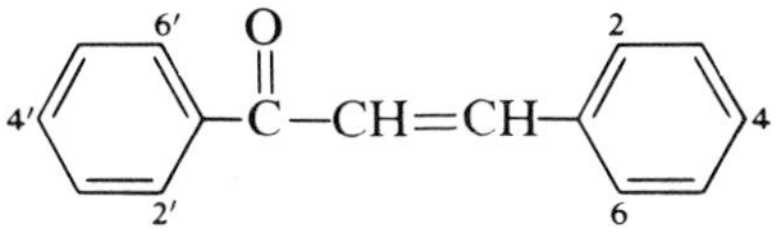

(100)

The metabolic fate of chalcone (100) derivatives is largely unknown. Formanek and Höller (1961) administered 2,4,4'-trihydroxy-, 2,4,2'-trihydroxy-, 4,2'-dihydroxy- and 4-hydroxychalcone orally or i.v. to rats. They reported that a series of unidentified compounds was rapidly excreted in the urine and bile.

G. ISOFLAVONOIDS

Isoflavonoids differ structurally from the flavonoids in having the phenyl ring (B-ring) attached at the 3- rather than at the 2-position of the heterocyclic ring. This is illustrated in the structure of isoflavone (101), deriva-

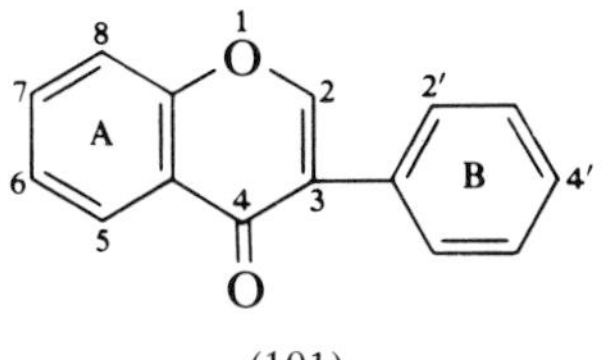

(101)

tives of which form the most common group of isoflavonoids. Our knowledge of isoflavonoid metabolism is largely limited to these isoflavones and owes its existence mainly to the fact that several of these compounds exert effects on the reproductive system of animals. As these isoflavones occur naturally in several species of forage legumes, knowledge of their metabolism as well as other biological properties is of practical interest. The isoflavonoid structure is also present in the rotenoids which are discussed in Section VI..

(102)

(a) Daidzein, R = OH
(b) Formononetin, R = OMe

(103)

(a) Genistein, R = OH
(b) Biochanin A, R = OMe

Most of the data on isoflavone metabolism deal with the compounds **daidzein** (102a), **formononetin** (102b), **genistein** (103a) and **biochanin A** (103b). The structural formulas show that formononetin and biochanin A are the 4'-O-methyl ethers of daidzein and genistein, respectively. The first published report on isoflavone metabolism indicated that O-demethylation could occur (Nilsson, 1961a). When biochanin A (103b) was incubated *in vitro* with rumen fluid from cattle or sheep, genistein (103a) was formed along with variable amounts of two unidentified metabolites. Subsequent reports confirmed the O-demethylation of biochanin A when incubated with rumen fluid (Batterham *et al.*, 1965; Braden *et al.*, 1967; Nilsson *et al.*, 1967) and the corresponding reaction with formononetin (102b) to give daidzein (102a) has also been demonstrated (Nilsson, 1962a; Batterham *et al.*, 1965; Nilsson *et al.*, 1967; Braden *et al.*, 1967). In contrast, the two methoxy compounds have been shown to be largely resistant to demethylation when incubated with microorganisms from rat intestine (Nilsson, 1961b; Griffiths and Smith, 1972a). However, administration of biochanin A (25 mg/kg, i.p.) to rats resulted in the excretion of unchanged compound and its demethylation product in the urine and faeces (Nilsson, 1961b). This reaction was shown to be carried out by the liver. Further studies in rats using ^{3}H-labelled biochanin A given i.p. showed that about 30% of the radioactivity appeared in the urine and 60% in the faeces (Nilsson, 1962b). Most of the urinary material was in the form of conjugates, but about 8% free biochanin A and 1% free genistein were detected. Faecal radioactivity was entirely unconjugated and consisted of about 40% unchanged compound and 14% genistein. These figures are explainable in terms of the biliary excretion of biochanin A and its metabolites. The isoflavone conjugates appear to be glucuronides and sulphates (Nilsson, 1962b). The conjugates, chiefly glucuronides, comprise 98–99% of the total plasma isoflavones in sheep fed on diets of oestrogenic clover (Shutt *et al.*, 1967). Labow and Layne (1972) demonstrated that biochanin A, genistein, formononetin and daidzein as well as the related compound equol (104) form monoglucuronides when incubated with rabbit liver microsomal fractions and UDP-glucuronic acid. The structures of the glucuronides were not determined but with formononetin (102b) the attachment must be at the 7-position which is the only available site. Interestingly, Labow and Layne also showed that these isoflavones were converted in low yield to monoglucosides when UDP-glucose was substituted for UDP-glucuronic acid in the incubates.

A comprehensive study of the liver microsomal system catalysing the O-demethylation of biochanin A, formononetin and other methoxylated isoflavones was carried out by Nilsson (1963). Appreciable demethylating activity was present in liver preparations from rabbits, rats and mice with lower activities found in swine, cattle and sheep. The lower activities in the preparations from the larger animals may, however, be due to technical factors rather than intrinsic species differences. The following isoflavones underwent O-demethylation when incubated with a cofactor-fortified supernatant fraction (14 000 g) from rabbit liver: **prunetin** (5,4'-dihydroxy-7-methoxyisoflavone), **prunusetin** (7,4'-dihydroxy-5-methoxyisoflavone), **muningin** (6,4'-dihydroxy-5,7-dimethoxyisoflavone), 7-methoxyisoflavone and 5,7,2'-trimethoxyisoflavone. These results indicate that O-demethylation may take place at sites in both the A and B rings. In addition, **ferreirin** (5,7,2'-trihydroxy-4'-methoxyisoflavanone) was demethylated by the rabbit liver preparation. Both biochanin A and formononetin undergo demethylation in sheep following their intraruminal or i.m. administration (Lindner, 1967).

From the foregoing it can be seen that methoxylated isoflavones are subject to demethylation both by tissue enzymes and by some intestinal bacteria. However, several other and far more profound metabolic changes also occur with isoflavones. Batterham *et al.* (1965) and Braden *et al.* (1967) reported that biochanin A and genistein, when given intraruminally to sheep, were degraded to p-ethylphenol which was excreted in the urine. This pathway appears to account for about 80% of the ingested isoflavone. Interestingly, i.m. injection of these isoflavones did not lead to an increased excretion of p-ethylphenol. Similar experiments in sheep employing [4-^{14}C]-biochanin A (Batterham *et al.*, 1971) showed that the p-ethylphenol excreted was not radioactive, indicating that this metabolite is derived from the isoflavone B-ring. The steps leading to p-ethylphenol are not known although it seems possible that scission of the isoflavone to a phenyl-α-methylbenzyl ketone intermediate followed by hydrolysis of the latter may be involved. This would also give rise to phenolic acids and, in fact, evidence is available indicating the formation of several such compounds, although the small amounts detected were insufficient to permit identification (Batterham et al., 1971). Evidence concerning the site of p-ethylphenol formation is conflicting. Braden et al. (1967) looked for but did not find degradation of biochanin A and genistein to simple phenols in incubation experiments using sheep rumen fluid. However, Griffiths and Smith (1972a) showed that genistein was extensively converted to p-ethylphenol in incubates of rat intestinal bacteria.

(104)   (105)

Formononetin and daidzein differ from biochanin A and genistein only by their lack of a 5-hydroxyl group. Nonetheless, this deficiency is sufficient to give rise to profound differences in their overall metabolic fates. When formononetin is given intraruminally to sheep, equol (7,4'-dihydroxy-isoflavan) (104) is excreted in the urine (Batterham et al., 1965; Braden et al., 1967). This does not occur when the isoflavone is given by i.m. injection. The metabolism of formononetin to equol in sheep has been confirmed by Shutt and Braden (1968) and Batterham et al. (1971) who also identified 2,4-dihydroxyphenyl-α-(4'-hydroxyphenyl) ethyl ketone (O-desmethyl angolensin) (105) as a urinary metabolite. These reactions are not confined to sheep, as guinea pigs, fed a diet containing formononetin, were found to have both of these metabolites in the blood (Shutt and Braden, 1968). Equol and O-desmethyl angolensin are evidently produced by different metabolic sequences, but none of these investigations has found simpler end-products such as p-ethylphenol which, as noted above, is formed from biochanin A and genistein. Incubation of formononetin or daidzein with sheep rumen fluid (Nilsson et al., 1967) or rat intestinal microorganisms (Griffiths and Smith, 1972a) results in the formation of equol.

H. NEOFLAVONOIDS

The term neoflavonoid is employed to describe a group of compounds possessing a 4-arylchroman skeleton (106). They are thereby closely

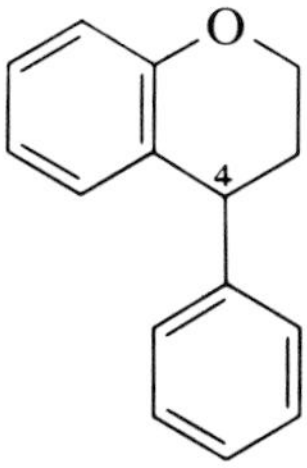

(106)

related to the flavanoids and isoflavanoids which are based upon 2-aryl- and 3-arylchroman, respectively. Examples of neoflavanoid metabolism are found in other sections and include the compounds calophyllolide (Section III.C) and haematoxylin (Section II.B).

V. Xanthones

Derivatives of xanthone (107) are not widespread in Nature and relatively little is known of their metabolism in animals. For this reason the metabolism of xanthone itself will be included although all of the naturally occurring compounds have a hydroxyl group at C1.

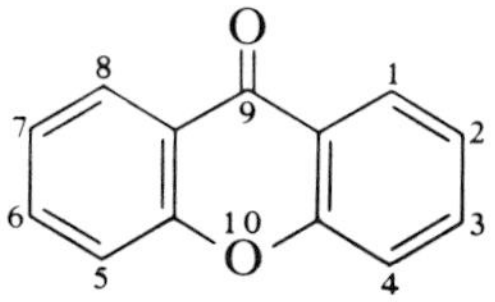

(107)

The metabolism of xanthone in rats was studied by Griffiths (1974) who found that the major unconjugated urinary metabolite was 4-hydroxy-xanthone, accounting for about one fourth of the 300 mg/kg dose. The other monohydroxy derivatives formed were 2-hydroxyxanthone (13%) and 3-hydroxyxanthone (6%). Trace amounts of five additional urinary metabolites were present, one of which was an acid-labile conjugate and another an o-dihydroxy derivative. Unchanged compound was not excreted. Conjugates of both 2- and 4-hydroxyxanthone were excreted in the urine and bile, the latter occurrence no doubt responsible for the finding that metabolite excretion was prolonged. Rupture of the heterocyclic ring, a reaction shown to occur with coumarin and its derivatives (Section III) and many hydroxylated flavonoids (Section IV), was not detected. This finding is similar to that seen with the structurally related compound flavone, which undergoes ring hydroxylation but not ring fission (Section IV.A).

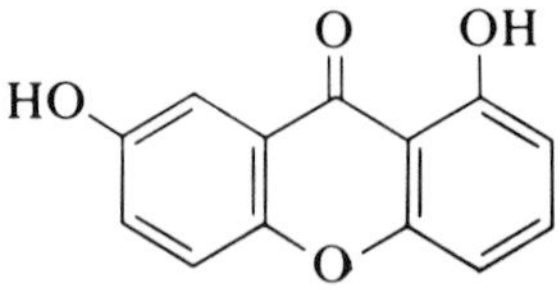

$$\text{HO} \quad \overset{\text{O}}{\big|} \quad \text{OH}$$

(108)

Euxanthone

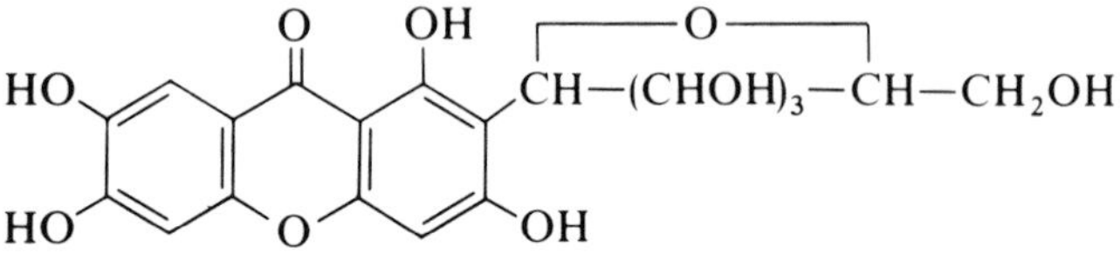

$$\text{HO} \quad \overset{\text{O}}{\big|} \quad \text{OH} \qquad \text{CH}-(\text{CHOH})_3-\text{CH}-\text{CH}_2\text{OH}$$

HO   O   OH

(109)

Mangiferin

Of the naturally occurring xanthones, only **euxanthone** (1,7-dihy-droxyxanthone) (108) and mangiferin (109) have been studied metabolic-ally. Very early experiments (Kostanecki, 1886; Külz, 1887) showed that oral administration of euxanthone to rabbits and dogs resulted in the urinary excretion of the 7-glucuronide derivative, euxanthic acid. It appears that the 1-glucuronide (isoeuxanthic acid) is not formed. The same metabolite, euxanthic acid, was shown by Wiechowski (1923) to be excreted in the urine of rabbits fed the crude yellow colouring matter from mango leaves. Subsequently, pure samples of **mangiferin** (109), the yellow principle from the leaves of *Mangifera indica*, were administered to ani-mals. Iseda (1956) showed that euxanthic acid was excreted in the urine of rabbits given mangiferin orally. Similar experiments were reported later by Krishnaswamy *et al.* (1971). No free mangiferin or 1,3,6,7-tetra-hydroxyxanthone was detected in the urine but both euxanthone (10%) and its glucuronide, euxanthic acid (12%), were found. This unusual metabolic sequence involves removal of two phenolic hydroxyl groups and the loss of the C-glucoside moiety. Other points of interest from this study are the findings that mangiferin was only slowly eliminated from the organism and that a relatively large amount of benzoic acid was excreted by the treated but not the control animals.

VI. Other Oxygen Heterocyclic Compounds

The rotenoids are a group of compounds, the most important being rotenone (110), possessing considerable insecticidal activity but short residual effects. Also, rotenone is highly toxic to fish and is therefore useful

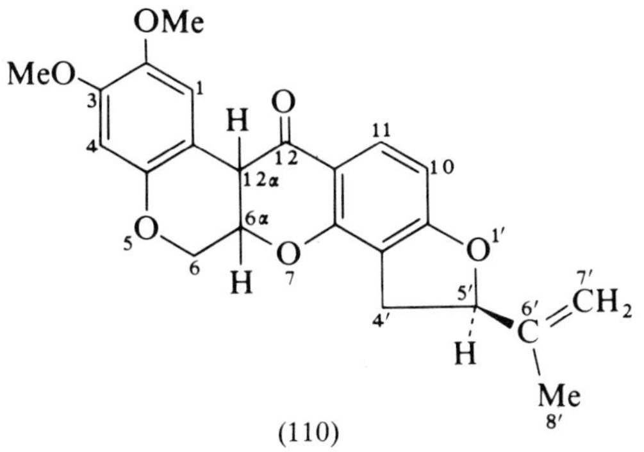

(110)

Rotenone

in the management of fish populations in fresh-water systems. On the other hand, rotenone shows only moderate toxicity in mammals.

Information on the metabolism of **rotenone** (110) has been obtained following the preparation of the compound labelled with ^{14}C, either at the 6α or the 3-methoxy position. With the former compound, Fukami et al. (1967) and Yamamoto (1969) described experiments which showed that eight metabolites were formed in vitro during incubation with the microsome-$NADPH_2$ system from rat or mouse liver (Table 7.3). While each of these is more polar than rotenone, the toxicity to mice of several of the metabolites was similar to that of rotenone. The major metabolities in vitro were compounds (111), (113), (114), (116) and (117) and, of these, metabolite (116) was the most important. Sequence studies showed the pathways of metabolism and these results are summarized in Table 7.3. Fukami et al. (1969) discussed differences in metabolic rate or route in relation to the selective toxicity of rotenone in various animal species.

In addition to the in vitro studies with rotenone labelled in the 6α-position, experiments in mice showed that compounds (116) and (117) are urinary metabolites (Fukami et al., 1967; Yamamoto, 1969). It is interesting to note that the radioactivity was excreted predominantly in the faeces following the intragastric administration of rotenone to mice. After 48 h, 20% of the ^{14}C was found in the urine, 0·3% in the expired air, 5% in the body and the remainder in the faeces. The nature of the faecal metabolites is not known.

Further information on the metabolism of natural rotenone ($5'$-β-rotenone) (110) was obtained using material labelled with ^{14}C in the 3-methoxy group (Unai et al., 1973). These experiments showed that extensive demethylation occurs. In mice, 27% of the radioactivity was found in the expired air within 50 h, after either oral or i.p. administration. The corresponding figure in rats was 12·5%. The finding that the extent of

TABLE 7.3

Metabolites of [6α-^{14}C]-rotenone formed by liver microsomes[a]

Compound	Name	Chemical change[b] at 12α	Chemical change[b] at 5'-isopropenyl	Precursor
(111)	Rotenolone I	OH	—	(110)
(112)	Rotenolone II	OH	—	(110)
(113)	8'-Hydroxyrotenone	—	$\diagdown C{=}CH_2$ CH_2OH	(110)
(114)	8'-Hydroxyrotenolone I	OH	$\diagdown C{=}CH_2$ CH_2OH	(111), (113)
(115)	8'-Hydroxyrotenolone II	OH	$\diagdown C{=}CH_2$ CH_2OH	(112), (113)
(116)	6',7'-Dihydro-6',7' -dihydroxyrotenone	—	CH_2OH $\diagdown C\diagdown$ OH Me	(110)
(117)	6',7'-Dihydro-6',7'- dihydroxyrotenolone I	OH	CH_2OH $\diagdown C\diagdown$ OH Me	(111), (116)
(118)	6',7'-Dihydro-6',7'- dihydroxyrotenolone II	OH	CH_2OH $\diagdown C\diagdown$ OH Me	(112), (116)

[a] Fukami *et al.* (1967); Yamamoto (1969).
[b] Refer to structure (110).

formation of $^{14}CO_2$ is similar following both routes of administration is of interest as it suggests that rotenone absorption from the gut is extensive and that the high faecal excretion of radioactivity noted above may be due to biliary excretion of rotenone and its metabolites.

References

Agurell, S., Nilsson, I. M., Ohlsson, A. and Sandberg, F. (1969). *Biochem. Pharmac.* **18**, 1195–1201.

Agurell, S., Nilsson, I. M., Ohlsson, A. and Sandberg, F. (1970). *Biochem. Pharmac.* **19**, 1333–1339.

Agurell, S., Nilsson, I. M., Nilsson, J. L. G., Ohlsson, A. and Widman, M. (1971). *Acta pharm. suec.* **8**, 698–699.

Agurell, S., Dahmén, J., Gustafsson, B., Johansson, U.-B., Leander, K., Nilsson, I., Nilsson, J. L. G., Nordqvist, M., Ramsay, C. H., Ryrfeldt, Å., Sandberg, F. and Widman, M. (1972). *In* "Cannabis and its Derivatives. Pharmacology and Experimental Psychology" (W. D. M. Paton and J. Crown, Eds), pp. 16–36. Oxford University Press, London.

Arora, R. B., Tahir, P. J., Krishnaswamy, N. R. and Seshadri, T. R. (1966). *Indian J. Biochem.* **3**, 58–60.

Barrow, A. and Griffiths, L. A. (1971). *Biochem. J.* **125**, 24P–25P.

Batterham, T. J., Hart, N. K., Lamberton, J. A. and Braden, A. W. H. (1965). *Nature* Lond. **206**, 509.

Batterham, T. J., Shutt, D. A., Hart, N. K., Braden, A. W. H. and Tweeddale, H. J. (1971). *Aust. J. agric. Res.* **22**, 131–138.

Ben-Zvi, Z. and Burstein, S. (1974). *Res. Commun. Chem. Path. Pharmac.* **8**, 223–229.

Ben-Zvi, Z. and Burstein, S. (1975). *Biochem. Pharmac.* **24**, 1130–1131.

Ben-Zvi, Z., Mechoulam, R. and Burstein, S. (1970). *J. Am. chem. Soc.* **92**, 3468–3469.

Ben-Zvi, Z., Bergen, J. R. and Burstein, S. (1974a). *Res. Commun. Chem. Path. Pharmac.* **9**, 201–204.

Ben-Zvi, Z., Burstein, S. and Zikopoulos, J. (1974b). *J. pharm. Sci.* **63**, 1173–1174.

Ben-Zvi, Z., Bergen, J. R., Burstein, S., Sehgal, P. K. and Varanelli, C. (1976). *In* "Pharmacology of Marihuana" (M. C. Braude and S. Szara, Eds), pp. 63–75. Raven Press, New York.

Bickoff, E. M., Livingston, A. L. and Booth, A. N. (1964). *J. pharm. Sci.* **53**, 1411–1412.

Booth, A. N. and DeEds, F. (1958). *J. Am. pharm. Ass.* (Sci. Edn) **47**, 183–184.

Booth, A. N. and Williams, R. T. (1963). *Biochem. J.* **88**, 66P–67P.

Booth, A. N., Murray, C. W., DeEds, F. and Jones, F. T. (1955). *Fedn Proc. Fedn Am. Socs exp. Biol.* **14**, 321.

Booth, A. N., Jones, F. T. and DeEds, F. (1956a). *Fedn Proc. Fedn Am. Socs exp. Biol.* **15**, 223.

Booth, A. N., Murray, C. W., Jones, F. T. and DeEds, F. (1956b). *J. biol. Chem.* **223**, 251–257.

Booth, A. N., Jones, F. T. and DeEds, F. (1958a). *J. biol. Chem.* **230**, 661–668.

Booth, A. N., Jones, F. T. and DeEds, F. (1958b). *J. biol. Chem.* **233**, 280–282.

Booth, A. N., Masri, M. S., Robbins, D. J., Emerson, O. H., Jones, F. T. and DeEds, F. (1959). *J. biol. Chem.* **234**, 946–948.

Boyland, E. and Chasseaud, L. F. (1967). *Biochem. J.* **104**, 95–102.

Braden, A. W. H., Hart, N. K. and Lamberton, J. A. (1967). *Aust. J. agric. Res.* **18**, 335–348.

Braun, W., Whittaker, V. P. and Lotspeich, W. D. (1957). *Am. J. Physiol.* **190**, 563–569.

Braymer, H. D. (1960). Ph.D. Thesis. University of Oklahoma.

Braymer, H. D., Shetlar, M. R. and Wender, S. H. (1960). *Biochim. biophys. Acta* **44**, 606–607.

Bülles, H., Bülles, J., Krumbiegel, G., Mennicke, W. H. and Nitz, D. (1975). *Arzneimittel-Forsch.* **25**, 902–905.

Burstein, S. H. (1973). *In* "Marihuana. Chemistry, Pharmacology, Metabolism and Clinical Effects" (R. Mechoulam, Ed.), pp. 167–190. Academic Press, New York and London.

Burstein, S. H. and Kupfer, D. (1971). *Chem.-Biol. Interact.* **3**, 316.

Burstein, S. and Varanelli, C. (1975). *Res. Commun. chem. Path. Pharmac.* **11**, 343–354.

Burstein, S. H., Menezes, F., Williamson, E. and Mechoulam, R. (1970). *Nature Lond.* **225**, 87–88.

Burstein, S., Rosenfeld, J. and Wittstruck, T. (1972). *Science N.Y.* **176**, 422–423.

Cayen, M. N. and Common, R. H. (1965). *Biochim. biophys. Acta* **100**, 567–573.

Cheng, K.-J., Jones, G. A., Simpson, F. J. and Bryant, M. P. (1969). *Can. J. Microbiol.* **15**, 1365–1371.

Christensen, H. D., Freudenthal, R. I., Gidley, J. T., Rosenfeld, R., Boegli, G., Testino, L., Brine, D. R., Pitt, C. G. and Wall, M. E. (1971). *Science N.Y.* **172**, 165–167.

Christiansen, J. and Rafaelsen, O. J. (1969). *Psychopharmacologia* **15**, 60–63.

Clark, W. G. and MacKay, E. M. (1950). *J. Am. med. Ass.* **143**, 1411–1415.

Creaven, P. J., Parke, D. V. and Williams, R. T. (1965). *Biochem. J.* **96**, 390–398.

Crew, M. C., Szpiech, J. M. and DiCarlo, F. J. (1976a). *Xenobiotica* **6**, 83–88.

Crew, M. C., Melgar, M. D., George, S., Greenough, R. C., Szpiech, J. M. and DiCarlo, F. J. (1976b). *Xenobiotica* **6**, 89–100.

Csallany, A. S., Draper, H. H. and Shah, S. N. (1962). *Archs Biochem. Biophys.* **98**, 142–145.

Das, N. P. (1969). *Biochim. biophys. Acta* **177**, 668–670.

Das, N. P. (1971). *Biochim. Pharmac.* **20**, 3435–3445.

Das, N. P. (1974). *Drug Metab. Disposit.* **2**, 209–213.

Das, N. P. and Griffiths, L. A. (1966). *Biochem. J.* **98**, 488–492.

Das, N. P. and Griffiths, L. A. (1968). *Biochem. J.* **110**, 449–456.

Das, N. P. and Griffiths, L. A. (1969). *Biochem. J.* **115**, 831–836.

Das, N. P. and Sothy, S. P. (1971). *Biochem. J.* **125**, 417–423.

Das, N. P., Scott, K. N. and Duncan, J. H. (1973). *Biochem. J.* **136**, 903–909.

DeEds, F. (1968). *In* "Comprehensive Biochemistry" (M. Florkin and E. H. Stotz, Eds), Vol. 20, pp. 127–171. Elsevier Publ. Co., Amsterdam, London and New York.

Demole, V. (1962). *Helv. physiol. Acta* **20**, 93–96.

DiCarlo, F. J., Herzig, D. J., Kusner, E. J., Schumann, P. R., Melgar, M. D., George, S. and Crew, M. C. (1976). *Drug Metab. Disposit.* **4**, 368–371.

Douglass, C. D. and Hogan, R. (1958). *J. biol. Chem.* **230**, 625–629.

Draper, H. H. and Csallany, A. S. (1969). *Fedn Proc. Fedn Am. Socs exp. Biol.* **28**, 1690–1695.

Drasar, B. S. and Hill, M. J. (1974). "Human Intestinal Flora", p. 63. Academic Press, London, New York, San Francisco.

Erdmann, G., Just, W. W., Thel, S., Werner, G. and Wiechmann, M. (1976). *Psychopharmacology* **47**, 53–58.

Estevez, V. S., Englert, L. F. and Ho, B. T. (1973). *Res. Commun. Chem. Path. Pharmac.* **6**, 821–827.

Fedurov, V. V. (1966). Fenol'nye Soedin. Ikh Biol. Funkts., Mater Vses, Simp., 1st (Publ. 1968) pp. 371–377. (*Chem. Abstr.* (1969) **71**, 11414e.)

Feuer, G. (1974). *In* "Progress in Medicinal Chemistry" (G. P. Ellis and G. B. West, Eds), Vol. 10, pp. 85–158. North-Holland Publishing Co., Amsterdam and New York.

Feuer, G., Golberg, L. and Gibson, K. I. (1966). *Fd Cosmet. Toxicol.* **4**, 157–167.

Fink, P. C. and Kerekjarto, B. V. (1966). *Hoppe-Seyler's Z. physiol. Chem.* **345**, 272–279.

Flatow, L. (1910). *Hoppe-Seyler's Z. physiol. Chem.* **64**, 367–392.

Foltz, R. L., Fentiman, A. F., Leighty, E. G., Walter, J. L., Drewes, H. R., Schwartz, W. E., Page, T. F. and Truitt, E. B. (1970). *Science N.Y.* **168**, 844–845.

Fonseka, L. and Widman, M. (1977). *J. Pharm. Pharmac.* **29**, 12–14.

Formanek, K. and Höller, H. (1961). *Sci. Pharm.* **29**, 102–107. (*Chem. Abstr.* (1961) **55**, 22619h).

Fujita, M. and Furuya, T. (1958a). *Chem. pharm. Bull.* Tokyo **6**, 517–520.

Fujita, M. and Furuya, T. (1958b). *Chem. pharm. Bull.* Tokyo **6**, 520–523.

Fukami, J.-.I., Yamamoto, I. and Casida, J. E. (1967). *Science N.Y.* **155**, 713–716.

Fukami, J.-.I., Shishido, T., Fukunaga, K. and Casida, J. E. (1969). *J. agric. Fd Chem.* **17**, 1217–1226.

Furuya, T. (1958a). *Chem. pharm. Bull.* Tokyo **6**, 696–700.

Furuya, T. (1958b). *Chem. pharm. Bull.* Tokyo **6**, 701–706.

Gangolli, S. D., Shilling, W. H., Grasso, P. and Gaunt, I. F. (1974). *Biochem. Soc. Trans.* **2**, 310–312.

Gau, W., Bieniek, D., Coulston, F. and Korte, F. (1974). *Chemosphere* **3**, 71–76.

Gautrelet, J. and Gravellat, H. (1906). *C. r. Séanc. Soc. Biol.* **61**, 134–135.

Gill, E. W. and Jones, G. (1971). *Acta pharm. suec.* **8**, 700–701.

Gill, E. W., Jones, G. and Lawrence, D. K. (1973). *Biochem. Pharmac.* **22**, 175–184.

Greene, M. L. and Saunders, D. R. (1974). *Gastroenterology* **66**, 365–372.

Griffiths, L. A. (1962). *Nature* Lond. **194**, 869–870.

Griffiths, L. A. (1964). *Biochem. J.* **92**, 173–179.

Griffiths, L. A. (1974). *Xenobiotica* **4**, 375–382.

Griffiths, L. A. and Barrow, A. (1972a). *Biochem. J.* **130**, 1161–1162.

Griffiths, L. A. and Barrow, A. (1972b). *Angiologica* **9**, 162–174.

Griffiths, L. A. and Smith, G. E. (1972a). *Biochem. J.* **128**, 901–911.

Griffiths, L. A. and Smith, G. E. (1972b). *Biochem. J.* **130**, 141–151.

Gugler, R., Leschik, M. and Dengler, H. J. (1975). *Eur. J. clin. Pharmac.* **9**, 229–234.

Gurny, O., Maynard, D. E., Pitcher, R. G. and Kierstead, R. W. (1972). *J. Am. chem. Soc.* **94**, 7928–7929.

Hanuš, L. and Krejčí, Z. (1974). *Acta Univ. Palacki. Olumuc., Fac. Med.* **71**, 253–264.

Harvey, D. J. and Paton, W. D. M. (1976). *Res. Commun. chem. Path. Pharmac.* **13**, 585–599.

Harvey, D. J. and Paton, W. D. M. (1977). *J. Pharm. Pharmac.* **29**, 498–500.

Harvey, D. J., Martin, B. R. and Paton, W. D. M. (1976). *Biochem. Pharmac.* **25**, 2217–2219.

Harvey, D. J., Martin, B. R. and Paton, W. D. M. (1977a). *Res. Commun. chem. Path. Pharmac.* **16**, 265–279.

Harvey, D. J., Martin, B. R. and Paton, W. D. M. (1977b). *J. Pharm. Pharmac.* **29**, 482–486.

Harvey, D. J., Martin, B. R. and Paton, W. D. M. (1977c). *J. Pharm. Pharmac.* **29**, 495–497.

Hawksworth, G., Drasar, B. S. and Hill, M. J. (1971). *J. med. Microbiol.* **4**, 451–459.

Ho, B. T., Estevez, V., Englert, L. F. and McIsaac, W. M. (1972). *J. Pharm. Pharmac.* **24**, 414–416.

Ho, B. T., Estevez, V. S. and Englert, L. F. (1973). *J. Pharm. Pharmac.* **25**, 488–490.

Hollister, L. E., Kanter, S. L., Moore, F. and Green, D. E. (1972). *Clin. Pharmac. Ther.* **13**, 849–855.

Hollister, L. E., Kanter, S. L., Board, R. D. and Green, D. E. (1974). *Res. Commun. chem. Path. Pharmac.* **8**, 579–584.

Honohan, T., Hale, R. L., Brown, J. P. and Wingard, R. E. (1976). *J. agric. Fd Chem.* **24**, 906–911.

Horwitt, M. K. (1933). *Proc. Soc. exp. Biol. Med.* **30**, 949–951.

Hunt, C. A. (1977). *Diss. Abstr. Int.* **37B**, 3356–3357.

Indahl, S. R. and Scheline, R. R. (1971). *Xenobiotica* **1**, 13–24.

Iseda, S. (1956). *Nippon Kagaku Zasshi* **77**, 1629–1630.

Joachimoglu, G., Kiburis, J. and Miras, C. (1967). *Prakt. Akad. Athenon* **42**, 161–167. (*Chem. Abstr.* (1969) **70**, 18585z.)

Jones, G., Pertwee, R. G., Gill, E. W., Paton, W. D. M., Nilsson, I. M., Widman, M. and Agurell, S. (1974a). *Biochem. Pharmac.* **23**, 439–446.

Jones, G., Widman, M., Agurell, S. and Lindgren, J.-E. (1974b). *Acta pharm. suec.* **11**, 283–294.

Just, W. W., Erdmann, G., Thel, S., Werner, G. and Wiechmann, M. (1975). *Naunyn-Schmiedeberg's Arch. Pharmac.* **287**, 219–225.

Kaighen, M. and Williams, R. T. (1961). *J. med. Chem.* **3**, 25–43.

Kallianos, A. G., Petrakis, P. L., Shetlar, M. R. and Wender, S. H. (1959). *Archs Biochem. Biophys.* **81**, 430–433.

Klausner, H. A. and Dingell, J. V. (1971). *Life Sci.* **10I**, 49–59.

Kostanecki, S. V. (1886). *Ber. Dtsch. chem. Gesellschaft* **19**, 2918–2920.

Kratz, F. and Staudinger, H. (1965). *Hoppe-Seyler's Z. physiol. Chem.* **343**, 27–34.

Krishnamurty, H. G., Cheng, K.-J., Jones, G. A., Simpson, F. J. and Watkin, J. E. (1970). *Can. J. Microbiol.* **16**, 759–767.

Krishnaswamy, N. R., Seshadri, T. R. and Tahir, P. J. (1971). *Indian J. exp. Biol.* **9**, 458–461.

Külz, E. (1887). *Z. Biol.* **23**, 475–485.

Labow, R. S. and Layne, D. S. (1972). *Bicohem. J.* **128**, 491–497.

Lang, K. and Weyland, H. (1955). *Biochem. Z.* **327**, 109–117.

Leighty, E. G. (1973). *Biochem. Pharmac.* **22**, 1613–1621.

Leighty, E. G., Fentiman, A. F. and Foltz, R. L. (1976). *Res. Commun. chem. Path. Pharmac.* **14**, 13–28.

Lemberger, L. (1972). *In* "Advances in Pharmacology and Chemotherapy" (S. Garattini, A. Goldin, F. Hawking and I. J. Kopin, Eds), Vol. 10, pp. 221–255. Academic Press, New York and London.

Lemberger, L. and Rubin, A. (1976). "Physiologic Disposition of Drugs of Abuse", Ch. 8. Spectrum Publications, New York.

Lemberger, L., Silberstein, S. D., Axelrod, J. and Kopin, I. J. (1970). *Science* N.Y. **170**, 1320–1322.

Lemberger, L., Axelrod, J. and Kopin, I. J. (1971a). *Ann. N.Y. Acad. Sci.* **191**, 142–152.

Lemberger, L., Tamarkin, N. R., Axelrod, J. and Kopin, I. J. (1971b). *Science* N.Y. *173*, 72–74.

Lindner, H. R. (1967). *Aust. J. agric. Res.* *18*, 305–333.

Martin, B. R., Harvey, D. J. and Paton, W. D. M. (1976a). *J. Pharm. Pharmac.* *28*, 773–774.

Martin, B. R., Nordqvist, M., Agurell, S., Lindgren, J-E., Leander, K. and Binder, M. (1976b). *J. Pharm. Pharmac.* *28*, 275–279.

Martin, B. R., Agurell, S., Nordqvist, M. and Lindgren, J-E. (1976c). *J. Pharm. Pharmac.* *28*, 603–608.

Martin, B. R., Harvey, D. J. and Paton, W. D. M. (1977). *Drug Metab. Disposit.* *5*, 259–267.

Masquelier, J., Claveau, P. and Colse, J. (1965). *Bull. Soc. Pharm. Bordeaux* *104*, 193–199. (*Chem. Abstr.* (1966) *65*, 9531g.)

Masri, M. S., Booth, A. N. and DeEds, F. (1959). *Archs Biochem. Biophys.* *85*, 284–286.

Maynard, D. E., Gurny, O., Pitcher, R. G. and Kierstead, R. W. (1971). *Experientia* *27*, 1154–1155.

McCallum, N. K., Yagen, B., Levy, S. and Mechoulam, R. (1975). *Experientia* *31*, 520–521.

Mead, J. A. R., Smith, J. N. and Williams, R. T. (1955). *Biochem. J.* *61*, 569–574.

Mead, J. A. R., Smith, J. N. and Williams, R. T. (1958a). *Biochem. J.* *68*, 61–67.

Mead, J. A. R., Smith, J. N. and Williams, R. T. (1958b). *Biochem. J.* *68*, 67–74.

Mechoulam, R. (1970). *Science* N.Y. *168*, 1159–1166.

Mechoulam, R. (1973). *In* "Marihuana. Chemistry, Pharmacology, Metabolism and Clinical Effects" (R. Mechoulam, Ed.), pp. 1–99. Academic Press, New York and London.

Mechoulam, R., McCallum, N. K. and Burstein, S. (1976). *Chem. Rev.* *76*, 75–112.

Mechoulam, R., Levy, S., Yagen, B. and Ben Zvi, Z. (1977). International Symposium on Drug Activity, Jerusalem. Abstracts, p. 36.

Mikes, F., Hofmann, A. and Waser, P. G. (1971). *Biochem. Pharmac.* *20*, 2469–2576.

Milić, B. Lj. (1972). *J. Sci. Fd Agric.* *23*, 1151–1156.

Milić, B. Lj. and Stojanović, S. (1972). *J. Sci. Fd Agric.* *23*, 1163–1167.

Murray, C. W., Booth, A. N., DeEds, F. and Jones, F. T. (1954). *J. Am. pharm. Ass.* (Sci. Edn) *43*, 361–364.

Nahas, G. G. (1973). "Marihuana—Deceptive Weed". Raven Press, New York.

Nakagawa, Y., Shetlar, M. R. and Wender, S. H. (1965). *Biochim. biophys. Acta* *97*, 233–241.

Nakazawa, K. and Costa, E. (1971). *Nature* Lond. *234*, 48–49.

Neumeyer, J. L. and Shagoury, R. A. (1971). *J. pharm. Sci.* *60*, 1433–1457.

Nilsson, A. (1961a). *Ark. Kemi* *17*, 305–310.

Nilsson, A. (1961b). *Nature* Lond. *192*, 358.

Nilsson, A. (1962a). *Ark. Kemi* *19*, 549–550.

Nilsson, A. (1962b). *Acta chem. scand.* *16*, 31–40.

Nilsson, A. (1963). *Ark. Kemi* *21*, 97–121.

Nilsson, A., Hill, J. L. and Lloyd Davies, H. (1967). *Biochim. biophys. Acta* *148*, 92–98.

Nilsson, I. M., Agurell, S., Nilsson, J. L. G., Ohlsson, A., Sandberg, F. and Wahlqvist, M. (1970). *Science* N.Y. *168*, 1228–1229.

Nilsson, I. M., Agurell, S., Leander, K., Nilsson, J. L. G. and Widman, M. (1971). *Acta pharm. suec.* *8*, 701.

Nilsson, I. M., Agurell, S., Nilsson, J. L. G., Ohlsson, A., Lindgren, J-E. and Mechoulam, R. (1973a). *Acta pharm. suec.* **10**, 97–106.
Nilsson, I., Agurell, S., Nilsson, J. L. G., Widman, M. and Leander, K. (1973b). *J. Pharm. Pharmac.* **25**, 486–487.
Nordqvist, M., Agurell, S., Binder, M. and Nilsson, I. M. (1974). *J. Pharm. Pharmac.* **26**, 471–473.
Oshima, Y. and Watanabe, H. (1958). *J. Biochem.* Tokyo **45**, 973–977.
Oshima, Y., Watanabe, H. and Isakari, S. (1958). *J. Biochem.* Tokyo **45**, 861–865.
Oshima, Y., Watanabe, H. and Kuwazuka, S. (1960). *Bull. agric. chem. Soc.* Japan **24**, 497–500.
Ozawa, H. (1951). *J. pharm. Soc.* Japan **71**, 1191–1194.
Paton, W. D. M. (1975). *In* "Annual Review of Pharmacology" (H. W. Elliott, R. George and R. Okun, Eds), Vol. 15, pp. 191–220. Annual Reviews Inc., Palo Alto, California.
Paton, W. D. M. and Pertwee, R. G. (1973). *In* "Marihuana. Chemistry, Pharmacology, Metabolism and Clinical Effects" (R. Mechoulam, Ed.), pp. 191–285. Academic Press, New York and London.
Pekker, I. and Schäfer, E.-A. (1969). *Arzneimittel-Forsch.* **19**, 1744–1745.
Perez-Reyes, M., Lipton, M. A., Timmons, M. C., Wall, M. E., Brine, D. R. and Davis, K. H. (1973). *Clin. Pharmac. Ther.* **14**, 48–55.
Petrakis, P. L., Kallianos, A. G., Wender, S. H. and Shetlar, M. R. (1959). *Archs Biochem. Biophys.* **85**, 264–271.
Rasmussen, A., Scheline, R., Solheim, E. and Hänsel, R. (1976). Unpublished observations.
Rennhard, H. H. (1971). *J. agric. Fd Chem.* **19**, 152–154.
Roseman, S., Huebner, C. F., Pankratz, R. and Link, K. P. (1954). *J. Am. chem. Soc.* **76**, 1650–1652.
Scheline, R. R. (1968a). *Acta pharmac. tox.* **26**, 325–331.
Scheline, R. R. (1968b). *Acta pharmac. tox.* **26**, 332–342.
Scheline, R. R. (1970). *Biochim. biophys. Acta* **222**, 228–230.
Schüller, J. (1911). *Z. Biol.* **56**, 274–308.
Shilling, W. H., Crampton, R. F. and Longland, R. C. (1969). *Nature* Lond. **221**, 664–665.
Shutt, D. A. and Braden, A. W. H. (1968). *Aust. J. agric. Res.* **19**, 545–553.
Shutt, D. A., Axelsen, A. and Lindner, H. R. (1967). *Aust. J. agric. Res.* **18**, 647–655.
Sieburg, E. (1921). *Biochem. Z.* **113**, 176–199.
Siemens, A. J. and Kalant, H. (1975). *Biochem. Pharmac.* **24**, 755–762.
Simon, E. J., Gross, C. S. and Milhorat, A. T. (1956a). *J. biol. Chem.* **221**, 797–805.
Simon, E. J., Eisengart, A., Sundheim, L. and Milhorat, A. T. (1956b). *J. biol. Chem.* **221**, 807–817.
Simpson, F. J., Jones, G. A. and Wolin, E. A. (1969). *Can. J. Microbiol.* **15**, 972–974.
Solheim, E. and Scheline, R. R. (1973). *Xenobiotica* **3**, 493–510.
Solheim, E. and Scheline, R. R. (1976). *Xenobiotica* **6**, 137–150.
Stelzig, D. A. and Ribiero, S. (1972). *Proc. Soc. exp. Biol. Med.* **141**, 346–349.
Tatematsu, A., Nadai, T., Yoshizumi, H., Sakurai, H., Furukawa, H. and Hayashi, M. (1972). *Shitsuryo Bunseki* **20**, 339–346. (*Chem. Abstr.* (1973) **78**, 119062x.)
Timberlake, C. F. and Bridle, P. (1975). *In* "The Flavonoids" (J. B. Harborne, T. J. Mabry and H. Mabry, Eds), pp. 214–266. Chapman and Hall, London.

Turk, R. F., Dewey, W. L. and Harris, L. S. (1973). *J. pharm. Sci.* **62**, 737–740.
Unai, T., Cheng, H-M., Yamamoto, I. and Casida, J. E. (1973). *Agric. biol. Chem.* **37**, 1937–1944.
Van Sumere, C. F. and Teuchy, H. (1971). *Arch. int. Physiol. Biochim.* **79**, 665–679.
Venkataraman, K. (1975). *In* "The Flavonoids" (J. B. Harborne, T. J. Mabry and H. Mabry, Eds), pp. 267–295. Chapman and Hall, London.
Wall, M. E. (1971). *Ann. N.Y. Acad. Sci.* **191**, 23–37.
Wall, M. E. (1975). *In* "Recent Advances in Phytochemistry" (V. C. Runeckles, Ed.), Vol. 9, pp. 29–61. Plenum Press, New York and London.
Wall, M. E., Brine, D. R., Brine, G. A., Pitt, C. G., Freudenthal, R. I. and Christensen, H. D. (1970). *J. Am. chem. Soc.* **92**, 3466–3468.
Wall, M. E., Brine, D. R., Pitt, C. G. and Perez-Reyes, M. (1972). *J. Am. chem. Soc.* **94**, 8579–8581.
Wall, M. E., Brine, D. R., and Perez-Reyes, M. (1976). *In* "Pharmacology of Marihuana" (M. C. Braude and S. Szara, Eds), pp. 93–113. Raven Press, New York.
Watanabe, A. and Oshima, Y. (1965). *Agric. biol. Chem.* **29**, 90–93.
Watanabe, H. (1959a). *Bull. agric. chem. Soc.* Japan **23**, 257–259.
Watanabe, H. (1959b). *Bull. agric. chem. Soc.* Japan **23**, 260–262.
Watanabe, H. (1959c). *Bull. agric. chem. Soc.* Japan **23**, 263–267.
Watanabe, H. (1959d). *Bull. agric. chem. Soc.* Japan **23**, 268–271.
Watanabe, M., Toyoda, M., Imada, I. and Morimoto, H. (1974). *Chem. pharm. Bull.* Tokyo **22**, 176–182.
Westlake, D. W. S., Talbot, G., Blakley, E. R. and Simpson, F. J. (1959). *Can. J. Microbiol.* **5**, 621–629.
Widman, M., Nilsson, I. M., Nilsson, J. L. G., Agurell, S. and Leander, K. (1971). *Life Sci.* **10II**, 157–162.
Widman, M., Nordqvist, M., Agurell, S., Lindgren, J-E. and Sandberg, F. (1974). *Biochem. Pharmac.* **23**, 1163–1172.
Widman, M., Dahmén, J., Leander, K. and Petersson, K. (1975a). *Acta pharm. suec.* **12**, 385–392.
Widman, M., Nordqvist, M., Dollery, C. and Briant, R. H. (1975b). *J. Pharm. Pharmac.* **27**, 842–848.
Wiechowski, W. (1923). *Arch. Exp. Path. Pharmak.* **97**, 462–488.
Willinsky, M. D., Kalant, H., Meresz, O., Endrenyi, L. and Woo, N. (1974). *Eur. J. Pharmac.* **27**, 106–119.
Wiss, O. and Gloor, H. (1966). *Vitams Horm.* **24**, 575–586.
Würsch, M. S., Otis, L. S., Green, D. E. and Forrest, I. S. (1972). *Proc. west Pharmac. Soc.* **15**, 68–73.
Yamamoto, I. (1969). *In* "Residue Reviews" (F. A. Gunther, Ed.), Vol. 25, pp. 161–174. Springer Verlag, Berlin, Heidelberg and New York.
Yang, C-H., Braymer, H. D., Petrakis, P. L., Shetlar, M. R. and Wender, S. H. (1958). *Arch. Biochem. Biophys.* **75**, 538–539.
Yisak, W-A., Widman, M., Lindgren, J-E. and Agurell, S. (1977). *J. Pharm. Pharmac.* **29**, 487–490.

8

METABOLISM OF AMINES, NITRILES, AMIDES AND NON-PROTEIN AMINO ACIDS

I. Amines

Plant amines fall largely into three groups: aliphatic monoamines, aliphatic polyamines and aliphatic amines with aromatic substituents. The majority of these compounds are primary amines, having the general formula R—NH$_2$, although some secondary amines (e.g. ephedrine), tertiary amines (e.g. hordenine) and quaternary amines are also encountered. The aliphatic monoamines are volatile compounds ranging from the simples homologue, methylamine, to n-hexylamine with several examples of branched chain compounds. In fact, isoamylamine ((Me)$_2$CH—CH$_2$—CH$_2$—NH$_2$) has widespread occurrence. The metabolic data on these simple amines derive mainly from older investigations, several having been carried out a century ago. This information on aliphatic monoamines, as well as that dealing with the polyamines, was conveniently summarized by Williams (1959, Chapter 6), from which much of the present summary is taken.

As pointed out in Chapter 1, Section I.A.15, monoamine oxidase catalyses the oxidation of aliphatic amines. The best substrates are primary amines (R—CH$_2$—NH$_2$), although mono- and dimethylated amino compounds are also deaminated. Alles and Heegaard (1943) showed that optimal rates of oxidation are achieved with higher homologues of alkylamines, however other data have shown that **methylamine** appears to be largely metabolized in rabbits, dogs and man as only a very small percentage of the dose can be recovered unchanged in the urine. In accordance with the general reaction for the oxidative deamination of primary alkylamines by monoamine oxidase and the subsequent metabolism of the products (Fig. 8.1), both the carboxylic acid (formic acid) and urea have been identified as metabolites of methylamine. **Dimethylamine**, however, is poorly metabolized, being excreted instead nearly quantitatively unchanged in the urine by man. **Ethylamine** appears to be partly metabolized and partly excreted unchanged. With higher homologues the latter pathway is reduced and both **n-propylamine** and **n-butylamine** are largely or entirely metabolized. Alles and Heegaard (1943) found n-

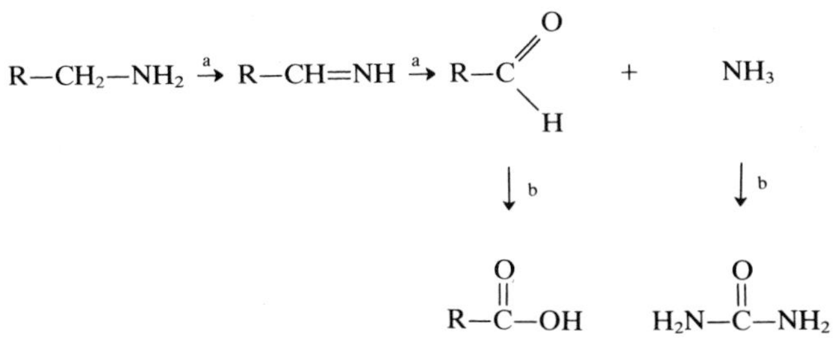

$$R-CH_2-NH_2 \xrightarrow{a} R-CH=NH \xrightarrow{a} R-C\overset{\displaystyle O}{\underset{\displaystyle H}{\big<}} \quad + \quad NH_3$$

$$\downarrow b \qquad\qquad\qquad \downarrow b$$

$$\underset{\displaystyle R-C-OH}{\overset{\displaystyle O}{\|}} \qquad \underset{\displaystyle H_2N-C-NH_2}{\overset{\displaystyle O}{\|}}$$

FIG. 8.1. Oxidative deamination (a) and further metabolism (b) of aliphatic amines.

butylamine to be a good substrate for monoamine oxidase *in vitro*, the rate of oxidation being about half of that observed with **n-amylamine** and **n-hexylamine**. Both of the latter compounds gave near maximal rates among the aliphatic amines tested. A branched side-chain may reduce metabolism slightly and **isobutylamine** ((Me)$_2$CH—CH$_2$—NH$_2$) undergoes some excretion unchanged in humans. With **isoamylamine** ((Me)$_2$-CH—CH$_2$—CH$_2$—NH$_2$), however, metabolism is complete as Richter (1938) found no unchanged amine in the urine of humans given 100 mg of the compound orally. Earlier work demonstrated the conversion of isoamylamine to the expected isovaleric acid and urea and Richter (1937) isolated isovaleraldehyde as its 2,4-dinitrophenylhydrazone from incubates of the amine with an monoamine oxidase preparation from guinea pig liver.

Aliphatic polyamines are often diamines having the general formula H$_2$N—(CH$_2$)$_n$—NH$_2$. **Putrescine** (n = 4) is of fairly widespread occurrence in plants and **cadaverine** (n = 5) is also sometimes found. Diamine oxidase (histaminase) oxidatively deaminates these compounds and, analogous to that described above with monoamines, the reaction products are an aldehyde and ammonia. Putrescine is an excellent substrate for this enzyme and quite large oral doses are required before some unchanged compound will escape metabolism and be excreted in the urine. Monoamine oxidase shows little activity towards putrescine and cadaverine. Another metabolic route for these two diamines involves cyclization by the intestinal bacteria (Asatoor, 1964). In this manner putrescine gives rise to pyrrolidine while cadaverine forms piperidine. Both of these heterocyclic amines are normal urinary constituents, although in this context their production is mainly related to the decarboxylation of the amino acids ornithine and lysine to the respective diamino compounds. Another widespread polyamine is the quanidine derivative **agmatine** (H$_2$N—(CH$_2$)$_4$—NHC(=NH)NH$_2$) which is metabolized by diamine oxidase. It is not a substrate for liver arginase which converts the quanidine group of arginine to urea.

The third group of plant amines consists of aliphatic amines containing aromatic substituents. It is in this area that a great number of metabolic investigations have been carried out, primarily because of the important physiologic and pharmacologic properties of many of these compounds. A point of interest with this group is the fact that several of these compounds occur both in plants and animals. This is true of histamine, noradrenaline and serotonin (5-hydroxytryptamine) and by far the greatest number of investigations with these compounds have dealt with them in the context of mammalian physiology. Accordingly, and because many other sources of information on their metabolism are available, the present summary will deal with compounds which are primarily considered to be of plant origin. A further noteworthy point is that the compounds to be covered are nearly exclusively aromatic (phenyl or indolyl) derivatives of ethylamine or occasionally isopropylamine. An exception to this statement is **benzylamine** which is now known to be the compound formerly termed moringin, however it seems to be of limited occurrence in plants. Its mammalian metabolism is straightforward according to earlier reports (see Williams, 1959 p. 130). These showed that it is converted to benzaldehyde which, in the dog, is further oxidized and then excreted in the urine as hippuric acid. Metabolizing capacity is large and an oral dose of 160 mg to humans does not result in the urinary excretion of unchanged compound. In the *in vitro* liver system used by Alles and Heegaard (1943), benzylamine was oxidized at about a third of the rate observed with the higher homologues of ω-phenylalkylamines. In addition to the oxidative deamination of benzylamine by the monoamine oxidase of liver mitochondria, this reaction can also be carried out by plasma amine oxidases. Considerable interest has been shown in the mechanisms of these copper-dependent enzymes (see Buffoni *et al.*, 1968: Taylor and Knowles, 1971).

The best represented group of plant alkylamines containing aromatic substituents is that of the β-phenylethylamines. Their metabolic fate is largely determined by the reactions of the amino group, however ring substituents may also be directly involved. In the first case, oxidative deamination leading via the aldehyde to a phenylacetic acid derivative is likely, possibly followed by conjugation of the acid with an amino acid. The amino group can also undergo acetylation or methylation. It is also possible that these pathways may not proceed readily, in which case the compound is excreted unchanged. Ring substituents include hydroxyl or methoxyl groups which may undergo conjugation or *O*-demethylation reactions, respectively. A further possibility involves the β-hydroxylase pathway found in the synthesis of catecholamines and which gives ethanolamine derivatives.

(1)

β-Phenylethylamine

The simplest member of this group is **β-phenylethylamine** (phenethyl-amine) (1) itself. An early study by Guggenheim and Löffler (1915) showed that when administered to rabbits or dogs it was converted to both phenylacetic acid and phenylethanol. Richter (1937) found that it was metabolized by guinea pig liver and intestine preparations to phenylacetic acid and ammonia. Similarly, Alles and Heegaard (1943) reported that β-phenylethylamine was rapidly oxidized by liver amine oxidase. *In vivo* experiments have also indicated rapid deamination and Richter (1938) found that no unchanged compound was excreted in the urine in man following an oral dose of 300 mg (approx. 4–5 mg/kg). Richter, noting that early work indicated the conversion of the amine to phenylacetic acid in rabbits, found that 62% of the dose was excreted in 4·5 h as the acid, a portion of which may have been conjugated with glutamine. Block (1953), using β-phenylethylamine labelled with ^{14}C in the α-position, showed that radioactivity was rapidly excreted in the urine (64% in 2 h) by mice given an i.p. dose of about 80 mg/kg. Total urinary excretion of radioactivity reached 80% and most of this was as phenylacetic acid. Only traces of $^{14}CO_2$ were detected in the respiratory air, indicating that decarboxylation of the acid or other metabolic pathways leading to C_6—C_1 compounds are not of importance. Recently, Wu and Boulton (1975) carried out an extensive study of the distribution and metabolism of injected β-phenyl-ethylamine in rats. The major metabolite in tissues and urine was phenyl-acetic acid and significant amounts of an unidentified metabolite were also detected. This compound was not the intermediate aldehyde, however, and no aldehydes were detected. A very small amount (0·1–0·3%) of phenyl-ethanolamine was detected and further indication that a β-hydrolase enzyme is also involved in the metabolism of β-phenylethylamine was the evidence for the formation of octopamine. This involves aliphatic hydroxylation at the β-position and aromatic hydroxylation at the p-posi-tion. Aromatic hydroxylation was also indicated by the finding of radioac-tivity associated with m- and p-tyramine. Creveling *et al.* (1962) described a β-oxidase capable of forming β-hydroxylated products with numerous phenylethylamines including β-phenylethylamine. Axelrod (1962) report-ed that enzyme preparations from several tissues, among which lung was most active, were able to transfer the methyl group of S-adenosyl-methionine to numerous amino compounds. This N-methylating enzyme showed moderate activity towards β-phenylethylamine, however the product thus formed has not been reported as an actual metabolite of β-phenylethylamine and it is likely that this pathway has little *in vivo* significance.

Tyramine (2) is β-*p*-hydroxyphenylethylamine and its metabolism is similar to that noted above with β-phenylethylamine. Ewins and Laidlaw (1910) carried out experiments *in vivo* in dogs and *in vitro* using perfused rabbit or cat livers which showed that *p*-hydroxyphenylacetic acid was formed. When dogs were given tyramine (approx. 60 mg/kg, p.o.) about 25% of the dose was isolated in the urine as the acid, however, other data indicated that the actual extent of oxidative deamination was about twice this amount. Other early studies showed that *p*-hydroxyphenylacetic acid and *p*-hydroxyphenylethanol were tyramine metabolites in rabbits and dogs (Guggenheim and Löffler, 1915), that guinea pig liver preparations metabolized tyramine to *p*-hydroxyphenylacetaldehyde and ammonia (Richter, 1937) and that active deaminating systems are present in the liver and kidneys of rats, guinea pigs, rabbits, cats and dogs (Bernheim and Bernheim, 1938). Similar results were reported by Blaschko and Philpot (1953).

As a result of several more recent studies using more sensitive and selective methods, our knowledge of the mammalian metabolism of tyramine has been greatly expanded (Lemberger *et al.*, 1966; Tacker *et al.*, 1970; Tacker *et al.*, 1972b). Indeed, these three studies which provide both the main findings in the present summary and a convenient source to the relevant literature, indicate that essentially complete quantitative assessments of the patterns of tyramine metabolism are available in rats, rabbits and man. The pathways involved are shown in Fig. 8.2 and the values for urinary metabolite excretion are listed in Table 8.1. These studies also showed that the radioactivity was rapidly excreted in the urine, essentially none being lost by faecal or respiratory routes. The most striking result is that which shows the extensive excretion of the phenylacetic acid derivative (4) resulting from the oxidative deamination of tyramine. This metabolite, mainly in the free form, accounts for more than 80% of the dose. Excretion of unchanged compound is a minor pathway. The results from rats and man showed that pathways involving β-hydroxylation and/or aromatic hydroxylation giving rise initially to octopamine, dopamine and noradrenaline and then to the acidic metabolites (8), (9) and (10), are not of importance. Similar findings were reported by Wiseman-Distler *et al.* (1965) who showed that neither metabolite (10) nor its 3-*O*-demethylated precursor (3,4-dihydroxyphenylacetic acid) was excreted in the urine of rats given tyramine (9·4 mg/kg, i.p.). Octopamine, which is readily metabolized to *p*-hydroxymandelic acid (8) by oxidative deamination (Hengstmann *et al.*, 1974), has been detected as a urinary metabolite of

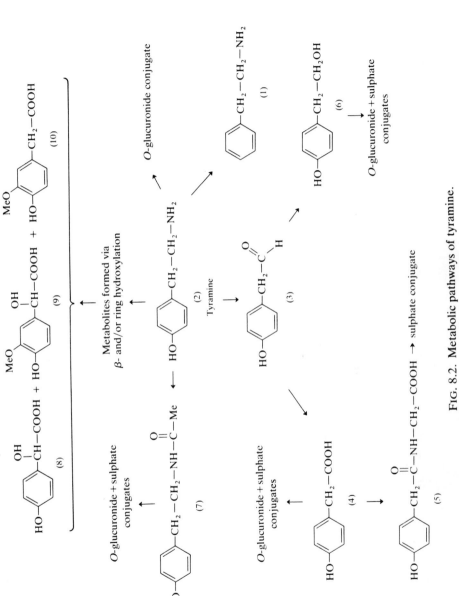

FIG. 8.2. Metabolic pathways of tyramine.

TABLE 8.1.

Urinary metabolites of tyramine (% of urinary radioactivity)

Compound	Structure (see Fig. 8.2)	Rat[a] (1·2 mg/kg, i.p.)	Rabbit[b] (5 mg/kg, i.p.)	Man[c] (1·3 mg, i.v.)
Tyramine	2	1·4		6·1
O-Glucuronide of (2)		0·6		0
p-Hydroxyphenylacetaldehyde	3	0·2		<0·5
p-Hydroxyphenylacetic acid	4	77	83	84
O-Glucuronide + sulphate of (4)		4·2		
p-Hydroxyphenylacetylglycine	5	10·5		0
p-Hydroxyphenylethanol, O-glucuronide + sulphate	6	0·7	4	0
N-Acetyltyramine	7	0.2		<0·5
O-Glucuronide + sulphate of (7)		2·9		0
p-Hydroxymandelic acid	8	0	7	0
3-Hydroxy-4-methoxymandelic acid	9	0	1	0
Homovanillic acid	10	0	0·5	0
Unidentified metabolites		1 (2 cpd.)	3 (2 cpd.)	approx. 10 (7 cpd.)

[a] Tacker *et al.* (1970).
[b] Lemberger *et al.* (1966).
[c] Tacker *et al.* (1972b).

tyramine in rats pre-treated with a monoamine oxidase inhibitor (Tacker *et al.*, 1972a). As shown in Table 8.1, these pathways involving hydroxylation reactions are responsible for the metabolism of at least 8–9% of the dose in rabbits. The β-oxidase system studied by Creveling *et al.* (1962) was very active in β-hydroxylating *p*-tyramine. Dehydroxylation appears to be a minor reaction of tyramine. Mosnaim *et al.* (1977) reported that this conversion to β-phenylethylamine (1) was carried out by the 10 000 g fraction from rabbit brain preparations.

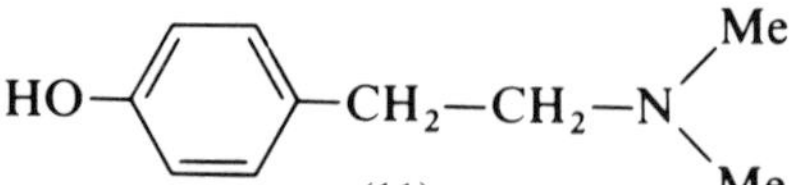

$$\text{HO} - \langle \rangle - \text{CH}_2 - \text{CH}_2 - \text{N} \overset{\text{Me}}{\underset{\text{Me}}{}}$$

(11)

Hordenine

Hordenine (11) is *N,N*-dimethyltyramine and was the subject of an early study by Ewins and Laidlaw (1910) who both administered it orally to dogs (approx. 125 mg/kg) and perfused it through cat liver. In both cases a relatively small amount of unchanged compound was recovered, however little *p*-hydroxyphenylacetic acid was found. The low recoveries of metabolites indicated that metabolism had taken place although the presence of the tertiary amino group clearly depressed the oxidative deamination pathway. That metabolism along this latter route can occur, however, was shown by Richter (1937) who identified dimethylamine in the incubates of hordenine and an amine oxidase preparation from guinea pig liver. An unidentified aldehyde was also detected in the incubates. The inhibiting influence of *N*-methyl groups on amine oxidase activity was reported by Alles and Heegaard (1943) in an extensive study of substrate specificity. A minor metabolic pathway in the metabolism of hordenine may involve ring hydroxylation as Wiseman-Distler *et al.* (1965) reported that nearly 2% of the dose (23 mg/kg, i.p.) was excreted in the urine as 3-hydroxy-4-methoxyphenylacetic acid (10). Daly *et al.* (1965) described a microsomal enzyme system which readily formed a catechol derivative with hordenine.

Mescaline (12), the active constituent of the Mexican cactus peyote, has been an object of continued interest since the scientific discovery of its hallucinogenic properties towards the end of the last century. This interest has accelerated in recent years due to the greatly increased usage of drugs of abuse. Of the many facets related to the study of mescaline, that of its metabolism has received a great deal of attention because of the possibility that it may be closely associated with the pharmacological effects, perhaps through the formation of one or more active metabolites. Metabolic studies with mescaline date from the investigation of Slotta and Müller (1936) who administered it orally to rabbits and dogs and intravenously to humans. Interestingly, the 24 h urine samples from the first two species contained roughly a third of the dose as 3,4,5-trimethoxyphenylacetic acid (13), whereas this metabolites was not found in human urine. In the latter case a monomethoxy derivative was detected. This early indication of species differences in the metabolism of mescaline has spurred the subsequent investigation, both *in vitro* and *in vivo*, of its fate in many animal species. In order to summarize this abundant and widespread material in a way that more clearly underlines the metabolic features, the results from numerous *in vivo* studies are tabulated in Table 8.2, which lists the metabolites found, their amounts and other relevant points. The pathways in the metabolism of mescaline are illustrated in Fig. 8.3.

The data in Table 8.2, in spite of the numerous differences in experimental details involved, allow several general conclusions to be drawn about the metabolism of mescaline. It is evident that this compound is more resistant to oxidative deamination than are, for example, β-phenylethylamine and tyramine. This fact is reflected in the generally high values for the excretion of unchanged compound, although some species differences do appear to exist, notably with the rather low values from rats compared with those from mice. However, the urinary pH values of the experimental urines in these studies were not reported and possible varia-

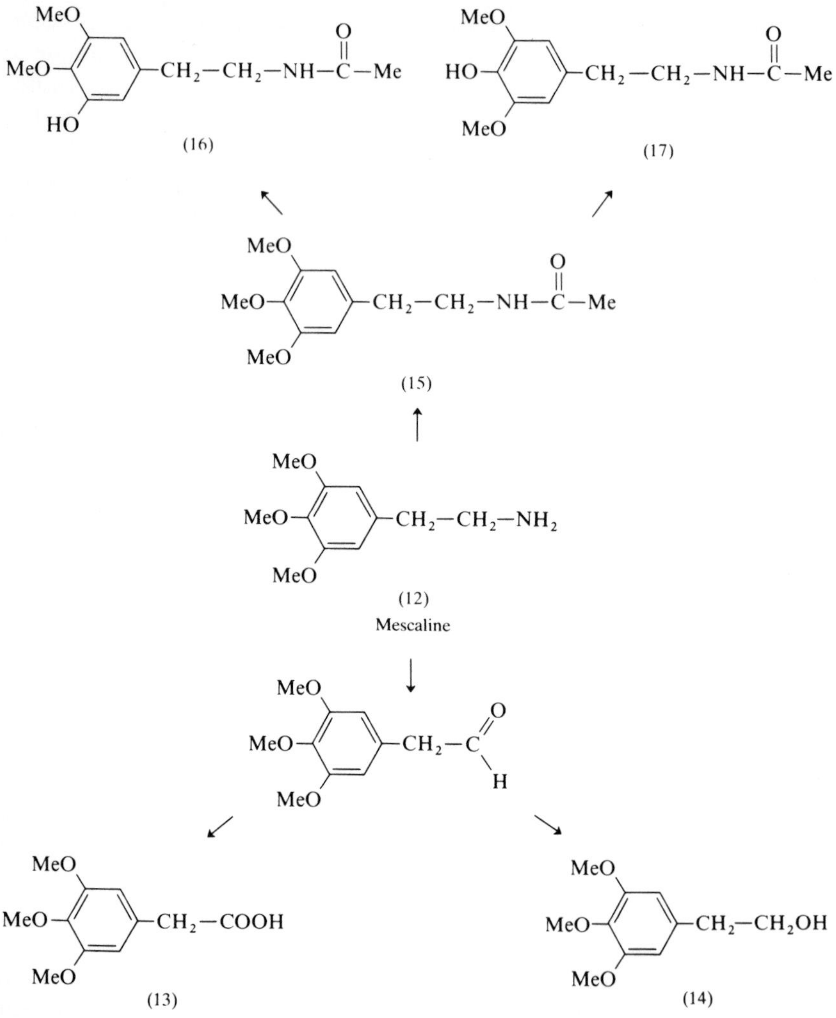

FIG. 8.3. Metabolic pathways of mescaline. See text for description of minor additional reactions.

tions in these values, influencing as they do the urinary excretion of the unchanged compound, make interpretation difficult. Another point of interest which emerges from the data shown in Table 8.2 is that 3,4,5-trimethoxyphenylacetic acid has been reported as a urinary metabolite in all of the species studied (mouse, rat, rabbit, cat, dog and man). This metabolite and unchanged compound generally furnish the bulk of the

TABLE 8.2

Urinary metabolites of mescaline

Species	Dose	Collection period (h)	Mescaline (12)	(13)	(14) (15) (16) (17)	Other	Unknown	References
Mouse	80 mg/kg, i.p.	40	79[a]	16			4	Block et al. (1952)
Mouse	50 mg/kg, i.p.	3	68[a]	20 (31 in 24 h)	8		5 (neutral compounds)	Shah and Himwich (1971)
Rat	24 mg/kg, i.p.	40	18[a]	72			9	Block et al. (1952)
Rat	0·4 mg/kg	24	+[b]	+	+			Goldstein et al. (1961)
Rat	0·2 or 40 mg/kg, i.p.	24	20[a]	42 (incl. (14))	1·7 14 15		6	Musacchio and Goldstein (1967)
Rabbit	340 mg, p.o.	24		30–50[c]		No other metabolites detected		Slotta and Müller (1936)
Cat	25 mg/kg, i.v.	0·5–6	+	+				Neff et al. (1964)
Dog	340 mg, p.o.	24		30–50[c]				Slotta and Müller (1936)
Dog	20 mg/kg, p.o., i.v., i.m.	24	28–46[c]	trace				Cochin et al. (1951)
Dog	20 mg/kg, i.p.	13	approx. 40[a]	approx. 60				Spector (1961)
Man	approx. 3 mg/kg, i.v., p.o.	24	52, 58[c]					Richter (1938)
Man	400 mg	24	approx. 35[c]	—		1–2 of glutamine conjugate of 3,4-dihydroxy-5-methoxyphenylacetic acid; Small amount of 3,4-dimethoxy-5-hydroxyphenylethylamine		Harley-Mason et al. (1958)
Man	400 mg, p.o.	5	+ (large amount)	7				Ratcliffe and Smith (1959)
Man	5 mg/kg, i.v.	6	31[c] (12–67)					Mokrasch and Stevenson (1959)
Man	350 mg, p.o.	12		26[c]				Charalampous et al. (1964)
Man	6 mg/kg	24	23[c]	18				Friedhoff and Hollister (1966)
Man	500 mg, p.o.	24	55–60[a]	27–30	0·1 5	A hydroxymethoxyphenylacetic acid	approx. 10 (5 cpds)	Charalampous et al. (1966)

Compound (see Fig. 8.3 for structures)

[a] % of radioactivity in urine.   [b] + = present, − = absent.   [c] % of dose.

excreted material with little contribution apparently being made by N-acetylation. However, the latter reaction has been reported in the mouse, rat and man (Table 8.2) and as a liver metabolite of mescaline in monkeys (Taska and Schoolar, 1972, 1973). Musacchio and Goldstein (1967) reported that N-acetyl derivatives accounted for about 30% of the urinary metabolites in rats. This study also showed that β-hydroxymescaline was not an excretory product of mescaline in rats. This metabolite or other β-hydroxylated derivatives were not reported in any of the other investigations summarized in Table 8.2 and it seems likely that β-hydroxylation is quantitatively of little importance in the metabolism of mescaline. The role of O-demethylation is more difficult to assess, however several metabolites of this type have been identified (Table 8.2) and an unidentified metabolite of this type was reported in the early work of Slotta and Müller (1936).

In addition to the urinary metabolites, those found in the tissues and especially the brain deserve attention. Ho et al. (1973) administered [^{14}C]-mescaline (3·5–25 mg/kg, i.v.) to rats and found that N-acetylmescaline (15) accounted for as much as a third of the radioactivity in the brain after 30 min. Unchanged mescaline was the major component, however small amounts (2–6% each) of the acid (13) and alcohol (14) were also present. Similar results were reported by Taska and Schoolar (1973) in monkeys (*Saimiri sciureus*) given mescaline (approx. 5 mg/kg, i.v.). The radioactivity in the liver during the first 6 h consisted mainly of unchanged compound but 10–20% was due to the N-acetyl derivative (15). Smaller amounts of (13) and (14) were also present. N-Acetylmescaline has also been identified in the brains of mice treated with mescaline (120 mg/kg, i.p.) labelled with ^{14}C in the α-position (Seiler and Demisch, 1974a). Interestingly, this investigation showed that 0·05% of the administered radioactivity appeared in the respiratory air. Furthermore, the expected metabolite resulting from side-chain decarboxylation, 3,4,5-trimethoxybenzoic acid, was identified in brain and liver. The small amounts of this benzoic acid derivative which were found confirm that this route is a very minor pathway in the metabolism of mescaline. Three other metabolites, shown to be non-demethylated amines but otherwise only partially characterized, were also identified in small amounts in the brain.

A development which has coincided with the *in vivo* studies of mescaline metabolism is the elucidation of the enzyme systems responsible for its metabolism and their tissues and species distribution. Several early studies indicated that the systems which readily attacked β-phenylethylamine or tyramine were often without significant effects towards mescaline. Pugh and Quastel (1937) showed that this was the case with the tyramine oxidase in rat brain slices and guinea pig kidney slices. Bernheim and Bernheim (1938) reported that liver and kidneys from most animals contain an active tyramine oxidase, however only the preparation from rabbit liver showed appreciable activity towards mescaline. Recently, Roth *et al.* (1977) found that mescaline-oxidizing activity is several times higher in rabbit lung homogenates than in similar preparations from liver and kidney. Blaschko (1944) concluded that the activity of rabbit liver was not due to amine oxidase. The liver amine oxidase studied by Alles and Heegaard (1943) and which oxidized many phenylethylamine derivatives, was without activity towards mescaline. Steensholt (1947) investigated the oxidation of mescaline by rat and rabbit liver homogenates. Little activity was found in the former case whereas appreciable oxidation occurred with the latter. However, it was concluded that mescaline oxidase was distinct from monoamine oxidase, although many similarities exist between the two. This problem was re-examined in detail by Zeller *et al.* (1958) who studied the mescaline oxidase activity of various tissues in seven mammalian species. This enzyme in rabbit liver was found to be present in the mitochondrial and microsomal fractions and inhibitor studies indicated that it is a typical diamine oxidase rather than a monoamine oxidase. This was confirmed by Huszti and Borsy (1966) using rabbit liver mitochondria. Zeller *et al.* showed that mescaline was also oxidized by the diamine oxidases of pig kidney cortex and sheep plasma. Nonetheless, examination of other tissue preparations from these species (e.g. mouse or pig liver mitochondria) indicated that the enzyme system involved in mescaline deamination had an inhibitor pattern typical for monoamine oxidase. Interestingly, Riceberg *et al.* (1975) recently reported that the enzymes oxidizing mescaline in rabbits are related to the copper-containing plasma amine oxidases and are not the flavin-containing mitochondrial monoamine oxidases. A monoamine oxidase was shown to be responsible for mescaline oxidation in mouse brain mitochondria (Seiler, 1965; Seiler and Demisch, 1971, 1974b). Guha and Mitra (1971) reported the presence of a dehydrogenase system localized in rat and guinea pig brain mitochondria which is capable of dehydrogenating mescaline.

In the section above on the *in vivo* metabolism of mescaline several minor metabolic pathways were noted. One of these included side-chain degradation giving small amounts of 3,4,5-trimethoxybenzoic acid. Recently, Demisch and Seiler (1975) reported that this metabolite is formed *in vitro* in mouse tissue preparations, especially those from brain. The system, which is localized in the nuclear and microsomal fractions, also forms the corresponding phenylacetic acid derivative (13). The *in vivo* metabolism of mescaline involves the formation of several *O*-demethylated derivatives, usually in small amounts. This subject was studied *in vitro* by Axelrod (1956) and Daly *et al.*(1962). Demethylation is effected by liver microsomes of many species and experiments using rabbit liver microsomes showed that the reaction takes place in both the 4- and 5-positions. The demethylated metabolites were formed in relatively small amounts and with somewhat more 5-hydroxy derivative being produced. The N-methylation of mescaline had not been reported in *in vivo* studies, however rabbit lung preparations are capable of transferring the methyl group of S-adenosylmethionine to numerous amines including mescaline (Axelrod, 1962). The absence of β-hydroxylation of mescaline *in vivo* was noted above. Goldstein and Contrera (1962) reported that mescaline is a weak substrate of phenylamine-β-hydroxylase *in vitro* and that O-methylation in the aromatic ring is a factor which decreases substrate affinity for the hydroxylating enzyme.

The number of studies on the metabolism of **l-ephedrine** (18) has grown considerably in recent years, mainly due to the widespread interest in the phenylisopropylamine drugs (amphetamines). As most members of this group, of which amphetamine itself is the simplest example, are synthetic compounds, a general discussion of their metabolism falls outside the scope of this book. However, the articles by Williams *et al.* (1974) and Caldwell (1976) and the monograph by Lemberger and Rubin (1976) cover the comparative aspects of the metabolism of this group of compounds and serve as supplements to the present summary of ephedrine metabolism. Ephedrine contains two asymmetric carbon atoms and can therefore exist in four isometric forms. These are *l*- and *d*-ephedrine and *l*- and *d*-pseudoephedrine, of which the first and the last are naturally occurring. Most of the metabolic studies have used *l*-ephedrine, however the racemic *dl*-form has been employed in a few cases. Some data is also available on *d*-ephedrine and *dl*-pseudoephedrine.

The major pathways in the metabolism of *l*-ephedrine are illustrated in Fig. 8.4 which shows that the main reactions are N-demethylation, aromatic hydroxylation and oxidative deamination which may lead to degradation of the aliphatic side-chain. Table 8.3 summarizes the quantitative aspects of urinary metabolite excretion in seven mammalian species. These data show that although all three pathways are often operative, their relative importance varies greatly from species to species. In fact the large species variations in *l*-ephedrine metabolism have provided perhaps the most interesting aspect of this subject. Perusal of the data in Table 8.3 shows that mouse and man excrete most of the dose unchanged in the urine. It should be noted that urinary pH influences the extent of excretion of the unchanged compound and that this may be reduced appreciably when the urine is alkaline (Wilkinson and Beckett, 1968). The rat is intermediate (roughly 30–50% of the dose) in the amount of compound which is excreted unchanged while guinea pigs, rabbits, dogs and horses all metabolize the compound extensively.

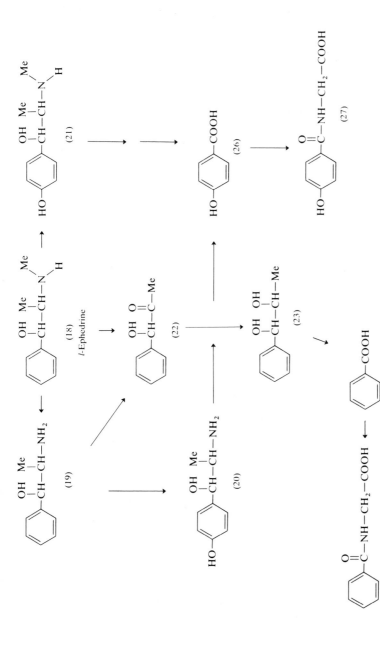

FIG. 8.4. Metabolic pathways of *l*-ephedrine.

TABLE 8.3

Urinary metabolites of l-ephedrine

Species	Dose	Collection period (h)	% of dose recovered in urine	l-Ephedrine (18)	(19)	(20)	(21)	(22)	(23)	(24)	(25)	(26)	(27)	Other	References
Mouse	5 mg/kg, s.c.	24	93	79[a]	2·7		2·1		0	0·4	1·0	0	0		Baba et al. (1972)
Rat	41 mg/kg, i.p.	24		32	7·5										Axelrod (1953)
Rat	i.p.	24	64	43	0·5		13			0·3	1·8				Nagase et al. (1967)
Rat	2 mg/kg, i.p.[b]	40	80	53			2			4					Bralet et al. (1968)
Rat	5 mg/kg, s.c.	24	71	40	3·3		20			0·3	1·9				Baba et al. (1972)
Rat	8 mg/kg, s.c.	24	70	33-37	39		22	<0·1							Kawai and Baba (1974)
Guinea pig	41 mg/kg, i.p.	24		2			11-14								Axelrod (1953)
Guinea pig	5 mg/kg, s.c.	24	83	1·3	43		0·9		7·2	19	2·3	0·4			Baba et al. (1972)
Guinea pig	1 mg/kg, i.v.	40	85	13	51					17				2·6 as p-hydroxy-compounds	Jacquot et al. (1974)
Rabbit	41 mg/kg, i.p.	24		0·1	1·8	1·9									Axelrod (1953)
Rabbit	5 mg/kg, p.o.	24		0·7[c]	1·3[c]	0	0·7[c]	0·8[c]	29·1[c]	19·1[c]	32·9[c]	1·3[c]	8·8[c]		Matsuda et al. (1971); Baba et al. (1972)
Rabbit	8 mg/kg, s.c.	24	95						13		31				Kawai and Baba (1974)
Rabbit	8 mg/kg, s.c.	24							14-17		25				Kawai and Baba (1975)
Rabbit	3 mg/kg, i.p.	24	71	0·5	1·1			0·2	3·3	14	21				Feller and Malspeis (1977)
Dog	i.p.	24			approx. 80	approx. 1	0·3								Axelrod (1952)
Dog	41 mg/kg, i.p.	24		6·5	58		1·5								Axelrod (1953)
Horse	approx. 0·4 mg/kg, s.c.	3[d]		6	7[e](15[f])										Karawya et al. (1968)
Horse	approx. 1 mg/kg, p.o., i.v.	24		0			3							mandelic acid not present	Nicholson (1970)
Horse	approx. 1 mg/kg, p.o.	24							19						Nicholson (1971)
Horse	i.v.	24	87			18		0							Chapman and Marcroft (1977)
Man	0·4-1·5 mg/kg, p.o.	48			80-100										Richter (1938)
Man	27 mg/kg, p.o.	24		79	4·3										Beckett and Wilkinson (1965)
Man	20-25 mg/kg, p.o.	24		27[e](88[f])	18[e](7[f])		3								Wilkinson and Beckett (1968)
Man	p.o.	24	88	77	4·0				1-10	tr[g]	6·3				Baba (1973)
Man	30 mg, p.o.	24		61[h]	13						4·6				Sever et al. (1975)
Man	0·82 mg/kg, p.o.	24		75	3·6					+[i]	3·5				Kawai and Baba (1976)

[a] Figures = % of dose unless otherwise indicated.
[b] Used dl-ephedrine.
[c] ...
[d] No amine excreted after 3 h.
[e] Alkaline urine.
[f] ...
[g] tr = trace.
[h] Urine pH 6·5-6·8.
[i] ... present

The summarized data show that N-demethylation occurs to an appreci-able extent in rats, guinea pigs and dogs with the demethylated metabolites (19) and (20) accounting for perhaps about 20% of the dose in rats, 40–50% in guinea pigs and 60–80% in dogs. The actual extent of N-demethylation in rabbits is masked by the extensive metabolism to phenylpropane derivatives and to benzoic and hippuric acids. However, it is not unreasonable to expect that the extent of N-demethylation may be considerably higher than that suggested by the small amounts of urinary N-demethylated compounds since it is known that norephedrine (19) undergoes extensive further metabolism in rabbits to compounds (22), (23) and (25) (Sinsheimer et al., 1973).

As seen in Table 8.3, aromatic hydroxylation of l-ephedrine is a prominent pathway only in the rat. On the other hand, that of oxidative deamination and side-chain shortening is a minor route in this species. The latter pathway is the major feature of l-ephedrine metabolism in rabbits and an important route in both guinea pigs and horses. It appears to account for about 10% of the dose in man.

Nagase et al. (1967) studied the excretion and metabolism of $[^{14}C]$-d-ephedrine in rats and found that the excretion rate was similar to that of the l-isomer but that the recovery of urinary radioactivity was about 20% less with the former. The N-demethylation and aromatic hydroxylation which occurred with l-ephedrine were not detected with the d-isomer. Feller and Malspeis (1977) compared the in vitro and in vivo metabolism of l- and d-ephedrine in rabbits. The l-isomer was found to be metabolized at a faster rate by liver microsomes and the rate of benzoic acid formation was about three times as large. The relative amounts of demethylated compound (19) and diol intermediate (23) formed from the two isomers were nearly identical. The in vivo data showed that both isomers are mainly metabolized by N-demethylation and oxidative deamination of the side-chain. However, some quantitative differences were noted in urinary metabolite excretion. About 91% of the administered dose of d-ephedrine was excreted in the urine in 24 h compared with the 71% found with the l-isomer (Table 8.3). Other differences with d-ephedrine were a higher excretion of benzoic acid and lower excretion of hippuric acid, although the total amounts of these metabolites excreted in the two sets of experiments were not very different. About five times as much unchanged compound and diol (23) were excreted when the d-isomer was given.

Jacquot et al. (1973) administered racemic **pseudoephedrine** (5 mg/kg, i.p.) labelled with ^{14}C to rats. About 80% of the radioactivity was eli-minated in the urine in 18 h and, of this, about 42% was unchanged compound, 44% was p-hydroxypseudoephedrine and 12% was benzoic and hippuric acids.

Norephedrine (19) is formed in widely varying amounts from ephedrine in various species and several investigations have dealt with its excretion and metabolism. *dl*-Norephedrine is a synthetic sympathomimetic agent which has been widely used as a nasal decongestant. Interestingly, the stereoisomer ***d*-pseudonorephedrine** (cathine) is a major constituent of khat, a stimulant drug obtained from the leaves of an East African plant. While its metabolism has not been studied, it is probable that this is similar to that seen with norephedrine. Axelrod (1953) found that *l*-norephedrine is mainly excreted unchanged in the dog. The same result was found in man (Beckett and Wilkinson, 1965; Wilkinson and Beckett, 1968). Sinsheimer *et al.* (1973) administered racemic norephedrine to rats, rabbits and man and confirmed that most (86%) of the dose (25 mg) was excreted unchanged in the urine by man. They identified 4-hydroxynorephedrine (20) and hippuric acid (25) as minor metabolites. Rabbits excreted less than 10% of the dose (12 mg/kg) unchanged. Oxidative deamination was the main pathway and about 80% of the dose was converted to metabolites (22), (23) and (25). At the same dose level, rats excreted about half unchanged and 28% as the *p*-hydroxylated metabolite (20). Thiercelin *et al.* (1976) also administered racemic [^{14}C]-norephedrine to rats (0·535 mg/kg, i.v.) and found that about 75% of the radioactivity was excreted in the urine in 18 h. This material consisted of unchanged compound and metabolite (20) (54% and 14% of the dose, respectively). This investigation also dealt with the kinetics of norephedrine distribution and elimination.

The cellular and enzymic aspects of ephedrine metabolism have been studied using a number of *in vitro* preparations. An early study (Alles and Heegaard, 1943) showed that the amine oxidase preparation from rabbit liver which effectively oxidized aliphatic amines and β-phenylethylamine and its derivatives, did not oxidize the racemic mixtures of ephedrine, pseudoephedrine or their demethylated derivatives. It was later shown that liver microsomes are able to carry out numerous metabolic reactions with ephedrine including *N*-demethylation and oxidative deamination. Axelrod (1955a) showed that rabbit liver microsomes which oxidized amphetamine to phenylacetone and ammonia actively metabolized *l*-ephedrine. Baba *et al.* (1969, 1971) used NADPH-fortified rabbit liver preparations or rabbit liver slices and showed conversion of *l*-ephedrine to the ketone (22) and its reduction product (23). These two metabolites were also reported by Beckett and Al-Sarraj (1973) using fortified 10 000 g fractions from rabbit liver. A subsequent extensive study of this reaction, using norephedrine as the substrate, was also carried out (Beckett *et al.*, 1974). The dehydrogenase system in brain mitochondria noted above in the section on mescaline was also found to dehydrogenate ephedrine (Guha and Mitra, 1971).

Microsomal *N*-demethylation of *l*-ephedrine and its isomers has been reported in several studies. In an important early investigation Axelrod (1955b) showed that rabbit liver microsomes in the presence of NADPH catalysed the conversion of *l*-ephedrine to norephedrine and formaldehyde. This reaction was also carried out by dog, guinea pig and rat liver preparations, although with lower relative activities. The high *N*-demethylase activity of rabbit liver microsomes was confirmed by Gaudette and Brodie (1959) in a study relating oxidation to lipid solubility. Mc-Mahon (1964) reported the *N*-demethylation of *l*-ephedrine by rat, guinea pig and rabbit liver microsomes and studied the effect of inhibitors on this reaction. Dann *et al.* (1971) and Feller *et al.* (1973) studied the rabbit microsomal *N*-demethylation of the four stereoisomers of ephedrine and noted small differences in the relative reaction rates. An investigation of the mechanism of *N*-dealkylation of several secondary amines including *l*-ephedrine and pseudoephedrine by a purified microsomal mixed function oxidase from pig liver was carried out by Ziegler *et al.* (1969). The reaction involves an initial *N*-oxidation followed by the breakdown of this unstable intermediate to the primary amine and formaldehyde. Beckett and Al-Sarraj (1973) and Beckett *et al.* (1974) studied the *N*-oxidation of ephedrine and norephedrine using 10 000 g rabbit liver preparations. Their results showed that in addition to the α-*C*-oxidation noted above which leads to the formation of metabolites (22) and (23), *N*-oxidation giving *N*-hydroxy compounds (hydroxylamines) also occurs.

A few *in vitro* studies have shown the formation of most of the known metabolites of ephedrine in the species studied. Baba *et al.* (1969) incubated *l*-ephedrine with an NADPH-fortified rabbit liver preparation and detected the demethylated compound (19), the α-*C*-oxidized products (22) and (23) and benzoic acid (24). With rabbit liver slices Baba *et al.* (1971) noted the formation of hippuric acid (25) in addition to the above four metabolites.

Another major group of amines containing aromatic substituents is that of the indoleamines. These compounds are usually derivatives of ethylamine (tryptamines), although simpler homologues are sometimes found. Gramine (3-(dimethylaminomethyl)-indole) is such an example, however its metabolism in mammals has not been investigated. The tryptamine derivatives which occur in higher plants include tryptamine (28) itself as well as *N*-methylated and/or 5-hydroxylated or methoxylated compounds. The more recent interest in the metabolism of these compounds has nearly exclusively been related to the well-known hallucinogenic properties of several of them.

Tryptamine (28) was the subject of early metabolic studies by Ewins and Laidlaw (1913) and Guggenheim and Löffler (1915). In the latter case, the amine was shown to be converted to indole-3-acetic acid (29) following its

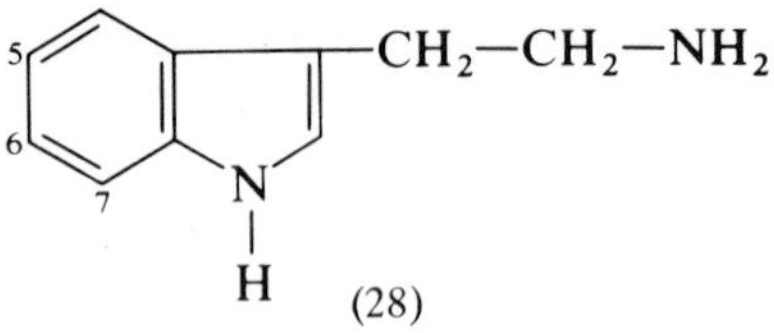

(28)

Tryptamine

administration (p.o. or i.v.) to rabbits and dogs. This metabolite was also detected by Ewins and Laidlaw in perfusion experiments with rabbit or cat liver. However, they found that 20–30% of the dose (50–100 mg/kg) was excreted in the urine of dogs as indoleaceturic acid (30), the glycine conjugate of (29). These results were confirmed by Erspamer (1955) in rats. Following a dose of about 16 mg/kg (s.c.), the urine in 11 h contained 25% and 59% as metabolites (29) and (30), respectively. These results agree well with those of Blaschko and Philpot (1953) who demonstrated the *in vitro* oxidation of tryptamine by liver and kidney preparations from several mammalian species.

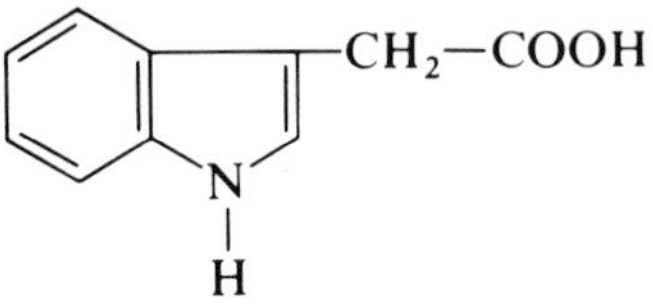

(29)

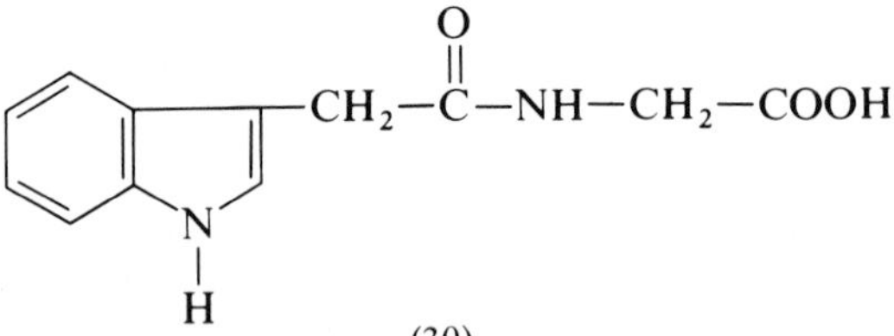

(30)

Another metabolic pathway of tryptamine involves aromatic hydroxylation. Jepson *et al.* (1962) reported that rabbit liver microsomes supplemented with TPNH and oxgyen hydroxylated tryptamine in the 6-position. The 5- and 7-hydroxytryptamines were not formed. However, this pathway appears to have little *in vivo* significance. An additional metabolic possibility involves N-methylation. Axelrod (1962) reported the presence of an enzyme system in the soluble supernatant fraction of several tissues from rabbits, but mainly in the lung, which utilized S-adenosylmethionine in forming mono- and di-methylated derivatives from several tryptamines including tryptamine itself. However, this enzyme is very species dependent, probably being present in human lung but not in that from mouse, rat, guinea pig, cat and monkey. Another enzyme system with a narrower range of substrate specificity was found in rat brain which formed the mono- and di-methylated derivatives of tryptamine (Morgan and Mandell, 1969).

The pathways of metabolism of the N-methyl and N,N-dimethyl derivatives of tryptamine are similar to those noted above with the corresponding primary amine. In experiments similar to that described above with tryptamine, Erspamer (1955) found that rats excreted metabolites (29) and (30) (see Fig. 8.5) in the urine after injection of **N-methyltryptamine** or **N,N-dimethyltryptamine** (31). As with tryptamine, most of these metabolites consisted of the glycine conjugate (30) but the total amount of oxidized metabolites ((29)+(30)) excreted decreased from about 85% with tryptamine to 36% and 18% for the secondary and tertiary amines, respectively. Compound (29) was reported to be the major excretory product of N,N-dimethyltryptamine in rats (Ahlborg et al., 1968) and in man (Szára, 1956). Conversion of the two methylated tryptamines to indole-3-acetic acid (29) in in vitro experiments has also been demonstrated. Fish et al. (1955) found that the conversion rate was higher with the secondary amine when using a mitochondrial preparation from mouse liver. When fortified mouse liver homogenates were employed, three additional metabolites including the N-oxide of the added N,N-dimethyltryptamine were formed. Experiments with the last metabolite indicated that it is not an intermediate in the oxidation of the parent amine by the mitochondrial monoamine oxidase. Szara and Axelrod (1959), using rabbit liver microsomes, found that N,N-dimethyltryptamine was demethylated to the monomethyl derivative but not further to the primary amine. However, rats given the dimethyl compound (10 mg, i.p.) excreted in the urine unchanged compound and both of the demethylated derivatives.

As noted above in the summary of tryptamine metabolism, hydroxylation of the indole nucleus may occur at the 6-position. This pathway appears to be of somewhat greater importance with the methylated tryptamines, which are less susceptible to oxidative deamination, than with the primary amine. Szara and Axelrod (1959) obtained evidence for the 6-hydroxylation of N,N-dimethyltryptamine in vitro using rabbit liver microsome preparations and in vivo in rats. In the latter experiments, the urines collected for 48 h after dosage (10 mg, i.p.) contained 6-hydroxy-N,N-dimethyltryptamine (32) and 6-hydroxyindole-3-acetic acid (33), as well as the N-demethylated metabolites noted above.

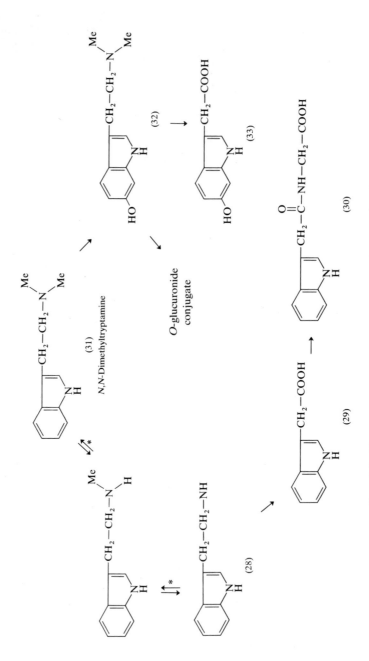

FIG. 8.5. Metabolic pathways of *N, N*-dimethyltryptamine (* see text).

It was pointed out in the summary of tryptamine metabolism given above that some tissue preparations from a few species effect the *N*-methylation of tryptamine to the mono- and dimethyl derivatives. Ahn *et al.* (1973) reported that this reaction occurred *in vivo* when *N*-methyl-tryptamine was administered intravenously to rabbits pretreated with a monoamine oxidase inhibitor. The dimethylated product was found in several tissues and especially in the lungs, however this reaction was not detected when rats were similarly treated. This species difference in metabolism must be kept in mind when viewing the metabolic pathways of *N,N*-dimethyltryptamine shown in Fig. 8.5. Also, the *N*-oxide derivatives of compounds (31) and (32), both of which are produced *in vitro* by liver microsomal preparations (Fish *et al.*, 1955; Szara and Axelrod, 1959), are not shown. Further information on the 6-hydroxylation pathway is available (Szara, 1961), however it appears to be less developed in man than in rodents (Szara, 1968).

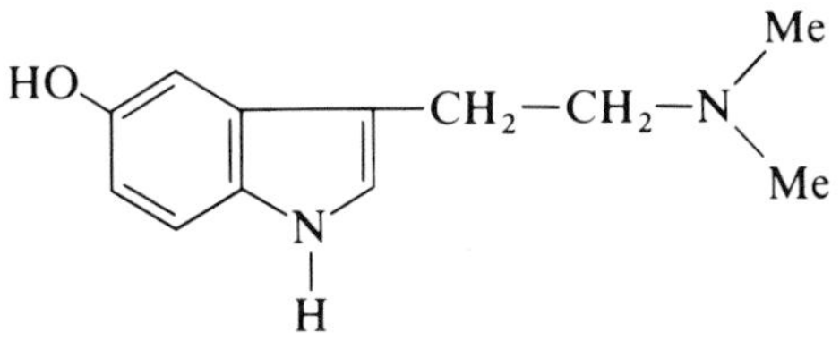

HO—[indole ring]—CH$_2$—CH$_2$—N(Me)(Me)

(34)

Bufotenine

HO—[indole ring]—CH$_2$—COOH

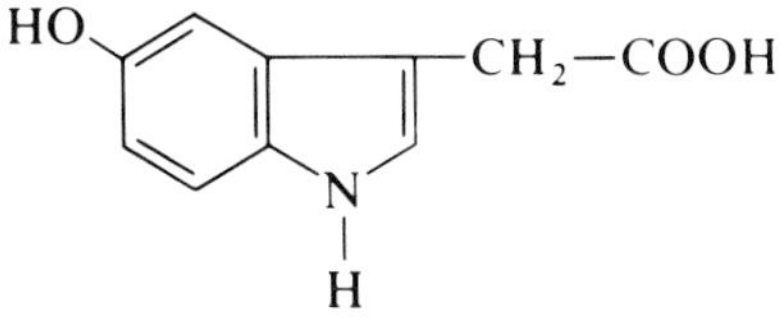

(35)

Several studies on the metabolism of **bufotenine** (5-hydroxy-*N,N*-dimethyltryptamine) (34) have been carried out in rats, with the result that its metabolism in this species is partially understood. Erspamer (1955) detected unchanged compound and 5-hydroxyindole-3-acetic acid (35) as urinary metabolites after small doses (1·2–1·5 mg/kg, s.c.) of bufotenine. Compound (35) was found to be a major urinary metabolite by Ahlborg *et al.* (1968). Quantitative data on its excretion were obtained by Gessner *et al.* (1960) and by Sanders and Bush (1967). In the former study, about 7% of a large dose (125 mg/kg, given s.c. in portions over 6 h) was excreted as the acid in 48 h. The latter study employed a much smaller dose (approx.

1 mg/kg, i.v.) and obtained a value of 14% of the dose for urinary 5-hydroxyindole-3-acetic acid. These low values for the excretion of the deaminated product are not unexpected in view of the reports by Blaschko and Philpot (1953) and Govier *et al.* (1953) showing that bufotenine is only slowly oxidized by monoamine oxidase. Sanders and Bush found that bufotenine glucuronide is the major metabolite, accounting for 35% of the dose. The earlier study by Gessner *et al.* indicated that the increase in urinary glucuronide output following dosing could account for up to 25% of the dose. These studies also confirmed the excretion of unchanged compound. With the higher dose this amounted to about 22% while 6% was excreted unchanged with the lower dose. This difference is hardly surprising in view of the difference in dosage of over two orders of magnitude. The data of Sanders and Bush indicated that bufotenine metabolism in the rat is both rapid and extensive. No evidence was obtained for the *O*-methylation of bufotenine, however a small amount of a metabolite tentatively identified as 5-methoxyindole-3-acetic acid was detected. It is conceivable that the latter metabolite may arise from methylation of 5-hydroxyindole-3-acetic acid.

These studies of the metabolism of bufotenine in rats have not accounted for the entire dose and the fate of the remainder is not known. However, unidentified metabolites have been reported by Erspamer (1955) and by Gessner *et al.* (1960). It was noted above that *N,N*-dimethyltryptamine undergoes aromatic hydroxylation and this route may also be involved in the metabolism of bufotenine. While this reaction has not been demonstrated *in vivo*, Daly *et al.* (1965) showed that a rabbit liver preparation containing microsomes, soluble supernatant and added *S*-adenosylmethionine had low but detectable activity to first hydroxylate bufotenine and then methylate the catechol formed. By analogy with results obtained using other indoles. the hydroxylation most likely occurred at the 6-position.

The above picture of bufotenine metabolism in rats has recently been extended to humans. Sanders-Bush *et al.* (1976) administered ^{14}C-labelled material (0·2 and 1 mg, i.v.) to two subjects and found that essentially quantitative excretion of the radioactivity in the urine occurred within the first 24 h. Most (68 and 74%) of this material was 5-hydroxyindole-3-acetic acid (35) and only 1 and 6% were found to be due to unchanged bufotenine. The remainder consisted of very polar metabolites (22 and 20%) and neutral metabolites (9 and <1%, respectively). It was believed that the alcohol analogue of compound (35), 5-hydroxytryptophol, occurred in the latter fraction.

When **5-methoxy-*N*-methyltryptamine** labelled with ^{14}C was administered to rats using a dose of 3 mg, about 90% of the radioactivity was excreted in the urine in 24 h (Taborsky and McIsaac, 1964). No unchanged compound was present and about 95% of the urinary material was due to 5-methoxyindole-3-acetic acid. In some cases, a few percent of the dose corresponded to a compound which was probably the O-glucuronide of the 6-hydroxylated derivative. An alternative metabolic pathway, however probably not in the rat, may be the N-methylation of 5-methoxy-N-methyltryptamine. Mandel and Walker (1974) reported on an indoleamine-N-methyltransferase from rabbit and human lung which carries out this reaction.

The results summarized above indicate that non-methylated tryptamines undergo rapid and extensive oxidative deamination. Di-methylated derivatives are more resistant in this regard with the result that the rate of metabolism is lower and extent of conversion to the indoleacetic acid derivative is less. This has been shown with **5-methoxy-N,N-dimethyltryptamine**. Ahlborg *et al.* (1968) found that about 50% of the radioactivity appeared in the urine in 12 h when rats were given the ^{14}C-labelled compound (10 mg/kg, i.p.). Most of this material was 5-methoxyindole-3-acetic acid. Agurell *et al.* (1969), in similar experiments using half the above dose level, found that 59–65% of the radioactivity was excreted during the first 24 h. The urinary material consisted of four compounds: 5-methoxyindole-3-acetic acid (33% of the dose), bufotenine (5%), bufotenine glucuronide (14%) and 5-hydroxyindole-3-acetic acid (8%). Several possible urinary metabolites were shown to be absent. These included the unchanged compound and its N-demethylated products, the mono- and di-demethylated products of bufotenine, 5-methoxyindoleaceturic acid and the 6-hydroxy derivative of the administered compound. The data indicated that 5-hydroxyindole-3-acetic acid is formed via bufotenine as 5-methoxyindole-3-acetic acid does not undergo demethylation.

The metabolism of a few miscellaneous amino compounds is discussed in other sections. Betaine is included in the section on aliphatic acids (Chapter 5, Section I.A) and the metabolism of isojuripidine, a triterpenoid compound containing an alicyclic amino group, is summarized in Chapter 6, Section III.

II. Nitriles

The nitriles or organic cyanides found in plants are mainly cyanogenetic glycosides. These compounds, which have the general formula shown in Fig. 8.6, are widely distributed and occur in several hundred genera belonging to dozens of plant families. Many of these plants have economic importance, partly as forages but also as human foods. The best-known cyanogenetic glycoside is **amygdalin** (36), the toxic principle of bitter almonds which is also found in the kernels of fruits including cherries, apricots and plums. Other common examples include **linamarin** (37) and **lotaustralin** (38), which are often found together (e.g. in white clover), dhurrin (39), a constituent of sorghum, and prunasin. The latter compound contains the same aglycone (mandelonitrile) (40) as amygdalin but the sugar moiety consists of a single glucose unit. All of these compounds are β-glycosides and will undergo hydrolysis when subjected to the actions β-glycosidases also present in the plants. Normally, glycoside and enzyme are separated in the plant tissue but damage by cutting, crushing or soaking in water permits their contact and results in the liberation of HCN as

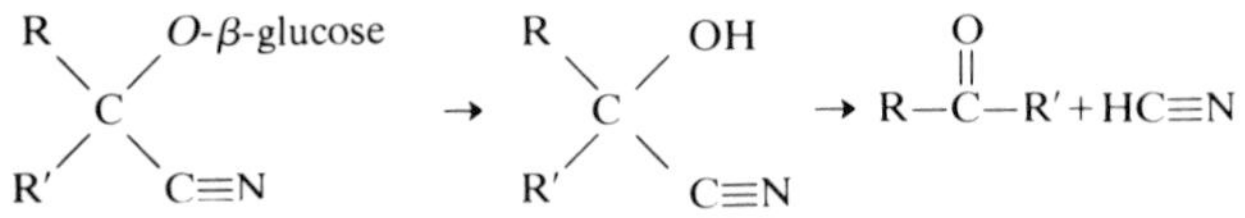

FIG. 8.6. Hydrolysis of cyanogenetic glycosides. R′ = H in many cases.

illustrated in Fig. 8.6. With amygdalin (36) this occurs via the sequential loss of two glucose units to give the aglycone mandelonitrile (40) followed by the decomposition of this unstable cyanohydrin (α-hydroxynitrile) to benzaldehyde and HCN.

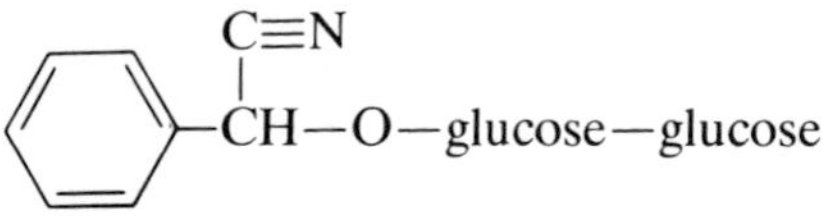

(36)

Amygdalin

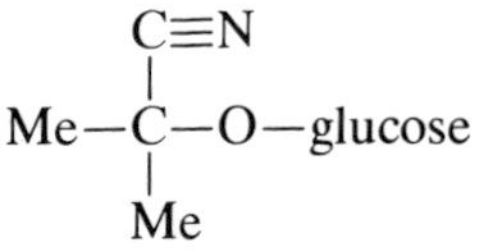

(37)

Linamarin

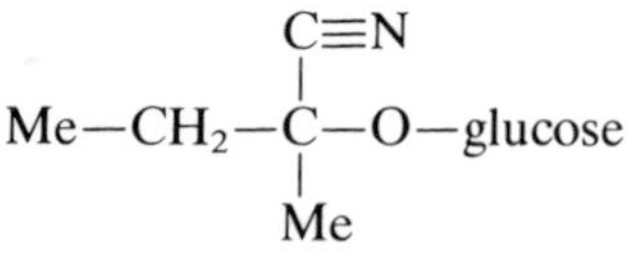

(38)

Lotaustralin

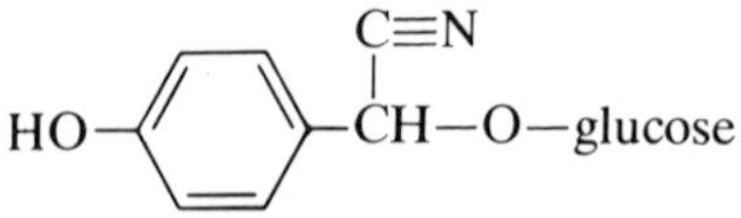

(39)

Dhurrin

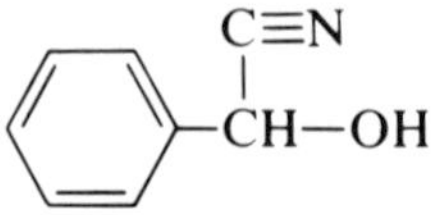

(40)

Considerable attention has been given to the subject of cyanide poisoning by cyanogenetic glycosides, however there appears to have been a good deal of confusion among earlier workers as to the nature of the hydrolysis process. The belief lingered that hydrolysis following ingestion of the glycosides by animals was dependent upon concomitant intake of the β-glycosidase. This subject was investigated by Coop and Blakley (1949) who, using a sheep with a rumen fistula, measured the formation of HCN from amygdalin (36), linamarin (37) or lotaustralin (38) administered intraruminally. They found that the rumen microflora can rapidly hydrolyse the glycosides and that plant enzyme is therefore not necessary. Two rumen microorganisms, a Gram positive diplococcus and a Gram negative bacillus, were isolated which readily hydrolysed lotaustralin *in vitro*. These findings are also useful in explaining why the cyanogenetic glycosides are somewhat less toxic than HCN or an equivalent amount of cyanide ion. In the latter case symptoms of poisoning appear very rapidly whereas Coop and Blakley reported that periods from five min to several hours elapsed following glycoside administration. In this case the HCN will be liberated over a certain period of time and thus allow the normal thiocyanate pathway of detoxification to more effectively remove the absorbed HCN. Coop and Blakley (1950) also showed that relatively large amounts of HCN or cyanogenetic glycoside are tolerated when they are given repeatedly in small divided doses. It is therefore important that the absorption rate of HCN, which is rapid, does not exceed the rate of conversion of cyanide to thiocyanate. It was also found that feeding which promotes the bacterial population of the rumen increases the toxicity of the glycosides whereas starvation reduces it.

Linamarin (37), the glucoside of acetone cyanohydrin, is found in several foodstuffs including cassava and lima beans. Barrett *et al.* (1977) found that an oral dose of about 500 mg/kg was lethal to rats but that one of about 300 mg/kg was not. In the latter case, nearly 20% of the administered linamarin was recovered unchanged in the urine within 24 h. A further 10% of the dose was accounted for as urinary thiocyanate ion. No unchanged glycoside was identified in the blood or faeces. Unfortunately, other routes of administration were not employed in this study and the data therefore do not furnish information on the site or sites of linamarin decomposition. It is not unreasonable to assume that this is the intestinal tract and Winkler (1958) reported that cooked lima beans released significant amounts of HCN when incubated with extracts of faeces or with *E. coli.* Further evidence of the key role of the gastrointestinal microflora in the metabolism of cyanogenetic glycosides was obtained by Smith (1971) who administered amygdalin (36) to mice. No toxicity was observed following intraperitoneal injection of doses as large as 5 g/kg whereas the oral LD$_{50}$ was 350 mg/kg. While the latter value was considerably higher than that given by potassium cyanide, the symptoms of toxicity were similar to those produced by inorganic cyanide. It was found that pretreatment of the mice with lactose, another β-glycoside, conferred protection against a lethal dose of amygdalin but not against cyanide. This protection was not achieved when an α-glycoside, maltose, was used. Thus, competition for the available β-glycosidase present in the intestine will reduce the rate of cyanide formation. That the source of the enzyme is the microflora was indicated by the fact that its suppression by pretreatment with the antibiotic kanamycin reduced amygdalin toxicity, especially in fasted mice. Finally, ten strains of enterobacteria and enterococci isolated from mouse intestine were isolated which degraded amygdalin to cyanide *in vitro*. Similar results using isolated strains of intestinal bacteria were reported by Drasar and Hill (1974) who also found that amygdalin hydrolysis could be carried out by some non-sporing anaerobes (e.g. bifidobacteria).

$$N\equiv C-CH_2-CH_2-NH-\overset{\overset{\displaystyle O}{\|}}{C}-CH_2-CH_2-\overset{\overset{\displaystyle NH_2}{|}}{CH}-COOH$$

(41)

$$H_2N-CH_2-CH_2-C\equiv N \qquad\qquad HOOC-CH_2-C\equiv N$$

(42)   (43)

Another type of nitrile is represented by the osteolathyrogen found in the seeds of *Lathyrus* species (vetches) including the sweet pea. The causative agent, which produces changes in bone and connective tissue structure, is β-(γ-L-glutamyl)-aminopropionitrile (41). Its metabolism has not been studied, however several investigations have dealt with β-aminopropionitrile (42) which also produces the symptoms of osteolathyrism (odoratism). The glutamyl residue is therefore not necessary for the biological activity of the compound (see Sarma and Padmanaban, 1969) and it seems reasonable to assume that the amino acid (41) undergoes hydrolysis of the amide bond in the body to the simple aminonitrile. An analogous reaction has been reported with the closely related amino acid theanine (56) (see Section III). Lipton *et al.* (1958a,b) reported that β-aminopropionitrile is metabolized in rats and rabbits to cyanoacetic acid (43). This metabolite does not share the toxic properties of the aminonitrile. This study indicated that about equal amounts of the administered compound and the metabolite were excreted in the urine of rats given a dose of approx. 135 mg/kg, i.p. Keiser *et al.* (1967), using a very small dose (7 μg/rat), reported that roughly six times as much cyanoacetic acid as β-aminopropionitrile was excreted in the urine. No other urinary metabolites were detected. Disappearance of the aminonitrile from the body was rather slow and appears to involve the participation of monoamine oxidase. This was shown in an experiment in mice in which administration of a monoamine oxidase inhibitor greatly reduced the rate of disappearance of the compound from the body. Recently, Fleischer *et al.* (1976) reinvestigated the metabolism of β-aminopropionitrile in rats. Following a large dose (400 mg/kg, i.p.), excretion of unchanged compound in the urine accounted for 30% of the dose in 12 h and 32% in 24 h, after which no further excretion occurred. Urinary excretion of cyanoacetic acid was 12 and 22% in 24 h and 48 h, respectively. When cyanoacetic acid itself was administered it was rapidly excreted. These findings, as well as that indicating the presence in liver and brain of the metabolite but not the administered nitrile after 24 h, suggest that the metabolically formed acid is only slowly released from tissues. Fleisher *et al.* assumed that monoamine oxidase is involved in the metabolism of β-aminopropionitrile, however *in vitro* experiments using rat liver homogenates showed a low capacity for conversion to cyanoacetic acid. Sievert *et al.* (1960) reported that rat liver homogenates carried out this conversion more slowly than did preparations from several other species (mouse, cotton rat, guinea pig, rabbit, cow and horse). Human liver homogenates showed the lowest activity of all while rabbit preparations were most active. Lipton *et al.* (1958b) reported that rabbits are much more efficient in converting β-aminopropionitrile to cyanoacetic acid *in vivo*. Unlike that noted above with rats, no unchanged compound was detected in the urine of rabbits given the nitrile. Further information on the metabolism and distribution of β-aminopropionitrile and its oxidatively deaminated metabolite, especially with regard to teratogenic effects, is available (Wilk *et al.*, 1972; Waddell *et al.*, 1974).

(44)

A further type of organic nitrile is that produced from the breakdown of glucosinolates. As pointed out in Chapter 10, Section VI, the hydrolysis of the glycosides can lead to the formation of nitriles ($R-C\equiv N$) as well as thiocyanates ($R-S=C=N$) and, more commonly, isothiocyanates ($R-N=C=S$). Various aliphatic and aromatic nitriles including benzyl cyanide (44) may thus be formed. Ohkawa *et al.* (1972) studied the metabolism of several compounds of this type employing mouse liver microsomes. It appears that this system hydroxylates the methylene position to produce an unstable cyanohydrin which breaks down to the aldehyde and HCN. This reaction sequence with benzyl cyanide is illustrated in

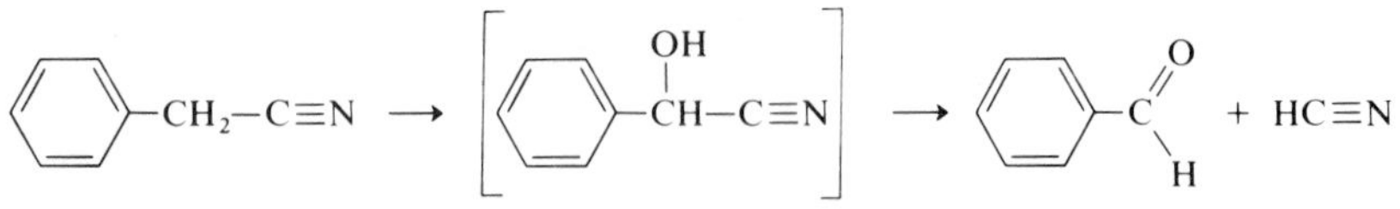

$$\text{C}_6\text{H}_5\text{—CH}_2\text{—C}{\equiv}\text{N} \longrightarrow \left[\text{C}_6\text{H}_5\text{—}\overset{\overset{\text{OH}}{|}}{\text{CH}}\text{—C}{\equiv}\text{N}\right] \longrightarrow \text{C}_6\text{H}_5\text{—C}\overset{\text{O}}{\underset{\text{H}}{\diagup\diagdown}} + \text{HC}{\equiv}\text{N}$$

(40)

FIG. 8.7. Metabolism of benzyl cyanide by mouse liver microsomes.

Fig. 8.7. Thus, the intermediate formed (mandelonitrile) (40) is identical to that produced from the hydrolysis of the cyanogentic glycoside amygdalin.

III. Amides

Plant compounds which are characterized primarily by the presence of an amide group appear to be quite restricted in their occurrence, except, of course, for peptides which fall outside the scope of this book. A few amides consisting of various amines linked to the γ-carboxyl group of L-glutamic acid have been reported. These include β-(γ-L-glutamyl)-amino-propionitrile (41) which is dealt with in Section II and the tea constituent theanine (56). As noted in Section IV, theanine undergoes hydrolysis in the body and serves as a source of the ethylamine excreted in the urine by tea drinkers. Other examples of this type include amides derived from other simple aliphatic amines or from aromatic amines (e.g. p-hydro-xyphenylamine, o- and p-hydroxybenzylamine), however the mammalian metabolism of these compounds has not been reported. Another group of amides deserving mention consists of derivatives of isobutylamine and various unsaturated fatty acids. These products have insecticidal properties and include pellitorine, a mixture of several isobutylamides which consists mainly of N-isobutyldeca-trans-2-trans-4-dienamide (45). The metabol-ism of this group has not been studied, however Kuhn et al. (1937) reported on the oxidation of amides of several related unsaturated fatty acids in rabbits. This investigation determined the extent of ω-oxidation of the compounds to the semi-amides of the corresponding dicarboxylic acids following large repeated doses ranging from 7 to 20 g. Based on the isolation of the corresponding ω-carboxy derivatives from urine, the yields of the metabolites were (% of dose): sorbamide $\text{Me—(CH{=}CH)}_2\text{—CO—NH}_2$ (32%) octatrienoic amide, $\text{Me—(CH{=}CH)}_3\text{—CO—NH}_2$ (42%) and deca-tetraenoic amide, $\text{Me(CH{=}CH)}_4\text{—CO—NH}_2$ (20%). The only N-substi-tuted amide, sorbic acid N-methyl amide, underwent ω-oxidation to an extent of 44% of the dose. Nonetheless, all of these examples involve allylic oxidation, a situation which will not be encountered in the ω-oxidation of the naturally occurring isobutylamides. Also, the extent of hydrolysis of the amide bond in the above compounds is undertermined and this is a possible route of metabolism with the isobutylamides.

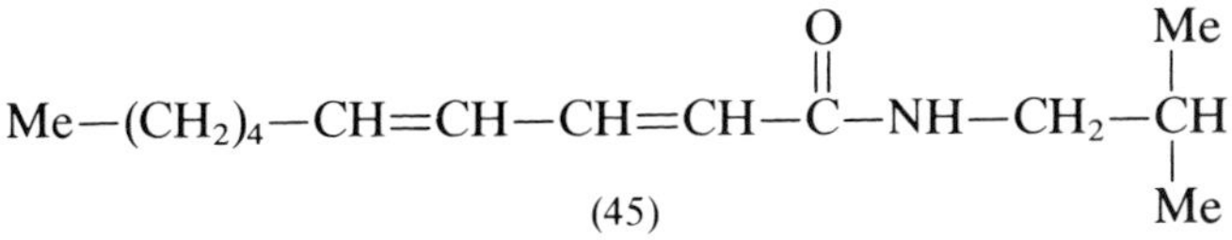

$$Me-(CH_2)_4-CH=CH-CH=CH-\overset{\overset{\displaystyle O}{\|}}{C}-NH-CH_2-\overset{\overset{\displaystyle Me}{|}}{\underset{\underset{\displaystyle Me}{|}}{CH}}$$

(45)

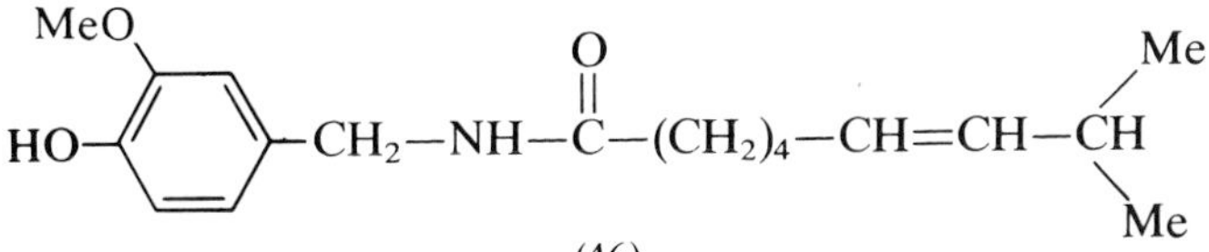

$$\text{MeO, HO}\diagdown\!\!\!\!\!\bigcirc\!\!\!\!\!-CH_2-NH-\overset{\overset{\displaystyle O}{\|}}{C}-(CH_2)_4-CH=CH-\overset{\diagup Me}{\underset{\diagdown Me}{CH}}$$

(46)

Capsaicin

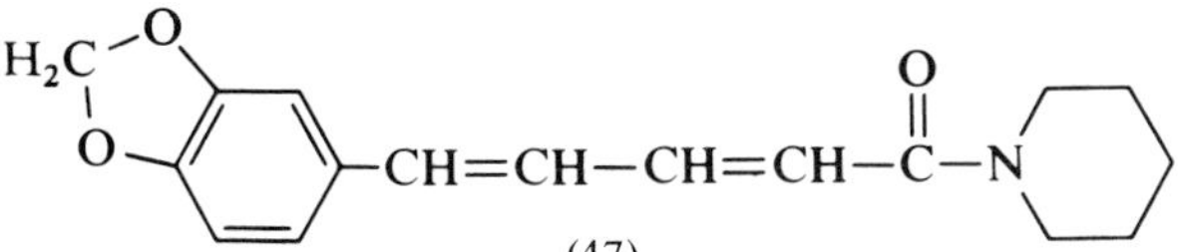

$$\text{H}_2\text{C}\diagdown\!\!\!\!\!\overset{O}{\underset{O}{\diagup}}\!\!\!\!\!\bigcirc\!\!\!\!\!-CH=CH-CH=CH-\overset{\overset{\displaystyle O}{\|}}{C}-N\!\!\!\bigcirc$$

(47)

Piperine

$$\text{H}_2\text{C}\diagdown\!\!\!\!\!\overset{O}{\underset{O}{\diagup}}\!\!\!\!\!\bigcirc\!\!\!\!\!-CH=CH-CH=CH-\overset{\overset{\displaystyle O}{\|}}{C}-NH-CH_2-COOH$$

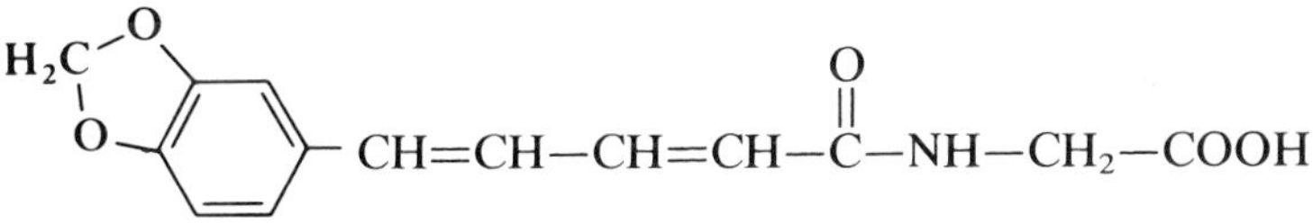

(48)

Two further amides worthy of note but for which metabolic data are lacking include capsaicin (8-methyl-*N*-vanillyl-6-nonenamide) (46), the pungent principle from various species of *Capsicum*, and piperine (1-piperoylpiperidine) (47), a constituent of black pepper and other species of *Piper*. However, Acheson and Atkins (1961) investigated the metabolism in rats of the acidic component of piperine, piperic acid (see Chapter 5, Section I.C), and also the amide of piperic acid and glycine, piperoylglycine (48). Following a total oral dose of 20 mg (approx. 200 mg/kg) of compound (48), the 48 h urines contained some unchanged compound as well as the glycine conjugates of the homologous acids, 3,4-methylenedioxybenzoic and 3,4-methylenedioxycinnamic acid. Although free C_6-C_1, C_6-C_3 or C_6-C_5 acids were not detected, the presence of the glycine conjugates of the first two compounds indicates that amide hydrolysis of the administered compound must have taken place.

(49)

Colchicine

The final compound to be included among this small group of amides is **colchicine** (49). This complex polymethoxylated alkaloid has a long history of use, especially in the treatment of acute attacks of gouty arthritis. Nonetheless, its metabolic fate is far from clear. As sporadic studies of its metabolism have been carried out over a relatively long period of time, it is natural that several different analytical methods have been employed. Earlier methods were based on chemical or biological assays and, using the former, Boyland and Mawson (1938) concluded that hydrolysis of colchicine to the C10-demethylated derivative colchiceine in the acidic environment of the stomach is not a significant factor in its metabolism. Most of the present knowledge concerning colchicine metabolism has been obtained subsequent to the availability of the [3]H- or ^{14}C-labelled compounds and the present summary of results deals largely with these more recent findings. A short review of earlier data is available in the article by Wallace *et al.* (1970).

The first study of the metabolism of radioactive colchicine was carried out by Walaszek *et al.* (1960) using a product either randomly labelled by biosynthetic means or labelled specifically in the acetyl moiety at C7. With the former compound given at dose levels of $0 \cdot 1–10$ mg/kg, they found that about 3–9% was excreted unchanged in the urine in 48 h in mice, rats, hamsters and guinea pigs. The corresponding faecal values were lower (1–2%) except for the rat which showed urinary and faecal excretion values for colchicine of about 3 and 6%, respectively. In addition, smaller amounts of chloroform-soluble metabolites were found in the excreta in most cases. Loss of radioactivity as respiratory [14]CO_2 was also noted. This was only 2 and 9% in mice and hamsters, respectively, whereas guinea pigs excreted 28% and rats 32% of the dose by this route. Administration of the preparation labelled in the *N*-acetyl moiety to a patient with gout resulted in excretion of radioactivity as [14]CO_2, indicating hydrolysis of the amide bond. Urinary excretion of unchanged colchicine in three humans averaged more than a fourth of the dose (2–3 mg), however Wallace and

Ertel (1969) reported that less than 10% of the administered compound or its metabolites was excreted in the urine in 24 h following i.v. injection.

An important factor in the metabolic disposition of colchicine appears to be its biliary excretion. This excretory route was recognized in rats in an early report by Brues (1942). Later, Fleischmann *et al.* (1965, 1968) reported that appreciable biliary excretion of colchicine occurs in the golden hamster as well as in the related rodent, the Mongolian gerbil (Fleischmann *et al.*, 1967). An extensive study of the biliary excretion of colchicine labelled with ^{3}H or ^{14}C in the methoxy group at C10 was carried out by Hunter and Klaassen (1975a). Initial experiments with rats receiving a dose of 0·2 mg/kg (i.v.) indicated appreciable biliary excretion as the faecal : urinary ratio of radioactivity was 5 : 1 for the initial period of three days. More than 80% of the administered radioactivity was excreted during this time. At a dose level of 2 mg/kg (i.v.), the biliary excretion of radioactivity in 2 h in rats, hamsters, rabbits and dogs was 50, 32, 16 and 20%, respectively. The percentages of this material excreted as unchanged colchicine were 53 in rats, 45 in hamsters, 72 in rabbits and 34 in dogs. In addition to unchanged colchicine, the bile of all four animal species contained metabolites including polar compounds, some of which was a glucuronide conjugate of a colchicine derivative formed by demethylation in the A ring. In all species except the rabbit, some free demethylated metabolite was also detected. This amounted to 15, 10 and 35% of the biliary material in rats, hamsters and dogs, respectively. There was no evidence indicating that colchiceine, formed by demethylation at C10, was a biliary metabolite or that *N*-deacetylated metabolites were excreted by this route. The lower capacity of newborn rats to conjugate and excrete colchicine in the bile was studied by Hunter and Klaassen (1975b). This deficiency was related to the higher toxicity of colchicine in very young rats than in older rats.

An early report by Axelrod (1956) on the mechanism of cleavage of aromatic ethers indicated that colchicine is a substrate for the liver microsomal system dependent on NADPH and O_2. This reaction has been studied in detail by Schönharting *et al.* (1974) who found that mouse, rat and hamster liver microsomes oxidatively demethylate colchicine at the C2 and C3 positions. Additionally, demethylation at C10 to colchiceine is carried out by mouse and rat preparations. The monodemethylated metabolites are not metabolized further by the microsomes except for glucuronidation which took place at the 2-position with rat microsomes and the 3-position with hamster microsomes. The three demethylated metabolites and a fourth derivative formed by rearrangement of the tropolone ring (C-ring) are formed by a modified Udenfriend system (Schönharting *et al.*, 1973).

IV. Non-protein Amino Acids

Of the more than 200 amino acids isolated from plants, only about 10% consist of those which are incorporated into the proteins of living matter. The remainder, of which most have been identified in recent years following the introduction of chromatographic separation techniques, are found as free amino acids in a variety of higher plants. These non-protein amino acids are often found in low concentrations, however in some cases they may accumulate to such a degree that they constitute several percent of the dry weight of the plant. Major areas of interest in this field have dealt with the biosynthetic pathways involved, the possible taxonomical correlations and the toxicological properties shown by some of these compounds. Neurotoxicity is an important consideration with regard to the latter point and this subject was reviewed by Johnston (1974). On the other hand, information describing the mammalian metabolism of non-protein amino acids is extremely limited, both in regard to the number of compounds studied and the amount of information available on specific compounds.

A toxicological problem of considerable practical importance is that of lathyrism. This disease is produced by the consumption of the seeds of the chick pea (*Lathyrus sativus*) and is characterized by extreme neurological symptoms including muscular rigidity and paralysis. It should be noted that this disease, neurolathyrism, is the classical form seen in humans whereas another form, osteolathyrism, is produced in animals ingesting vetches (*Lathyrus* sp.). The skeletal changes produced in animals by feeding the sweet pea (*Lathyrus odoratus*) are due to a nitrile, the metabolism of which is discussed above in Section II. Although the causative agents of neurolathyrism are not known with certainty, it appears that some acidic amino acids are involved. One of these is **β-N-oxalyl-L-α,β-diaminopropionic acid** (50). Cheema *et al.* (1971), using ^{14}C-labelled compound, found that 50–70% of the amino acid was excreted in the urine within 24 h after i.p. injection. In 1 h experiments, most of the radioactivity was located in the kidneys and liver with about 70% found in unchanged form. About 5–10% of the radioactivity in the kidneys was in a keto form and this transamination product was shown to be N-oxalyl-β-aminopyruvic acid. This indicates that the α-amino group undergoes transamination while the N-oxalyl moiety remains intact.

$$\underset{\text{HOOC}-\overset{\displaystyle O}{\overset{\displaystyle \|}{C}}-\text{NH}-\text{CH}_2-\overset{\displaystyle \text{NH}_2}{\overset{\displaystyle |}{\text{CH}}}-\text{COOH}}{}$$

(50)

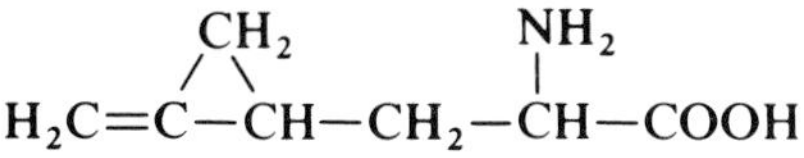

$$\underset{(51)}{\overset{\overset{\displaystyle CH_2}{\underset{\displaystyle /\ \backslash}{}}}{H_2C=C-CH-CH_2}}\underset{}{\overset{\displaystyle NH_2}{\underset{\displaystyle |}{CH}}}-COOH \qquad \underset{(52)}{\overset{\overset{\displaystyle CH_2}{\underset{\displaystyle /\ \backslash}{}}}{H_2C=C-CH-CH_2}}\overset{\displaystyle O}{\underset{\displaystyle ||}{C}}-COOH$$

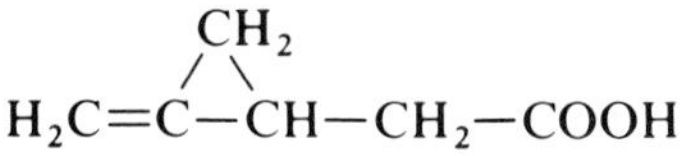

(51) Hypoglycin

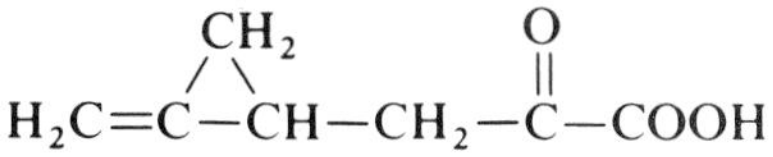

$$\underset{(53)}{\overset{\overset{\displaystyle CH_2}{\underset{\displaystyle /\ \backslash}{}}}{H_2C=C-CH-CH_2-COOH}}$$

The fruit of the tropical tree *Blighia sapida* contains a hypoglycaemic toxin known as **hypoglycin**. This compound is β-(methylenecyclopropyl) alanine (51) and its metabolism in rats and by rat liver homogenates was studied by Von Holt *et al.* (1964) and Von Holt (1966). The results indicated that the first step in the metabolism of hypoglycin involved transamination to the keto acid (52). When rats are given [^{14}C]-hypoglycin by injection, about 60% of the radioactivity is lost within 2 h as respiratory $^{14}CO_2$ and the data indicated that the decarboxylating enzyme is associated with the mitochondria. The decarboxylation product, which was detected in the homogenates and in the livers of rats treated with hypoglycin, was identified as methylenecyclopropaneacetic acid (53).

The legume *Leucaena glauca* is used as a forage plant, however excessive consumption leads to several toxic effects of which loss of hair is the most dramatic and characteristic. The toxic principle is **mimosine** (β-(3-hydroxy-4-oxo-1-pyridyl)alanine) (54). Hegarty *et al.* (1964) showed that little degradation of the amino acid occurs in sheep following intravenous or intra-abomasal administration and that most of the compound is excreted unchanged in the urine. Following intraruminal administration,

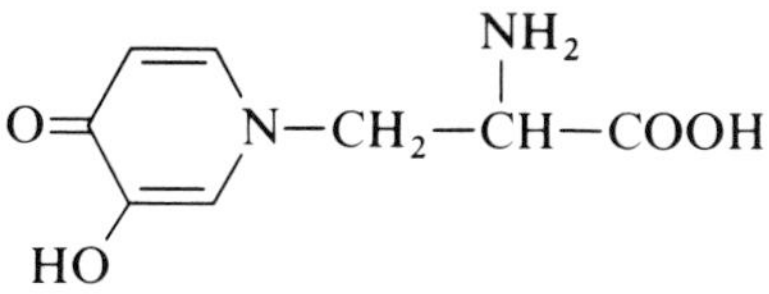

(54) Mimosine

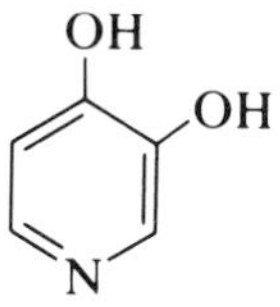

(55)

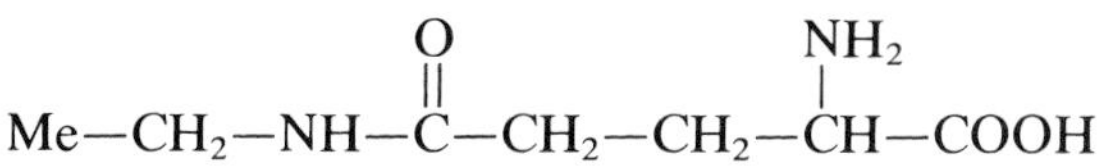

$$Me-CH_2-NH-\overset{\displaystyle O}{\underset{\displaystyle ||}{C}}-CH_2-CH_2-\overset{\displaystyle NH_2}{\underset{\displaystyle |}{CH}}-COOH$$

(56) Theanine

however, only 5–10% of the dose was recovered in the urine as unchanged compound and a metabolite, 3,4-dihydroxypyridine (55), was excreted which accounted for over 40% of the dose. The results indicated that metabolite (55) is the end product of mimosine metabolism in the rumen and that this degradation is probably carried out by the microflora.

Theanine (*N*-ethyl-γ-glutamine) (56), a tea constituent, was found by Asatoor (1966) to be a source of urinary ethylamine in humans. Ethylamine was also excreted in the urine of rats given tea extract orally or by s.c. injection. The latter finding, as well as that showing that neomycin did not have a marked effect on the excretion of the amine in humans, suggests that theanine hydrolysis occurs in the tissues rather than in the gut as a result of bacterial metabolism.

$$\text{HOOC}-\text{CH}_2-\text{CH}_2-\text{S}-\text{CH}_2-\overset{\overset{\displaystyle \text{NH}_2}{|}}{\text{CH}}-\text{COOH}$$
(57)

$$\text{HOOC}-\overset{\overset{\displaystyle \text{NH}_2}{|}}{\text{CH}}-\text{CH}_2-\text{S}-\text{CH}_2-\text{S}-\text{CH}_2-\overset{\overset{\displaystyle \text{NH}_2}{|}}{\text{CH}}-\text{COOH}$$
(58)

Djenkolic acid

Several non-protein amino acids which contain sulphur are known, however information on their mammalian metabolism is scanty. Some data on the metabolism of *S*-methylcysteine and *S*-methylcysteine sulphoxide are available but this is more conveniently covered in Chapter 10 together with other sulphur compounds. The metabolism of two related amino acids, **S-2-carboxyethyl-L-cysteine** (57) and **djenkolic acid** (58) was studied by Binkley (1950) using an enzyme preparation from rat liver capable of cleaving thioethers. With the former compound, very little cleavage of the carbon–sulphur bond of the cysteine moiety was observed whereas the symmetrical molecule of djenkolic acid was readily split to form cysteine.

References

Acheson, R. M. and Atkins, G. L. (1961). *Biochem. J.* **79**, 268–270.
Agurell, S., Holmstedt, B. and Lindgren, J. E. (1969). *Biochem. Pharmac.* **18**, 2271–2781.
Ahlborg, U., Holmstedt, B. and Lindgren, J-E. (1968). *In* "Advances in Pharmacology" (S. Garattini and P. A. Shore, Eds), Vol. 6, Part B, pp. 213–229. Academic Press, New York and London.

Ahn, H. S., Walker, R. W., VandenHeuvel, W. J. A., Rosegay, A. and Mandel, L. R. (1973). *Fedn Proc. Fedn. Am. Socs exp. Biol.* **32**, 511.

Alles, G. A. and Heegaard, E. V. (1943). *J. biol. Chem.* **147**, 487–503.

Asatoor, A. M. (1964). Proc. Ass. Clin. Biochemists, E. J. King Memorial Symposium, August, 1964, pp. 82–87.

Asatoor, A. M. (1966). *Nature* Lond. **210**, 1358–1360.

Axelrod, J. (1952). *J. Pharmac. exp. Ther.* **106**, 372.

Axelrod, J. (1953). *J. Pharmac. exp. Ther.* **109**, 62–73.

Axelrod, J. (1955a). *J. biol. Chem.* **214**, 753–763.

Axelrod, J. (1955b). *J. Pharmac. exp. Ther.* **114**, 430–438.

Axelrod, J. (1956). *Biochem. J.* **63**, 634–639.

Axelrod, J. (1962). *J. Pharmac. exp. Ther.* **138**, 28–33.

Baba, S. (1973). *Nippon Aisotopu Kaigi Hobunshu* **11**, 171–176. (*Chem. Abstr.* (1976) **84**, 83892a.)

Baba, S., Matsuda, A. and Nagase, Y. (1969). *J. pharm. Soc.* Japan **89**, 833–836 (*Chem. Abstr.* (1969) **71**, 100055f).

Baba, S., Matsuda, A., Nagase, Y. and Kawai, K. (1971). *J. pharm. Soc.* Japan **91**, 584–586.

Baba, S., Enogaki, K., Matsuda, A. and Nagase, Y. (1972). *J. pharm. Soc.* Japan **92**, 1270–1274.

Barrett, M. D., Hill, D. C., Alexander, J. C. and Zitnak, A. (1977). *Can. J. Physiol. Pharmac.* **55**, 134–136.

Beckett, A. H. and Al-Sarraj, S. (1973). *J. Pharm. Pharmac.* **25**, 169P–170P.

Beckett, A. H. and Wilkinson, G. R. (1965). *J. Pharm. Pharmac.* **17**, 107S–108S.

Beckett, A. H., Jones, G. R. and Al-Sarraj, S. (1974). *J. Pharm. Pharmac.* **26**, 945–951.

Bernheim, F. and Bernheim, M. L. C. (1938). *J. biol. Chem.* **123**, 317–326.

Binkley, F. (1950). *J. biol. Chem.* **186**, 287–296.

Blaschko, H. (1944). *J. Physiol.* Lond. **103**, 13P–14P.

Blaschko, H. and Philpot, F. J. (1953). *J. Physiol.* Lond. **122**, 403–408.

Block, W. (1953). *Z. Naturf.* **8b**, 440–444.

Block, W., Block, K. and Patzig, B. (1952). *Hoppe-Seyler's Z. physiol. Chem.* **290**, 160–168.

Boyland, E. and Mawson, E. H. (1938). *Biochem. J.* **32**, 1204–1206.

Bralet, J., Cohen, Y. and Valette, G. (1968). *Biochem. Pharmac.* **17**, 2319–2331.

Brues, A. M. (1942). *J. clin. Invest.* **21**, 646–647.

Buffoni, F., Della Corte, L. and Knowles, P. F. (1968). *Biochem. J.* **106**, 575–576.

Caldwell, J. (1976). *Drug. Metab. Rev.* **5**, 219–280.

Chapman, D. I. and Marcroft, J. (1977). Unpublished observations cited by M. S. Moss *in* "Drug Metabolism—from Microbe to Man" (D. V. Parke and R. L. Smith, Eds), pp. 263–280. Taylor and Francis, London.

Charalampous, K. D., Orengo, A., Walker, K. E. and Kinross-Wright, J. (1964). *J. Pharmac. exp. Ther.* **145**, 242–246.

Charalampous, K. D., Walker, K. E. and Kinross-Wright, J. (1966). *Psychopharmacologia* **9**, 48–63.

Cheema, P. S., Padmanaban, G. and Sarma, P. S. (1971). *Indian J. Biochem. Biophys.* **8**, 16–19.

Cochin, J., Woods, L. A. and Seevers, M. H. (1951). *J. Pharmac. exp. Ther.* **101**, 205–209.

Coop, I. E. and Blakley, R. L. (1949). *N.Z. Jl. Sci. Technol.* **30A**, 277–291.

Coop, I. E. and Blakley, R. L. (1950). *N.Z. Jl. Sci. Technol.* **31A**, 44–58.
Creveling, C. R., Daly, J. W., Witkop, B. and Udenfriend, S. (1962). *Biochim. biophys. Acta* **64**, 125–134.
Daly, J., Axelrod, J. and Witkop, B. (1962). *Ann. N.Y. Acad. Sci.* **96**, 37–43.
Daly, J., Inscoe, J. K. and Axelrod, J. (1965). *J. med. Chem.* **8**, 153–157.
Dann, R. E., Feller, D. R. and Snell, J. F. (1971). *Eur. J. Pharmac.* **16**, 233–236.
Demisch, L. and Seiler, N. (1975). *Biochem. Pharmac.* **24**, 575–580.
Drasar, B. S. and Hill, M. J. (1974). "Human Intestinal Flora" p. 63. Academic Press, London, New York, San Francisco.
Erspamer, V. (1955). *J. Physiol.* Lond. **127**, 118–133.
Ewins, A. J. and Laidlaw, P. P. (1910). *J. Physiol.* Lond. **41**, 78–87.
Ewins, A. J. and Laidlaw, P. P. (1913). *Biochem. J.* **7**, 18–25.
Feller, D. R. and Malspeis, L. (1977). *Drug. Metab. Disposit.* **5**, 37–46.
Feller, D. R., Basu, P., Mellon, W., Curott, J. and Malspeis, L. (1973). *Archs int. Pharmacodyn. Thér.* **203**, 187–199.
Fish, M. S., Johnson, N. M., Lawrence, E. P. and Horning, E. C. (1955). *Biochim. biophys. Acta* **18**, 564–565.
Fleischmann, W., Price, H. G. and Fleischmann, S. K. (1965). *Med. Pharmacol. exp.* **12**, 172–176.
Fleischmann, W., Price, H. G. and Fleischmann, S. K. (1967). *Med. Pharmacol. exp.* **17**, 323–326.
Fleischmann, W., Price, H. G. and Fleischmann, S. K. (1968). *Pharmacology* **1**, 48–52.
Fleischer, J. H., Arem, A. J., Choapil, M. and Peacock, E. E. (1976). *Proc. Soc. exp. Biol. Med.* **152**, 469–474.
Friedhoff, A. J. and Hollister, L. E. (1966). *Biochem. Pharmac.* **15**, 269–273.
Gaudette, L. E. and Brodie, B. B. (1959). *Biochem. Pharmac.* **2**, 89–96.
Gessner, P. K., Khairallah, P. A., McIsaac, M. W. and Page, I. H. (1960). *J. Pharmac. exp. Ther.* **130**, 126–133.
Goldstein, M. and Contrera, J. F. (1962). *J. biol. Chem.* **237**, 1898–1902.
Goldstein, M., Friedhoff, A. J., Pomeranz, S., Simmons, C. and Contrera, J. F. (1961). *J. Neurochem.* **6**, 253–254.
Govier, W. M., Howes, B. G. and Gibbons, A. J. (1953). *Science* N.Y., **118**, 596–597.
Guggenheim, M. and Löffler, W. (1915). *Biochem. Z.* **72**, 325–350.
Guha, S. R. and Mitra, C. (1971). *Biochem. Pharmac.* **20**, 3539–3542.
Harley-Mason, J., Laird, A. H. and Smythies, J. R. (1958). *Confin. neurol.* **18**, 152–155.
Hegarty, M. P., Schinckel, P. G. and Court, R. D. (1964). *Aust. J. agric. Res.* **15**, 153–167.
Hengstmann, J. H., Konen, W., Konen, C., Eichelbaum, M. and Dengler, H. J. (1974). *Naunyn-Schmiedeberg's Arch. Pharmac.* **283**, 93–106.
Ho, B. T., Pong, S. F., Browne, R. G. and Walker, K. E. (1973). *Experientia* **29**, 275–277.
Hunter, A. L. and Klaassen, C. D. (1975a). *J. Pharmac. exp. Ther.* **192**, 605–617.
Hunter, A. L. and Klaassen, C. D. (1975b). *Drug Metab. Disposit.* **3**, 530–535.
Huszti, Z. and Borsy, J. (1966). *Biochem. Pharmac.* **15**, 475–480.
Jacquot, C., Bralet, J. and Cohen, Y. (1973). *C.r.Séanc. Soc. Biol.* **167**, 1789–1794.
Jacquot, C., Rapin, J. R., Wepierre, J. and Cohen, Y. (1974). *Archs int. Pharmacodyn. Thér.* **207**, 298–309.

Jepson, J. B., Zaltzman, P. and Udenfriend, S. (1962). *Biochim. biophys. Acta* **62**, 91–102.

Johnston, G. A. R. (1974). *In* "Neuropoisons. Their Pathophysiological Actions" (L. L. Simpson and D. R. Curtis, Eds), Vol. 2, pp. 179–205. Plenum Press, New York and London.

Karawya, M. S., El-Keiy, M. A., Wahba, S. K. and Kozman, A. R. (1968). *J. Pharm. Pharmac.* **20**, 650–652.

Kawai, K. and Baba, S. (1974). *Chem. pharm. Bull.* Tokyo **22**, 2372–2376.

Kawai, K. and Baba, S. (1975). *Chem. pharm. Bull.* Tokyo **23**, 289–293.

Kawai, K. and Baba, S. (1976). *Chem. pharm. Bull.* Tokyo **24**, 2728–2732.

Keiser, H. R., Harris, E. D. and Sjoerdsma, A. (1967). *Clin. Pharmac. Ther.* **8**, 587–592.

Kuhn, R., Köhler, F. and Köhler, L. (1937). *Hoppe-Seyler's Z. physiol. Chem.* **247**, 197–220.

Lemberger, L. and Rubin, A. (1976). *In* "Physiologic Disposition of Drugs of Abuse", pp. 31–63, Spectrum Publications, New York.

Lemberger, L., Klutch, A. and Kuntzman, R. (1966). *J. pharmac. exp. Ther.* **153**, 183–190.

Lipton, S. H., Lalich, J. J. and Strong, F. M. (1958a). *J. Am. chem. Soc.* **80**, 2022–2023.

Lipton, S. H., Lalich, J. J., Garbutt, J. T. and Strong, F. M. (1958b). *J. Am. chem. Soc.* **80**, 6594–6596.

Mandel, L. R. and Walker, R. W. (1974). *Life Sci.* **15**, 1457–1463.

Matsuda, A., Baba, S. and Nagase, Y. (1971). *J. pharm. Soc.* Japan **91**, 542–545.

McMahon, R. E. (1964). *Life Sci.* **3**, 235–241.

Mokrasch, L. C. and Stevenson, I. (1959). *J. nerv. ment. Dis.* **129**, 177–183.

Morgan, M. and Mandell, A. J. (1969). *Science* N.Y. **165**, 492–493.

Mosnaim, A. D., Edstrand, D. L., Wolf, M. E. and Silkaitis, R. P. (1977). *Biochem. Pharmac.* **26**, 1725–1728.

Musacchio, J. M. and Goldstein, M. (1967). *Biochem. Pharmac.* **16**, 963–970.

Nagase, Y., Baba, S. and Matsuda, A. (1967). *J. pharm. Soc.* Japan **87**, 123–128.

Neff, N., Rossi, G. V., Chase, G. D. and Rabinowitz, J. L. (1964). *J. Pharmac. exp. Ther.* **144**, 1–7.

Nicholson, J. D. (1970). *Archs int. Pharmacodyn. Thér.* **188**, 375–386.

Nicholson, J. D. (1971). *Archs int. Pharmacodyn. Thér.* **192**, 291–301.

Ohkawa, H., Ohkawa, R., Yamamoto, I. and Casida, J. E. (1972). *Pestic. Biochem. Physiol.* **2**, 95–112.

Pugh, C. E. M. and Quastel, J. H. (1937). *Biochem. J.* **31**, 2306–2321.

Ratcliffe, J. and Smith, P. (1959). *Chemy Ind.* 925.

Riceberg, L. J., Simon, M., Van Vunakis, H. and Abeles, R. H. (1975). *Biochem. Pharmac.* **24**, 119–125.

Richter, D. (1937). *Biochem. J.* **31**, 2022–2028.

Richter, D. (1938). *Biochem. J.* **32**, 1763–1769.

Roth, R. A., Roth, J. A. and Gillis, C. N. (1977). *J. Pharmac. exp. Ther.* **200**, 394–401.

Sanders, E. and Bush, M. T. (1967). *J. Pharmac. exp. Ther.* **158**, 340–352.

Sanders-Bush, E., Oates, J. A. and Bush, M. T. (1976). *Life Sci.* **19**, 1407–1412.

Sarma, P. S. and Padmanaban, G. (1969). *In* "Toxic Constituents of Plant Foodstuffs" (I. E. Liener, Ed.), pp. 267–291. Academic Press, New York and London.

Schönharting, M., Pfaender, P., Rieker, A. and Siebert, G. (1973). *Hoppe-Seyler's Z. physiol. Chem.* **354**, 421–436.

Schönharting, M., Mende, G. and Siebert, G. (1974). *Hoppe-Seyler's Z. physiol. Chem.* **355**, 1391–1399.
Seiler, N. (1965). *Hoppe-Seyler's Z. physiol. Chem.* **341**, 105–110.
Seiler, N. and Demisch, L. (1971). *Biochem. Pharmac.* **20**, 2485–2493.
Seiler, N. and Demisch, L. (1974a). *Biochem. Pharmac.* **23**, 259–271.
Seiler, N. and Demisch, L. (1974b). *Biochem. Pharmac.* **23**, 273–287.
Sever, P. S., Dring, L. G. and Williams, R. T. (1975). *Eur. J. clin. Pharmac.* **9**, 193–198.
Shah, N. S. and Himwich, H. E. (1971). *Neuropharmacology* **10**, 547–556.
Sievert, H. W., Lipton, S. H. and Strong, F. M. (1960). *Archs Biochem. Biophys.* **86**, 311–316.
Sinsheimer, J. E., Dring, L. G. and Williams, R. T. (1973). *Biochem. J.* **136**, 763–771.
Slotta, K. H. and Müller, J. (1936). *Hoppe-Seyler's Z. physiol. Chem.* **238**, 14–22.
Smith, R. L. (1971). *In* "A Symposium on Mechanisms of Toxicity" (W. N. Aldridge, Ed.), pp. 229–247. Macmillan, London.
Spector, E. (1961). *Nature* Lond. **189**, 751–752.
Steensholt, G. (1947). *Acta physiol. scand.* **14**, 356–362.
Szára, S. (1956). *Experientia* **12**, 441–442.
Szára, S. (1961). *Fedn Proc. Fedn. Am. Socs exp. Biol.* **20**, 885–888.
Szára, S. (1968). *In* "Advances in Pharmacology" (S. Garattini and P. A. Shore, Eds), Vol. 6, Part B, pp. 230–231. Academic Press, New York and London.
Szára, S. and Axelrod, J. (1959). *Experienta* **15**, 216–217.
Taborsky, R. G. and McIsaac, W. M. (1964). *Biochem. Pharmac.* **13**, 531–534.
Tacker, M., McIsaac, W. M. and Creaven, P. J. (1970). *Biochem. Pharmac.* **19**, 2763–2773.
Tacker, M., McIsaac, W. M. and Creaven, P. J. (1972a). *J. Pharm. Pharmac.* **24**, 245–246.
Tacker, M., Creaven, P. J. and McIsaac, W. M. (1972b). *J. Pharm. Pharmac.* **24**, 247–248.
Taska, R. J. and Schoolar, J. C. (1972). *J. Pharmac. exp. Ther.* **183**, 427–432.
Taska, R. J. and Schoolar, J. C. (1973). *Archs int. Pharmacodyn. Thér.* **202**, 66–78.
Taylor, C. and Knowles, P. F. (1971). *Biochem. J.* **122**, 29P–30P.
Thiercelin, J. F., Jacquot, C., Rapin, J. R. and Cohen, Y. (1976). *Archs int. Pharmacodyn. Thér.* **220**, 153–163.
Von Holt, C. (1966). *Biochim. biophys. Acta.* **125**, 1–10.
Von Holt, C., Chang, J., Von Holt, M. and Böhm, H. (1964). *Biochim. Biophys. Acta* **90**, 611–613.
Waddell, W. J., Wilk, A. L., Pratt, R. M. and Steffek, A. J. (1974). *Teratology* **9**, 211–216.
Walaszek, E. J., Kocsis, J. J., Leroy, G. V. and Geiling, E. M. K. (1960). *Archs int. Pharmacodyn. Thér.* **125**, 371–382.
Wallace, S. L. and Ertel, N. H. (1969). *Bull. rheum. Dis.* **20**, 582–587.
Wallace, S. L., Omokoku, B. and Ertel, N. H. (1970). *Am. J. Med.* **48**, 443–448.
Wilk, A. L., King, C. T. G., Horigan, E. A. and Steffek, A. J. (1972). *Teratology* **5**, 41–48.
Wilkinson, G. R. and Beckett, A. H. (1968). *J. Pharmac. exp. Ther.* **162**, 139–147.
Williams, R. T. (1959). "Detoxication Mechanisms". Chapman and Hall, London.
Williams, R. T., Caldwell, J. and Dring, L. G. (1974). *Biochem. Pharmac.* Suppl. 2, 765–770.

Winkler, W. O. (1958). *J. Ass. off. agric. Chem.* **41**, 282–287.
Wiseman-Distler, M. H., Sourkes, T. L. and Carabin, S. (1965). *Clinica chim. Acta* **12**, 335–339.
Wu, P. H. and Boulton, A. A. (1975). *Can. J. Biochem.* **53**, 42–50.
Zeller, E. A., Barsky, J., Berman, E. R., Cherkas, M. S. and Fouts, J. R. (1958). *J. Pharmac. exp. Ther.* **124**, 282–289.
Ziegler, D. M., Mitchell, C. H. and Jollow, D. (1969). *In* "Microsomes and Drug Oxidations" (J. R. Gillette, A. H. Conney, G. J. Cosmides, R. W. Estabrook, J. R. Fouts and G. J. Mannering, Eds), pp. 173–187. Academic Press, New York and London.

9

METABOLISM OF NITROGEN HETEROCYCLIC COMPOUNDS

I. Pyrrolidine Derivatives

Alkaloids based on pyrrolidine are usually subdivided into simple derivatives of pyrrolidine (tetrahydropyrrole) (1) and more complex types containing the pyrrolidine or pyrrole ring systems. Of the first type, only a few compounds are known (e.g. hygrine, stachydrine and cuscohygrine) but no reports on their metabolic fate are available. Nicotine and several related tobacco alkaloids contain the pyrrolidine ring but this group is included among the pyridine derivatives (Section II.B). However, it should be noted that among those tobacco alkaloids which contain a pyrrolidine ring, it is usually this moiety which is the site of greatest metabolic activity. The more complex types of pyrrolidine derivatives include the tropane, indole and pyrrolizidine alkaloids which are treated separately below in Sections III, IV and VII, respectively.

(1)

II. Piperidine, Pyridine and Pyrazine Derivatives

A. Piperidine Derivatives

Pipecolic acid (2) is of metabolic interest because it is an intermediate in the pathway from lysine to α-aminoadipic acid (3) (Rothstein and Miller, 1954; Rothstein and Greenberg, 1959) and therefore a normal constituent found in blood and urine. In addition, pipecolic acid is present in many plants, often in the fruits and seeds. Intraperitoneal administration of the DL-compound to rats resulted in its excretion unchanged in the urine, no metabolites being detected (Boulanger and Osteux, 1960). However, small amounts of unidentified compounds were formed when it was incubated with rat liver homogenates. Other *in vitro* experiments employing guinea pig brain and kidney homogenates have indicated that pipecolic acid can undergo decarboxylation to piperidine (Kasé *et al.*, 1967).

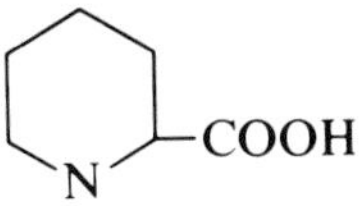

(2)

Pipecolic acid

(3)

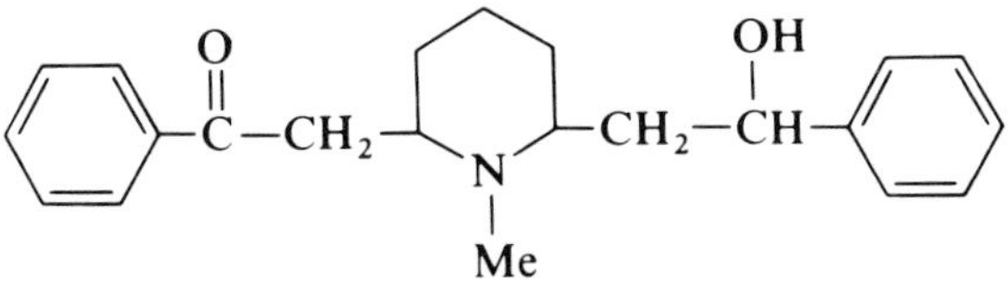

(4)

Lobeline

Uehleke (1963) reported that **lobeline** (4) disappeared rapidly from the blood of rats following i.v. administration and that some unchanged compound was excreted in the urine. In addition, two unidentified metabolites including one having phenolic properties were excreted.

The metabolism of **arecoline** (5) and **arecaidine** (6), two of the betel alkaloids, was investigated in rats by Boyland and Nery (1969) and Nery (1971). These findings are summarized in Fig. 9.1 which shows that arecoline is metabolized along three pathways: 1. ester hydrolysis to arecaidine (6); 2. N-oxide formation with or without ester hydrolysis giving metabolites (7) and (8); and 3. conjugation of the α,β-unsaturated moiety with glutathione followed by subsequent conversion to the mercapturic acid (9). Conjugation with glutathione occurs spontaneously. Arecoline was shown to be rapidly hydrolysed to arecaidine by rat liver homogenates (Nieschulz and Schmersahl, 1968). Rats given ^{3}H-labelled arecoline orally excrete little radioactivity in the faeces and about 20% of the dose in the urine within 18 h (Nery, 1971). This material is mainly unchanged compound (5–6%) and metabolite (8) (about 4%). The remaining metabolites and also an unidentified metabolite each accounted for 1–2% of the dose. However, it was felt that the actual extent of production of N-oxides was greater due to the ease of the *in vivo* reduction of these compounds to the tertiary amines.

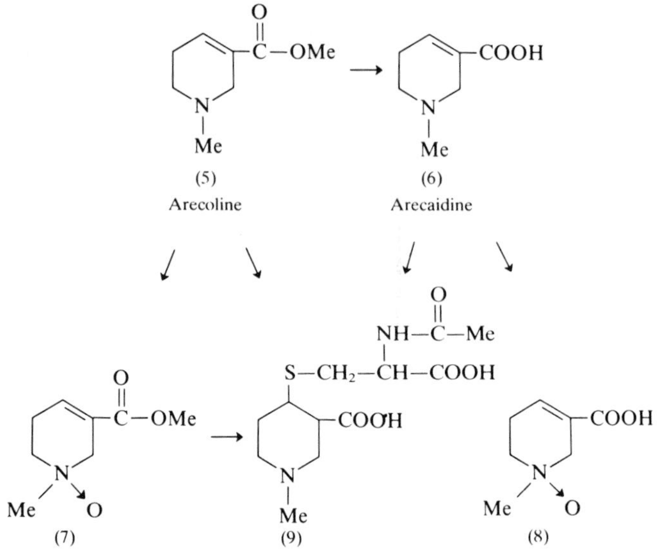

O
‖
C—OMe   ⟶   COOH

N   N
| |
Me   Me
(5)   (6)
Arecoline   Arecaidine

O
‖
NH—C—Me
|
S—CH₂—CH—COOH

O
‖
C—OMe   ⟶   COOH   COOH

N   N   N
Me   O   Me   Me   O
(7)   (9)   (8)

FIG. 9.1. Metabolic pathways of arecoline and arecaidine.

The metabolism of some tobacco alkaloids containing the piperidine moiety (ababasine, methylanabasine) is discussed in the following section.

B. Pyridine Derivatives

The pyridine derivatives covered in this section are largely the tobacco alkaloids. These compounds typically contain a pyridine ring attached at the β-position to the α-position of a pyrrolidine or piperidine ring. An example of the former type is seen with (−)-**nicotine** (10) (see Fig. 9.2), the chief alkaloid in tobacco and also a compound found widely distributed in the plant world. Great interest has been attached to the elucidation of the metabolic fate of nicotine and this has resulted in an extensive literature extending roughly over the past three decades. Detailed coverage of this wealth of data is obviously beyond the scope of the present summary which may be advantageously supplemented by several chapters in the book edited by von Euler (1965), the review by Gorrod and Jenner (1975) or the appropriate sections of the exhaustive monographs written by Larson *et al.* (1961) and Larson and Silvette (1968, 1971, 1975).

It is well established that only a minor part of the elimination of nicotine from the body can be ascribed to the urinary excretion of unchanged compound (see Larson and Haag, 1942; Finnegan *et al.*, 1947). However, the extent of this excretion is dependent upon the pH value of the urine

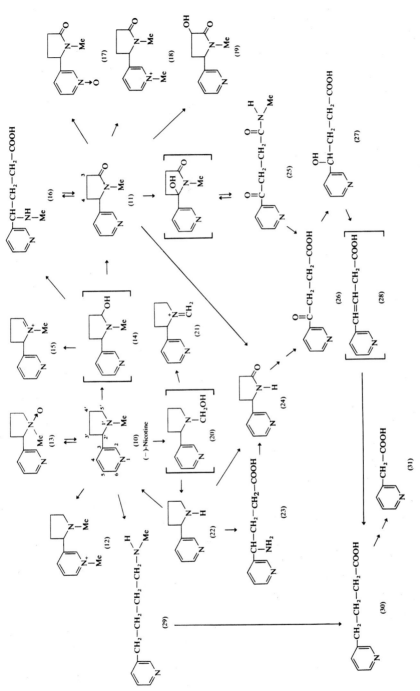

FIG. 9.2. Metabolic pathways of (−)-nicotine. See Table 9.1 for chemical names of metabolites. Proposed intermediates in brackets.

(Haag and Larson, 1942; Beckett *et al.*, 1971). In the former study nicotine excretion was about four times as great when the urine was acidic as when it was alkaline. The latter investigation indicated that when the urinary pH was raised to a range of $7 \cdot 5$–$8 \cdot 5$, the excretion of nicotine fell to zero. Another variable affecting the excretion of unchanged nicotine is the dose. Finnegan *et al.* (1947) found a linear increase with increasing dosage. Nonetheless, a range of 5–10% of the dose may be considered as reasonably normal for the urinary excretion of nicotine. Adir *et al.* (1976) carried out a pharmacokinetic study of the nicotine plasma levels following i.v. injection of the alkaloid in rats.

The pathways in the mammalian metabolism of $(-)$-nicotine (10) are shown in Fig. 9.2. The compounds illustrated are metabolites identified in a large number of *in vivo* and *in vitro* studies in which not only nicotine, but also cotinine (11) and several other metabolites were employed. The experiments upon which Fig. 9.2 is based are summarized in Table 9.1. Many of these investigations have led to the suggestion of hypothetical intermediates in the formation of the metabolites shown in Fig. 9.2, however some of these have been omitted in the figure.

A central pathway in the metabolism of nicotine is its oxidation to cotinine (11), a reaction which has been repeatedly demonstrated in many studies with a large number of animal species (Table 9.1). The urinary excretion of this metabolite was reported to be about 10% of the dose in rabbits (Hucker *et al.*, 1959) and, in man, 10% (Bowman *et al.*, 1959) or 5–20% (Beckett *et al.*, 1971) of the dose. Adir *et al.* (1976) studied the pharmacokinetics of its plasma concentrations in rats given nicotine by i.v. administration. According to Fig. 9.2, cotinine formation is a result of C-oxidation at the 5'-position, possibly due to lactamization of γ-(3-pyridyl)-γ-methylaminobutyric acid (16) (Bowman *et al.*, 1959). Hucker *et al.* (1960) proposed that nicotine is initially oxidized at the α-position of the pyrrolidine ring to form 5'-hydroxynicotine. The existence of this pathway is supported by the findings which indicate that an iminium ion (15) is formed from nicotine (Murphy, 1973). This metabolite is thought to arise from the loss of water from the 5'-hydroxynicotine (14) initially formed by the mono-oxygenase system. Dehydrogenation of the 5'-hydroxy metabolite to cotinine may be affected by an aldehyde oxidase (Hucker *et al.*, 1960).

The N-demethylation of nicotine to nornicotine (22) is well documented (Table 9.1) and further evidence for this reaction has been obtained in studies using [*methyl*-^{14}C]-nicotine. Radioactive CO_2 was produced *in vitro* with mouse tissue slices (Hansson *et al.*, 1964) and *in vivo* in mice and rats (Hansson and Schmiterlöw, 1962; McKennis *et al.*, 1962b). However, in none of these particular investigations was nornicotine specifically detected and $^{14}CO_2$ formation is also possible at other stages in the metabolism of nicotine, e.g. the demethylation of cotinine (11) to demethylcotinine (24) or in the further metabolism of methylamine formed when the amide (25) is hydrolysed to the acid (26). The formation of methylamine from nicotine was reported by Werle and Meyer (1950) using liver slices and McKennis et al. (1962b) noted that the methylamine produced would be rapidly metabolized and eliminated as respiratory CO_2, rather than appearing as a urinary metabolite. The absence of methylamine in the urine of dogs given nicotine has been reported by Owen and Larson (1958). Nguyen et al. (1976) recently obtained evidence for the demethylation of nicotine via the intermediate carbinol (20). They showed that nicotine is metabolized to an iminium ion (21), a reaction involving the hydroxylated intermediate (20) in an analogous manner to that seen in the conversion of nicotine to cotinine (11) in which the 5'-hydroxynicotine intermediate (14) is converted to the isomeric iminium ion (15).

As seen in Fig. 9.2, nicotine undergoes both C- and N-oxidations, the former leading to cotinine and further products while the latter route results in N-oxides. Gorrod et al. (1971) obtained results from various enzyme inhibition studies in vitro which indicated that these two oxidative pathways are mediated by different processes. Formation of nicotine-1'-N-oxide (13), which accounts for 3–4% of the dose of nicotine in man (Beckett et al., 1971), gives both the cis and trans diastereoisomers (Booth and Boyland, 1970, 1971; Beckett et al., 1971; Jenner and Gorrod, 1973). Their formation is stereospecific, the ratios of isomers produced varying in different tissues and animal species. Marked species differences in the in vitro metabolism of both (−)-nicotine and (+)-nicotine to their diastereoisomeric N-oxides have been encountered (Jenner et al., 1973c). The stereochemistry of the diastereoisomers of nicotine-1'-oxide formed from the (−)- and (+)-isomers of nicotine was investigated by Beckett et al. (1973) and a reappraisal of some of the data was given by Testa et al. (1976). A further point of metabolic interest with nicotine-1'-N-oxide is that it can be reduced to nicotine. The reducing conditions essential for this reaction are found in the lower intestine (Beckett et al., 1970; Jenner et al., 1973a) and in vitro studies using rat intestinal contents have demonstrated considerable reduction of the N-oxide to nicotine (Dajani et al., 1975b). The last investigation, which also utilized germ-free rats, pointed also to the involvement of the liver and/or other tissues in the reduction of nicotine-1'-N-oxide. Other reports have shown that many tissues, and especially the liver, can reduce the N-oxide to nicotine under anaerobic conditions (Booth and Boyland, 1971; Dajani et al., 1972, 1975a). The metabolism of cotinine-N-oxide (17) in rabbits and dogs was studied by Yi

TABLE 9.1
Metabolites of (−)-nicotine

Precursor[a,b]	Metabolite[c]	Structure[a]	Species and conditions	References
Cotinine		(11)	Mouse, liver *in vitro*	Stålhandske (1970) Jenner and Gorrod (1973)
			Mouse, liver kidney and lung *in vitro*	Hansson *et al.* (1964)
			Mouse, brain *in vivo*	Appelgren *et al.* (1962)
			Mouse, urine	Hansson and Schmitterlöw (1962) Stålhandske (1970)
			Rat, liver *in vitro*	Jenner and Gorrod (1973)
			Rat, liver *in vivo*	De Clercq and Truhaut (1962)
			Rat, urine	Truhaut and De Clercq (1959) Harke *et al.* (1970)
			Hamster, liver *in vitro*	Harke *et al.* (1970) Jenner and Gorrod (1973)
			Hamster, urine	Harke *et al.* (1970)
			Guinea pig, liver *in vitro*	Jenner and Gorrod (1973)
			Guinea pig, liver kidney and lung *in vitro*	Booth and Boyland (1971)
			Rabbit, liver *in vitro*	Hucker *et al.* (1959, 1960) Papadopoulos and Kintzios (1963) Decker and Sammeck (1964) Jenner and Gorrod (1973)
			Rabbit, liver and kidney *in vivo*	Papadopoulos (1964a)
			Rabbit, urine	Hucker *et al.* (1959) Truhaut and De Clercq (1959) Papadopoulos (1964a) Decker and Sammeck (1964)
			Cat, blood and liver *in vivo*	Turner (1969, 1971)

No.	Compound	Tissue/preparation	Reference
(11)		Cat, brain *in vivo*	Appelgren *et al.* (1962)
		Cat, urine	Turner (1969, 1971)
		Dog, perfused lung	Turner *et al.* (1975)
		Dog, urine	McKennis *et al.* (1957, 1958, 1961)
		Pig, urine	Harke *et al.* (1974a)
			Harke and Frahm (1976)
		Rhesus monkey, hepatocytes	Poole and Urwin (1976)
		Man, blood *in vivo*	Armitage *et al.* (1974)
		Man, urine	Bowman *et al.* (1959)
			Beckett *et al.* (1971)
(12)	Isomethylnicotinium ion	Dog, urine	McKennis *et al.* (1963)
(18)	Cotinine methonium ion	Dog, urine	McKennis *et al.* (1963)
		Man, urine	McKennis *et al.* (1963)
(15)	Nicotine $\Delta^{1'(5')}$iminium ion	Rabbit, liver *in vitro*	Murphy (1973)
(22)	Nornicotine	Rat, urine	Harke *et al.* (1970)
		Hamster, urine	Harke *et al.* (1970)
		Rabbit, liver *in vitro*	Papadopoulos and Kintzios (1963)
		Rabbit, liver and kidney *in vivo*	Decker and Sammeck (1964)
		Rabbit, urine	Papadopoulos (1964a)
			Papadopoulos (1964a)
			Decker and Sammeck (1964)
		Cat, liver *in vivo*	Turner (1969, 1971)
		Cat, urine	Turner (1969, 1971)
		Pig, urine	Harke *et al.* (1974a,b)
			Harke and Frahm (1976)
(21)	Nicotine methyleniminium ion	Rabbit, liver *in vitro*	Nguyen *et al.* (1976)
(13)	Nicotine, 1'-*N*-oxide	Mouse, liver *in vitro*	Booth and Boyland (1970)
			Jenner and Gorrod (1973)
		Rat, liver *in vitro*	Jenner and Gorrod (1973)
		Hamster, liver *in vitro*	Booth and Boyland (1970)
			Jenner and Gorrod (1973)
		Guinea pig, liver *in vitro*	Jenner *et al.* (1973b)

TABLE 9.1—continued

Precursor[a,b]	Metabolite[c]	Structure[a]	Species and conditions	References
			Guinea pig, liver, kidney and lung *in vitro*	Booth and Boyland (1970, 1971)
			Rabbit, liver *in vitro*	Papadopoulos (1964b)
				Booth and Boyland (1970)
				Jenner and Gorrod (1973)
			Rabbit, liver and kidney *in vivo*	Papadopoulos (1964a)
			Cat, liver *in vivo*	Turner (1969, 1971)
			Cat, urine	Turner (1969, 1971)
			Dog, perfused lung	Turner et al. (1975)
			Pig, urine	Harke et al. (1974a)
				Harke and Frahm (1976)
			Rhesus monkey, hepatocytes	Poole and Urwin (1976)
			Man, urine	Booth and Boyland (1970)
				Beckett et al. (1971)
(11)	Cotinine-*N*-oxide	(17)	Rhesus monkey, urine	Dagne and Castagnoli (1972b)
	γ-(3-Pyridyl)-γ-methylamino butyric acid	(16)	Dog, urine	McKennis et al. (1957, 1958, 1961)
				Owen and Larson (1958)
	γ-(3-Pyridyl)-γ-amino-butyric acid	(23)	Dog, urine	Wada et al. (1961)
(11)	Demethylcotinine	(24)	Mouse, urine	Bowman et al. (1964)
(11)			Rat, urine	McKennis et al. (1962b)
(11)				Morselli et al. (1967)
(11)			Hamster, urine	Harke et al. (1970)
			Rabbit, liver *in vitro*	Harke et al. (1970)
				Papadopoulos and Kintzios (1963)
			Rabbit, liver and kidney *in vivo*	Decker and Sammeck (1964)
				Papadopoulos (1964a)

No.	Compound	Source	Reference
(11)		Rabbit, urine	Papadopoulos (1964a)
(22)		Cat, liver *in vivo*	Turner (1969, 1971)
		Cat, urine	Turner (1969, 1971)
		Dog, perfused lung	Turner *et al.* (1975)
		Dog, urine	McKennis *et al.*(1959)
			McKennis *et al.* (1961)
(11)		Pig, urine	Wada *et al.* (1961)
			Harke *et al.* (1974a,b)
			Harke and Frahm (1976)
			Harke *et al.* (1974b)
		Man, urine	Bowman *et al.* (1959)
(25)	γ-(3-Pyridyl)-γ-oxo-N-methylbutyramide	Mouse, liver *in vitro*	Stålhandske (1970)
		Mouse, liver, kidney and lung *in vitro*	Hansson *et al.* (1964)
(11)		Mouse, urine	Stålhandske (1970)
(11)		Rat, urine	McKennis *et al.* (1962b)
			Morselli *et al.* (1967)
		Dog, urine	McKennis *et al.* (1960, 1961)
(19)[d]	Hydroxycotinine[d]	Mouse, liver *in vitro*	McKennis *et al.* (1962a,b)
		Mouse, liver, kidney and lung *in vitro*	Stålhandske (1970)
(11)			Hansson *et al.* (1964)
		Mouse, liver *in vitro*	Stålhandske (1970)
		Mouse, urine	Hansson and Schmitterlöw (1962)
			Stålhandske (1970)
(11)		Rat, urine	Bowman *et al.* (1964)
(11)		Dog, urine	McKennis *et al.* (1962b)
			McKennis *et al.* (1961)
(11)		Man, urine	McKennis *et al.* (1959)
(11)			Bowman *et al.* (1959)
(19)	3-Hydroxycotinine	Rhesus monkey, urine	Bowman and McKennis (1962)
(11)			Dagne and Castagnoli (1972a)
(11)			Dagne *et al.* (1974)

TABLE 9.1—continued

Precursor[a,b]	Metabolite[c]	Structure[a]	Species and conditions	References
(11), (25)	γ-(3-Pyridyl)-γ-oxo-butyric acid	(26)	Rat, urine	Morselli et al. (1967)
(11), (25)				Schwartz and McKennis (1963)
(24)				Schwartz and McKennis (1964)
(24)				McKennis et al. (1964)
(11), (25)			Dog, urine	Schwartz and McKennis (1963)
(24)				McKennis et al. (1964)
(24)	γ-(3-Pyridyl)-γ-hydroxy-butyric acid	(27)	Rat, urine	Schwartz and McKennis (1964)
(24)			Dog, urine	McKennis et al. (1964)
(24)	Dihydrometanicotine	(29)	Rat, liver in vivo	McKennis et al. (1964)
				De Clercq and Truhaut (1962)
(29)	γ-(3-Pyridyl)butyric acid	(30)	Rat, urine	Meacham et al. (1972)
			Dog, urine	Meacham et al. (1972)
(11)	3-Pyridylacetic acid	(31)	Mouse, urine	Schwartz and McKennis (1964)
(29)				Meacham et al. (1972)
			Cat, liver in vivo	Turner (1969, 1971)
			Cat, urine	Turner (1969, 1971)
			Dog, urine	McKennis et al. (1961)
(11)				McKennis et al. (1961)
(29)			Man, urine	Meacham et al. (1972)
(11)				McKennis et al. (1964)

[a] Numbers refer to structural formulas in Fig. 9.2.
[b] Compound employed or administered is nicotine unless otherwise indicated.
[c] Some excretion of unchanged compound occurs. See text.
[d] Position of hydroxyl group uncertain but usually assumed to be at C3.

et al. (1977) who found that it was excreted in the urine partly unchanged but also as cotinine and several further metabolites of cotinine.

While the metabolic pathways illustrated in Fig. 9.2 reveal that nicotine is subject to extensive metabolism along numerous routes, it seems reasonable to expect that further unidentified metabolites exist. Some of these will no doubt clarify known metabolic pathways while others may reveal hitherto unknown routes. Recent examples of the latter possibility are seen in the reports of Shen and Van Vunakis (1974a,b) showing that nicotine under *in vitro* conditions using DPNases or liver microsomal fractions can replace nicotinamide in NAD and NADP. Also, further investigation of metabolites which have been reported in earlier investigations but whose identification is generally deemed to be tentative may also prove fruitful. Examples of these include β-nicotyrine (32) (Werle *et al.*, 1950; Werle and Meyer, 1950), the dehydrogenation product of nicotine, and also a phenylalanine conjugate of a nicotine metabolite reported by Truhaut and de Clercq (1959). Further discussion of these and other possible metabolites of nicotine is found in the article by McKennis (1965).

(32)

β-Nicotyrine

(33)

Metanicotine

In addition to nicotine, tobacco contains small quantities of structurally similar alkaloids. Those for which some metabolic information is available include (−)-nornicotine (22), dihydrometanicotine (29), metanicotine (33), (−)-methylanabasine (34), (−)-anabasine (35) and β-nicotyrine (32). Hucker and Larson (1958) administered **nornicotine** (22) to dogs and found that the extent of its excretion unchanged in the urine was several times greater than that seen with nicotine. As shown in Fig. 9.2, nornicotine can undergo methylation to nicotine. Axelrod (1962b) described an enzyme system capable of transferring the methyl group from *S*-adenosylmethionine to nornicotine. In studies using rabbits, *N*-methylating activity was found in several tissues, with the highest activity occurring in lung tissue.

Dihydrometanicotine (29) is another minor tobacco alkaloid which is implicated in the metabolism of nicotine (see Fig. 9.2). Meacham *et al.* (1972) found that its administration to rats and dogs resulted in the urinary excretion of the pyridyl acids (30) and (31). **Metanicotine** (33), the dehydro derivative of compound (29), undergoes similar metabolism. When the *trans* isomer (15 mg/kg) was administered orally to dogs, the urine was found to contain trans-4-(3-pyridyl)-3-butenoic acid (28) and 3-pyridyl-acetic acid (31) (Fig. 9.2) (Meacham *et al.*, 1973).

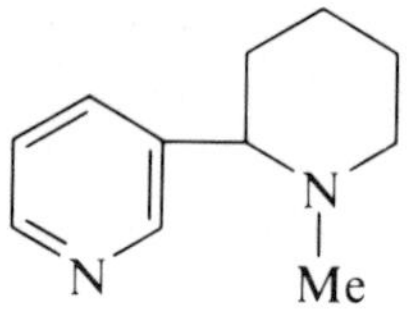

(34)
(−)-Methylanabasine

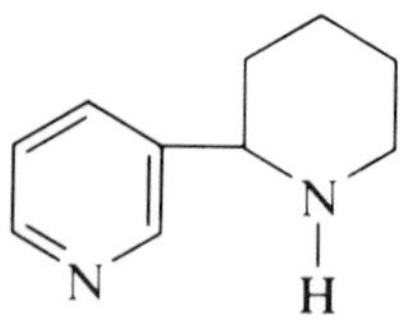

(35)
(−)-Anabasine

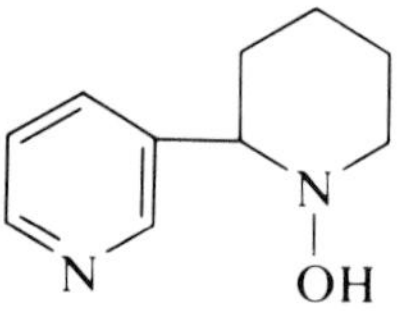

(36)

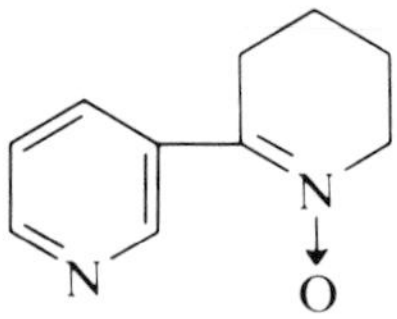

(37)

Methylanabasine (34) was reported by Gvishiani (1960) to be metabolized by the liver following its i.v. administration to cats, however the metabolic products were not investigated. Beckett and Sheikh (1973) studied the metabolism of both (−)-**anabasine** (35) and (−)-methyl-anabasine in liver and lung homogenates from rats, rabbits and guinea pigs. The main metabolites of (−)-methylanabasine in both tissue preparations were its two diastereoisomeric *N*-oxides. Also formed, especially with the liver preparations, were some *N*-demethylated compound, anabasine, and its *N*-oxidation products 1′-hydroxyanabasine (36) and anabasine-1′Δ-nitrone (37). Metabolites (36) and (37) were also formed when (−)-anabasine was used as the substrate in the incubations. Similar experiments with the *N*-hydroxy compound (36) showed that it was converted to the nitrone (37) whereas the latter compound was not metabolized. The metabolism of (−)-methylanabasine (34) by liver preparations from mice, rats, hamster, guinea pigs and rabbits was also studied by Jenner and Gorrod (1973). The formation of *N*-oxides was small in all species except the guinea pig. Furthermore, no evidence was obtained for the *N*-demethylation of methylanabasine to anabasine or for the α-*C* oxidation of the substrate to a cotinine-like metabolite. The former finding thus differs from that reported by Beckett and Sheikh (1973).

The *in vitro* metabolism of **β-nicotyrine** (32) was studied by Jenner and Gorrod (1973) in experiments similar to those noted above with (−)-methylanabasine. Extensive, and in some cases complete, metabolism of this compound occurred in all species. However, the nature of the metabolic products was not ascertained.

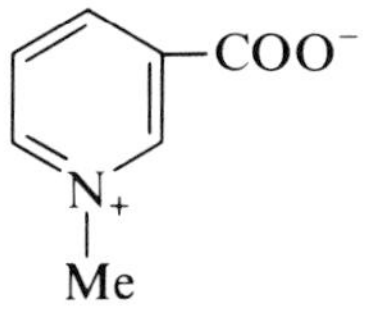

(38)

Trigonelline

Trigonelline (38), of fairly widespread occurrence in plants, is also a urinary metabolite of nicotinic acid. An early report by Kohlrausch (1912) indicated that trigonelline is excreted unchanged in the urine of rabbits and cats given the compound by injection. This was confirmed in rabbits by Chattopadhyay et al. (1953). The latter investigation also showed that trigonelline was not excreted in the faeces and that it was not demethylated to nicotinic acid. However, as only about half of the injected trigonelline was recovered in the urine, it was assumed that some of it was metabolized.

C. PYRAZINE DERIVATIVES

The interest in pyrazine compounds has expanded greatly during the past decade as their flavour importance in many foods has become apparent. As indicated in the review by Maga and Sizer (1973), these compounds are mainly found in roasted or cooked foods although several derivatives have been detected in fresh vegetables. The high odour potency of these compounds is especially noteworthy. The only report on the metabolic fate of this group of compounds is that of Hawksworth and Scheline (1975) who investigated the metabolism of four alkyl- and two alkoxy-derivatives of pyrazine (39) in rats. When **2-methyl-, 2,3-dimethyl, 2,5-dimethyl-** or **2,6-dimethylpyrazine** were administered orally at a dose level of 100 mg/kg it was found that they were oxidized to monocarboxylic acids which were then excreted in the urine as such or sometimes partly as their glycine conjugates. This oxidative pathway accounted for nearly all of the dose with all of these pyrazines except the 2,3-dimethyl derivative. In the latter case 10–15% of the dose was oxidized to 2-methylpyrazine-3-carboxylic acid. Ring hydroxylation, a reaction which was not detected with the three other alkyl pyrazines, resulted in nearly 40% of the dose being excreted as conjugates of 2,3-dimethyl-5-hydroxypyrazine.

The two methoxypyrazines investigated were 2-methoxypyrazine and **2-isobutyl-3-methoxypyrazine** (40), the latter compound being the major

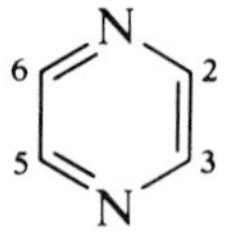

(39) 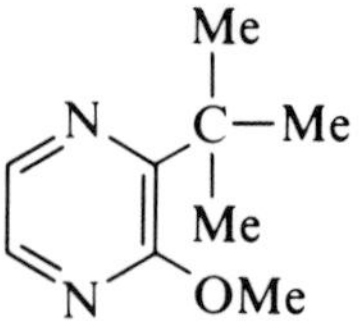   (40)

2-Isobutyl-3-methoxypyrazine

characteristic aroma component of bell peppers. While less precise quantitative information was obtained with these compounds, O-demethylation
was noted in both cases. With the 2-methoxy compound, about 20% of the
identified urinary metabolites consisted of 2-hydroxypyrazine while the
remainder was a single monohydroxy derivative of 2-methoxypyrazine.
Both metabolites were excreted as glucuronide and/or sulphate conjugates. With the bell pepper compound, the demethylated product was a
major urinary metabolite. Aliphatic side-chain oxidation to 2-methoxy-3-
(2-carboxypropyl)pyrazine was also noted, but this reaction was not
extensive. In this regard the compound resembled 2,3- rather than 2,5-
or 2,6-dimethylpyrazine. Three unidentified urinary metabolites were
detected but these accounted for only a minor part of the dose. These
appeared to be compounds in which the isobutyl group had undergone
modification but in which the methoxyl group was intact and ring hydroxylation was lacking.

III. Tropane Derivatives

Plant compounds based on tropane (41) are not numerous but are usually
divided into two main groups, the solanaceous alkaloids and the coca
alkaloids. The most common compounds in the solanaceous group are
$(-)$-**hyoscyamine** (42) and its 6,7-epoxide derivative, $(-)$-scopolamine
(hyoscine). These compounds are esters of the amino alcohols tropine (43)
and scopine (44), respectively, and tropic acid (45). Racemization of the
esters occurs fairly readily and dl-hyoscyamine, known as **atropine**, is
therefore generally used rather than the $(-)$-isomer.

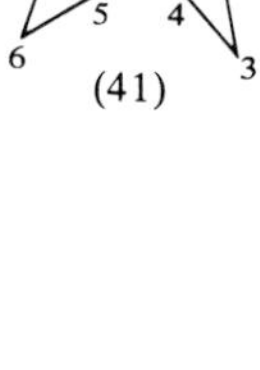

(41)

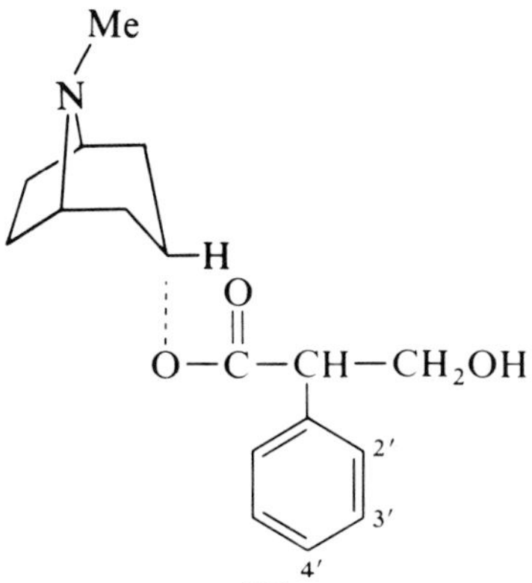

(42)

(−)-Hyoscyamine

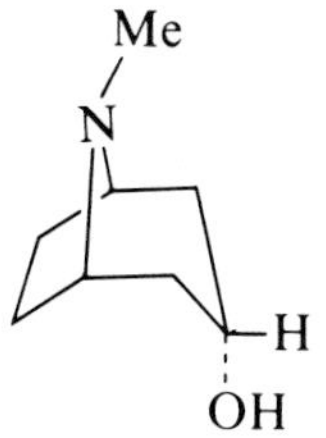

(43) 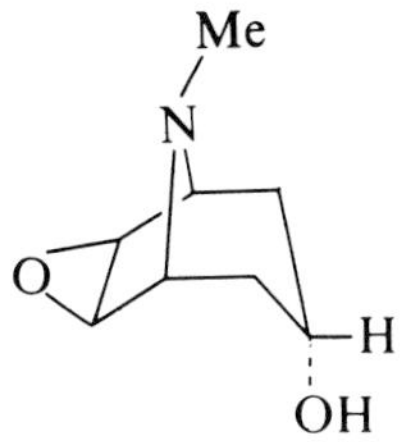   (44) 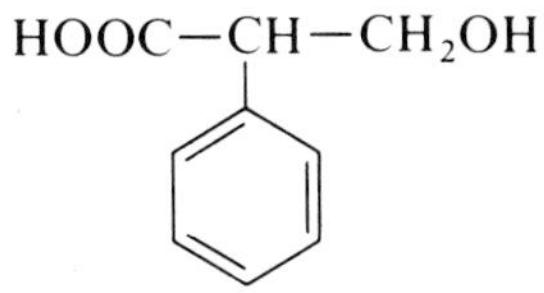   (45)

Information on the metabolic fate of atropine derives from a fairly large number of studies carried out over a lengthy period of time. Several general points stand out from this body of information. Firstly, a relatively high proportion of the administered compound is excreted in the urine in unchanged form. Secondly, the patterns of metabolism vary considerably among different animal species and it is therefore not possible to summarize the metabolism of atropine in a scheme which has general applicability. Thirdly, knowledge of the precise nature of the metabolites of atropine is often scanty.

The appreciable urinary excretion of unchanged atropine was registered at an early date as Wiechowski (1901) found that its i.p. administration to dogs (approx. 30 mg/kg) resulted in the excretion of 33% of the dose unchanged. Subsequent investigations in the dog with a much lower dose (0·5 mg/kg) given by s.c. injection indicated that the urinary excretion of unchanged compound was about 23% within 2 h (Albanus et al., 1968b) or 20–35% within 4 h (Winbladh, 1973). Appreciable urinary excretion of unchanged atropine is commonly seen in other animal species also. In the mouse, the reported values for the amount of atropine excreted unchanged in the urine are about 25% in 48 h (Gosselin *et al.*, 1955; Gabourel and Gosselin, 1958) and about 18% in 2 h (Albanus *et al.*, 1968a). Similar values (23–28%) have been obtained in rabbits (Tønnesen, 1950). The latter investigation showed that atropine excretion in rats varied from 10% (s.c.) to 33% (i.v.) and also that the values were reduced to roughly 0·5–1% of the dose in rats and rabbits following oral administration of the same amount of atropine. Kalser *et al.* (1957) reported that the urinary excretion of what appeared to be unchanged atropine accounted for 39% of the dose (i.v.) in rats and 25% in guinea pigs. They also found a considerable but undetermined amount of unchanged atropine in the urine of cats. The situation in man does not appear to be appreciably different from that observed in laboratory animals. Tønnesen (1950) found 20% of a s.c. dose excreted unchanged. However, the data of Gosselin *et al.* (1960) indicate that about half of the atropine given by i.m. injection appeared unchanged in the urine. Recognition must be given to the influence of urinary pH on the excretion of atropine and Albanus *et al.* (1968b) showed that alkalinization of the urine to pH 8 completely abolished the net tubular transport of atropine and more than halved the amount excreted by dogs.

As noted above, appreciable species differences in the metabolism of atropine have been registered and this, together with the fact that our knowledge of the identity of metabolites is often fairly sketchy, makes it difficult to summarize schematically the pathways of atropine metabolism generally or, in fact, in most animal species. Most of our knowledge of the nature of atropine metabolites has been obtained in studies using mice. In this species the following reactions have been established: 1. N-demethylation to noratropine; 2. ester hydrolysis to tropine (43) and tropic acid (45); 3. conjugation with glucuronic acid; and 4. aromatic hydroxylation followed by conjugation with glucuronic acid. The use of $[N\text{-}methyl\text{-}^{14}C]$-atropine has shown that 7–10% of the dose is excreted as respiratory CO_2 (Eling, 1968; Werner and Schmidt, 1968). This extent of N-demethylation was confirmed in the former investigation by showing that the nortropine formed by hydrolysis of the urinary noratropine and closely related metabolites also accounted for about 10% of the dose. In the mouse, ester hydrolysis to tropic acid and tropine is not extensive. It is known that the administration of tropic acid leads to the excretion of essentially all of the compound unchanged in the urine (Gosselin *et al.*, 1955). Therefore, the extent of hydrolysis of atropine to these compounds should be reflected in the amount of tropic acid excreted in the urine. While the evidence regarding tropic acid formation in atropine-treated mice is conflicting, it does appear that this pathway is of minor importance. Gosselin *et al.* (1955) reported its absence as a urinary metabolite while Gabourel and Gosselin

(1958) found that it comprised about 1% of the dose. Its presence as a urinary metabolite has also been noted by Eling (1968) and Werner and Schmidt (1968). In contrast to the limited formation of the metabolites noted above, it appears that conjugation of atropine or its hydroxylation products with glucuronic acid is a major metabolic pathway in the mouse (Gabourel and Gosselin, 1968; Albanus *et al.*, 1968a; Werner and Schmidt, 1958). A major hydroxylated product is 4'-hydroxyatropine but Gabourel and Gosselin (1958) obtained evidence for excretion of large amounts (40–50% of the dose) of glucuronides of 3',4'-dihydroxyatropine.

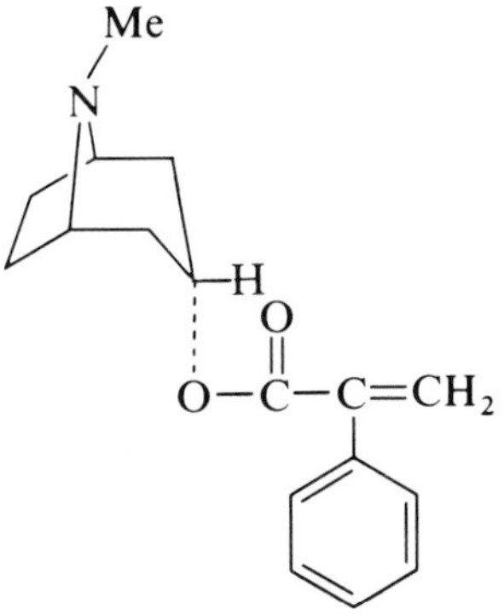

(46)

With the above outline of atropine metabolism in mice as a guideline, the main similarities and differences found in other animal species can be summarized. Gabourel and Gosselin (1958) found that rats and mice metabolized atropine similarly and Werner (1961) also found a fairly close similarity between these two species. Truhaut and Yonger (1967) reported that atropine was converted by rat liver *in vitro* to noratropine, apoatropine (46) and hydroxylated derivatives, especially 2'-hydroxyatropine. Franklin (1965) reported that atropine was demethylated when incubated with a rat liver microsomal preparation. Activity was seen with several *N*-methyl compounds of plant or synthetic origin but not with endogenous mammalian substrates. In the rat, as in the mouse (Evertsbusch and Geiling, 1956), biliary excretion of atropine metabolites has been shown to be especially extensive (Kalser *et al.*, 1965a,b). As much as 50% of an injected dose appears in the bile within 4 h, entirely as metabolites more polar than atropine. Bernheim and Bernheim (1938) found that rat liver was not very active in hydrolysing atropine. On the other hand, this activity was pronounced in guinea pig liver. Several other investigations have noted the ease of atropine hydrolysis in this species (Langecker and Lewit, 1938; Kalser *et al.*, 1957; Werner, 1961). Liver preparations from guinea pigs hydrolysed atropine, a reaction which was not observed in similar experiments using rat liver (Truhaut and Yonger, 1967). Otherwise, the oxidation products noted above for the rat (noratropine, apoatropine and hydroxylated derivatives) were also formed *in vitro* by guinea pig liver. Phillipson *et al.* (1976) incubated atropine with microsomal preparations from guinea pig liver and found that noratropine was formed. Additionally, these incubates contained the two isomeric *N*-oxides of atropine. When noratropine was used as a substrate, this secondary amine underwent *N*-oxidation to give the hydroxylamine derivative.

The metabolism of atropine in rabbits is of interest because of the presence of a genetically determined serum esterase in some animals. Bernheim and Bernheim (1938) found the enzyme to be lacking in about one third of the rabbits they investigated. Other studies gave values of about 70% (Glick and Glaubach, 1941), slightly over 50% (Godeaux and Tønnesen, 1949) and about 40% (Ellis, 1948) for the proportions of animals which did not hydrolyse atropine. It is known that this esterase is not identical with that responsible for the hydrolysis of cocaine in rabbits (Ammon and Savelsberg, 1949). A clear difference in the patterns of urinary atropine metabolites between the two different types of rabbits is shown in the data of Werner (1961). Information on the metabolism of atropine in cats is both scanty and apparently contradictory. While Bernheim and Bernheim (1938) found that its hydrolysis by cat liver occurred slowly, Langecker and Lewit (1938) concluded that cats could readily hydrolyse the alkaloid. The latter finding was confirmed by Godeaux and Tønnesen (1949). However, the data given by Werner (1961) on the urinary metabolites of [14C]-atropine in six different animal species indicate that the amount of urinary products is lowest in the cat. Knowledge of the metabolic fate of atropine in dogs is very limited. Albanus *et al.* (1968b) concluded that the pattern in dogs is qualitatively similar to that seen in mice. They detected four urinary metabolites of atropine, two of which were probably tropic acid and tropine.

Considerable uncertainty regarding the metabolism of atropine also exists in man. In order to overcome the technical problems associated with the detection and identification of small quantities of metabolites arising from the relatively small dose of atropine tolerated by man, the use of radioactive alkaloid seems to be essential. Gosselin *et al.* (1960) employed a preparation labelled with ^{14}C in the tropic acid moiety. When 2 mg was given by i.m. injection, most of the radioactivity appeared in the urine within 24 h, a trace was found in the faeces and, not unexpectedly, none was detected in the expired air. As mentioned above, about half of the dose was excreted in the urine unchanged while about one third was found to consist of unidentified metabolites and less than 2% as tropic acid (45).

The unidentified metabolites appeared to be esters of tropic acid but neither aromatic hydroxylation nor glucuronide formation could be demonstrated, indicating that these metabolites are not the same as those found in rat or mouse urine. A subsequent study was carried out by Kalser and McLain (1970) using two different preparations labelled with ^{14}C either in the ring or the N-methyl group of the tropine moiety. With the latter compound, N-demethylation was detected and about 3% of the dose was recovered as $^{14}CO_2$, although it was considered possible that as much as 5–10% may be lost by this route. In contrast to that mentioned above, this study indicated the excretion of considerable amounts of a glucuronide conjugate. In addition, evidence was obtained for the excretion of four other metabolites. Of these, one was not characterized, one was tentatively identified as tropine and the others were converted upon alkaline hydrolysis to a compound resembling tropine.

The investigations noted above utilized atropine, the racemic form of hyoscyamine, but a few reports have appeared in which the optically active (−)-hyoscyamine was studied. In regard to the amount of alkaloid excreted unchanged in the urine, Tønnesen (1950) reported values similar to those for atropine in rabbits and man but not in rats. Similar values have been obtained for the hydrolysis of (−)-hyoscyamine and atropine in the perfused rabbit liver (Godeaux and Tønnesen, 1949) and rabbit serum (Ammon and Savelsberg, 1949). However, Bernheim and Bernheim (1938) reported the preferential hydrolysis of the (−)-isomer by guinea pig liver *in vitro* and the data of Glick and Glaubach (1941) suggest a slightly greater hydrolysis of the (−)-isomer using rabbit serum. An esterase found in rabbit serum and liver which hydrolyses (−)-hyoscyamine has been studied by Werner (1961) and Werner and Brehmer (1963, 1967).

(−)-Scopolamine, the 6,7-epoxide of (−)-hyoscyamine and the other medically important solanaceous alkaloid, is also a substrate of the esterase present in rabbits. Glick and Glaubach (1941) found a slightly lower rate of hydrolysis with scopolamine than with (−)-hyoscyamine in rabbit serum and Godeaux and Tønnesen (1949) reported similar rates in the perfused rabbit liver. However, Bernheim and Bernheim (1938) claimed that scopolamine was usually not hydrolysed by liver preparations, even those from guinea pigs which otherwise showed high esterase activity. Tønnesen (1950) determined the extent of excretion of unchanged scopolamine in the urine. In rats and rabbits the values following various routes of administration were generally similar to those found with atropine or (−)-hyoscyamine but in man the values were much lower with scopolamine. Ziegler *et al.* (1969) isolated scopolamine-N-oxide from an incubation mixture of the alkaloid and a purified amine oxidase preparation. Interestingly, one isomeric N-oxide of (−)-scopolamine (N-

oxide moiety in equatorial position) was formed when (−)-scopolamine was incubated with microsomal preparations from guinea pig liver (Phillipson et al., 1976). These preparations also formed norscopolamine and the latter compound, when itself used as a substrate, underwent N-oxidation to give the hydroxylamine derivative. It seems likely that scopolamine is also metabolized by several other routes which are seen with atropine. These pathways include aromatic hydroxylation and dehydration to an apo derivative similar to compound (46). This suggestion is supported by the findings of Sano and Hakusui (1974) who showed that the quaternary methyl derivative of scopolamine undergoes appreciable hydroxylation in rats to the 4'-hydroxy derivative and that small amounts of aposcopolamine are also excreted. Another possible reaction of scopolamine is conversion of the epoxide moiety to a diol by the action of epoxide hydratase. In vitro studies with this enzyme indicated, however, that this does not occur (Oesch, 1977).

The second major group of tropane derivatives is that of the coca alkaloids. These consist of several types including those based on ecgonine, tropine and hygrine. Ecgonine derivatives furnish the principal type and, among these, **cocaine** (47) (see Fig. 9.3) is the most important and most widely studied. Until very recently, the amount of information on the metabolic fate of cocaine was quite limited. This is perhaps surprising in view of the fact that this drug, in its pure form, has been used in one way or another for nearly a century. However, this deficiency is now being corrected as the increasing illicit use of cocaine makes a better understanding of its biological disposition necessary. Early studies of this subject were largely devoted to the question of the extent of urinary excretion of cocaine. Wiechowski (1901) administered cocaine i.p. to rabbits and dogs and found that about 5% of the dose could be recovered in the 48 h urine in the latter species. However, no cocaine was detected in rabbit urine. Oelkers and Vincke (1935) studied the urinary excretion of cocaine following its injection in mice, rats, guinea pigs, rabbits, cats and dogs. In general, excretion values of 1–5% were obtained but as much as 16% of the dose was found excreted unchanged in rabbits when the urine was acidic. A low level of cocaine excretion in the rat was confirmed by Nayak et al. (1974) who found 0·75% of the dose (20 mg/kg, s.c.) in the 96 h urine. Faecal excretion during this period was a mere 0·25%. Likewise, Woods et al. (1951) detected 1–12% of the dose (10–15 mg/kg, s.c. or i.v.) unchanged in the 24 h urine in dogs. It was also noted that much less cocaine is absorbed following oral administration. The latter publication is useful also for its summary of the early work on cocaine metabolism and excretion. In studies using rabbits, Woods et al. found no or at most, traces of cocaine excreted in the urine. Information on the urinary excretion of unchanged cocaine in man has been obtained mainly from addicts, although Woods *et al.* (1951) referred to unpublished results which gave values of 1–21% in non-addicts. In addicts, the excretion has been reported to be 1–9% (Fish and Wilson, 1969), 4–9% (Sánchez, 1957) and 7–21% (Ortiz, 1952). Fish and Wilson (1969) obtained the highest excretion values when the urinary pH was reduced.

From the foregoing it is apparent that cocaine is largely metabolized in animals. It contains two ester linkages and most of the information on cocaine metabolism deals with its hydrolysis at these metabolically active sites. Langecker and Lewit (1938) were the first to propose that the probable metabolic sequence involves the initial hydrolysis of the methyl ester group to give benzoylecgonine (48) followed by further hydrolysis to ecgonine (49) and benzoic acid. They also showed that the hydrolysis products were less toxic that cocaine. However, the results obtained on the excretion of these hydrolysis products have not always been consistent. Wiechowski (1901) did not find ecgonine in the urine of rabbits given cocaine i.p. whereas Ortiz (1966) detected both ecgonine and benzoylecgonine as urinary metabolites in rats. Interestingly, it was noted that only a small proportion of the dose (45 mg/kg, i.p.) was excreted as these metabolites in rats. This has also been noted by Sánchez (1957) who found only 1–3% of the 90–220 mg daily dose taken orally by addicts to be excreted in the urine as ecgonine. Valanju *et al.* (1973) noted that whenever ecgonine was present in the urine of addicts, benzoylecgonine was also found. However, the latter metabolite was often present in the absence of ecgonine. Comprehensive studies of the metabolism of cocaine in rats (Nayak *et al.*, 1976) and dogs (Misra *et al.*, 1976b) have recently clarified many aspects of this subject. Some of these findings are summarized in Fig. 9.3 which shows the general pathways of cocaine metabolism. Of course, the quantitative picture may vary from species to species and, in some cases, certain pathways may be inconsequential. Firm evidence for the formation of hydroxylated metabolites is, for example, only available from the experiments using rats. When rats were given cocaine (20 mg/kg, s.c.), a phenolic and two non-phenolic hydroxylated metabolites were detected as urinary metabolites (Nayak *et al.*, 1976). These are the *p*-hydroxylated metabolite (54) and compounds which are presumably hydroxylated in the 6- and/or 7-positions of the pyrrolidine ring. The other urinary metabolites were benzoylecgonine (48), ecgonine (49), ecgonine methyl ester (50) and benzoylnorecgonine (52). The likely precursor of benzoylnorecgonine, norcocaine (51), was not detected. Neither was its possible hydrolysis product norecgonine (53). It is also possible that the nor compound (52) might be formed via an alternative pathway from benzoylecgonine (48). The results of Misra *et al.* (1975) suggest that

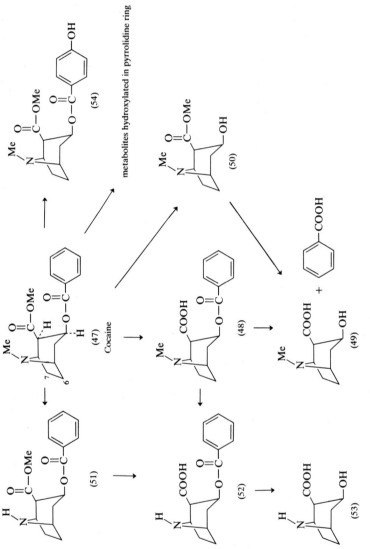

FIG. 9.3. Metabolic pathways of cocaine.

this may not be the case as the major urinary products of benzoylecgonine in rats were found to be unchanged compound and ecgonine, its hydrolysis product, together with minor amounts of a phenolic metabolite and its glucuronide. Nevertheless, Mulé et al. (1976) recently reported that benzoylnorecgonine is present in the brains of rats given benzoylecgonine. When the nor compound was given, however, no further metabolism was detected. This confirms the earlier result that benzoylnorecgonine is not metabolized further to norecgonine in rats. Misra et al. (1976a) administered benzoylnorecgonine to rats and found that it was rapidly and extensively excreted unchanged in the urine. A small amount of an unidentified compound was also detected, however this was not norecgonine. On the other hand, both benzoylnorecgonine and norecgonine were detected in brain when rats were given norcocaine (51) (Misra et al., 1976c). These differences may be related to the variations in lipophilicity of the different compounds employed. Formation of norecgonine appears not to occur via N-demethylation of ecgonine (49). Misra et al. (1974b) found that the latter compound was rapidly and extensively excreted in the urine of rats without undergoing metabolism. The routes of excretion of cocaine and its metabolites in rats were also studied by Nayak et al. (1976). At the dose level noted above, only about 1% was excreted unchanged in the urine and faeces. However, about 22% of the ^{3}H-labelled material was lost by the faecal route. This finding is explained by the considerable excretion of cocaine metabolites in the bile. Following an intravenous dose of 5 mg/kg, approx. 36% was excreted by this route as metabolites in 3·5 h. The amount of unchanged cocaine excreted in the bile was very small.

The biological disposition of cocaine in dogs was studied by Misra et al. (1976b). Using ring-labelled [^{3}H]-cocaine (5 mg/kg, i.v.), 58% of the radioactivity was excreted in the urine in 48 h. The urinary metabolites and amounts excreted were: cocaine (47) (2·9% of dose), benzoylecgonine (48) (2·5%), ecgonine (49) (37·1%), ecgonine methyl ester (50) (0·1%), norcocaine (51) (0·8%), benzoylnorecgonine (52) (9·1%), norecgonine (53) (0·9%) and unidentified polar compounds (5·6%). The detection of these compounds as urinary metabolites indicates that the metabolism of cocaine in dogs is qualitatively similar to that seen in rats. However, some additional metabolites are also detected. These include the demethylated intermediate norcocaine (51) and the fully hydrolysed and demethylated norecgonine (53). Nevertheless, the amount of the latter metabolite found was quantitatively small compared with that of ecgonine (49). Rapid metabolism of cocaine was also indicated in dogs which had high levels of radioactivity in the bile shortly after administration of the compound.

Several studies on the enzymic hydrolysis of cocaine have been carried out. Many of these dealt with the activity present in serum and Glick and

Glaubach (1941) showed that hydrolysis took place with rabbit serum. Glick *et al.* (1942) noted that horse and human serum do not possess this activity. Blaschko *et al.* (1955) confirmed this lack of cocainesterase in horse serum and human plasma. Glick and Glaubach believed that the cocainesterase in rabbit serum was not identical with atropinesterase, however Glick *et al.* were not able to separate the enzymes by electrophoretic means and all of the enzymes were found mainly in the α- and β-globulin fractions. Blaschko *et al.* (1947) confirmed the separate entities of cocainesterase and tropinesterase (atropinesterase) in rabbit serum. The same result was reported by Ammon and Savelsberg (1949) who found that about one third of the rabbits tested were able to hydrolyse cocaine. This activity was also seen when ecgonine methyl ester (50) was used as the substrate. The usual hydrolytic pathway is via benzoylecgonine (48) to ecgonine (49), however Werner and Brehmer (1963) showed the presence of a cocainesterase in rabbit serum which produced ecgonine via the alternative pathway forming ecgonine methyl ester (50). Taylor *et al.* (1976) studied the *in vitro* hydrolysis of cocaine by human serum and found that hydrolysis to ecgonine occurred via both hemiesters, benzoylecgonine (48) and ecgonine methyl ester (50). The reason for the earlier negative results with human serum is not clear.

The ability of other tissues to hydrolyse cocaine has also been studied. Heim and Haas (1950) investigated its hydrolysis to benzoylecgonine in guinea pig tissues and found that the liver had greater activity than kidney, brain or muscle tissue. Mikami (1951) studied the formation of benzoic acid from cocaine in the liver *in vitro* and found the activity to decrease in the following order: rabbit, guinea pig, cat, mouse, dog and rat. Cocaine hydrolysing activity was also studied by Severi *et al.* (1967) who also found a decreasing order of activity in rabbits, guinea pigs and rats. Iwatsubo (1965) studied the hydrolysis of various esters by liver microsomes solubilized with sodium deoxycholate. Preparations from rats, guinea pigs and rabbits showed good activity with several simple, synthetic esters, however cocaine hydrolysis was seen only with rabbit liver microsomes which had weak activity. Leighty and Fentiman (1974) reported that the observed hydrolysis of cocaine to benzoylecgonine by rat liver microsomes accounted for only a small part of the cocaine metabolized in these incubates. On the other hand, Estevez *et al.* (1977) found that the *in vitro* and *in vivo* hydrolysis of cocaine by rate liver was significantly reduced by SKF-525A, a microsomal enzyme inhibitor.

Although the excretion of norcocaine (51) as a urinary metabolite of cocaine has only recently been demonstrated, considerable evidence is available which indicates the existence of this pathway of N-dealkylation. Axelrod and Cochin (1957) showed it to be carried out by rat liver microsomes and similar results were obtained by Ramos-Aliaga and Chiriboga (1970) using mouse, rat, guinea pig and dog liver microsomes. Estevez *et al.* (1977) reported a reduction in the amounts of demethylated metabolites (norcocaine and benzoylnorecgonine) in the livers of rats given the inhibitor SKF-525A prior to cocaine administration. Leighty and Fentiman (1974) also demonstrated norcocaine formation by rat liver microsomes, however this compound and benzoylecgonine, the only metabolites identified made up together less than 20% of the cocaine metabolized. Using [N-*methyl*-^{14}C]-cocaine, Werner (1961) found that 45% of the injected radioactivity was recovered as respiratory $^{14}CO_2$ in mice. This suggests that N-demethylation may be a prominent metabolic pathway in this species. Also, norcocaine has been identified as a metabolite of cocaine in the brains of rats (Misra *et al.*, 1974c) and dogs (Nayak *et al.*, 1976) and in the brains and blood plasma of monkeys (Hawks *et al.*, 1974).

Very little information is available on the metabolism of other coca alkaloids. Based on the excretion of the unchanged compounds in cats, Gruhn (1925) suggested that a larger proportion of pseudococaine (55) than *l*-cocaine (47), the usual isomer, was metabolized. When pseudococaine was administered to rats by intravenous injection, the brains contained the pseudo isomers of metabolites (48), (49), (51) and (52) (Misra *et al.*, 1976c). Glick and Glaubach (1941) found that **tropacocaine** (56) was hydrolysed by rabbit and horse serum and believed that this

(55)	(56)
Pseudococaine	Tropacocaine

activity may be different from that hydrolysing atropine, cocaine and choline esters. Blaschko *et al.* (1947) confirmed the presence of this enzyme in horse serum and showed that the hydrolysis of tropacocaine was not inhibited by several other esters including atropine and cocaine. They concluded that the tropacocainesterase of horse serum is not identical with pseudocholinesterase. Other studies of the enzymic hydrolysis of tropacocaine (Seiler *et al.*, 1968a) and nortropacocaine (Seiler *et al.*, 1968b) indicated that the esterase activity involved is not identical with that responsible for the hydrolysis of benzoylcholine.

IV. Indole Derivatives

Indole (57) (see Fig. 9.4) is the parent of an extensive and sometimes complex group of nitrogen heterocyclic compounds. In fact, perhaps as many as one fourth of all known alkaloids contain the indole moiety. One important type of indole compound consists of indolethylamines (tryptamine derivatives). However, their metabolism is governed more by the alkylamine than the indole moiety and these compounds are therefore covered in Chapter 8, Section I. In contrast to that seen with the parent compounds of most of the other classes included in this chapter, indole itself occurs in a number of essential oils (see Opdyke, 1974). Although attempts to determine its metabolic fate were made in the nineteenth century, it is only recently that an adequate understanding of indole metabolism has been obtained. Earlier investigations, the results of which are summarized by Williams (1959, pp. 668–669), showed mainly that indole is hydroxylated to 3-hydroxyindole (indoxyl) (58) which is then excreted in the urine of several animal species partly as the glucuronide (59b) but mainly as the sulphate (59a) conjugate (indican). King *et al.* (1966) carried out a detailed investigation of the metabolism of [2-^{14}C]-indole in rats. Employing oral doses of 60–70 mg/kg they found that about 80% of the radioactivity was excreted in the urine in two days, with only traces of unchanged compound appearing. The major urinary metabolite was indoxyl sulphate (59a) which accounted for 50% of the dose. The metabolites and the amounts excreted are shown in Fig. 9.4 which indicates that two metabolic pathways exist. The major route involves hydroxylation at the 3-position giving indoxyl (58), its conjugates and further oxidation products leading to N-formylanthranilic acid (61) and anthranilic acid (62). As the indole used was labelled in the 2-position, scission of the pyrrole ring to give metabolites (61) and (62) leads to loss of the label from the latter metabolite. No measure of the urinary anthranilic acid excretion was therefore obtained but a value of about 2%, that found for the amount of ^{14}CO$_2$ lost in the expired air, appears reasonable. *In vitro* experiments indicated that 3-hydroxylation occurred under aerobic conditions, suggesting that this reaction is carried out by the microsomal hydroxylating enzymes. This conversion of indole to indoxyl (58) was also demonstrated by Posner *et al.* (1961) and Beckett and Morton (1966a) using the microsomal hydroxylating system from rabbit liver. All of these reports noted the rapid further oxidation of indoxyl to indigo blue (indigotin), although Beckett and Morton found this reaction to take place with rabbit liver preparations but not with those from rat liver which formed oxindole (63) instead. Rimington (1946) noted the occasional occurrence of bis-indole pigments in normal and pathological urines.

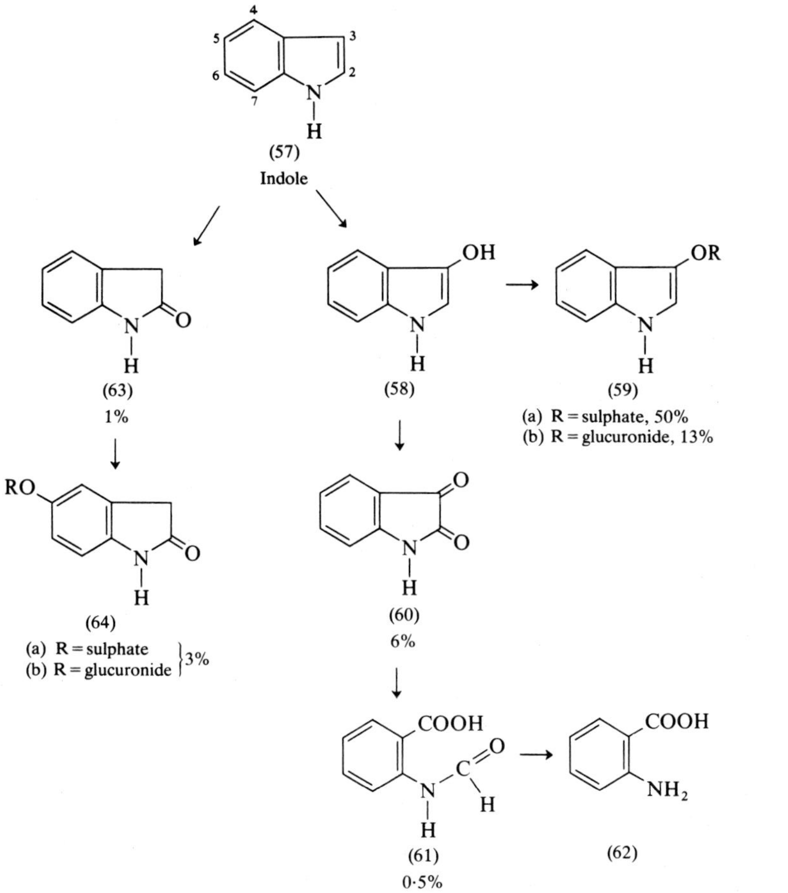

FIG. 9.4. Metabolic pathways of indole in rats

The second metabolic pathway of indole involves the initial oxidation at the 2-position to form oxindole (63). As noted above, Beckett and Morton (1966a) showed that this metabolite was formed when rat liver microsomes were used and King *et al.* (1966) found it in aerobic and anaerobic incubates of rat liver preparations. The available data suggest that this pathway is of minor quantitative importance and Beckett and Morton did not detect oxindole in the urine of rats given indole intraperitoneally. However, they did find small amounts of conjugates of 5-hydroxyoxindole. King *et al.* showed that the main urinary metabolites resulting from this reaction sequence are the sulphate (64a) and glucuronide (64b) conjugates of 5-hydroxyoxindole. Beckett and Morton found that microsomal preparations from rat, guinea pig and rabbit liver carried out the 5-hydroxylation of oxindole. Interestingly, indole itself does not undergo direct hydroxylation at the 5-position (see Posner *et al.*, 1961). The results of King *et al.* indicated that biliary excretion of indole and its metabolites is not extensive in the rat (about 5% of the dose) and that about one tenth of the dose was excreted in the faeces, mainly as unidentified metabolites.

Whereas metabolic indican is indoxyl sulphate (59a), the indican found in plants (plant indican) is **indoxyl-β-D-glucoside**. Li *et al.* (1963) gave oral doses of plant indican to rabbits, dogs and humans and found that the 24 h urines contained unchanged compound, indoxyl sulphate and indoxyl glucuronide as well as some highly polar compounds. In the urine of rabbits only, some anthranilic acid (62) and its glycine conjugate were also detected.

(65)

Several groups of indole alkaloids are derivatives of β-carboline (65). Among the simpler types are the closely related harmala alkaloids, **harmine** (66), **harmaline** (67) and **harmalol** (69). The metabolic fate of these compounds is not complex, involving primarily *O*-demethylation followed by conjugation (Fig. 9.5). Zetler *et al.* (1974) studied their pharmacokinetics in the rat and a kinetic study of the liver microsomal *O*-demethylation of harmine in six mammalian species was reported by Burke and Upshall (1976). Very little harmine is excreted unchanged in the urine following its injection in rats or man (Villeneuve and Sourkes, 1966; Slotkin and DiStefano, 1970a,b; Slotkin *et al.*, 1970). In the rat, nearly 75% of the dose (5 mg/kg, i.v.) was excreted in the bile and the remainder in the urine (Slotkin and DiStefano, 1970b). This study showed that the same metabolites were found in both urine and bile and that their relative proportions were fairly similar. Urinary excretion of metabolites in non-cannulated rats was about 30% (Slotkin *et al.*, 1970), which suggests that reabsorption of the biliary metabolites is not extensive. In man, the urinary excretion of harmine metabolites amounts to about a third of the dose. These investigations have also shown that the major metabolic product is harmol (68) which is excreted in small amounts free but mainly as glucuronide and sulphate conjugates. In man, somewhat more harmol is excreted in the urine conjugated with glucuronate than with sulphate whereas, in rats given the same or a larger dose (0·5 or 5 mg/kg), excretion of the sulphate conjugate is 2·5–3 times greater than excretion of the glucuronide (Slotkin *et al.*, 1970). Conjugate formation of harmol is discussed below together with that of harmalol.

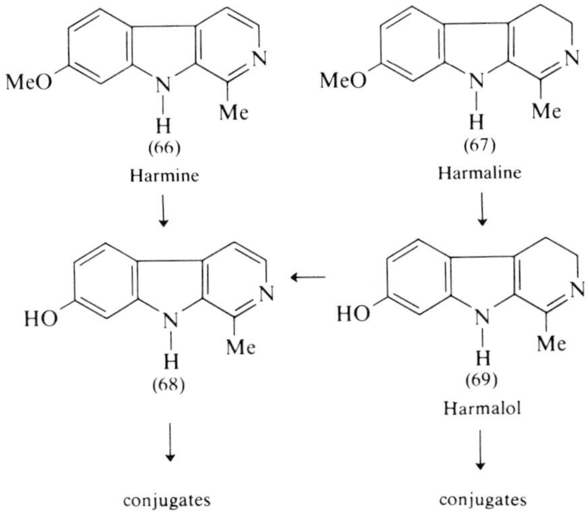

FIG. 9.5. Metabolic pathways of harmine, harmaline and harmalol.

The metabolic pathways shown in Fig. 9.5 indicate that harmaline (67) is subject to the same metabolic alterations as is harmine but, additionally, dehydrogenation at the 3,4-position occurs giving rise to the fully aromatic derivative harmol (68). This reaction was suggested by the results of Villeneuve and Sourkes (1966) and Ho *et al.* (1971b) obtained evidence which indicated that the dehydrogenation reaction occurs in rats with harmol but not with harmaline. This contradicts an early report (Flury, 1911) which stated that harmaline is metabolized to harmine in rabbits and dogs. The last investigation also reported the degradation of harmaline to harminic acid (6-azaindole-2,3-dicarboxylic acid), but this metabolite was not detected in rat urine (Ho *et al.*, 1971b). Villeneuve and Sourkes (1966) found that harmaline was metabolized more slowly in rats than was harmine and that little unchanged compound was excreted, a finding in agreement with that of Flury (1911) in rabbits. However, Ho *et al.* (1971b) found nearly 10% of the dose (40 mg/kg, s.c.) of harmaline excreted unchanged in the urine of rats within 8 h. Similar to that noted above for harmol excretion following harmine administration, very little free harmalol or its dehydrogenation product harmol was found in the urine of the harmaline-treated rats. Again, the bulk of the urinary metabolites, which amounted to slightly more than 60% of the dose in four days, was conjugated material. Interestingly, this was mainly the glucuronide conjugate rather than the sulphate which is the case with harmol. This subject was subsequently studied in more detail by Mulder and Hagedoorn (1974). They used both harmol and harmalol and found that i.v. administration to rats resulted in extensive excretion of conjugates in the urine and bile. In agreement with the earlier results, harmol was found to be excreted as the sulphate (70%) and the glucuronide (30%) whereas harmalol was excreted mainly as the glucuronide, only a trace of the sulphate being found. In addition, *in vitro* studies showed that both substrates were readily glucuronidated by UDP-glucuronyltransferase and that harmol was a good substrate of phenol sulphotransferase. However, harmalol was only poorly converted to its sulphate conjugate by this enzyme. The glucuronidation of harmol and harmalol using UDP-glucuronic acid and the glucuronyltransferase enzyme from guinea pig liver microsomes has been studied by Wong and Sourkes (1967, 1968). Other closely related investigations have dealt with the conjugation and biliary excretion of harmol *in vivo* and in the perfused liver (Mulder *et al.*, 1975), with the inhibition of conjugation and biliary excretion of harmol (Mulder and Pilon, 1975) and with the differences in the biliary and urinary excretion of the sulphate and glucuronide conjugates of harmol in the mouse, rat, guinea pig, rabbit and cat (Mulder and Bleeker, 1975). The last study showed that large species varriations occur with respect to the proportions of the two conjugates which are excreted.

Daly *et al.* (1965) reported that harmalol is an excellent substrate for the hydroxylating system in rabbit liver microsomes. The hydroxylated product was not identified, however it was readily converted by the catechol-*O*-methyltransferase also present in the preparation to a hydroxymethoxy derivative. Oxidation has therefore taken place at either the 6- or 8-position and from the knowledge that indoles are commonly hydroxylated at the 6-position (Daly and Witkop, 1963), it seems likely that the product is the 6,7-dihydroxy derivative which is then methylated, probably to a mixture of the two possible monomethyl ethers. This reaction does not appear to have been demonstrated with harmalol and related compounds *in vivo*.

The following types of alkaloids to be discussed include those based on the ajmaline, corynantheine and yohimbine skeletons. Except for the latter group which includes reserpine and related compounds, very little metabolic information is available. Iven (1977) studied the distribution of **ajmaline** (70) in mice and found that it was highly accumulated in lung, liver and heart. It underwent extensive metabolism and only 5% of the dose (10 mg/kg, i.v.) was excreted unchanged in the urine. The quaternary propyl hydrogen tartrate derivative of ajmaline, an antiarrhythmic drug, was shown by Schaumlöffel (1974) to be well absorbed after oral administration to rats and man. It was metabolized to identical metabolites in both species and metabolite excretion occurred mainly in the bile.

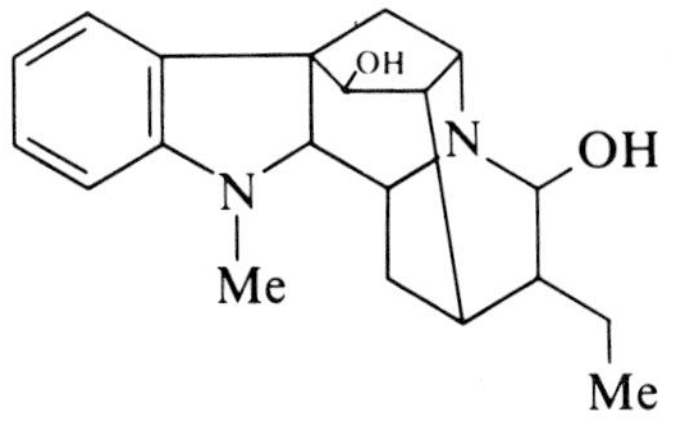

Ajmaline

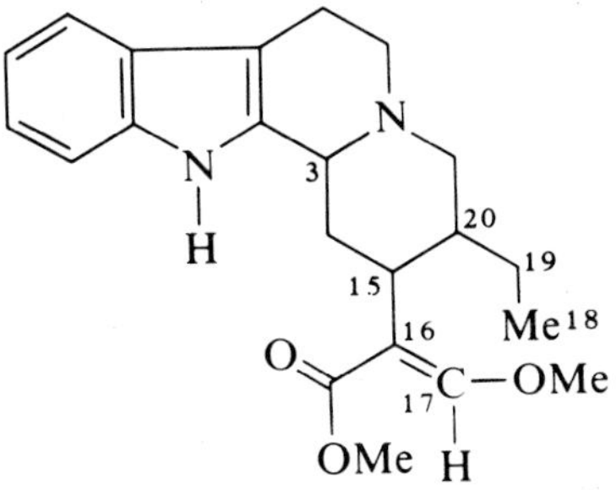

(71)

Corynantheidine

Corynantheidine (71) differs from **corynantheine** in lacking a double bond at C18,19. Three other diastereoisomeric forms of (71), **dihydrocorynantheine, isocorynantheidine** and **hirsutine**, have different configurations at the C3, C15 and C20 positions. The metabolism of these four mitragyna alkaloids by rabbit liver microsome preparations was studied by Beckett and Morton (1966b). They found that the major metabolic reaction *in vitro* with corynantheidine, dihydrocorynantheine and isocorynantheidine was O-demethylation at C17. This metabolic route accounted for most of the metabolism observed, especially with the first two compounds, and no other metabolic products were detected. With hirsutine, however, O-demethylation accounted for only about 5% of the total metabolism and another route is therefore mainly involved in the metabolism of this compound. These variations in the extent of O-demethylation are explained in terms of the preferred conformations of the various diastereoisomers, O-demethylation being greater in those cases where the indole and piperidine ring systems lie in the same plane.

The metabolism of **yohimbine** (72) in mice was investigated by Ho et al. (1971a) using ³H-labelled material. The compound (10 mg/kg, i.p.) was found to be metabolized rapidly although some unchanged yohimbine was excreted in the urine. Of the 22% of the radioactivity excreted in the urine in 24 h, about 35–40% (i.e. 8% of the dose) was unchanged compound. Several unidentified metabolites were detected in liver and spleen extracts and two metabolites in addition to yohimbine were found in the urine at 3 and 24 h. Interestingly, distribution studies of the radioactivity showed that large amounts of activity were found in the intestine. Shortly after injection this consisted of metabolites, including yohimbic acid which results from

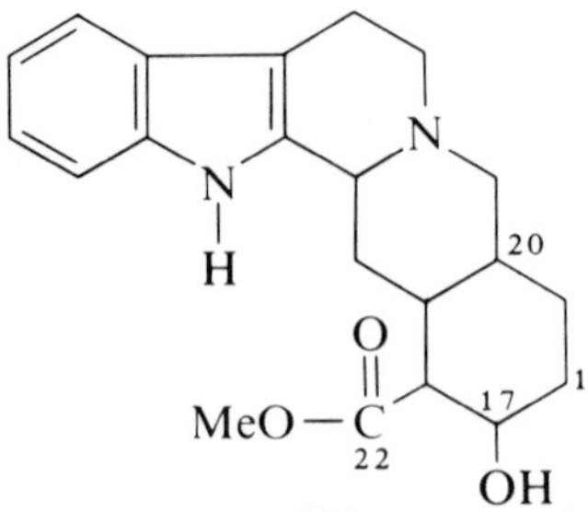

(72)

Yohimbine

thé hydrolysis of the ester group at C22. However, at 0·75 and 1·5 h but not from 3 h onwards most of the intestinal radioactivity was due to yohimbine itself. While faecal excretion of radioactivity was not studied, the above results and the finding that activity in the tissues after 24 h was low suggest that this is an important excretory route for yohimbine in mice.

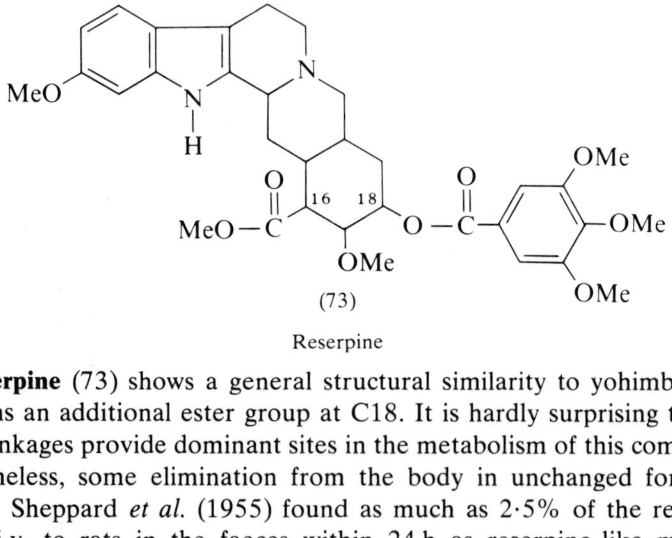

(73)

Reserpine

Reserpine (73) shows a general structural similarity to yohimbine but contains an additional ester group at C18. It is hardly surprising that the ester linkages provide dominant sites in the metabolism of this compound. Nonetheless, some elimination from the body in unchanged form also occurs. Sheppard *et al.* (1955) found as much as 2·5% of the reserpine given i.v. to rats in the faeces within 24 h as reserpine-like material. Roughly similar values were found in guinea pigs (Sheppard and Tsien, 1955; Sheppard *et al.*, 1957) and the latter report also noted that nearly 10% of the administered reserpine was recovered unchanged in the faeces of a rat only 6 h after dosing. In mice, as much as 35% of the dose (i.v.) may be excreted unchanged in the faeces within 24 h (Numerof *et al.*, 1955). Maass *et al.* (1969) reported that an average of 62% of the dose, mainly as reserpine, was recovered in the faeces of man within four days after administering 0·25 mg reserpine orally. However, they detected little unchanged compound in the urine from these subjects. Maronde *et al.* (1963) also reported little or no urinary excretion of free reserpine in man. Similar results showing amounts of urinary reserpine ranging from nil to a few tenths of percent of the dose have been reported in mice (Numerof *et al.*, 1955), rats (Sheppard *et al.*, 1955; Glazko *et al.*, 1956) and guinea pigs (Sheppard and Tsien, 1955; Sheppard *et al.*, 1957).

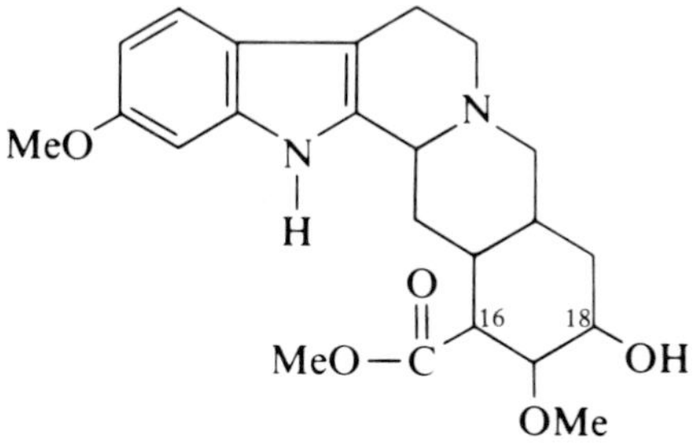

(74)

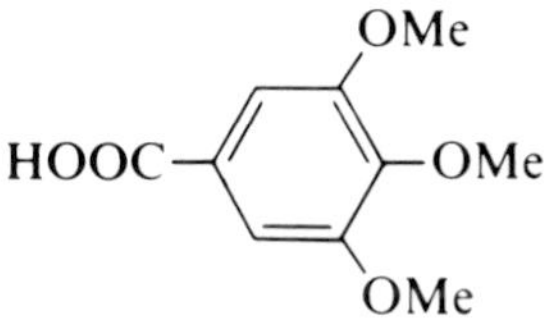

(75)

Although our present understanding of the metabolism of reserpine is far from complete, it appears that hydrolysis of the ester linkage at C18 is the dominating metabolic step. This leads to the formation of methyl reserpate (74) and 3,4,5-trimethoxybenzoic acid (75), both of which have been detected in various organs or excreta in numerous investigations. Dhar *et al.* (1955, 1956) reported detecting both metabolites in various organs, urine and faeces from rats given reserpine by both the oral and parenteral routes. Sheppard *et al.* (1955), using reserpine labelled with ^{14}C in the 4-methoxy group of the trimethoxybenzoic acid moiety, found that 7–14% of the radioactivity was excreted in the urine of rats as tri-methoxybenzoic acid-like material within 24 h of injection. The corresponding figures in guinea pigs were 5–24%, of which the bulk of the radioactivity was metabolite (75) or similar material (Sheppard and Tsien, 1955). Similar results were reported when reserpine labelled with ^{14}C in the carbonyl group of the trimethoxybenzoic acid moiety was employed (Sheppard *et al.*, 1957). Using the latter labelled compound in mice, Numerof *et al.* (1955) found that essentially all of the radioactivity excreted in the urine in 24 h (40–70% of the dose) consisted of trimethoxybenzoic acid. Maggiolo and Haley (1964) detected the latter metabolite and methyl reserpate (74) in the brains of mice given reserpine and they also showed that the concentration of metabolites increased during the five day experimental period whereas that of the unchanged compound decreased. In man given radioactive reserpine orally, trimethoxybenzoic acid can be detected in the blood, where it appears to be a major metabolite shortly after dosing (Numerof *et al.*, 1958). Also, it is the main metabolite found in the urine 24 h following dosing (Maass *et al.*, 1969). Maronde *et al.* (1963) calculated that about 6% of the daily oral dose was converted to urinary methyl reserpate.

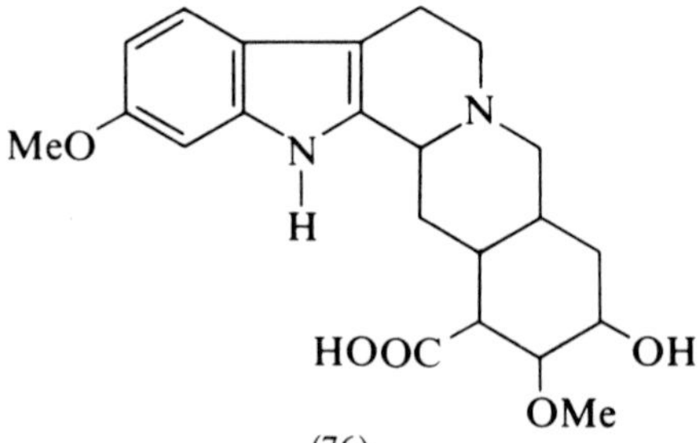

(76)

In contrast to the abundant documentation showing that reserpine is hydrolysed at the C18 ester linkage, no reports are available which show that the monoester resulting from hydrolysis at C16 is formed. However, this reaction may occur as reserpic acid (76) has been reported to be formed from reserpine, although the data on this point are conflicting. Dhar *et al.* (1955, 1956) claimed that reserpic acid was found in several organs of reserpine-treated rats as well as in intestinal contents, faeces and urine. In addition, both it and methyl reserpate (74) were formed when reserpine was incubated with rat liver slices. However, Glazko *et al.* (1956) found no trace of reserpic acid in the urine of reserpine-treated rats.

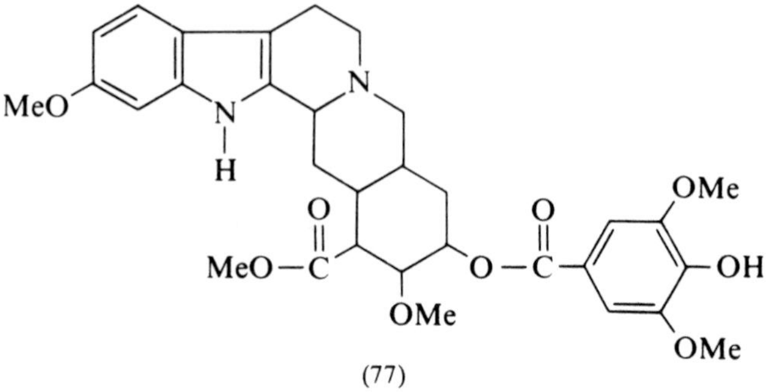

(77)

OMe

HOOC—⟨benzene ring⟩—OH

OMe

(78)

An important consideration in the metabolism of reserpine involves the enzyme systems responsible for its hydrolysis. Interestingly, Glazko *et al.* (1956) found that much higher levels of urinary methyl reserpate (74) were found in rats after oral than after parenteral administration. This hydrolysis was carried out *in vitro* by rat intestinal mucosa but not by similar preparations from dogs and monkeys. The two latter species also formed relatively small amounts of methyl reserpate *in vivo*. It is also known that the liver is active in hydrolysing reserpine in several animal species (Sheppard and Tsien, 1955). Stitzel *et al.* (1972) and Stawarz and Stitzel (1974) carried out a detailed investigation of the *in vitro* metabolism of reserpine by liver preparations. They found that most of the esterase activity was associated with the microsomal fraction and also that the mechanism is dependent upon the mono-oxygenase system, as indicated by the requirement for NADPH and O_2 as well as by inhibition of the reaction by inhibitors of this system. These results suggest that the hydrolysis of reserpine is preceded by an oxidative step, most likely *O*-demethylation of methyl ether group in the 4-position of the trimethoxybenzoyl moiety. This reaction sequence will form syringoyl methyl reserpate (77) and then syringic acid (78), although the articles do not make it clear if these metabolites were actually formed. The report of Sheppard *et al.* (1957) is pertinent in this regard as they found that rat liver slices converted significant amounts of reserpine to syringoyl methyl reserpate. This ester was also excreted in the faeces of reserpine-treated rats. Further information supporting the existence of this metabolic pathway has been obtained using reserpine labelled with ^{14}C in the 4-methoxyl group of the trimethoxybenzoyl moiety (Sheppard *et al.*, 1955; Sheppard and Tsien, 1955). They showed that as much as 20% of the dose was converted to [14]CO_2 *in vitro* using rat liver slices, with lesser amounts in preparations from dogs, mice and guinea pigs. Also, nearly a fourth of the radioactivity was recovered in the respiratory CO_2 within 6 h after giving rats the labelled compound by i.v. injection. Nonetheless, Sheppard *et al.* (1957) found that the urinary trimethoxybenzoic acid-like material in guinea pigs and rats was trimethoxybenzoic acid with only a small contribution by syringic acid being noted in the case of rats. At the present time, therefore, our understanding of the steps involved in the metabolism of reserpine is incomplete and it seems evident that future investigations must give adequate consideration to the unequivocal identification not only of reserpine and 3,4,5-trimethoxybenzoic acid but also to their closely related demethylation products, syringoyl methyl reserpate and syringic acid.

Some information on the metabolism of **raubasine** (ajmalicine), another alkaloid from *Rauwolfia serpentina* showing structural similarity to reserpine, has been reported by Löhr and Bartsch (1975) in a report dealing primarily with its detection in biological fluids and its absorption and excretion in rats. Less than 0·1% of the compound was excreted unchanged in the urine following oral dosage. Both raubasine and some unidentified metabolites were excreted in the bile.

Studies on the biological disposition of the experimental anti-cancer drug **ellipticine** (79) have been carried out in rats and dogs (Chadwick *et al.*, 1971) and in mice (Hardesty *et al.*, 1972). When given an i.v. dose of 6 mg/kg, dogs excreted less than 5% of the material in the urine in 24 h whereas rats excreted 11 and 13% in 24 and 96 h, respectively. In both cases, about one tenth of the excreted material consisted of unchanged compound and the remainder of one or more unidentified metabolites. As much as 70% of the dose underwent biliary excretion in rats and less than 10% of this material was unchanged ellipticine. The results of distribution studies indicated that biliary excretion was also important in ellipticine elimination in dogs. The data from mice suggest a similar situation as less than 1% of an i.p. dose (250 mg/kg) was recovered in the urine in five days. Faecal excretion was the main route of elimination, its rate and extent depending on the formulation of the preparation given. Reinhold *et al.* (1975) found that biliary excretion is the major (60% of the dose in 24 h) route of metabolite elimination in rats given ellipticine. Five components of this material were detected, two of which provided more than half of the total. Both of these metabolites were identified as polar conjugates of 9-hydroxyellipticine. Subsequently, Reinhold and Bruni (1976) reported on the characterization of this hydroxylated metabolite of ellipticine and also on the mechanism of its formation *in vivo* . An i.v. dose of 6 mg/kg was also used in this study which concluded that hydroxylation occurs without involving an arene oxide/NIH shift mechanism. In another study of ellipticine metabolism, Lesca *et al.* (1976) reported that rat liver microsomes metabolized ellipticine to 9-hydroxyellipticine and that this metabolite as well as an unidentified compound hydroxylated on the aromatic ring of the indole moiety are excreted in the bile of rats given ellipticine.

Ibogaine (80) is excreted in unchanged form in the urine of rabbits and man following oral doses of 20 mg/kg and 5 mg, respectively (Cartoni and Giarusso, 1972). An unidentified urinary metabolite of ibogaine was also

Me

Meo

N

Me

N
H
Me

N
H

Me

(79)   (80)

Ellipticine   Ibogaine

detected in the unhydrolysed fraction. It seems reasonable to expect that most of the urinary metabolites will be found as conjugates, however this fraction was not investigated.

Although the metabolic fate of **strychnine** (81a) has been the subject of sporadic investigations over a long period of time, relatively little is actually known on this subject. An important early development in this area was the investigation of Hatcher and Eggleston (1917), whose report is also useful as a guide to previous extensive and often conflicting literature. They found that only small amounts of unchanged compound were excreted in the urine 24 to 48 h after administering strychnine to guinea pigs, cats and dogs and that none was excreted in the faeces. Their evidence pointed to a rapid destruction of the compound in the body, especially in the case of guinea pigs. Perfusion experiments using guinea pig or dog liver showed rapid metabolism of strychnine by this organ. No unchanged strychnine was excreted in the bile of dogs given the compound i.v.

The *in vitro* metabolism of strychnine was studied by Adamson and Fouts (1959) using liver preparations from several species. With homogenates, increasing activity was found in dog, rat, mouse, rabbit and guinea pig liver, the activity in rabbits being about ten times greater than that in dogs. The activity was localized in the microsomal fraction, required TPNH and oxygen and was inhibited by SKF 525 A, the well-known inhibitor of the mono-oxygenase system. It was found that strychnine metabolism was associated with a loss in biological activity but the nature of the metabolic products was not studied. Kato *et al.* (1962) showed that the prior administration of inducing substances, especially phenobarbital, resulted in increased strychnine metabolism in rats together with increased tolerance to its toxic effects. The metabolism of strychnine has also been shown to increase in rabbits following treatment with phenobarbital (Tsukamoto *et al.*, 1964a). The latter investigation also studied the metabolic products formed when the alkaloid was incubated with rabbit liver slices or 9 000 g supernatant fractions. Four metabolites were detected, a minor one being 2-hydroxystrychnine which was found to possess only about 1% of the toxicity shown by strychnine itself. The demonstration that hydroxylation occurs at this position agrees with the results of Johns and Wright (1964) who found that the structurally related compound carbazole (82) is hydroxylated at the 3-position in rats and rabbits. The identity of the two major *in vitro* metabolites of strychnine was not determined, but it was shown that they were not strychnine, 18-oxystrychnine, 16-hydroxystrychnine or strychnine-N-oxide.

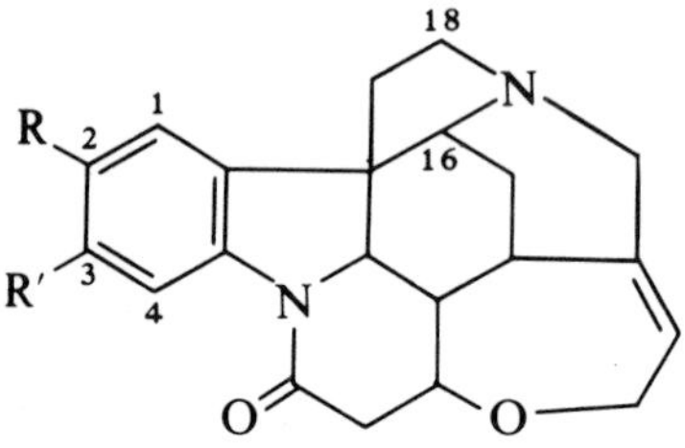

(81)

(a) Strychnine, R = R' = H
(b) Brucine, R = R' = OMe

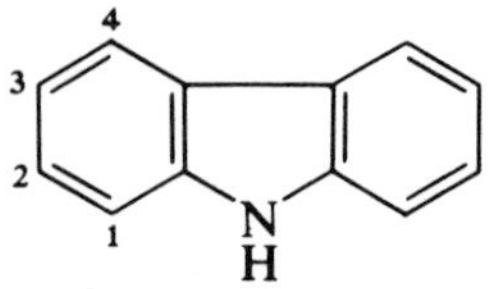

(82)

Brucine (81b) is 2,3-dimethoxystrychnine and its metabolism was studied in regard to its demethylation. Tsukamoto *et al.* (1964b) reported that brucine undergoes selective O-demethylation in the rabbit, giving mainly 3-hydroxy-2-methoxystrychnine which is excreted in the urine largely as its glucuronide conjugate. In addition, small amounts of the isomeric 2-hydroxy-3-methoxy compound, a non-phenolic base and unchanged brucine were also detected in the urine. Faecal excretion of all of these metabolites occurred, but to a very minor extent. A similar metabolic pattern showing preferential demethylation *meta* to the nitrogen atom was found when brucine was incubated with the 9 000 g supernatant fraction from rabbit liver (Watabe *et al.*, 1964).

The vinca alkaloids include some of the chemically most complex types of alkaloids covered in this section on indole derivatives. Examples include vinblastine and vincristine. The few reported investigations on the metabolic fate of vinca alkaloids have dealt with the metabolic disposition of these compounds rather than the metabolic alterations occurring and no metabolites have been characterized. A series of studies by Beer *et al.* (1964a,b), Beer and Richards (1964) and Greenius *et al.* (1968) described the fate in rats of injected **vinblastine** labelled with [3]H. Urinary excretion of radioactivity accounted for about 5% of the dose within 12 h and only very small amounts were lost thereafter by this route. Only about 1% of the dose was found in the urine as unchanged vinblastine. On the other hand, biliary excretion of about 20% of the radioactivity in the bile occurred within 24 h, about a tenth of which was unchanged compound.

Creasey and Marsh (1973) carried out similar experiments in dogs. Following a dose of 0·15 mg/kg (i.v.), the urinary excretion of radioactivity levelled off at 12–17% of the dose after three days. Faecal excretion was more prolonged, amounting to 30–36% after nine days and still increasing. Extensive metabolism of vinblastine in the dog was indicated by the rapidly diminishing proportions of unchanged compound excreted in the urine and faeces during this time period. The bile is the major excretory route of the injected vinblastine in dogs.

The metabolic disposition of tritiated **vincristine** was studied by Mellett and El Dareer (1971) in mice, rats, dogs and rhesus monkeys. The amounts of unchanged compound found in the urine of these animals were 7, 7, 14 and 8%, respectively, whereas the values for urinary metabolites were 8, 7, 14 and 7%. Biliary excretion appeared to be important in all of these species. A similar study in rats and dogs was reported recently by Castle *et al.* (1976). Again, the bile was found to be the major excretory route. When ^{3}H-labelled vincristine was administered (0·1 mg/kg, i.v.) to rats the cumulative excretion of radioactivity in 72 h was 15–17% (urine) and 70% (faeces). Relatively little (<10%) of the radioactivity in the bile and urine was due to metabolites. These were not identified but the urinary metabolites appeared to be more polar than those found in the bile. Most of the radioactivity found in the plasma is due to metabolites, including those detected in the urine and bile. Nonetheless, it was concluded that vincristine is not metabolized extensively in the rat.

Vincamine is a vinca alkaloid of much less structural complexity than the aforementioned compounds. Ezer and Szporny (1967), using ^{14}C-labelled material, found that most of injected compound was eliminated by rats within 24 h in the urine, partly unchanged and partly as unidentified metabolites.

V. Quinoline and Isoquinoline Derivatives

A. Quinoline Derivatives

The only examples in this group are provided by the cinchona alkaloids. The most important of these compounds are the stereoisomeric pairs cinchonine and cinchonidine with structure (83a) and their 6'-methoxyl derivatives quinidine and quinine with structure (83b). Most of the information on the metabolism of cinchonine and cinchonidine derives from studies in man. Earle (1946) and Earle *et al.* (1948) were able to account for about 85% of the daily dose of **cinchonine** in the urine as unchanged compound (4–5%), 2'-hydroxycinchonine (cinchonidine

(83)

(a) Cinchonine and cinchonidine, R = H
(b) Quinine and quinidine, R = OMe

carbostyril) (55%) and a further oxidation product in which an additional oxygen atom is found in the quinuclidine moiety. Renal clearance of the unchanged compound but not the 2'-hydroxy derivative was depressed upon alkalization of the urine. It was noted that cinchonine is metabolized much more rapidly in man than the other cinchona alkaloids, a feature also discussed by Taggart *et al.* (1948). The results pointed clearly to the initial step in cinchonine metabolism being oxidation at the 2'-position, the metabolite having considerably reduced antimalarial effect. This metabolic sequence, first to the 2'-hydroxy derivative which is subsequently hydroxylated in the quinuclidine moiety, was also demonstrated by Brodie *et al.* (1951). While the position of the second hydroxyl group was not determined, it was believed to be in the 2-position (however, see quinine and quinidine metabolism below).

The pattern of urinary metabolites of **cinchonidine** in man is similar to that described above for its stereoisomer. No quantitative data are available but the 2'-hydroxy product (carbostyril) is the major metabolite, being about five times as abundant as unchanged compound and the dihydroxy metabolite which are excreted in roughly equal amounts (Brodie *et al.*, 1951). In addition to these *in vivo* experiments in man, *in vitro* studies employing the quinoline oxidizing enzyme from rabbit liver (see below) were carried out by Knox (1946). Cinchonidine was oxidized much more rapidly than the other three cinchona alkaloids, the rates for the (−)-isomers being 15–25 times greater than those for the corresponding (+)-isomers.

Metabolic studies of quinine and quinidine (83b), although more numerous than those with cinchonine and cinchonidine, have nonetheless mainly been carried out in rabbits and especially rabbit liver preparations and in man. The older literature on **quinine** metabolism, sometimes conflicting, was summarized by Williams (1959, pp. 657–659). In short, the most noteworthy features of these early investigations were the claims that quinetine and haemoquinic acid are urinary metabolites. The former compound differs from quinine in having the vinyl group at C3 replaced with a carboxyl group while the latter compound was believed to be 6-methoxyquinoline-4-keto-carboxylic acid. Williams considered that their being true metabolites of quinine was an open question. A recent investigation by Watabe and Kiyonaga (1972) suggested that quinetine is not a urinary metabolite of quinine in rabbits.

Although no unchanged quinine was detected in the urine of rabbits following oral dosage (Watabe and Kiyonaga, 1972), this has been found to occur in man (Haag *et al.*, 1943; Brodie *et al.*, 1951; Schütz and Hempel, 1974). The results of the first study showed that this amounted to about 17% of the dose (0·5 g, p.o.) when the urine was acidic, decreasing to roughly half of this following alkalinization of the urine. Schütz and Hempel reported that no further quinine was excreted three days after an oral dose of 45 mg.

Information on the patterns of quinine metabolism as deduced from its urinary metabolites is available from experiments in man (Brodie *et al.*, 1951) and in rabbits (Watabe and Kiyonaga, 1972). In man, the major urinary product is not the 2'-hydroxy derivative (carbostyril), as is the case with cinchonine and cinchonidine, but a metabolite with a single hydroxyl group in the quinuclidine moiety. About three times as much of this compound was present as was the carbostyril derivative and some of it appeared to undergo an additional hydroxylation in the quinuclidine group. Similarly to that noted above with the two cinchona alkaloids lacking the 6'-methoxyl group, the location of the quinuclidine hydroxyl group was not determined, although it also was believed to be at the 2-position. However, the results of Watabe and Kiyonaga (1972) suggest that a more likely site is the 3-position, i.e. allylic hydroxylation. They detected six urinary metabolites of quinine (200 mg/kg, p.o.) in rabbits, mainly in conjugated form. All were hydroxylated in the 2'-position and 2'-hydroxyquinine itself was the major urinary metabolite. The spectral data indicated that the latter metabolite exists in the 2-quinolone form. The two other main metabolites were identified as 2',3-dihydroxyquinine and 2',6'-dihydroxycinchonidine. Quantitative measurements indicated that conversion to these metabolites accounted for 35 and 6% of the dose, respectively. The identification of the latter metabolite shows that *O*-demethylation has occurred. These data do not indicate if this reaction occurred before or after oxidation at the 2'-position, but no direct evidence was obtained which indicated that the *O*-demethylated product of quinine, 6'-hydroxycinchonidine, was excreted in rabbit urine. However, it was not excluded that this metabolite could be one of the three unidentified minor metabolites. Axelrod (1956a) reported the *O*-demethylation of quinine by a rabbit liver preparation containing microsomes and soluble fraction, but the nature of the product or products was not determined. An additional metabolic pathway of quinine in man is oxidation to an N-oxide derivative (Jovanović et al., 1976), however its quantitative importance is not known.

Regardless of the uncertainty of the exact identity of the major quinine metabolite in human urine, the results summarized above clearly show that the metabolism of quinine differs in man and rabbits, 2'-hydroxylation being much more prominent in the latter species. In fact, the quinine oxidase present in rabbit liver which forms quinine carbostyril has been the focal point in a number of metabolic investigations. While quinine oxidase activity is found in many tissues from numerous animal species, Kelsey and Oldham (1943), whose article is also useful as a guide to the early literature on this subject, found that rabbits contained the greatest amount of enzyme of all the species studied (mouse, rat, guinea pig, rabbit, cat, dog, sheep, pig, steer, monkey and man), with the liver being the most abundant source. Using this rabbit liver enzyme, Kelsey et al. (1944) isolated the metabolite which was soon shown by Mead and Koepfli (1944) to be 2'-hydroxyquinine. Oldham and Kelsey (1943) reported that little or no enzyme activity was present in rabbit liver before birth, after which it increased to adult levels some time after the age of six weeks. Also, quinine oxidase activity was reduced in rabbits during late pregnancy and the early post-partum period. Several subsequent investigations have dealt with the properties and substrate specificity of rabbit liver quinine oxidase. Knox (1946) showed that oxidation is limited to the α-position of quinoline and derivatives, the rate of oxidation correlating with the activity of the α-hydrogen. The enzyme has properties similar to, and is associated with, the flavoprotein liver aldehyde oxidase. Lang and Keuer (1957) found that quinine oxidase activity was localized mainly in the microsome plus soluble fraction with little activity being found in the mitochondrial fraction of rabbit liver. Führ and Kaczmarczyk (1955) demonstrated considerable quinine oxidase activity in vitro using human liver and also showed that quinine metabolism was inhibited by the simultaneous administration of N-methylnicotinamide, another excellent substrate of the enzyme. Pertinent related data on quinoline oxidation in vitro by rabbit liver preparations have been given by Sax and Lynch (1964).

The metabolism of **quinidine** (83b), the stereoisomer of quinine, has received relatively little attention. Brodie et al. (1951) reported that both isomers showed a similar qualitative pattern of metabolism in man, however the excretion of unchanged compound relative to that of its metabolites was greater with quinidine than with quinine. Iven (1977) reported that about 5% of the dose (11 mg/kg, i.v.) was excreted in the urine unchanged by mice. Brodie et al. demonstrated that the major metabolite in man of quinidine, like quinine, is a compound hydroxylated in the quinuclidine moiety. The 2'-hydroxy (carbostyril) derivative was also excreted, but in smaller amounts than with quinine. Palmer *et al.* (1969) reinvestigated the fate of quinidine in man. Following oral administration (0·66 g daily), numerous metabolites were detected which were shown to exist mainly in the unconjugated state. In addition, some metabolites were shown to be excreted as glucuronides, however no conjugate of quinidine itself was found, indicating that no conjugation takes place at the 9-position. Two metabolites were isolated which were found to be poly-oxygenated compounds having lower molecular weights than quinidine, however the two major characterized metabolites were derivatives containing oxygen in the 2'-position or in the quinuclidine moiety. The former metabolite was shown to be 2'-quinidinone, i.e. the keto form of 2'-hydroxyquinidine. Carroll *et al.* (1974), employing ^{13}C-nuclear magnetic resonance techniques, confirmed this structure and also showed that the second metabolite was 3-hydroxyquinidine. Beermann *et al.* (1976) also reported the urinary excretion of the 3-hydroxy derivative by man follow-ing quinidine administration. Oxidation at the 3-position furnishes an example of allylic hydroxylation.

B. Isoquinoline Derivatives

Studies on the metabolism of **papaverine** (84) extend over a period of about 60 years, although useful data on the transformations taking place

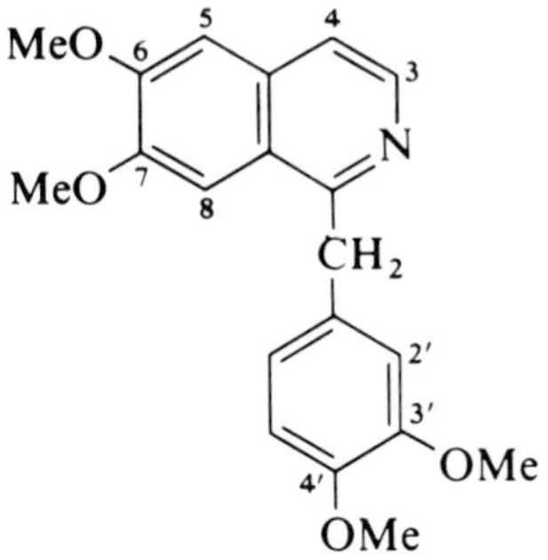

(84)
Papaverine

are of fairly recent date. Zahn (1915) was not able to detect unchanged compound or metabolites in the excreta of rabbits, cats and dogs given papaverine i.p. Only traces (<1% of the dose) of unchanged compound were found in the urine of rabbits 24 h after i.v. administration (20 mg/kg) (Lévy, 1945). In man, Elek and Bergman (1953) failed to find any unchanged papaverine in the urine following oral or parenteral dosage (approx. 200–600 mg), although one or more probable derivatives were noted. In another investigation, less than 1% of the dose was recovered unchanged in the 24 h urine of subjects receiving 3 mg/kg i.v. or p.o. (Axelrod *et al.*, 1958). This value for the urinary excretion of unchanged papaverine in man has been confirmed by Guttman *et al.* (1974) using other analytical methods. Additional excretion values are given in Table 9.2.

TABLE 9.2

Urinary papaverine and metabolites 48 h after administering papaverine hydrochloride[a]

Species	Dose	Papaverine %	Free phenolic metabolites %	Conjugated phenolic metabolites %
Rat	50 mg/kg, i.p.	0·5	2·9	45
Guinea pig	50 mg/kg, i.p.	0·7	18	56
Dog	15 mg/kg, i.v.	0·5	1·6	7·1
Man	10 mg/kg, p.o.	0·5	0·5	64

[a] Axelrod *et al.* (1958).

Axelrod (1956a) reported that papaverine was a good substrate for the O-demethylating enzyme found in a rabbit liver preparation containing the microsomal and soluble fractions. Using tissue preparations from guinea pigs and rabbits, Axelrod *et al.* (1958) subsequently found that the liver is the most active site of papaverine metabolism and that this activity is found in the microsomal plus soluble fraction, requiring TPNH and oxygen. Furthermore, one mole of formaldehyde was formed for every mole of substrate metabolized. They also showed that when guinea pigs received papaverine (100 mg/kg, i.p.), the 24 h urine contained three phenolic metabolites, the amounts of which increased following acid hydrolysis. One of these was identified as the $4'$-O-demethylated product, the so-called $4'$-hydroxypapaverine (85a) which accounted for about 70% of the phenolic material. The results of their experiments in rats, guinea pigs, dogs and man are summarized in Table 9.2. This table shows that rats, guinea pigs and man are fairly similar with respect to the amounts of conjugated phenolic material excreted, but that guinea pigs excrete much more unconjugated phenolic metabolites than is the case in the other species. Dogs, on the other hand, appear to rely on another pathway to metabolize papaverine. As with the guinea pig, the major urinary phenolic metabolite in man was the $4'$-O-demethylated compound.

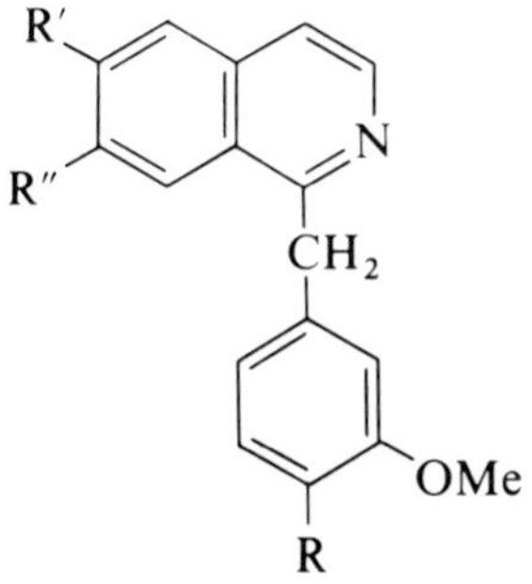

(85)

(a) R = OH, R′ = R″ = OMe
(b) R′ = OH, R = R″ = OMe
(c) R″ = OH, R = R′ = OMe
(d) R = R′ = OH, R″ = OMe

Our knowledge of papaverine metabolism has been extended by a series of detailed investigations by Belpaire and Bogaert. Using material labelled with ^{3}H, they showed that most of the radioactivity was eliminated within 24 h in the faeces of rats given doses of 5 mg/kg, p.o., i.m. or i.v. (Belpaire and Bogaert, 1972, 1973). After four days, the total faecal excretion of radioactivity was 68–76% whereas 11–18% was found in the urine, most in the first day. About 70% of an i.v. or i.m. dose of papaverine was excreted in the bile in 6 h. This finding, together with that showing that absorption of radioactivity does not occur when the biliary papaverine metabolites are administered duodenally, indicates that entero-hepatic circulation is not an important feature in the biological disposition of papaverine in rats. Hydrolysis of the biliary metabolites with a pre-paration containing β-glucuronidase and sulphatase gave rise to four metabolites, the structures of which have now been determined (Belpaire et al., 1975). These are the three monodemethylated derivatives, 4′-desmethyl- (85a), 6-desmethyl- (85b) and 7-desmethylpapaverine (85c) and the didemethylated compound, 4′,6-didesmethylpapaverine (85d). The studies on the excretion of the papaverine metabolites have been extended to other species including guinea pigs, rabbits, cats and dogs (Belpaire and Bogaert, 1974, 1975a). [^{3}H]-Papaverine (5 mg/kg) was given i.v. to these animals and rats and the biliary and urinary excretion of radioactivity measured for 5–6 h. Biliary excretion of ^{3}H, which accounted for about 80% of the dose in rats in these experiments, averaged about 50% in the other species. The average values for urinary ^{3}H excretion were: rat, 8%, guinea pig, 29%, rabbit, 45%, cat, 7% and dog, 10%. These values are mostly lower than those shown in Table 9.2, except with the dog, but this is likely a result, in part at least, of the shorter collection periods. Little or no unchanged papaverine was found in this material, the bulk in both the bile and urine samples being conjugates of the four phenolic metabolites. However, quantitative differences were found, metabolite (85b) being most important in cats and metabolite (85a) in the other species. An unidentified fifth metabolite was detected in the bile of dogs and, especially, cats. A further study by Belpaire and Bogaert (1975b) investigated the metabolism of papaverine to the monodemethylated metabolites by the 9 000 g fraction of rat liver. The O-demethylation reaction was stimulated by the inducing agent phenobarbital and inhibited by the compound SKF 525 A. Rosazza *et al.* (1977) found that the three desmethyl metabolites (85a), (85b) and (85c) were formed when papaverine was incubated with phenobarbital-induced rat liver microsomes. The 7-desmethyl derivative (85c) was the major metabolite in these studies, however about equal amounts of metabolites (85a) and (85b) were formed when liver microsomes from guinea pigs were used.

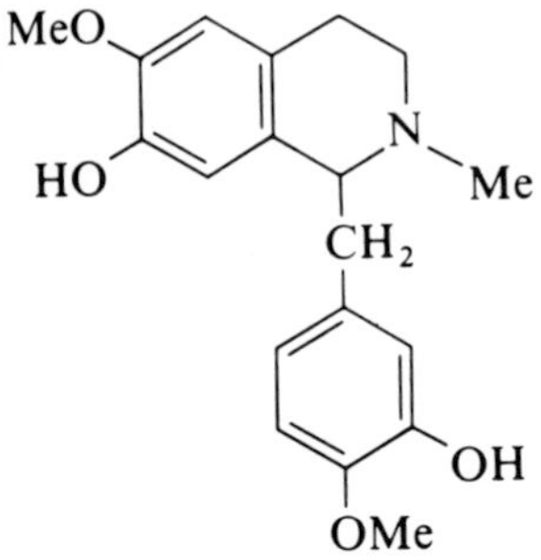

(86)
Reticulene

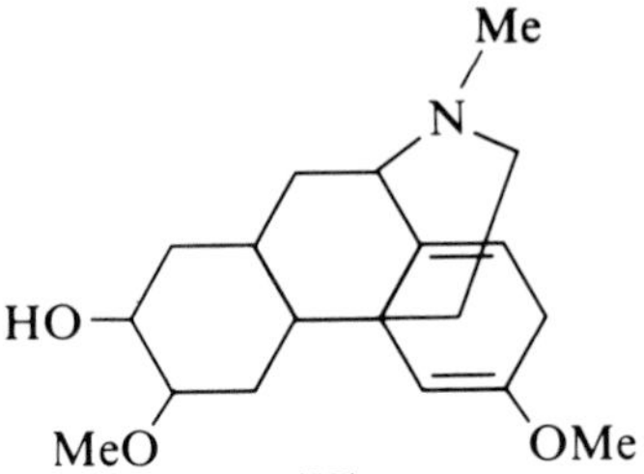

(87)

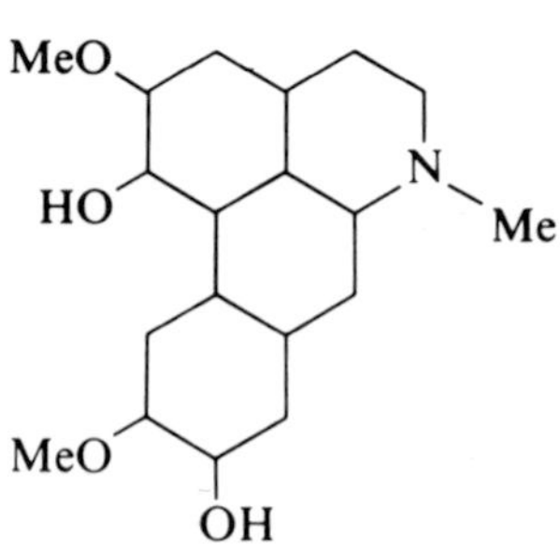

(88)

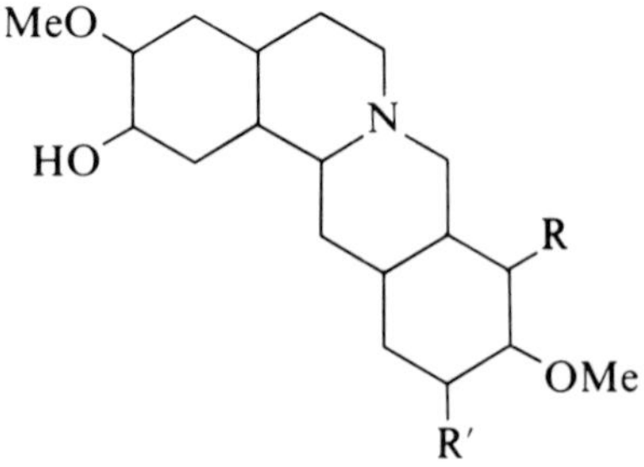

(89)
(a) R = H, R' = OH
(b) R = OH, R' = H

($\pm$)-**Reticulene** (86) is an opium alkaloid showing close structural similarity to papaverine (84). Its metabolism in rat liver homogenates was studied by Kametani *et al.* (1977) who reported that cyclization reactions occurred which led to the formation of several more complex alkaloids. The metabolites identified as a result of this process were the morphinandienone alkaloid pallidine (87), the aporphine alkaloid isoboldine (88) and the two protoberberine alkaloids coreximine (89a) and scoulerine (89b). Formation of the alkaloidal metabolites was stimulated by NAD, NADP and NADPH. An experiment with deuterium-labelled reticulene indicated that the *N*-methyl group did not appear in the protoberberines (89a) and (89b).

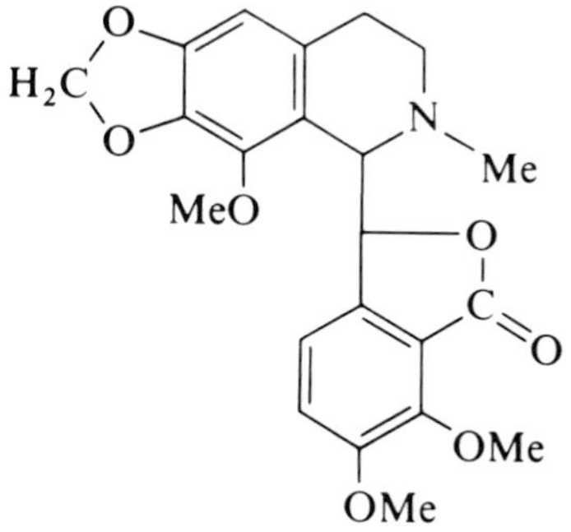

(90)

Noscapine

Noscapine (90) is another opium alkaloid related chemically to papaverine. It was formerly called narcotine, an unsuitable name considering that its main pharmacological effect is an antitussive action and that its narcotic properties are insignificant. Little definite information is available on the metabolic changes of noscapine and most of the published data deal with other aspects of its biological disposition. However, it seems that little unchanged compound is excreted from the body. Cooper and Hatcher (1934) reported a value of 0·3–0·5% of a total i.m. dose of 200–300 mg/kg for the urinary excretion of unchanged noscapine in 4–5 days in cats. Essentially no unchanged compound is excreted in the faeces and less than 1% of the dose (15 mg/kg, i.v.) is found in the urine of rabbits in 24 h (Nayak *et al.*, 1965). In man, the urinary excretion of free and conjugated noscapine did not exceed 1% over a 6 h period following a 500 mg oral dose (Vedsö, 1961). It was also shown that noscapine undergoes a very rapid tissue uptake and this was thought to indicate concentration of compound at these sites and also a low rate of metabolism. Subsequent reports have confirmed this rapid tissue uptake of noscapine, but they also showed that it is quickly lost from these areas and metabolized at a rapid rate (Nayak *et al.*, 1965; Nayak, 1966; Idänpään-Heikkilä, 1968). Nayak (1966) showed that rabbit liver and kidney slices metabolized noscapine and also that two glucuronide conjugates of unidentified phenolic derivatives were excreted in the urine in rats and rabbits. However, these accounted for less than 5% of the dose. A third highly fluorescent metabolite was detected which gave negative tests with alkaloidal, phenolic and glucuronide reagents. Further information on the nature of the metabolites of noscapine in rabbits was reported by Tsunoda *et al.* (1976). Using oral doses (85–150 mg/kg), they found that the 24 h urine contained three free and three conjugated compounds in the basic fraction. The former were shown to be unchanged compound and two mono-*O*-demethylated metabolites while the latter were conjugates of the two monodemethylated compound and of a di-*O*-demethylated metabolite. The major metabolite was the free and conjugated forms of one of the monodemethylated compounds, however the amount of total urinary metabolites detected was small. Idänpään-Heikkilä (1968) carried out an extensive study of the fate of [^{3}H]-noscapine in mice and rats. In the latter species, the same pattern of excretion of radioactivity was found after both p.o. and i.v. dosage. In the urine, about one fourth of the dose (0·5 mg/kg) was excreted in 24 h, increasing only slightly during the next few days. Faecal excretion of ^{3}H was 60–63% of the dose in four days. The bile was a major excretory route, accounting for about 70% of the dose in 24 h, of which about one fourth was reabsorbed from the intestine. Two major, although unidentified, biliary metabolites appeared to be glucuronide conjugates of phenolic derivatives of noscapine. It therefore appears that the metabolism of noscapine is quite similar to that of the closely related papaverine, however phenolic metabolites may also be formed as a result of scission of the methylenedioxy ring.

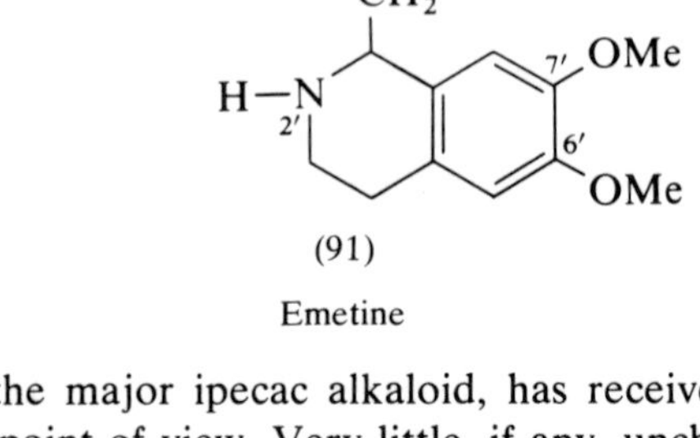

(91)

Emetine

Emetine (91), the major ipecac alkaloid, has received little attention from a metabolic point of view. Very little, if any, unchanged compound was found in the urines of rats and dogs receiving the alkaloid (1–2 mg/kg) by injection or in the urine of man after ten daily oral doses totalling 600 mg (Gimble *et al.*, 1952). Their results pointed also to a prolonged storage of the compound in the body. Schwartz and Herrero (1965), using [^{14}C]-emetine, found that 95% of the injected radioactivity was excreted by guinea pigs in the faeces and only 5% in the urine. No information was obtained concerning the nature of the possible metabolites, but most of the faecal material appears to be present as metabolites as only 40% of the radioactivity was extracted when methods used for the determination of emetine were employed.

A more recent investigation employed (±)-2,3-dehydroemetine, a less toxic synthetic analogue of the naturally occurring (−)-emetine, labelled with ^{14}C in the 3'-position (Johnson *et al.*, 1971; Johnson and Jondorf, 1973). They used bile duct-cannulated rats to which 5 mg/kg of the compound was given by i.v. injection. Nor surprisingly, faecal excretion of radioactivity was low (1% in 72 h). At this time, about 35% of the dose was found in the bile and urine, in approximately equal amounts. Interestingly, 30–40% of the urinary radioactivity appeared to consist of unchanged 2,3-dehydroemetine while the remainder included several polar metabolites not susceptible to hydrolysis with β-glucuronidase and sulphatase. In the bile, however, only a trace of the excreted material was unchanged compound and most of these metabolites appeared to be glucuronide conjugates. One of these released 2,3-dehydroemetine upon hydrolysis, suggesting that it may be a 2'-*N*-glucuronide derivative. Other aglycones more polar than the original compound were detected but it is not clear if these are derivatives formed by *O*-demethylation or ring hydroxylation. The former possibility appears reasonable by analogy with the metabolism of other methoxyisoquinolines, however incubation of 2,3-dehydroemetine with the rat liver microsomal drug metabolizing system failed to show appreciable *O*-demethylation.

Berberine is probably mainly metabolized before excretion since relatively small amounts of unchanged compound are excreted in the urine. Schein and Hanna (1960) found about 1% of the dose in the urine of rats 24 h after a s.c. injection (50 mg/kg). Chin *et al.* (1965) recovered about 15% of the dose (60 mg/kg, i.p.) in the urine of rabbits in 48 h. They believed that this material was unchanged berberine, however related compounds may have contributed to the value as their assay method was not specific for berberine. They also found that the alkaloid was metabolized by liver slices. The ability of the liver to metabolize berberine has also been reported by Furuya (1956) in perfusion experiments. The metabolite was an unidentified oxidation product which was also isolated from the urines of rabbits and dogs given berberine HCl by s.c. injection.

(92)

(93)

Sanguinarine

A single report on the metabolism of the argemone alkaloid **sanguinarine** (92) described its conversion to 3,4-benzacridine (93) (Hakim *et al.*, 1961). The metabolite was detected in the blood in isolated rat liver experiments, in the milk of lactating rabbits and in rabbit urine after s.c. injection of the alkaloid (approx. 2 mg/kg). This is a most unusual and profound metabolic alteration, however no further reports have appeared which might clarify the metabolic pathways of sanguinarine.

d-Tubocurarine and related curarines are relatively large molecules which possess two quaternary ammonium groups. The resultant ionized character of these compounds is the major determinant with regard to their biological disposition, as well as the feature which dictates the use of the i.v. route when administering the compounds to effect muscular relaxation. Waser *et al.* (1954) reported that the i.v., s.c. and p.o. doses of *C*-curarine required to produce the same degree of paralysis in cats increased in the ratio 1 : 10 : 100. For understandable reasons, the majority of the studies on the biological disposition of these compounds have given most attention to their distribution, plasma levels and, perhaps also, their excretion. These topics fall largely outside the scope of this book and will therefore not be covered, however most of the pertinent information is available in the articles of Mahfouz (1949), Marsh (1952), Foldes (1957), Kalow (1959), Chagas (1962), Cohen *et al.* (1965, 1968) and Crankshaw and Cohen (1975).

Investigations on the metabolism of *d*-**tubocurarine** and its relatives are not numerous and the findings are rather limited. The main reason for the latter fact is, simply, that these alkaloids appear to be metabolized to a very limited degree. The results from several early investigations (Mahfouz, 1949; Marsh, 1952; Kalow, 1953), which indicated that only a fraction of the injected dose was excreted in the urine, led to the suggestion that *d*-tubocurarine is metabolized. However, more recent data obtained with radioactive compounds suggest that this is not the case. Cohen *et al.* (1967) administered [³H]-*d*-tubocurarine (0·3 mg/kg, i.v.) to dogs and recovered about 75% of the radioactivity in the urine and 11% in the bile in 24 h. No metabolites were detected in the urine, even after hydrolysis with β-glucuronidase, although a demethylated derivative, accounting for 0·1–1% of the dose, was found in the bile together with the unchanged compound. Biliary excretion of tubocurarine was also demonstrated by Robelet *et al.* (1964) using the perfused, isolated rat liver. The material excreted was probably unchanged compound as it did not show any loss in biological activity. Meijer *et al.* (1976) carried out similar experiments and found no evidence for the metabolism of d-tubocurarine or its methylated derivative, trimethyltubocurarine. The latter compound, formerly called dimethyltubocurarine, was found by Chagas (1962) to be excreted to the extent of 50–70% in the urine and 2% in the bile in 3–6 h when radioactive material was given by injection to dogs. Chromatography failed to show the presence of any metabolized compounds. Waser and Lüthi (1968), using material labelled with ^{14}C in the N-methyl group, found that both *C*-curarine and *C*-toxiferine were mainly excreted in the urine and that no N-demethylation occurred in cats which received i.v. doses of 100–200 µg/kg. About 10% of the dose of the former compound was excreted in the bile.

VI. Morphinan Derivatives

The morphinan skeleton (94) is found in several of the principal opium alkaloids. The most important of these are morphine (95), codeine (96) and, to a lesser extent, thebaine (97). As morphine was the first alkaloid to

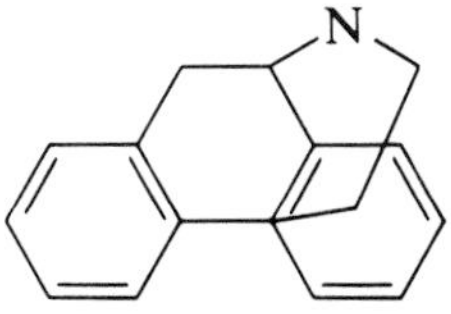

(94)

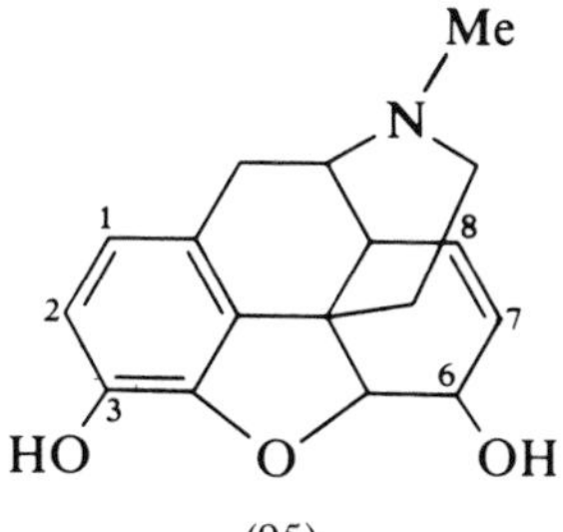

(95)

Morphine

be isolated in pure form as well as one of the few reliable and effective drugs available prior to the present century, it is hardly surprising that interest in its biological properties spans a relatively long period of time. In fact, studies of its metabolic fate in animals were carried out nearly 100 years ago. Many of these studies have their origin in the desire to clarify the enigma of the addiction and tolerance which result from repeated use of morphine and several of its relatives. It is therefore hardly surprising that

Me

Me

MeO   O   OH   MeO   O   OMe

(96)   (97)

Codeine   Thebaine

there exists a vast literature on the biological disposition of these opium alkaloids. In the following discussion, the pathways of their metabolism will be outlined with emphasis being placed on investigations carried out during the last two decades. It has been mainly during this period that specific information has been obtained on the identity of the metabolites and their pathways of formation. Further information on the metabolism and other aspects of the biological disposition of these opium alkaloids is available in the reviews by Way and Adler (1961, 1962), Mulé (1971), Scrafani and Clouet (1971), Misra (1972), Boerner et al. (1975) and Lemberger and Rubin (1976, Chapter 5).

It is of interest to note that injections are nearly always employed when studying the metabolism of **morphine** (95). This practice is related to the well-known unpredictability of effects when morphine is given orally. This has long been considered to be due to erratic absorption from the intestine, however an increasing body of evidence indicates that its absorption from this site is nearly complete and that the low and variable effects observed are more likely due to a first-pass effect. References to the literature on this subject are given in the publication of Iwamoto and Klaassen (1976) who studied the phenomenon in rats. They recorded an overall first-pass effect of more than 80% of the dose (5 mg/kg). Of this, the ratio of the extent of extraction and/or metabolism by the intestine compared with the liver was found to be about 2:1.

Although most of the administered morphine undergoes metabolic transformation prior to its excretion from the body, numerous investigations in many animal species have indicated that some excretion of unchanged compound also occurs, mainly in the urine. The reported values for free urinary morphine generally fall in the range of 5–30% of the dose. Species differences are not pronounced but it appears that free morphine excretion in rats and cats is slightly higher than that found in other species. Zauder (1952) reported a value of 28% in the 24 h urine of rats given morphine (12 mg/kg, s.c.). Values of 24% in 48 h following a similar dose

(Misra *et al.*, 1961b), 17% daily for Wistar rats and 24% daily for Gunn rats following repeated daily injections of 25–35 mg/kg (Abrams and Elliott, 1974) and 20% after a dose of 10 mg/kg (Oguri *et al.*, 1970) have also been found. On the other hand, Klutch (1974) reported that 7–11% of repeated i.p. doses (10 mg/kg) were excreted unchanged each day. Similarly, 10–11% of a single s.c. dose of morphine (20 mg/kg) was excreted in the urine of rats as such, mainly in the first 24 h (Milthers, 1962a). Yeh *et al.* (1977) recorded a value of 9% in similar experiments. This investigation of Yeh *et al.* also reported the extent of excretion of free morphine in several other species including the mouse which gave values of 6·5–10% in 144 h following a s.c. dose of 75 mg. Guinea pigs (25 mg/kg, s.c.) excreted slightly more than 3% in 24 h. Axelrod and Inscoe (1960) found 7% of the dose (90 mg/kg, i.p.) excreted unchanged in the 24 h urine in guinea pigs. Values of 3–12% in 48 h were obtained in rabbits given large (100–200 mg/kg) injections of morphine (Keeser *et al.*, 1933) and the newer study by Yeh *et al.* recorded values of approximately 4–5% in 24 h following a s.c. dose of 25 mg/kg. In cats, Yosikawa (1940) reported values of roughly 11–27% and Yeh *et al.* (1971) found that 19% of a 20 mg/kg injection was excreted as free morphine in 48 h. More recently, Yeh *et al.* (1977) reported 24 h values of 20–27% in similar experiments. Dogs have often been employed in studies of the biological disposition of morphine and abundant data are therefore available on its urinary excretion in this species. A comprehensive summary of this work has been made by Way and Adler (1961) who noted that about 15% of the dose is excreted as free urinary morphine within 24 h. Misra *et al.* (1970) subsequently reported that about 14% of a 20 mg/kg dose injected s.c. was excreted in the urine of dogs in 12 h and Yeh *et al.* (1977) recorded a 24 h value of 20% in similar experiments. In the monkey (*Macaca mulatta*) only 3–7% of the dose of morphine (30 mg/kg, s.c.) was excreted unchanged in the urine (Mellet and Woods, 1956). A later study using a smaller dose (2 mg/kg) gave slightly higher values (6–14%) (Mellet and Woods, 1961). The data on morphine excretion are undoubtedly most extensive in the case of man. These results, from both addicts and non-addicts, have been tabulated by Way and Adler (1961) and recently by Boerner *et al.* (1975). Their summaries indicate that there is fairly good agreement among these results which show that an average of 7% of the dose is excreted in the urine as free morphine. The range of values reported is 1–14%. A recent investigation by Yeh (1975a) in morphine-dependent subjects gave a value of 10%. These studies have also shown that the percentage excretion of free morphine does not vary greatly with varying doses and also that the differences between addicts and non-addicts are not significant.

In addition to the urinary excretion of morphine and its metabolites, some loss from the body occurs in the faeces. The data summarized by Way and Adler (1961) indicate that most species including man excrete from a few to roughly 10% of the dose by this route. Subsequently, data on cats appeared which showed that 2–20% of the dose was found as total morphine in 48 h in the faeces (Yeh *et al.*, 1971). The nature of the faecal metabolites has not been investigated in detail but it appears that they consist mainly of free and conjugated morphine. It seems clear that the faecal excretion of morphine is dependent upon its excretion in the bile, mainly as morphine glucuronide (see Smith, 1973). Interestingly, species variations in the biliary excretion of morphine and its metabolites are large, in contrast to that found with the faecal excretion. Roughly 60% of the dose (5 mg/kg, s.c.) was found in the bile of rats in 6 h (March and Elliott, 1954). Similar results were obtained in this species by Smith *et al.* (1973). More than 35% of the injected morphine is excreted in conjugated form in the bile of dogs (Woods, 1954). Biliary excretion is less extensive in the cat, the reported values being not more than 10% (Chernov and Woods, 1965), 5–23% in 3 h (Yeh *et al.*, 1971) and 14% in 3 h (Smith *et al.*, 1973). In monkeys (*Macaca mulatta*), 10–20% of the dose can be recovered within a few hours in the bile, mainly in conjugated form (Mellet and Woods, 1956). The biliary route is of less importance in man, the results of Elliott *et al.* (1954) suggesting a value of about 7%. These findings showing that large amounts of morphine, mainly in conjugated form, are sometimes excreted in the bile whereas much less, usually in the free form, is found in the faeces indicates that an enterohepatic circulation of the compound takes place. Walsh and Levine (1975) studied the enterohepatic circulation of morphine in rats.

As noted above, most of the morphine administered is metabolized. The two major general metabolic routes consist of synthetic reactions, of which glucuronide formation is usually the most important, and oxidative reactions, of which *N*-demethylation is the most important. Suggestions that morphine is excreted in the urine in what was called a bound form were made in some of the earliest reports on morphine metabolism. However, these beliefs were first placed on a firm foundation by the investigations of Gross and Thompson (1940) in dogs and Oberst (1940) in man. The former study showed that the substance released upon acid hydrolysis of the urine was actually morphine. The results in man indicated that acidic hydrolysis of the urine increased the amount of morphine present by factors of 3 to 36. Oberst (1941) found that the glucuronic acid concentration in urine increases proportionally with increasing doses of morphine and concluded therefore that the bound form is a glucuronide. This was later confirmed when several groups demonstrated conclusively that the bound form, or at least the major bound form, is morphine-3-glucuronide. Woods (1954) isolated this conjugate from the urine and bile of morphine-treated dogs and further confirmation was provided by the results of Seibert *et al.* (1954) and Fujimoto and Way (1954, 1957). The main pathway in the glucuronidation of morphine in mammals leads to the 3-glucuronide, although this is a minor reaction in cats which form instead the ethereal sulphate (see below). Capel *et al.* (1974) did not detect morphine-3-glucuronide in the urine 24 h following the i.p. administration of morphine to cats, however Yeh *et al.* (1971) reported that 1–2% of the dose was excreted in this form in roughly similar experiments. In addition to the 3-glucuronide, it is now clear that the 6-glucuronide is also formed in small amounts. This minor metabolite was detected in the urine of morphine-treated rabbits (Yoshimura *et al.*, 1969) and dogs (Misra *et al.*, 1970) and subsequently in the urine of mice, rats, guinea pigs, rabbits and man (Oguri *et al.*, 1970). The last study showed that the ratio of 3-glucuronide to 6-glucuronide found in guinea pigs, rabbits and man is about 100:1. The metabolism of these glucuronides in rats (Ida *et al.*, 1975a) and in mice, guinea pigs and rabbits (Ida *et al.*, 1975b) has been studied. In general, the patterns of urinary metabolites vary according to the route of administration of the glucuronides in the latter two species, whereas this is seen to a much lesser degree in rats and mice due to their greater ability to excrete the conjugates in the bile and thus expose them to the action of bacterial β-glucuronidase in the lower intestine.

The extent of urinary excretion of morphine glucuronide has been determined in many investigations. In their summary of earlier data, Way and Adler (1961) reported values of 11–45% of the dose in man. In a study using addicts, Yeh (1975a) reported a value of about 65% for the excretion of urinary morphine glucuronide. Depending on dosage route, extrahepatic conjugation may contribute to the formation of the glucuronide. Thus, Brunk and Delle (1974) obtained evidence which indicated that morphine glucuronide is rapidly formed in the cells of the intestinal mucosa following oral administration of the alkaloid to man. In the monkey (*Macaca mulatta*) the value for the urinary excretion of morphine glucuronide is 70–80% (Mellett and Woods, 1956). Typical values in dogs fall in the range of 38–75% (Way and Adler, 1961). Morphine-3-glucuronide is a major urinary metabolite of morphine in rabbits (Fujimoto and Haarstad, 1969), however quantitative data are lacking. Guinea pigs, following a relatively large i.p. dose (90 mg/kg), excreted 38% of the dose as glucuronide conjugated material (Axelrod and Inscoe, 1960). In rats, values between 20 and 50% are typical in earlier investigations (Way and Adler, 1961) although Misra *et al.* (1961b) reported that 14% of the dose was excreted as conjugated morphine in Sprague–

Dawley rats. A study by Abrams and Elliott (1974) using Gunn and Wistar rats showed no differences in the amounts of urinary glucuronide, both strains excreting about 9% of the repeated daily doses in this form. Relatively few studies have been reported on the *in vitro* formation of the glucuronide conjugates of morphine. Strominger *et al.* (1954) showed that morphine served as an acceptor in a glucuronide-synthesizing system consisting of calf or guinea pig liver microsomes plus supernatant fraction, UDPG and DPN, the latter substance being required in the oxidation of UDPG to UDPGA. A subsequent study by Axelrod and Inscoe (1960) showed that morphine glucuronide was formed when the alkaloid was incubated with a microsomal preparation from guinea pig liver and UDPGA. Sanchez and Tephly (1973) obtained results which indicated that different glucuronyltransferases may catalyse the conjugation of morphine and *p*-nitrophenol in rat liver microsomes. Details on the enzymic formation of glucuronides are covered in Chapter 1, Section I.D.1.a. The hepatic UDP-glucuronyltransferase activity, which in Gunn rats shows varying degrees of deficiency towards various substrates, was measured by Abrams and Elliott (1974) who found that conjugation with morphine was not significantly different from that shown by preparations from Wistar rats.

In the mammalian species studied, ethereal sulphate conjugation of morphine appears to be a major synthetic reaction only in cats. Morphine ethereal sulphate was first isolated from the urine of morphine-treated cats by Woods and Chernov (1966) and this conjugate was shown by Fujimoto and Haarstad (1969) to be morphine-3-O-sulphate. This metabolite was also shown to be present in the urine of cats given injections of morphine by Mori *et al.* (1972) who found no evidence for the excretion of the 6-O-sulphate conjugate. Quantitative data on the excretion of morphine-3-O-sulphate in cats have been reported by Yeh *et al.* (1971) and Capel *et al.* (1974). In the former study, about 48% of the dose (20 mg/kg, s.c.) was excreted in the urine in 48 h as conjugated material, mainly the 3-O-sulphate. Values of roughly 70% for this metabolite were recorded in the latter investigation, which employed a dose of 5 mg/kg and a 24 h collection period. Also, no morphine glucuronide was found but another sulphate conjugate, possibly morphine-6-O-sulphate, which accounted for about 1% of the dose was detected.

Other species in which ethereal sulphate conjugation of morphine has been demonstrated include the dog (Misra *et al.*, 1970) and man (Yeh, 1973). The latter study indicated that the ratio of urinary morphine ethereal sulphate to morphine glucuronide was about 1:4.

Morphine is converted to codeine (96) as a result of O-methylation at the 3-position. This reaction was first demonstrated by Elison and Elliott (1964) who found small amounts of the metabolite in the urine of rats given morphine. Increased formation of codeine was found when rats deficient in liver glucuronide conjugating ability were used. Codeine formation was also noted *in vitro* using rat liver preparations, however the nature of the methyl donor was unclear as [*methyl-*^{14}C]-S-adenosyl-methionine added to the incubates did not result in the formation of radioactive product. An investigation of the metabolism of morphine in both Gunn and Wistar rats failed to detect codeine in the urine, thereby raising doubts as to the significance of this metabolic pathway (Abrams and Elliott, 1974). Disagreement has also been registered on the formation of codeine from morphine in man. Börner and Abbott (1973) reported that 0·7–0·9% of the dose of morphine given p.o. or i.v. to non-tolerant subjects was recovered as urinary codeine. The codeine excretion was significantly increased in tolerant subjects. Subsequently, Boerner et al. (1974) identified norcodeine in the urine of non-tolerant and tolerant subjects. Yeh (1974) detected only very small amounts of codeine in the urine of subjects receiving morphine chronically and concluded that these were more likely due to the small amounts of codeine present in the morphine used than to O-methylation of the latter. Also, Yeh (1975b) has questioned the interpretation that norcodeine is a metabolite of morphine, suggesting instead that it arises from the metabolism of the small amounts of codeine present in morphine samples. This viewpoint is not, however, accepted by Boerner and coworkers (Boerner and Roe, 1975; Boerner et al., 1975).

(98)

Little is known regarding the possibility that morphine may be metabolized to other conjugated derivatives. However, Misra et al. (1970) detected chromatographically two minor metabolites in the urine of morphine-treated dogs which gave purple colours when sprayed with nin-hydrin reagent. Misra and Woods (1970) obtained evidence suggesting that morphine and codeine can form a covalently linked conjugate with glu-tathione under non-enzymic conditions, particularly in the presence of ferrous ions.

The most important and widely studied oxidative reaction occurring with morphine is N-demethylation. The initial evidence for its oxidative demethylation to normorphine (98) was obtained using [N-methyl-^{14}C]-morphine. March and Elliott (1952, 1954) found that the s.c. injection of labelled morphine (5 mg/kg) in rats resulted in loss of some of the radio-activity as respiratory ^{14}CO$_2$. When male rats were used, about 5% of the radioactivity was excreted in this manner while, with female rats, the values were only about 10% as large. Also, rat liver slices metabolized morphine to ^{14}CO$_2$ and the preparations showed the sex-related differences. Similar in vivo studies in man confirmed the N-demethylation reaction, with 3·5–6% of the radioactivity being lost as respiratory ^{14}CO$_2$ in a 24 h period following the i.m. injection of morphine (15 mg/kg). Another early but different approach to this problem was taken by Axelrod (1956b) who found that enzyme systems were present in the livers of several mammalian species which are capable of demethylating morphine to normorphine and formaldehyde. These enzyme systems are located in the liver microsomal fraction and require TPNH, O$_2$ and various cofactors. Franklin (1965) studied this system in more detail and found that increased rates of N-demethylation occurred when the concentration of nicotinamide in the incubate was increased. Normorphine has also been shown to be formed from morphine in the brain of rats in vivo (Milthers, 1962b). Gutierrez and Flaine (1958) detected normorphine when morphine was incubated with rat liver and brain homogenates. Using morphine labelled with ^{14}C in the N-methyl group, Elison and Elliott (1963) found that aerobic incubation with rat brain slices resulted in the formation of ^{14}CO$_2$, however relatively low activity was present compared with that present in the liver. Recently, Fishman et al. (1976) carried out in vivo studies in rats which indicated that the N-demethylation of morphine in brain occurs at sites which have a high content of opiate receptors.

Numerous investigations have demonstrated that normorphine is a metabolite of morphine in many mammalian species. In mice, Maggiolo and Huidobro (1967) and Oguri et al. (1970) obtained evidence for its formation and Adler (1967) recovered 1–2% of the dose as respiratory ^{14}CO$_2$ in 4 h following s.c. injection of [N-methyl-^{14}C]-morphine (5 mg/kg). In rats, many investigations including those noted above have demonstrated the N-demethylation of morphine, although Abrams and Elliott (1974) did not find evidence for normorphine excretion in the urine of Gunn and Wistar rats given repeated daily injections of morphine (20–35 mg/kg). However, several other studies indicate that values of a few to perhaps ten per cent of single or repeated doses of roughly this size can be expected (Milthers, 1962a; Heimans et al., 1971; Misra et al., 1973; Klutch, 1974). Further evidence for the conversion of morphine to normorphine in rats has been obtained by Misra et al. (1961b) and Oguri et al. (1970). The latter group obtained similar results with rabbits and normorphine was detected both in the liver (Tampier and Penna-Herreros, 1966) and urine (Yeh et al., 1971) of cats dosed with morphine. In dogs Misra et al. (1970) obtained evidence which suggested the urinary excretion of normorphine glucuronide. It appears that the N-demethylation of morphine in this species may be rather limited as Mellett and Woods (1961) found that only about 0·2% of the dose (2 mg/kg, s.c.) of [N-methyl-^{14}C]-morphine was excreted as respiratory $^{14}CO_2$. They obtained a value of about 1% in similar experiments in monkeys. Oguri et al. (1970) detected normorphine in the urine of a human given three 10 mg injections of morphine and several other investigations have confirmed and extended this finding. Yeh (1973, 1975a) reported that morphine-dependent subjects receiving four 60 mg injections daily excreted from 0·5–1% of the dose as free and about 3% as conjugated normorphine. Boerner et al. (1974) estimated that about 5% of large doses of morphine given chronically to a patient was excreted in the urine as normorphine. Yeh et al. (1977) investigated the urinary excretion of normorphine in several mammalian species following the acute or chronic administration of morphine. Free and conjugated normorphine was found in all species (mouse, rat, guinea pig, rabbit, cat, dog, monkey and man), however the total excretion values were not large (approximate range of 0·5–7%).

The above studies amply demonstrate that morphine undergoes N-demethylation in the body to normorphine (98) which is then excreted in the urine, partly free and partly in conjugated form (see also Misra et al., 1961a). Additionally, normorphine may be metabolized in other ways, especially by N-methylation which regenerates morphine. This capability is pronounced in rabbit lung, as shown by Axelrod (1962a,b) using an in vitro system incorporating S-adenosylmethionine. Clouet (1962, 1963) reported that normorphine injected into the rat brain was N-methylated and also that this reaction was carried out in vitro using rat liver and brain preparations containing S-adenosylmethionine. The N-methyl transferase involved was studied by Clouet et al. (1963). On the other hand, Elison and Elliott (1964) did not find evidence for the N-methylation of normorphine in vitro in rat liver preparations or for the urinary excretion of morphine in rats dosed with normorphine.

Another metabolite of morphine formed by an oxidative reaction is morphine-N-oxide (99), although its appearance seems to usually require special circumstances. Woo et al. (1968) identified morphine-N-oxide in the urine from patients receiving morphine together with amiphenazole (2,4-diamino-5-phenylthiazole) or tacrine (1,2,3,4-tetrahydro-9-amino-acridine). The metabolite was not detected when these compounds were given alone. It is possible that the metabolite is formed chemically since it was produced when morphine and amiphenazole were mixed in ammoni-

O   Me

N

HO   O   OH

(99)

acal methanol, although the authors believed that its excretion may be due to inhibition of its further metabolism or to inhibition of an alternate pathway in the metabolism of morphine. Heimans *et al.* (1971) found that 46% of the total urinary opiates consisted of morphine-N-oxide when rats were given morphine and tacrine. Also, the excretion of normorphine was greatly reduced compared with that seen in rats receiving morphine alone. Misra and Mitchell (1971) found no evidence indicating that morphine is converted to its N-oxide in the central nervous system of rats. Morphine-N-oxide has been detected as a metabolite of morphine under *in vitro* conditions. Misra *et al.* (1973) found that small amounts of the N-oxide were formed when morphine was incubated aerobically with rat liver homogenates fortified with NADPH. Also, Ziegler *et al.* (1969) showed that a purified microsomal amine oxidase from pig liver converted morphine to its N-oxide. When morphine-N-oxide is administered to rats, it is excreted in the urine to a large extent as unchanged compound but mainly as morphine and morphine-3-glucuronide together with some normorphine (Misra and Mitchell, 1971; Heimans *et al.*, 1971).

Daly *et al.* (1965) investigated a microsomal hydroxylating system from rabbit liver that converted phenols to catechols. This activity was assayed by converting the latter compounds to an O-methyl derivative using [*methyl*-^{14}C]-S-adenosylmethionine and catechol-O-methyltransferase. They reported that morphine was a substrate for this system, although the activity was low compared with that seen with many compounds. Misra *et al.* (1973) found that rat brain and liver homogenates were able to hydroxylate morphine to a derivative having the properties of a 2,3-catechol structure (2-hydroxymorphine) (100). In the urine of morphine-treated rats, however, a different metabolite was detected and this was tentatively identified as the 2,3-dihydrodiol derivative (101).

Another recently described novel metabolite of morphine is dihydromorphinone (102). Klutch (1974) isolated this compound from the urine of rats injected with repeated daily doses (25 mg/kg) of morphine. Quantitative data indicated that about 4% of the dose was excreted as

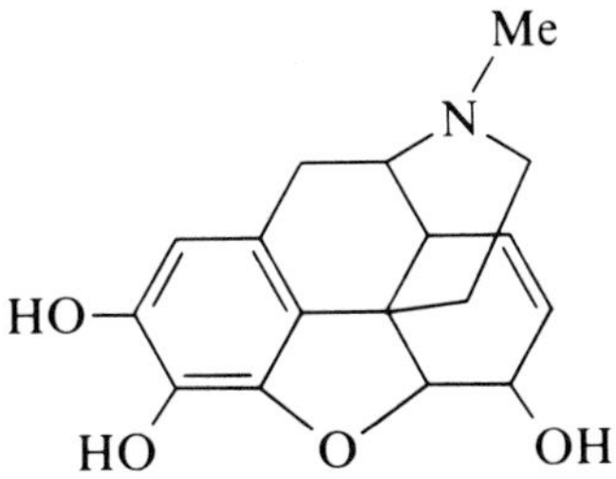

Me

N

HO

HO   O   OH
          (100)

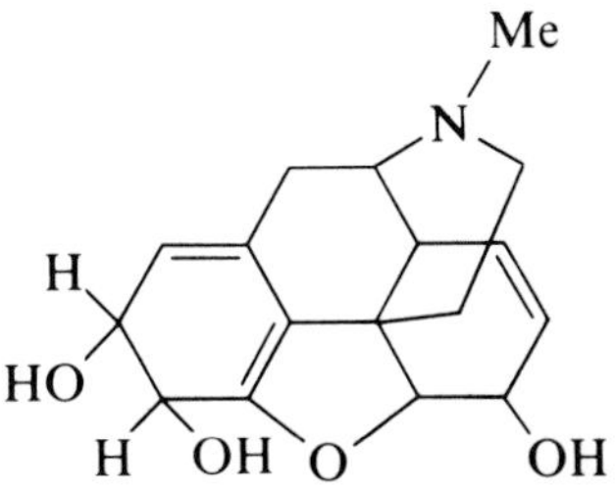

Me

N

H
HO

H   OH   O   OH
          (101)

dihydromorphinone. Yeh *et al.* (1977) studied its urinary excretion in several mammalian species given morphine by acute or chronic administration. No free dihydromorphinone was detected, however all acid-hydrolysed samples except those from dogs and morphine-dependent humans contained this metabolite. The amounts found were generally approx. 1% or less of the administered dose, however a value of 6–7% was obtained using chronically-treated monkeys. The excretion of dihydromorphinone in morphine-treated guinea pigs was not increased following chronic administration. Reduction of dihydromorphinone (102) to the corresponding 6-hydroxy derivative (dihydromorphine) was observed in rabbits (Roerig *et al.*, 1973) and in guinea pigs (Yeh, 1976).

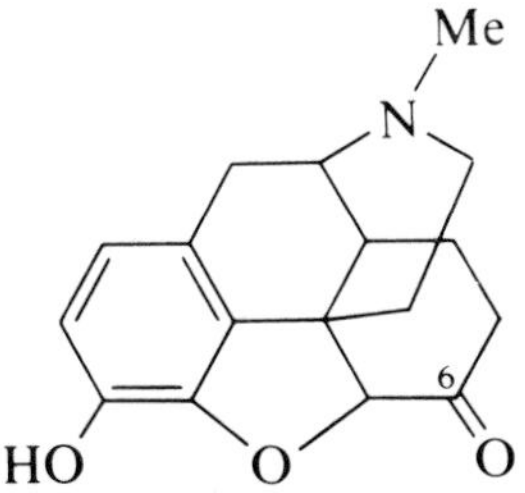

Me

N

HO   O   O
          6
          (102)

Pseudomorphine (2,2'-bimorphine) is an oxidation product of morphine which has long been suggested as a possible morphine metabolite. However, little evidence is available to support this view (see Way and Adler, 1961). More recently, Misra and Mulé (1972) found no evidence for the formation of pseudomorphine *in vivo* in the brain of rats injected with [³H]-morphine (10 mg/kg). Pseudomorphine itself was excreted mainly unchanged in the urine.

As with morphine, the metabolism of **codeine** (96) has been the subject of a large number of investigations. A comprehensive review of the earlier studies on its biological disposition was published by Way and Adler (1962) and other summaries have appeared more recently (Scrafani and Clouet, 1971; Misra, 1972). The general picture of codeine metabolism is one of the predominant excretion of the alkaloid and its metabolites in the urine, other excretory routes being of minor importance in most species. However Yeh and Woods (1969) found that 13–14% of the administered dose (2 mg/kg, s.c.) was excreted in the faeces in rats. With respect to the excretion of free codeine in the urine of animals, the results summarized by Way and Adler (1962) indicate that values in the range of a few to 10–15% of the dose are typical. Thus, in rats about 3% of the dose (33 mg/kg, s.c.) was excreted free and this finding agrees reasonably well with the 6–7% reported by Yoshimura *et al.* (1970) and Yeh and Woods (1971), using doses of 10 and 20 mg/kg, respectively. Values of 8–9% were recorded in guinea pigs (Axelrod and Inscoe, 1960; Yoshimura *et al.*, 1970) with widely different doses (90 and 10 mg/kg, respectively). Rabbits were found to excrete 3·5% of the dose (10 mg/kg) and 5–7% of the dose (20 mg/kg) as free urinary codeine (Yoshimura *et al.*, 1970; Yeh and Woods, 1971) and the latter investigation also recorded a figure of roughly 30% for cats. Excretion of free urinary codeine in dogs has been given as 4–11% of the dose (20 mg/kg, s.c.) (Woods *et al.*, 1956), 9% in a bile duct cannulated female dog and about 23% in normal male dogs given the same dose (Yeh and Woods, 1971). Values for the rhesus monkeys (*Macaca mulatta*) given a similar dose are 3–10% (Woods *et al.*, 1956) and a range of 3–17% was found in man in the earlier work summarized by Way and Adler (1962). In another study Nariyuki and Asaki (1959) noted a value of 14% following an oral dose of 100 mg of codeine phosphate in man.

The above results demonstrate that codeine, like morphine, is largely metabolized prior to its excretion from the body. Not unexpectedly, the main pathways of metabolism seen with morphine are also seen with codeine. Thus, glucuronide conjugation and N-demethylation are important metabolic reactions but, additionally, codeine can be O-demethylated. The bound or conjugated material excreted in the urine of animals receiving codeine may consist of derivatives of either codeine itself or its O-demethylated metabolite morphine (see below). Yeh and Woods (1970a) isolated codeine-6-glucuronide from the urine of codeine-treated dogs, thus identifying one of these conjugated metabolites. This article is also a useful source to the earlier investigations on the urinary excretion of bound codeine in various animal species. These workers subsequently showed that male dogs excreted 36% and female dogs 78% of the dose (20 mg/kg, s.c.) as codeine-6-glucuronide (Yeh and Woods, 1971). Similar previous experiments indicated a urinary excretion of 42–58% of the dose as conjugated codeine (Woods *et al.*, 1956) and it is evident that this route is the principal metabolic pathway for codeine in dogs. The data summarized by Way and Adler (1962) indicate that large amounts of conjugated codeine (20–50% of the dose) are excreted by man. Woods *et al.*

(1956) reported values of roughly 30–40% with the rhesus monkey. The values found in the other animal species studied are usually much lower, especially in cats which excrete very little codeine-6-glucuronide (1–2% of the dose) (Yeh and Woods, 1971) and in rats (1% of the dose) (Yoshimura et al., 1970). The latter workers also reported values of 11% in rabbits and 13% in guinea pigs, the latter result agreeing well with the value of 10% reported earlier by Axelrod and Inscoe (1960). Of course, those species showing a minor urinary excretion of codeine-6-glucuronide may have a correspondingly higher excretion of conjugated morphine. This appears to be the case in rabbits (Yeh and Woods, 1971) and rats (Yeh and Woods, 1969). The main glucuronide excreted in the urine and bile of codeine-treated rats was shown to be morphine-3-glucuronide (Yeh and Woods, 1970b).

(103)

In the same manner as that described above with morphine, the initial investigations which indicated that codeine is metabolized by oxidative N-demethylation utilized [N-methyl-^{14}C]-codeine. Thus, Adler (1952) found that about 13% of the radioactivity appeared in the respiratory $^{14}CO_2$ of rats within 30 h of dosing (40 mg/kg). In experiments in man, the demethylated product, norcodeine (103), was detected in the urine both free and conjugated. The amount of norcodeine excreted in the urine was later found to be about 14% of the dose, although the values for respiratory $^{14}CO_2$ were calculated to be only about half of this (Adler et al., 1955). The results did not differ appreciably when the dose (20–30 mg/kg) was given by different routes of administration (p.o., s.c., i.m.). Also, urinary excretion of metabolites predominated with only about 1% of the dose being found in the faeces within 24 h. A study in man by Ebbighausen et al. (1973) using 10–20 mg doses of codeine gave an estimate of about 9% of the dose being excreted as total urinary norcodeine.

Following the initial experiments in rats and man, several other investigations using various animal species have shown that N-demethylation of codeine to norcodeine is a general metabolic reaction. Many of these studies have used rats and an early report by Axelrod (1956b) described the system in rat liver microsomes which converted codeine to norcodeine and formaldehyde in the presence of TPNH, oxygen and other cofactors. Kuhn and Friebel (1962) found that urine of codeine-treated rats contained norcodeine and also small amounts of the fully demethylated metabolite, normorphine (98). Similar results were obtained by Yeh and Woods (1970b) using both rat urine and bile. The findings of Elison and Elliott (1964) pointed to the existence of different enzymes in rat liver being responsible for the N-demethylation and O-demethylation (see below) of codeine. The N-demethylation of codeine in mice appears to be fairly pronounced as 27–31% of the radioactivity was recovered in 24 h as respiratory $^{14}CO_2$ at a dose level of 33 mg/kg (Adler, 1967). Quantitative data in the guinea pig are not available but norcodeine has been detected as a urinary metabolite of codeine in this species (Kuhn and Friebel, 1962). Yeh and Woods (1971) investigated the urinary excretion of norcodeine in rabbits, cats and dogs following s.c. administration of codeine (20 mg/kg). The ability to N-demethylate codeine varied appreciably among these three species. It was most pronounced in cats, which excreted 48% of the dose as free and none as conjugated norcodeine in the 48 h urine. The values for rabbits were 1–3% and 7–14%, respectively, while the least N-demethylation was found in dogs which excreted about 4% of the dose in each fraction. Also, 9–14% of the dose in rabbits was excreted as conjugated normorphine. This fully demethylated metabolite was not detected in the experiments with cats. A further related point of interest is that norcodeine can be N-methylated by preparations of rat liver and brain (Clouet, 1962) and of rabbit lung (Axelrod, 1962a).

Codeine, which contains both N-methyl and O-methyl groups, can undergo oxidative demethylation at two sites. The first direct evidence showing that the latter type of demethylation takes place was obtained using codeine labelled with ^{14}C in the 3-methoxyl group. Adler and Latham (1950) and Latham and Elliott (1951) found that 50% of the radioactivity was lost as respiratory $^{14}CO_2$ in 31 h in rats given this material (40 mg/kg, s.c.). In vitro experiments in the former study showed that this conversion takes place mainly in the liver. While the identity of the alkaloidal metabolite remaining after demethylation was not determined in these investigations, Adler and Shaw (1952) soon showed that it is morphine. They isolated and characterized the metabolite following its formation in vitro using rat liver slices, which were able to convert a large proportion of the added codeine to morphine in these experiments. Axelrod (1955, 1956a) studied the in vitro O-demethylation of codeine using various liver preparations from several animal species. The reaction, which produced morphine and formaldehyde, was catalysed by an enzyme system located in the liver microsomes and required TPNH, oxygen and other cofactors. Greatest activity was found in rabbit liver preparations, intermediate activity in rat liver preparations, low activity in those from guinea pigs and dogs and none in mouse liver. However, Takemori and Mannering (1958) reinvestigated this reaction in microsomal preparations from the livers of several species including the mouse and found high activity in this species. It was noted that this apparent discrepancy is due to marked differences in demethylating ability of the livers of various strains of mice. These results also suggested that the N- and O-demethylating systems may be different. This view is also supported by the results of Elison and Elliott (1964) who carried out a kinetic study of N- and O-demethylation of codeine by rat liver preparations.

The results summarized above show clearly that codeine undergoes metabolic O-demethylation and indicate also that differences in this ability are seen among various animal species. The above results also suggest that rats and rabbits utilize this pathway to a relatively large extent and this has subsequently been confirmed in several investigations. Yeh and Woods (1969) found that rats given s.c. doses of codeine (2 mg/kg) excreted 3–8% of the dose as free morphine and 19–44% as conjugated morphine in the urine in 24 h. Yoshimura *et al.* (1970) reported values of 7% and 19% for free morphine and morphine-3-glucuronide, respectively, in a similar experiment employing a dose of 10 mg/kg. Yeh and Woods (1970b) isolated morphine-3-glucuronide from the urine and bile of codeine-treated rats. The 48 h urinary metabolites of codeine (20 mg/kg, s.c.) in rabbits included free morphine (1–2%) and conjugated morphine (47–51%) (Yeh and Woods, 1971). In another investigation with rabbits using half this dose, Yoshimura *et al.* (1970) found that the 24 h excretion values of morphine and morphine-3-glucuronide were 5% and 29%, respectively. They also showed that this pathway is less important in guinea pigs, which gave corresponding values of only 2% and 4·5%. The conversion of codeine to morphine in guinea pigs has also been reported by Kuhn and Friebel (1962). In view of the appreciable excretion of respiratory $^{14}CO_2$ following the administration of [3-O-*methyl*-^{14}C]-codeine to mice, it appears that this species also is readily able to O-demethylate this alkaloid. Adler (1967) reported a value of about 25% of the dose (33 mg/kg) for $^{14}CO_2$ in 24 h. Both cats and dogs appear to be particularly deficient in their ability to O-demethylate codeine. Using a s.c. dose of 20 mg/kg, Yeh and Woods (1971) found only about 1% free morphine and no conjugated morphine in the urine in 48 h in the former species while, with dogs, the corresponding values were about 1% and under 0·1%. In an earlier study, Woods *et al.* (1956) found neither free nor conjugated morphine in the plasma, urine, faeces or bile from dogs dosed with codeine (20 mg/kg, s.c.). Pærregaard (1958) recorded a value of about 2% for free urinary morphine in dogs following the administration of codeine in doses of 2–40 mg/kg. The remaining species studied, monkey and man, seem to occupy an intermediate position with regard to their ability to O-demethylate codeine. Woods et al. (1956) found that the rhesus monkey excreted in the urine about 0·5–2% of the dose as free and 6–10% as conjugated morphine following s.c. injection of codeine (20–40 mg/kg). The initial report on the O-demethylation of codeine in man stated that, following the s.c. injection of 30 mg of material labelled with ^{14}C in the 3-methoxyl group, 15% of the radioactivity was recovered in 24–30 h as respiratory $^{14}CO_2$ (Adler, 1954). Mannering et al. (1954) isolated and identified morphine in the urine following the oral administration of 130 mg of codeine sulphate to three subjects. They estimated that the amount of total morphine present corresponded, at most, to about 3% of the dose. However, Adler et al. (1955) reported a value of about 10% of the dose (20 mg, i.m. or p.o.) being excreted as total urinary morphine in 24 h. Redmond and Parker (1963) found this excretion to account for 14% of the dose (45 mg, p.o.) in 32 h. Pærregaard (1958) reported a value of 4% (24 h urine) in a similar experiment, but noted that this probably represents the actual conversion of nearly 10% of the dose into morphine. Confirmation of this general level of excretion of morphine in the urine of codeine-treated subjects has been made by Ebbighausen et al. (1973). Following administration of an oral dose of 20 mg, they found that urinary total morphine accounted for about 10% of the dose. In addition, the fully demethylated metabolite, normorphine, was detected and calculated to account for less than 4% of the dose.

The third and final compound to be discussed in this section dealing with morphinan alkaloids is **thebaine** (97). Although Axelrod (1955) reported that thebaine is demethylated to an unidentified phenolic metabolite when incubated with a preparation containing rabbit liver microsomes, no information on its metabolic fate has appeared until recently. In a detailed study with [^{3}H]-thebaine, Misra et al. (1974a) elucidated its metabolism in rats. They found that the compound (5 mg/kg, s.c.) was excreted unchanged in the urine and faeces in 96 h to an extent of 17% and 4% of the dose, respectively. The metabolism of thebaine in the rat was extensive, with the glucuronides of normorphine (98) and norcodeine (103) being identified as the major urinary metabolites. The minor free metabolites which were identified were codeine, norcodeine, morphine, normorphine and 14-hydroxycodeinone (104). Three further metabolites were detected, one of which appears to be the 3-O-demethylated compound, oripavine (105).

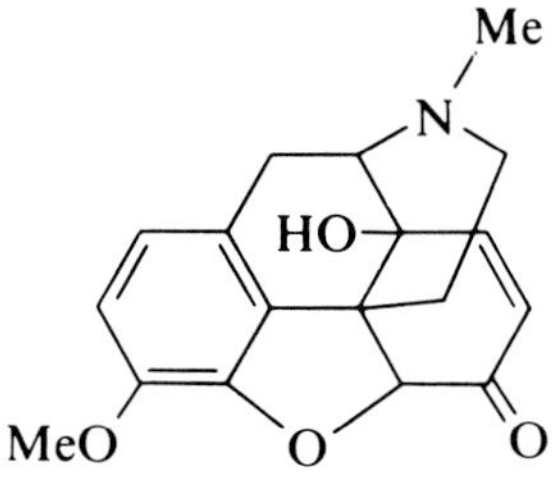

(104)

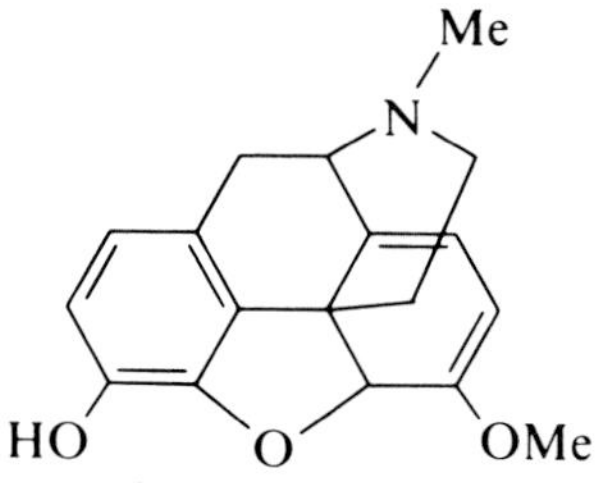

(105)

VII. Pyrrolizidine Derivatives

Plants containing pyrrolizidine alkaloids have long been the subject of considerable interest, mainly because of the poisonous nature of many of these when ingested by domestic animals. A contributing factor to this problem is the widespread geographical distribution of these compounds in many genera of several unrelated plant families. The chronic form of pyrrolizidine alkaloidosis seen in grazing animals often manifests itself as severe liver damage although, in humans, an acute type of toxicity involving veno-occlusive disease has been reported. These subjects are well covered in the monograph of Bull *et al.* (1968) and the review of McLean (1970).

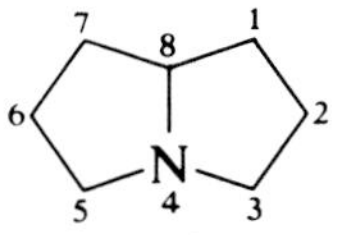

(106)

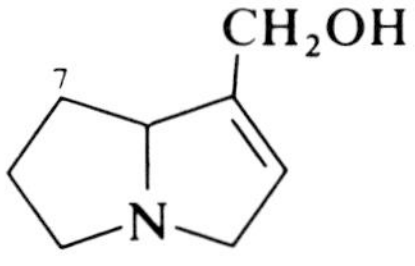

(107)

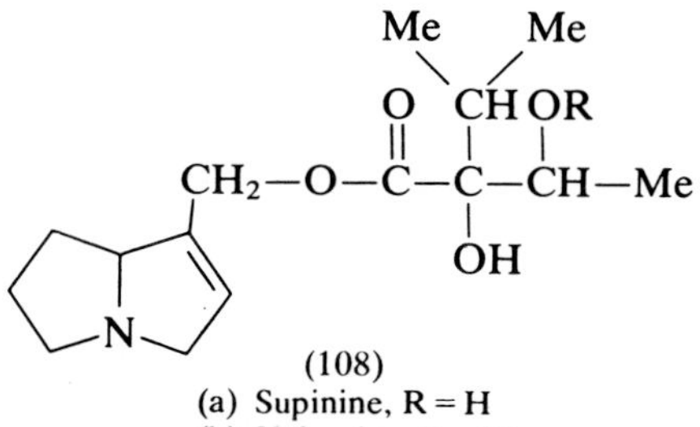

(108)
(a) Supinine, R = H
(b) Heleurine, R = Me

Pyrrolizidine (106) is the compound from which this group of alkaloids derives its name, however, all of the toxic compounds as well as most of those studied metabolically are esters of 1-hydroxymethyl-1,2-dehy-

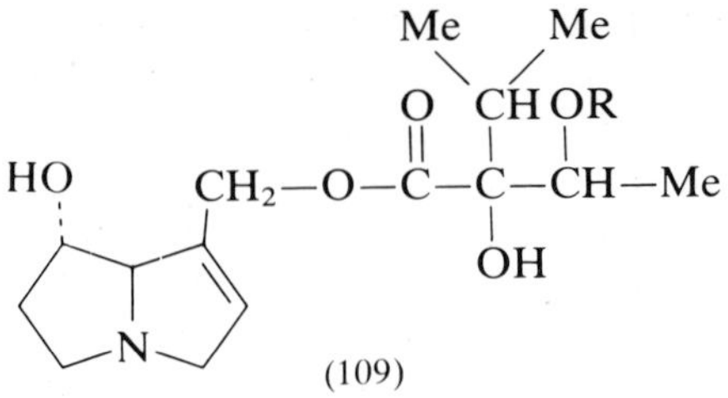

Me   Me
 \ /
  O CH OR
  || |
HO   CH₂—O—C—C—CH—Me
          |
          OH

(109)

(a) Heliotrine, R = Me
(b) Heliotridine trachelanthate, R = H

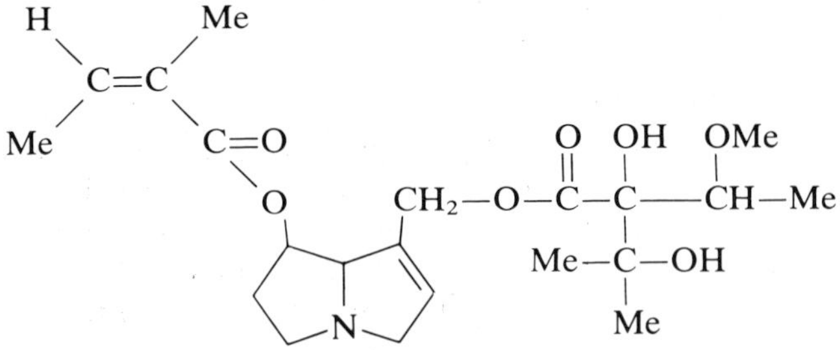

H   Me
 \   /
  C=C
 /   \
Me   C=O   O   OH OMe
         \ || | |
          O   CH₂—O—C—C———CH—Me
                        |
                     Me—C—OH
                        |
                        Me

(110)

Lasiocarpine

         Me
         |
H   CH₂———CH   OH
 \   /   \ /
  C=C   C—CH₂OH
 /   \ O |
Me   C   C=O
        \ |
         O   CH₂—O

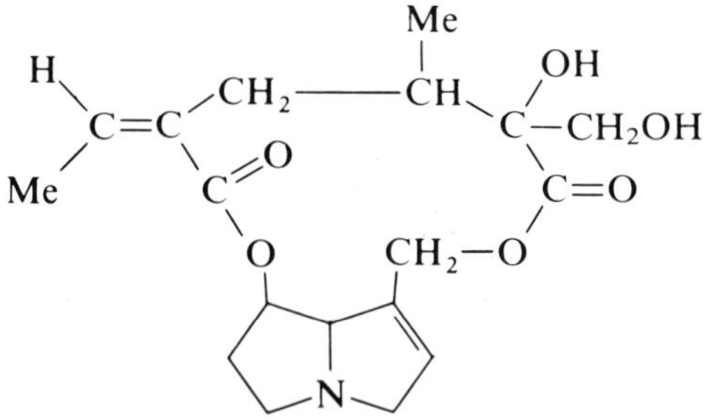

(111)

Retrorsine

dropyrrolizidine (107) or its 7-hydroxy derivative. An example of the simpler type based on (107) is **supinine** (108a) whereas a commonly studied 7-hydroxy derivative is exemplified by **heliotrine** (109a). Many pyrrolizidine alkaloids are diesters, either of the type seen with **lasiocarpine** (110) or cyclic forms exemplified by **retrorsine** (111). The heterocyclic nitrogen atom of the pyrrolizidine alkaloids is readily oxidized and the N-oxide derivatives of these alkaloids often contribute significantly to the total alkaloidal content in plants. The monograph by Bull *et al.* (1968)

provides a convenient source of information on the chemical properties and also the molecular structures of pyrrolizidine alkaloids.

Studies on the metabolism of pyrrolizidine alkaloids offer an interesting and rewarding area for research for several reasons, however, the most fundamental aspect involves its role in the elucidation of the mechanisms of toxicity of these compounds. As summarized below, metabolic alterations which lead to toxic pyrrolizidine metabolites are now well known in addition to the numerous reactions which produce detoxication products. Additionally, species differences in metabolism have been documented and these have been helpful in understanding corresponding differences in toxicity. Finally, gastrointestinal microorganisms are known to be partly responsible for the breakdown of pyrrolizidines in some species. These aspects, in view of the present state of knowledge, do not themselves provide the best framework for summarizing pyrrolizidine metabolism and the present discussion, following a short introduction on the disposition of these alkaloids, is therefore based upon the different types of metabolic reactions known to occur. Within this framework the relevant points concerning metabolite toxicity, species differences in metabolism and gut microfloral metabolism will be covered. The metabolism of pyrrolizidine alkaloids has previously been reviewed by Bull *et al.* (1968, pp. 215–220), McLean.(1970) and Mattocks (1972a).

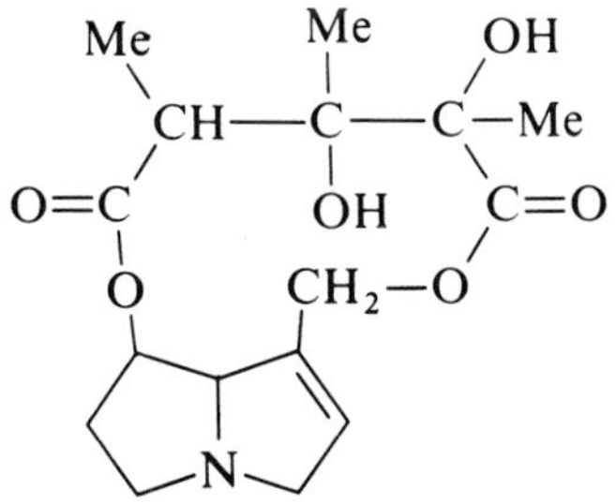

(112)

Monocrotaline

Hayashi (1966) studied the fate of non-labelled and ^{3}H-labelled **monocrotaline** (112) in rats. Following a s.c. injection (60 mg/kg), from 50–70% of the dose was detected in the urine as unchanged compound as determined by a methyl orange method. A fairly extensive excretion of this compound in the unchanged form may also be inferred from the data presented by Bull *et al.* (1968, p. 34, p. 217) who showed that monocrotaline has a relatively low partition coefficient (lipid phase : aqueous phase) and that, with heliotrine (109a) and several related compounds, increasing proportions of the dose are excreted unchanged as the hydrophilicity of the alkaloids increases. Thus, heliotrine which has a slightly greater partition coefficient than monocrotaline was found to be excreted unchanged in the urine to the extent of 30% of the dose in 16 h following i.p. injection to rats. The corresponding value for lasiocarpine (110), which has a partition coefficient more than 20-fold higher, was only 1–1·5%. Using [^{3}H]-monocrotaline, Hayashi found that about 30% of the radioactivity was excreted as an unknown metabolite in the bile of rats, only a trace of the unchanged alkaloid being detected. The chemical assay method for the alkaloid failed to detect significant amounts present in the tissues after 24 h, although radioactivity was present in the liver 72 h after injection of [^{3}H]-monocrotaline. Bull *et al.* (1968, pp. 213–214) studied the fate of randomly-labelled [^{14}C]-lasiocarpine (110) injected i.p. in rats. They recovered about 9% of the radioactivity in the expired CO_2 in slightly more than 4 h. In addition to these data obtained using rats, Jago *et al.* (1969) gave heliotrine (109a) by slow or rapid i.v. or slow duodenal routes to sheep. In the first instance, only 10–14% of the dose was recovered in the urine after 48 h as unchanged alkaloid or metabolites. Biliary excretion was much lower, accounting for only a few tenths of a percent of the dose in 24 h. Increased excretion (24% in 6 h) was observed following rapid i.v. dosage whereas only 6–9% was recovered, mainly in the urine, when the alkaloid was given by slow duodenal infusion. The generally low recoveries of pyrrolizidine compounds were interpreted to indicate either extensive metabolism resulting in modification of the pyrrolizidine ring or binding to tissues, or both.

A conspicuous chemical feature of the pyrrolizidine alkaloids is the presence of one or two ester linkages. Mattocks (1972a) reported that these compounds were generally fairly resistant to hydrolysis by esterases from rat liver or serum *in vitro*. However, Bull *et al.* (1968, p. 217) found that rats hydrolysed heliotrine (109a) and lasiocarpine (110) to small extent. Roughly 3% of the dose (i.p.) was found as heliotridine (113) in the urine in 16 h. The latter metabolite has also been identified in the urine of sheep given heliotrine i.v. or duodenally (Jago *et al.*, 1969), although this metabolite in this species also appears to account for only a few percent of the administered dose. As the hydrolysis products heliotridine (113) and heliotric acid (114) are not cytotoxic, this reaction is one of detoxication.

HO   CH$_2$OH

(113)

Me   Me
\   /
CH OMe
| |
HOOC—C—CH—Me
|
OH

(114)

(115)   (116)

An interesting alternative pathway resulting in cleavage of the ester linkage involves reductive fission. This reaction has been found to be carried out *in vivo* and *in vitro* by sheep rumen microorganisms. Dick *et al.* (1963) postulated that sheep, which may ingest appreciable quantities of heliotrine and yet show relatively slight liver damage, are likely to destroy much of the alkaloid in the rumen. They found that the rumen of these animals contained, in addition to some unchanged compound and its hydrolysis product heliotridine (113), 7α-hydroxy-1-methylene-8α-pyrrolizidine (*l*-goreensine) (115) as the major metabolite. This compound was devoid of acute hepatotoxic properties. Heliotrine was also converted to compound (115) *in vitro* by rumen microorganisms and the conversion rate was increased when the vitamin B_{12} concentration in the medium was increased to 1 μg/ml. Under the latter conditions several other pyrrolizidine alkaloids were metabolized to 1-methylene derivatives. Thus lasiocarpine (110) was converted to 7α-angelyloxy-1-methylene-8α-pyrrolizidine and supinine (108a) and **heleurine** (108b) gave 1-methylenepyrrolizidine. Russell and Smith (1968) isolated a small Gram-negative coccus from sheep rumen contents which was able to carry out this reaction. They found that one mole of heliotrine was reduced by one mole of hydrogen gas to give one mole each of compound (115) and heliotric acid (114). Formate could also act as the hydrogen donor, in which case stoichiometric amounts of CO_2 were formed. Further investigation of this detoxication reaction by Lanigan and Smith (1970) indicated that the 1-methylene derivatives are not the ultimate metabolic products but are further reduced to 1-methyl compounds. Thus, *in vitro* incubation of heliotrine with sheep rumen microorganisms resulted in the formation of 7α-hydroxy-1α-methyl-8α-pyrrolizidine (116) and *in vivo* experiments showed that this metabolite was formed from compound (115). In addition, the fully reduced metabolite (116) was also the final metabolic product of lasiocarpine (110). Lanigan (1970) discovered that the ability of sheep rumen fluid to effect reductive fission of pyrrolizidine alkaloids varied according to the diet employed. Adaptation of the rumen microflora resulting in greater rates of alkaloid metabolism occurred when the animals were fed plants containing the alkaloids and this property was rapidly lost when these plants were removed from the diet.

The heliotric acid moiety of heliotrine contains a methyl ether. This group has been shown to undergo cleavage in rats (Bull *et al.*, 1968, pp.

215–217) and sheep (Jago et al., 1969). In the former species 10% of the dose (i.p.) was excreted in the urine in 16 h as the corresponding alcohol, heliotridine trachelanthate (109b). A further 5% was recovered as the N-oxide of the latter metabolite and, together, these demethylated products accounted for about one fourth of the recovered dose. In sheep, the total urinary recovery was only 24% after i.v. injection and the quantity of heliotridine trachelanthate excreted was not determined. McLean (1970) reported unpublished observations by Jago which showed that cleavage of the methyl ether can be carried out by rat liver microsomes in vitro. This reaction is one of detoxication as the alcoholic derivatives are less toxic than heliotrine.

As noted above, Bull et al. (1968) found that some of the demethylated heliotrine was converted to the corresponding N-oxide. While very little of the corresponding derivative of heliotrine itself was detected, the N-oxide of heliotridine (113) accounted for 15% of the dose. Mattocks (1968) reported that about 14% of the dose (60 mg/kg, i.p.) of retrorsine (111) was excreted in the urine as the N-oxide mainly within 3–4 h. The rapid urinary excretion of these oxidation products was noted to occur generally with several pyrrolizidine alkaloids, not an unexpected finding in view of the higher water solubility of N-oxides. The oxidation of pyrrolizidine alkaloids to N-oxides has also been noted in sheep, although in this species heliotrine-N-oxide was the sole metabolite of this type detected following heliotrine administration (Jago et al., 1969). This metabolic reaction furnishes a further example of pyrrolizidine alkaloid detoxication as the N-oxide derivatives seem to show considerably less toxicity than that produced by the parent alkaloids (see Mattocks, 1972a). However, this situation can be greatly influenced by the route of administration when the N-oxides themselves are administered. Oral administration may expose these compounds to the reducing effects of the gut flora which generate the alkaloidal base, thus reducing or eliminating any difference in toxicity seen between the latter compound and its N-oxide. This situation was found by Mattocks (1971, 1972b) who also reported that the compound employed (retrorsine-N-oxide) was rapidly reduced to the base when incubated in vitro with rat gut contents. Lanigan (1970) found that pyrrolizidine N-oxides are rapidly reduced to the tertiary bases by sheep rumen fluid. This reaction was noted earlier with heliotrine-N-oxide by Dick et al. (1963).

In the search for a better understanding of the mechanisms involved in pyrrolizidine alkaloid toxicity, the possibility was raised that metabolically formed 1,2-epoxide derivatives might be implicated (Bull et al., 1968, pp. 214–215; Culvenor et al., 1969). This view was developed further by Schoental (1970) who felt that the epoxides are the proximal active forms of the alkaloids. However, these metabolites do not appear to have been detected as metabolic products. Furthermore, Culvenor *et al.* (1969, 1971) prepared the α- and β-epoxides of monocrotaline (112) and found them to be devoid of relevant biological activity.

A key step forward in the understanding of the events involved in production of toxicity by pyrrolizidine alkaloids was the demonstration by Mattocks (1968) of a new type of metabolite containing a pyrrole structure. Much of the subsequent work in this field has concerned itself with the formation and properties of these metabolites. Mattocks (1968) found that numerous pyrrolizidine alkaloids were converted to compounds termed metabolic pyrroles. This reaction was observed *in vitro* using rat liver slices and the metabolites were found in the livers and urine of rats injected with the alkaloids. These metabolites are strongly bound in the liver, especially in the microsomal fraction and solid debris of fractionated liver. Values for the urinary excretion (24 h) of metabolic pyrroles in rats fell into the range of roughly 5–15% of the dose (60 mg/kg, i.p.). Formation of pyrroles from heliotrine and lasiocarpine using rat liver homogenates and microsomal preparations was also reported by Culvenor *et al.* (1969) and Jago *et al.* (1970). They identified the main water-soluble pyrrolic derivative, which was also one of the main urinary metabolites, as dehydro-heliotridine (6,7-dihydro-7α-hydroxy-1-hydroxymethyl-5H-pyrrolizine) (117a). Formation of pyrroles was stimulated 4–5 fold when livers from phenobarbital pre-treated rats were used. Pyrrole derivatives containing the original ester groups were not detected, however this is not surprising in view of their lability (Mattocks, 1968). It was therefore suggested that heliotrine is first converted by liver microsomal enzymes to dehydro-heliotrine (118), the active alkylating species. Evidence was obtained which indicates that this species reacts with unchanged heliotrine to form the quaternary compound (119). Additionally, alkylation of proteins and perhaps nucleic acids near the site of formation occurs and some of the active metabolite may undergo hydrolysis to dehydroheliotridine (117a) which is stable and thus detected. A similar sequence was suggested by the results obtained with lasiocarpine and supinine. Interestingly, the *N*-oxides of these three alkaloids formed very small amounts of pyrrolic derivatives in similar experiments. This suggests that these compounds are not intermediates in the metabolic sequence. Mattocks and White (1971b) carried out an extensive investigation of this subject *in vitro* using rat liver microsomes and numerous pyrrolizidine alkaloids. They also concluded that *N*-oxides are not intermediates in the formation of pyrroles. Similar conclusions have been reached with monocrotaline and its *N*-oxide in a study involving the stimulation and inhibition of metabolism by rat liver microsomes (Chesney *et al.*, 1974) and in an *in vivo* and *in vitro* study with rats and guinea pigs (Chesney and Allen, 1973). The bile is also an excretory route for the pyrrolic metabolites. White (1977) gave rats retrorsine (40 mg/kg, i.v.) and found that about 25% of the dose was excreted in the bile in this form in 7 h. Excretion was greatest during the first hour and negligible after 7 h. On the other hand, biliary excretion of the parent compound was minor. Further experiments indicated that re-absorption of material excreted in the bile is very small and that entero-hepatic circulation is therefore of only minor importance in the elimination of retrorsine metabolites.

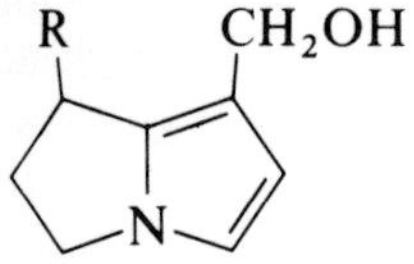

R   CH$_2$OH

(117)

(a) R = ···OH
(b) R = —OH

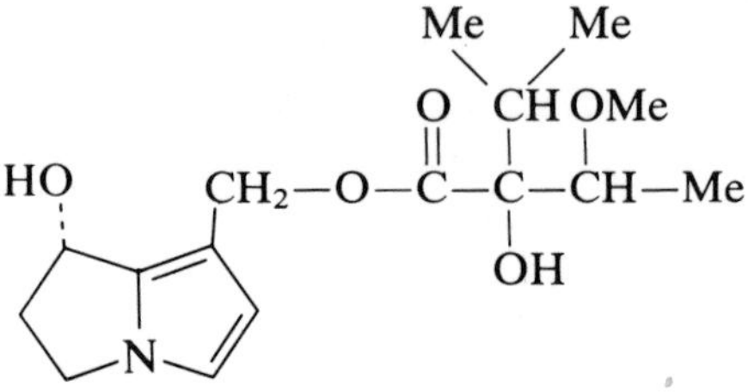

Me   Me

O   CH OMe

HO   CH$_2$—O—C—C—CH—Me

OH

(118)

Me   Me

O   CH OMe

HO   CH$_2$—O—C—C—CH—Me

OH

HO   CH$_2$

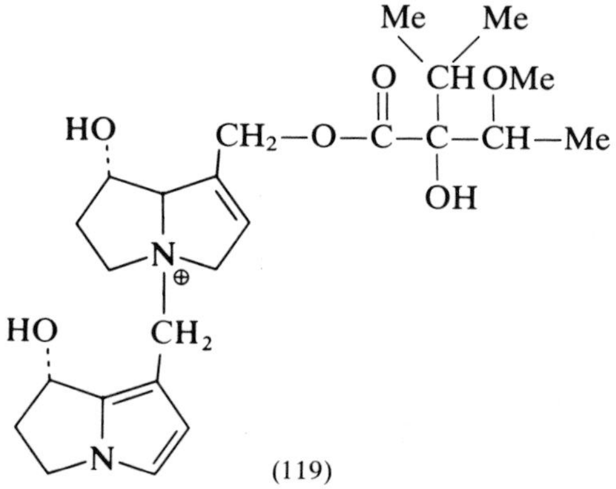

(119)

Mattocks (1972b) investigated the relationships between acute hepato-toxicity and pyrrolic metabolite formation in rats given pyrrolizidine alkaloids. Liver pyrrole levels were found to be proportional to the dose of alkaloid given 2 h earlier and in many cases the acute toxicity correlated with the amounts of pyrroles formed. However, it was pointed out that not all of the toxic metabolite formed may be involved in the production of toxicity as alternate pathways (e.g. hydrolysis, reaction with glutathione or amino acids) are available. Mattocks (1970) showed earlier that the acid moieties of the pyrrolizidine alkaloids influence their toxicity. In alkaloids having a common base moiety, esterification with different acids results in alkaloids showing both quantitative and qualitative differences in toxicity. These findings suggested that these differences are due to differences in the amounts and stabilities of pyrrolic metabolites formed. In view of the fact that liver microsomal enzyme activity is low in new-born animals, Culvenor *et al.* (1971) and Mattocks and White (1973) studied the ability of new-born and young rats to convert pyrrolizidine alkaloids to pyrrolic derivatives *in vitro*. This capacity was very low in new-born rats but increased rapidly. The former group found that the rate of conversion of lasiocarpine in day old rats was about 60% of that seen in adults and the latter study showed that the conversion of retrorsine reached very high levels within five days. In the case of human embryo tissue, liver but not lung was found to metabolize lasiocarpine, retrorsine and fulvine to pyrroles *in vitro* (Armstrong and Zuckerman, 1970).

Species differences in the metabolism of pyrrolizidine alkaloids have received some attention and Chesney and Allen (1973) reported that the guinea pig, which is known to be relatively resistant to the toxicity of these compounds, is much less able than rats to form pyrrolic derivatives. This was shown with monocrotaline (112) using both intact animals and *in vitro* preparations. Both animal species were highly susceptible to the effects of the pyrrolic metabolites when these were given by i.v. administration. The abilities to convert monocrotaline to its *N*-oxide derivative were similar in both species. Recently, Shull *et al.* (1976) studied the ability of liver microsomes from several species to produce pyrroles from monocrotaline or from tansy (*Senecio jacobaea*) alkaloids. In most cases, this ability appeared to be directly related to the species differences in susceptibility to pyrrolizidine alkaloids.

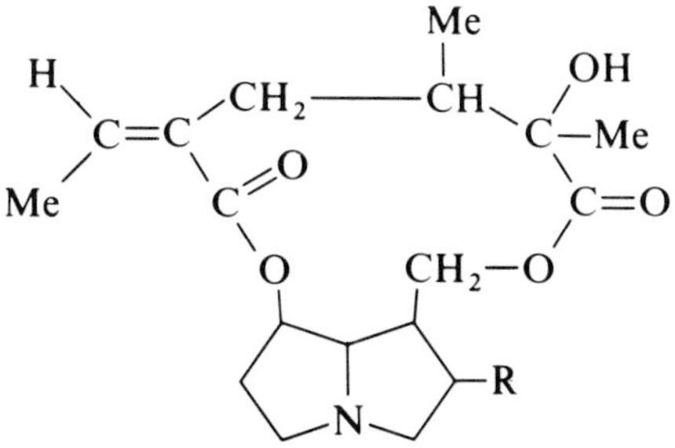

(120)
(a) Platyphylline, R = H
(b) Rosmarinine, R = OH

The metabolism of toxic pyrrolizidine alkaloids results in the formation of active pyrrole metabolites of the type shown with compound (118). The metabolism of the non-toxic saturated alkaloids **platyphylline** (120a) and **rosmarinine** (120b) also results in pyrroles, however these are of different type (Mattocks, 1968; Culvenor *et al.*, 1969). Mattocks and White (1971a),

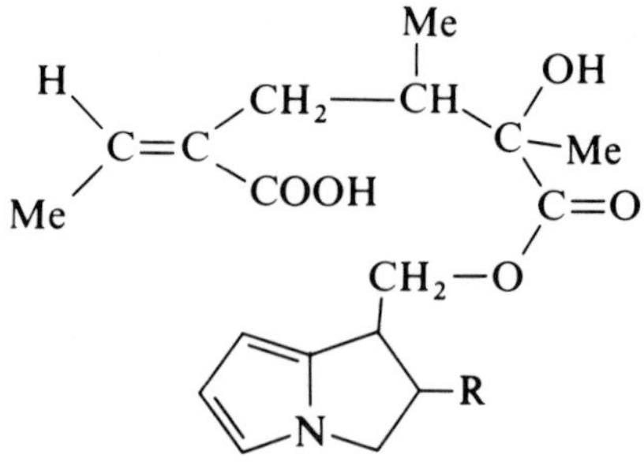

$$\begin{array}{c}
\text{Me} \\
| \\
\underset{\text{Me}}{\overset{\text{H}}{\diagdown}}\text{C}=\text{C}\underset{\text{COOH}}{\overset{\text{CH}_2-\text{CH}}{\diagup}}\overset{\text{OH}}{\underset{\text{C}=\text{O}}{\text{C}-\text{Me}}}
\end{array}$$

$$\text{CH}_2-\text{O}$$

(121)

(a) R = H
(b) R = OH

employing rat liver slices or microsomes, obtained evidence indicating that those alkaloids were dehydrogenated in the other ring. Thus, the metabolites formed were suggested to be compound (121a) from platyphylline and compound (121b) from rosmarinine. These pyrroles are not reactive.

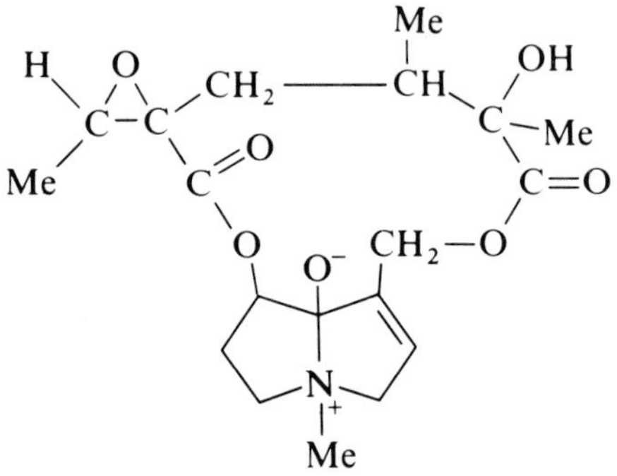

$$\begin{array}{c}
\text{Me} \\
| \\
\text{CH}_2 \longrightarrow \text{CH}
\end{array}$$

(122)

Otosenine

A few of the pyrrolizidine alkaloids are *N*-methyl derivatives and the metabolism of one of these, **otosenine** (122), was studied by Culvenor *et al.* (1971) using rat liver microsomes. A major pyrrolic metabolite was identified as dehydroretronicine (117b). This finding indicates that the *N*-methyl derivatives are subject to demethylation.

VIII. Imidazole Derivatives

The best known alkaloid containing an imidazole ring is **pilocarpine** (123), the only compound in this group for which metabolic data is available. This information is very limited, however, Koelle (1975) stated that it is partly destroyed in the body but mainly excreted in the urine in combined form.

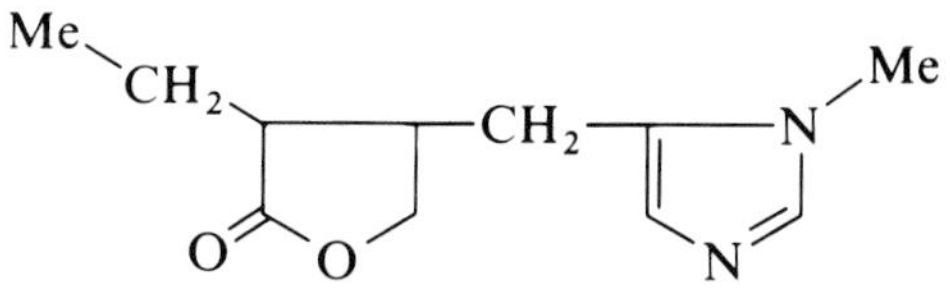

(123)

Pilocarpine

A likely centre of metabolism of pilocarpine is the ester linkage in the lactone ring and most of the available data deals with the hydrolysis, mainly in serum, of this group. Lavallee and Rosenkrantz (1966) found that the alkaloid was rapidly metabolized by rat serum *in vitro* to one or more metabolites which no longer gave positive reactions for the lactone or imidazole ring systems. Serum from rabbits, dogs, monkeys and humans also metabolized pilocarpine *in vitro*. Preliminary *in vivo* data using rats (10 mg/kg, i.v.) and dogs (3 mg/kg, i.v.) suggested that pilocarpine rapidly disappeared from the blood and that small amounts (3–4% in 2 h) were excreted unchanged in the urine. Extensive degradation of pilocarpine was also reported by Otorii (1969) who measured its degradation to NH_3 and CO_2 *in vitro* by rabbit serum.

The enzymic hydrolysis of pilocarpine by rabbit and human serum, ocular tissues and liver was studied by Schonberg and Ellis (1969) and Ellis *et al.* (1972). Pilocarpine inactivation in the samples was not due to binding with serum proteins but to an esterase which is not cholinesterase. This enzyme shows high stability and its activity can be prevented by the addition of a number of chelating agents. The enzyme levels in rabbit serum are considerably higher than those in human serum. Contrariwise, Newsome and Stern (1974) reported that pilocarpine is not destroyed enzymically by rabbit serum or several ocular tissues but instead bound to serum and tissue components. Release of active material was effected by heating. Ocular tissue uptake of pilocarpine was also reported by Lazare and Horlington (1975) who found no evidence for its metabolism, at least in the short term.

IX. Purine Derivatives

Derivatives of purine (124) are constituents of nucleotides and nucleic acids and therefore ubiquitous in plants. However, a discussion of the metabolism of these types falls outside the scope of this book, which limits itself to compound classed as alkaloidal purines. These alkaloids consist of methylated derivatives of xanthine (125) and uric acid (126), metabolic data being available on three compounds of the former type. These are theophylline (1,3-dimethylxanthine) (127), theobromine (3,7-dimethyl-xanthine) (128) and caffeine (1,3,7-trimethylxanthine) (129). They command attention both by virtue of their pharmacological properties, which may include a diuretic effect and stimulatory actions on cardiac muscle and the central nervous system, and their widespread occurrence in plants, especially coffee, tea and cocoa. These factors led, not unexpectedly, to an early interest in the metabolic fate of methylxanthines, especially caffeine, and it was reported as early as the middle of the nineteenth century that no unchanged caffeine could be detected in the urine following oral administration of the alkaloid. This finding was generally borne out in later experiments which also indicated that demethylated products were formed. However, the analytical methods available earlier were often unsatisfactory by modern standards and the data obtained, often conflicting, are therefore not treated in detail in the present summary. The publications of Myers and Wardell (1928) and Buchanan et al. (1945) are useful sources for references to and discussion of this early work.

(124) (125) (126)

(127) (128) (129)
Theophylline Theobromine Caffeine

Theophylline (127) was found by Myers and Hanzal (1946) to be metabolized mainly to 1,3-dimethyluric acid (131d) in the Dalmatian dog, a species which resembles man in its excretion of uric acid. This result confirmed the pathway suggested earlier by Myers and Wardell (1928) and

Buchanan *et al.* (1945) to be present in man. The latter authors felt that about half of a 3 g dose of theophylline was converted to methyluric acids, of which about half might be 1-methyluric acid (131a). Myers and Hanzal (1946) also left open the possibility that some of the methylated uric acid excreted could be this monomethyl derivative. Further work has clarified this situation and Weinfeld and Christman (1953) found that 1-methyl-(131a) and 1,3-dimethyluric acid (131d) are urinary metabolites of theophylline in rats, rabbits and man. In a quantitative study in man, Brodie *et al.* (1952) showed that as much as 50% of an oral dose (750 mg) was excreted in the urine as 1,3-dimethyluric acid. The excretion value for unchanged compound was about 10% but little or no monomethyluric acid was detected. Similarly, Cornish and Christman (1957) found 9–11% of the oral dose (1 g) of theophylline excreted unchanged by man in the urine during 48 h and, more recently, Jenne *et al.* (1976) reported a value of 8% in patients receiving roughly similar doses on a daily basis. These studies have also confirmed that 1,3-dimethyluric acid is a major urinary metabolite of theophylline. Values of 32–38% were found in the earlier study and Jenne *et al.* reported about 40%. 1-Methyluric acid was also reported and, in contrast to the results of Brodie *et al.* (1952) noted above, the values found were about 19% and 17%, respectively. The remaining recorded urinary metabolite of theophylline is 3-methylxanthine (130b) which accounted for 11–16% of the dose in the experiments of Cornish and Christman and 36% in those of Jenne *et al.* Pharmacokinetic data from the latter investigation suggested that demethylation at the 1-position to give metabolite (130b) is the dominant reaction determining the serum concentration of theophylline. An inverse relationship between the proportions of metabolites (130b) and (131d) excreted in the urine was noted. This finding suggests the competition between two reactions leading to the formation of

(130)

(a) R = Me, R' = R″ = H
(b) R' = Me, R = R″ = H
(c) R″ = Me, R = R' = H
(d) R = R″ = Me, R' = H

(131)

(a) R = Me, R' = R″ = H
(b) R' = Me, R = R″ = H
(c) R″ = Me, R = R' = H
(d) R = R' = Me, R″ = H
(e) R = R″ = Me, R' = H
(f) R' = R″ = Me, R = H
(g) R = R' = R″ = Me

these metabolites from a common precursor, theophylline. 3-Methyl-xanthine was not found to be a metabolite of theophylline in rats (Lohmann and Miech, 1976). Unpublished results of Moss and Horner (Moss, 1977) showed that 80% of the radioactivity was excreted in the urine following the oral administration of $[^{14}C]$-theophylline in horses. Excretion was complete within four days and about 21% of the dose was excreted as unchanged compound and 52% presumably as 1,3-dimethyluric acid (131d).

Until recently, little was known about the enzymes involved in the transformations of theophylline summarized above. Brodie *et al.* (1952) showed that its oxidation at C8 is not catalysed *in vitro* by milk xanthine oxidase. No N-demethylation of the alkaloid was observed in an *in vitro* system using rabbit liver microsomes (Gaudette and Brodie, 1959) although Mazel and Henderson (1965) reported slight demethylation activity when mouse liver microsomes were used. However, Lohmann and Miech (1976) carried out an extensive investigation of this subject in rats and found that theophylline is metabolized by the liver microsomal system. On the other hand, it is not a substrate for liver xanthine oxidase or aldehyde oxidase, the latter result confirming an earlier report by Krenitsky *et al.* (1972). Lohmann and Miech found that, among the many types of tissue slices used, only those from liver were able to metabolize theophylline. Fractionation studies localized this activity in the microsomal fraction and the results from stimulation and inhibition experiments were in accordance with the other data implicating the liver microsomal system. Furthermore, they found that this system converted theophylline to 1,3-dimethyluric acid (131d), which is a major urinary metabolite of the alkaloid, and apparently to 1-methylxanthine (130a), which is not found in the urine. The latter metabolite is normally metabolized further by xanthine oxidase to 1-methyluric acid (131a). However, 1-methylxanthine was found to accumulate in liver slice preparations containing the xanthine oxidase inhibitor allopurinol.

The metabolism of **theobromine** (128) differs from that described above with theophylline (127) in that oxidation at C8 to give uric acid derivatives is a minor reaction. It seems likely that this reaction is hindered by the methyl group at C7. Cornish and Christman (1957) found that no 3,7-dimethyluric acid (131f) was excreted in the urine of humans following the ingestion of 1 g doses of theobromine. However, about 4% of the dose was excreted as 7-methyluric acid (131c). Buchanan *et al.* (1945) showed earlier that theobromine is not metabolized to uric acid (126) or 3-methyluric acid (131b) to any significant extent in humans. The small amount of 7-methyluric acid formed probably arises from 7-methyl-xanthine (130c) which is a major urinary metabolite of theobromine in man, accounting for about 28% of the dose (1 g) (Cornish and Christman, 1957). In addition, they found that the isomeric 3-methylxanthine (130b) was excreted to the extent of about 20% of the dose. An additional 12% was excreted unchanged, leaving about a third of the dose unaccounted for. The finding that theobromine is mainly metabolized to monomethylxanthines in man confirms earlier work by Krüger and Schmidt (1901) who reported the urinary excretion, in roughly a 2 : 1 ratio, of the 7- and 3-methyl isomers. Both of these methylxanthines were excreted in the urine of rabbits and dogs following large doses of theobromine (Krüger and Schmidt, 1899, 1901). In rabbits the ratio of urinary 7-methylxanthine : 3-methylxanthine was about 15 : 1 whereas it was about 1 : 5 in dogs, which also converted less of the dose to methylxanthines than did rabbits or humans. Appreciable excretion of unchanged theobromine occurs in horses. Moss and Horner (Moss, 1977) gave ^{14}C-labelled compound orally and found that 75% of the radioactivity was recovered in the urine in four days. This material contained theobromine (42% of the dose) and a metabolite (21% of the dose) which was presumed to be 3,7-dimethyluric acid. Few studies have been carried out on the enzyme systems involved in the metabolism of theobromine. Similarly to that noted above with theophylline Gaudette and Brodie (1959) found no demethylation of theobromine *in vitro* by rabbit liver microsomes and Mazel and Henderson (1965) reported very slight demethylation activity when mouse liver microsomes were used. However, the recent results of Lohmann and Miech (1976) implicating the rat liver microsomal system in the 3-demethylation of theophylline suggest that this system may also be active in the formation of the monomethylxanthines from theobromine.

Caffeine (129) is the most important xanthine derivative from several points of view and it is therefore not suprising that it has received the greatest amount of attention with regard to its metabolic fate in animals. These studies have a long history, however many areas of uncertainty have accompanied this subject and we still do not have a thorough understanding of the metabolism of caffeine. Burg (1975) has comprehensively reviewed the literature dealing with the physiological disposition of caffeine. Caffeine is readily and extensively absobed from the gastrointestinal tract of many animal species including the mouse (Burg and Stein, 1972; Burg and Werner, 1972), rat (Czok *et al.*, 1969), pig (Cunningham, 1970) rhesus and squirrel monkey (Burg *et al.*, 1974) and man (Axelrod and Reichenthal, 1953). These studies have also shown that metabolism of the absorbed compound is generally rapid (plasma half-lives of 3 h are common), although slower metabolism was noted in the pig and especially in the squirrel monkey. Furthermore, caffeine is extensively metabolized with relatively little unchanged compound being excreted in the urine. In man, this value is about 1% of the dose following oral or intravenous administration of 0·3 to 1 g of caffeine (Axelrod and Reichenthal, 1953; Cornish and Christman, 1957; Schmidt and Schoyerer, 1966). The urinary excretion values of unchanged compound at various dose levels (generally in the range of 5–50 mg/kg) in other species are: 3–6% in mice (Burg and Stein, 1972); 2% (Usanova and Shnol, 1955), 4% (Arnaud, 1976b), 7% (Burg, 1975) and 9% (Khanna et al., 1972) in rats, although the lowest value covered only a 3-h collection period; 13% in rabbits (Weinfeld, 1952; Burg, 1975); 5% (Fisher et al., 1949) and about 7% (Krüger, 1899) in dogs; 6% in pigs (Cunningham, 1970); 1% (Moss, 1977) and 3% (Fisher et al., 1949) in horses; 3% in squirrel monkeys (Burg et al., 1974).

In general, the metabolism of caffeine shows greater similarity to that of theophylline than of theobromine as both N-demethylation and oxidation at C8 occur. Table 9.3 lists the urinary metabolites of caffeine together with information on species, dosage and amount excreted. Although the data on which this table is based are not always easily interpreted, several points of interest emerge. Firstly, all of the possible mono- and dimethylxanthines have been detected as caffeine metabolites, although no single species except the rat has been found to form them all. The rate of formation of theophylline from caffeine, as measured by the blood levels of the metabolite, was studied in man by Sved et al. (1976). Secondly, uric acid (126) is not a metabolite of caffeine. This point was the subject of much disagreement in the earlier literature until clarified by Buchanan et al. (1945). Thirdly, most of the data suggest that oxidation at C8 to uric acid derivatives is hindered when a methyl group is present at the 7-position. However, investigations in rats by Khanna et al. (1972) and Rao et al. (1973) have raised doubts about this point as urinary 1,3,7-tri-methyldihydrouric acid (132) was found to account for about 11% of the dose (approx. 35 mg/kg, i.p.). They proposed that metabolite (132) undergoes dehydrogenation to give 1,3,7-trimethyluric acid (131g). The latter compound which is not a prominent urinary metabolite, can then follow two pathways, one of which being of minor importance and leading to demethylated derivatives, e.g. 3-methyluric acid (131b). Myers and Hanzal (1946) obtained evidence which indicated that compound (131g) undergoes demethylation at the 7-position in rats and the Dalmatian dog. However, the major pathway for compound (131g) was suggested to involve scission of the pyrimidine ring to give 3,6,8-trimethylallantoin (133), 1,6,8-trimethylallantoic acid (134) and then glyoxylic acid, N-methylurea and N,N'-dimethylurea. The metabolites formed along this route, being highly polar and difficult to extract, may make up that portion of the dose not accounted for by previously identified metabolites. This interpretation

TABLE 9.3

Urinary metabolites of caffeine

Metabolite[a]	Structure	Species	Dose	Excretion[b]	Reference
1,3-Dimethylxanthine (theophylline)	(127)	Mouse	25 mg/kg, p.o.	—	Burg and Stein (1972)
		Rat	approx. 35 mg/kg, i.p.	1·2	Khanna et al. (1972)
		Rat	0·5–10 mg/kg, p.o.	5–6	Arnaud (1976b)
		Dog	chronic, p.o.	approx. 7	Krüger (1899)
		Horse	?	approx. 6	Moss (1977)
1,7-Dimethylxanthine (paraxanthine)	(130d)	Mouse	25 mg/kg, p.o.	11	Burg and Stein (1972)
		Mouse	?	+	Kamei et al. (1975)
		Rat	approx. 35 mg/kg, i.p.	9	Khanna et al. (1972)
		Rat	0·5–10 mg/kg, p.o.	9	Arnaud (1976b)
		Dog	chronic, p.o.	approx. 1	Krüger (1899)
		Squirrel monkey	?	+	Burg (1975)
		Man	600 mg daily, p.o.	+	Weissmann et al. (1954)
		Man	300 mg, p.o.	+	Schmidt and Schoyerer (1966)
		Man	1000 mg, p.o.	2–7	Cornish and Christman (1957)
3,7-Dimethylxanthine (theobromine)	(128)	Mouse	25 mg/kg, p.o.	—	Burg and Stein (1972)
		Mouse	?	+	Kamei et al. (1975)
		Rat	approx. 35 mg/kg, i.p.	5	Khanna et al. (1972)
		Rat	0·5–10 mg/kg, p.o.	6–7	Arnaud (1976b)
		Dog	chronic, p.o.	approx. 2	Krüger (1899)
		Horse	?	approx. 6	Moss (1977)
		Man	300 mg, p.o.	+	Schmidt and Schoyerer (1966)

TABLE 9.3—*continued*

Metabolite[a]	Structure	Species	Dose	Excretion[b]	Reference
1-Methylxanthine	(130a)	Mouse	25 mg/kg, p.o.	—	Burg and Stein (1972)
		Rat	chronic, i.p.	+	Otomo (1959)
		Rat	approx. 35 mg/kg, i.p.	—	Khanna et al. (1972)
		Rat	0·5–10 mg/kg, p.o.	4	Arnaud (1976b)
		Man	600 mg daily, p.o.	+	Weissmann et al. (1954)
		Man	300 mg, p.o.	+	Schmidt and Schoyerer (1966)
		Man	1000 mg, p.o.	18–20	Cornish and Christman (1957)
3-Methylxanthine	(130b)	Mouse	25 mg/kg, p.o.	approx. 2	Burg and Stein (1972)
		Mouse	?	+	Kamei et al. (1975)
		Rat	approx. 35 mg/kg, i.p.	—	Khanna et al. (1972)
		Rat	0·5–10 mg/kg, p.o.	0·5	Arnaud (1976b)
		Dog	chronic, p.o.	approx. 5	Krüger (1899)
7-Methylxanthine	(130c)	Mouse	25 mg/kg, p.o.	approx. 5	Burg and Stein (1972)
		Mouse	?	+	Kamei et al. (1975)
		Rat	approx. 35 mg/kg, i.p.	—	Khanna et al. (1972)
		Rat	0·5–10 mg/kg, p.o.	tr.	Arnaud (1976b)
		Man	600 mg daily, p.o.	+	Weissmann et al. (1954)
		Man	1000 mg, p.o.	4–9	Cornish and Christman (1957)
1,3,7-Trimethyl-dihydrouric acid	(132)	Rat	approx. 35 mg/kg, i.p.	11	Khanna et al. (1972), Rao et al. (1973)
		Rat	0·5–10 or 21 mg/kg, p.o.	20	Arnaud (1976a,b)
		Man	?	+	Arnaud (1976b)
1,3,7-Trimethyluric acid	(131g)	Rat	approx. 35 mg/kg, i.p.	tr.	Khanna et al. (1972)
		Rat	chronic, i.p.	+	Otomo (1959)
		Rat	0·5–10 mg/kg, p.o.	4	Arnaud (1976b)
1,3-Dimethyluric acid	(131d)	Mouse	25 mg/kg, p.o.	11	Burg and Stein (1972)
		Rat	0·5–10 mg/kg, p.o.	1·1	Arnaud (1976b)
		Rabbit	67 mg/kg, p.o.	+	Weinfeld and Christman (1953)
		Squirrel monkey	?	+	Burg (1975)

Compound		Species	Dose	Value	Reference
		Man	1000 mg, p.o.	5–14	Cornish and Christman (1957)
1,7-Dimethyluric acid	(131e)	Rat	0·5–10 mg/kg, p.o.	1·7	Arnaud (1976b)
3,7-Dimethyluric acid	(131f)	Rat	0·5–10 mg/kg, p.o.	tr.	Arnaud (1976b)
1-Methyluric acid	(131a)	Mouse	25 mg/kg, p.o.	9	Burg and Stein (1972)
		Rat	0·5–10 mg/kg, p.o.	6–7[c]	Arnaud (1976b)
		Rabbit	67 mg/kg, p.o.	+(?)	Weinfeld and Christman (1953)
		Squirrel monkey	?	+	Burg (1975)
		Man	13 mg/kg, p.o.	+	Weinfeld and Christman (1953)
		Man	1000 mg, p.o.	25–30	Cornish and Christman (1957)
3-Methyluric acid	(131b)	Rat	approx. 35 mg/kg, i.p.	tr.	Khanna et al. (1972)
		Rat	chronic, i.p.	+	Otomo (1959)
		Rat	0·5–10 mg/kg, p.o.	5[d]	Arnaud (1976b)
3,6,8-Trimethylallantoin	(133)	Rat	approx. 35 mg/kg, i.p.	1·3	Khanna et al. (1972), Rao et al. (1973)
		Rat	0·5–10 mg/kg, p.o.	14[e]	Arnaud (1976b)
N-Methylurea		Rat	0·5–10 mg/kg, p.o.	0·3	Arnaud (1976b)
N,N'-Dimethylurea		Rat	0·5–10 mg/kg, p.o.	0·2	Arnaud (1976b)
α-[7-(1,3-Dimethyl-xanthinyl)] methyl methyl sulphoxide[f]	(135b)	Mouse	?	+	Kamei et al. (1975)
		Rat	?	+	Kamei et al. (1975)
		Rabbit	?	+	Kamei et al. (1975)
		Horse	?	+	Kamei et al. (1975)

[a] Some excretion of the unchanged compound occurs. See text.
[b] Symbols: tr. = trace, — = looked for but not detected, + = present, numbers = % of dose.
[c] Includes unidentified compounds.
[d] Includes 7-methyluric acid and unidentified compounds.
[e] Includes an isomer or precursor.
[f] Corresponding sulphide (135a) and sulphone (135c) also identified in mouse urine.

was recently questioned by Arnaud (1976a) who found that 1,3,7-tri-methyldihydrouric acid (132) is the major (20% of the dose) urinary metabolite of caffeine in rats. When compound (132), labelled with ^{14}C in the 1-methyl group, was itself given, from 75–90% of the radioactivity was excreted in the urine and unchanged compound was the sole product present. This finding was interpreted to indicate that 1,3,7-tri-methyldihydrouric acid is an end product of caffeine matabolism and not an intermediate which forms 1,3,7-trimethyluric acid (131g) via dehy-drogenation. Arnaud (1976b) reported that 3,6,8-trimethylallantoin (133), together with an isomer or precursor, accounted for 14% of the dose of caffeine. It was presumed to be formed via this unidentified metaboite from caffeine and/or 1,3,7-trimethyluric acid. Further degradation of this ring-opened metabolite probably occurs as both N-methyl- and N,N'-dimethyl-urea were excreted in the urine. However, this latter pathway was of minor quantitative importance.

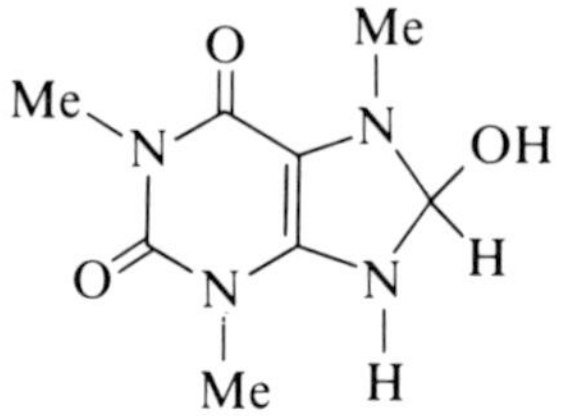

(132)

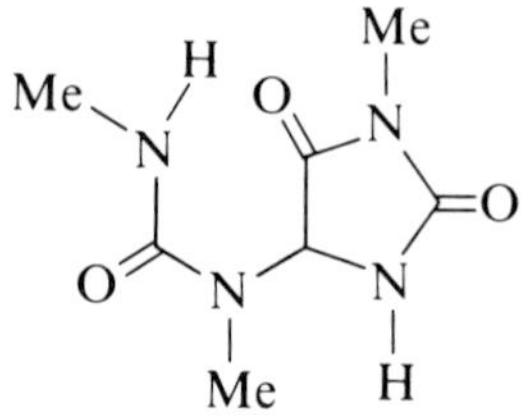

(133)

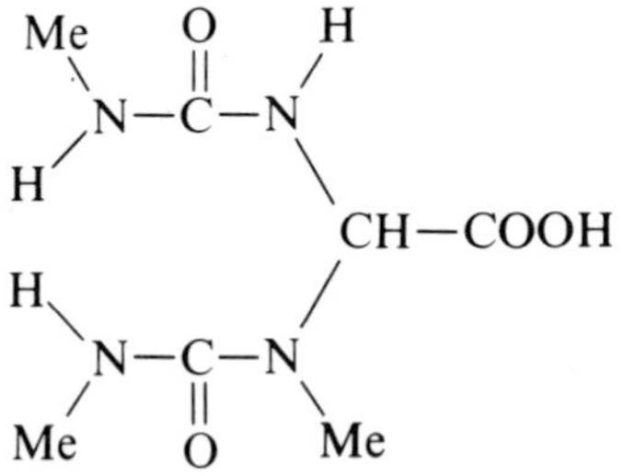

(134)

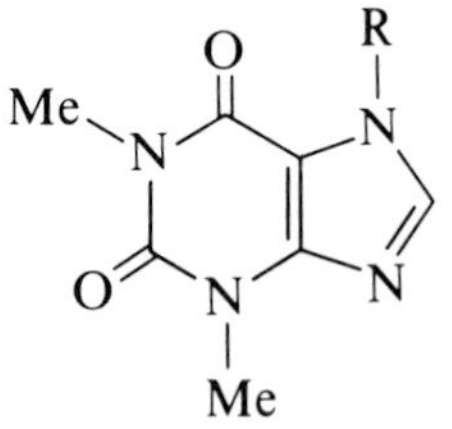

(135)

(a) R = CH_2—S—Me

(b) R = CH_2—$\overset{O}{\overset{\uparrow}{S}}$—Me

(c) R = CH_2—$\overset{O}{\underset{O}{\overset{\uparrow}{S}}}$—Me

An interesting report by Kamei et al. (1975) described the formation of several sulphur-containing metabolites of caffeine. These compounds have structures (135a), (135b) and (135c) and involve, in effect, the addition of a thiomethyl moiety to the 7-methyl group followed by oxidation of the sulphur to the corresponding sulphoxide and sulphone derivatives. The sulphoxide (135b) was identified as a urinary metabolite of caffeine in the mouse, rat, rabbit and horse and the other two metabolites were identified in mouse urine.

Several studies have been carried out on the N-demethylation of caffeine by liver microsomes. Low activity by mouse liver microsomes was reported by Mazel and Henderson (1965) whereas Gaudette and Brodie (1959) using those from rabbits and Franklin (1965) those from rats found moderate rates of activity (roughly 20% of that shown by the best substrates). Burg and Stein (1972) reported that mouse liver microsomes as well as the cytosol-microsome fraction from mouse kidney failed to demethylate caffeine. They also confirmed an earlier report of Bergmann and Dickstein (1956) that caffeine is not oxidized to its uric acid derivative by bovine xanthine oxidase. The rapid metabolism of caffeine *in vivo* therefore remains to be correlated with the activities of particular enzyme systems. Recently, Goth and Cleaver (1976) showed that caffeine is rapidly metabolized by demethylation in mouse and human cells in culture. The methyl groups appear to be used in the synthesis of thymine, guanine and adenine in nucleic acids.

X. Other Alkaloids

Betanin (136), the glucoside of betanidin, is the red pigment of beetroot. Watson (1964), who noted that excretion of the pigment in the urine even after consuming large amounts of beetroot is not observed in the majority

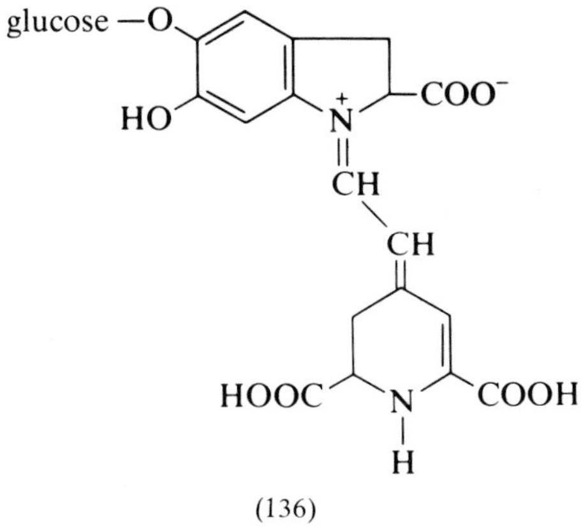

(136)

Betanin

of the normal population, investigated its urinary excretion after injection (5–40 mg, i.v.). In normal individuals, the urinary excretion of betanin is very rapid. The maximum rate is reached within minutes and diminishing excretion is seen for 3–10 h. Recovery ranged from 45–75% but it is not known if the material unaccounted for may be due to metabolites. Rats given the pigment (5–10 mg/kg, i.v.) also rapidly excreted it unchanged in the urine.

The metabolism of **acronine** (acronycine) (137a) labelled with ^{14}C in the N-methyl group was investigated in rats and humans by Sullivan *et al.* (1969). They found that radioactive metabolites were excreted mainly in the bile in rats. Five hydroxylated biliary metabolites were shown to be compounds (137b), (137c), (137d), (137e) and (137f). All of these metabolites except (137e) were excreted in the urine of patients receiving acronine.

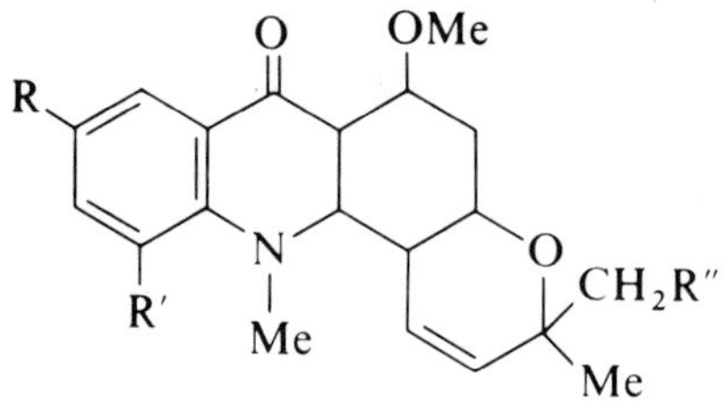

(137)

(a) Acronine, R = R' = R'' = H
(b) R = OH, R' = R'' = H
(c) R' = OH, R = R'' = H
(d) R'' = OH, R = R' = H
(e) R = R' = OH, R'' = H
(f) R' = R'' = OH, R = H

Eichelbaum *et al.* (1975) found that **sparteine** (138) is metabolized in man mainly by N-oxidation whereas routes involving C-oxidation are of minor quantitative importance. Some (15–30%) of the sparteine is also excreted unchanged in the urine. However, four individuals from a group of 100 healthy volunteers were found to differ from this normal pattern. They failed to convert the alkaloid to its N-oxide and excreted more than 90% of the dose in the urine as unchanged compound. A single oxidation

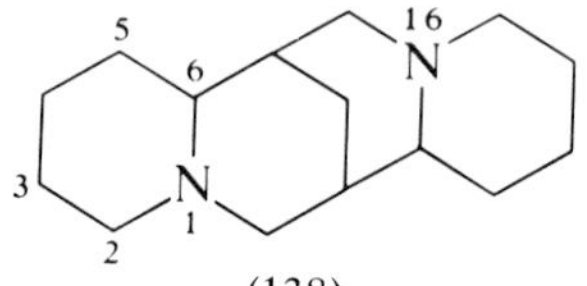

(138)

Sparteine

product, sparteine-1-N-oxide, is formed, however this primary metabolite is readily dehydrated to 2,3-dehydrosparteine (Δ_2-sparteine) and 5,6-dehydrosparteine (Δ_5-sparteine) (Dengler *et al.*, 1977).

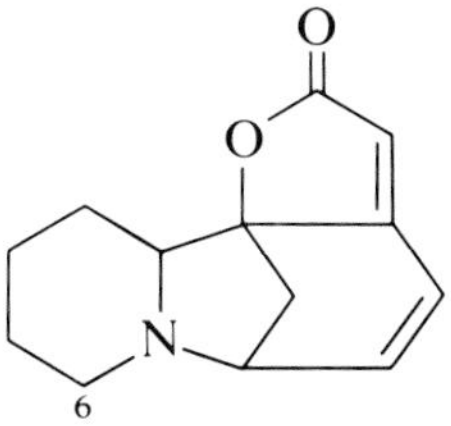

(139)

Securinine

The biological disposition of **securinine** (139) in several animal species was studied by Yao and Sung (1975). Disappearance of the alkaloid from the tissues was rapid following oral or parenteral administration. Very little unchanged compound was excreted in the urine and none was detected in the faeces. Securinine was found to be rapidly metabolized by rat liver slices or homogenates and also by the erythrocytes in this and several other species. The metabolic reaction involved was shown not to be simple cleavage of the lactone ring. Paper chromatography of the urine from rats treated with securinine indicated the presence of four metabolites in addition to some unchanged compound. Two of these metabolites were identified as the 6-hydroxy and 6-keto derivatives of securinine. These two metabolites were also detected in the urine of monkeys, however both were absent from human urine.

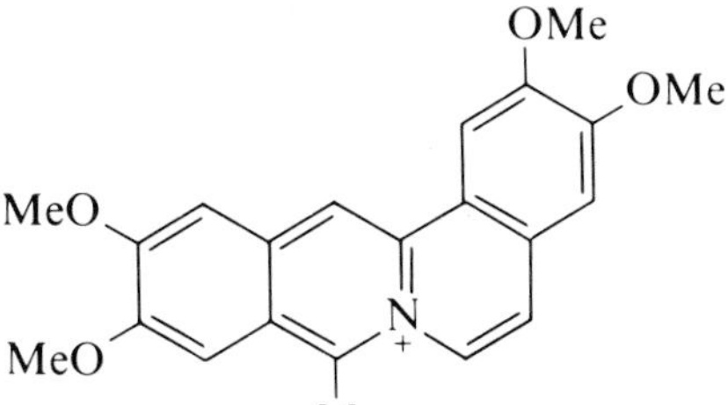

(140)

Coralyne

Plowman *et al.* (1976) studied the disposition in mice and rats of the sulphoacetate salt of the quaternary nitrogen alkaloid **coralyne** (140). Following administration (i.p. or i.v.) of the ^{14}C-labelled compound, the major route of excretion of radioactivity was the bile. Rats excreted 18–42% of the ^{14}C in the bile and only 1–2% in the urine in 24 h. In non-cannulated animals the cumulative excretion of urinary radioactivity after 96 h was 4–7% (rats) and 9–12% (mice) whereas the corresponding values for faecal radioactivity were roughly 40–50% and 70%, respectively. Unchanged coralyne was identified as the major component excreted in the bile, however a minor metabolite which is less polar than coralyne was also detected.

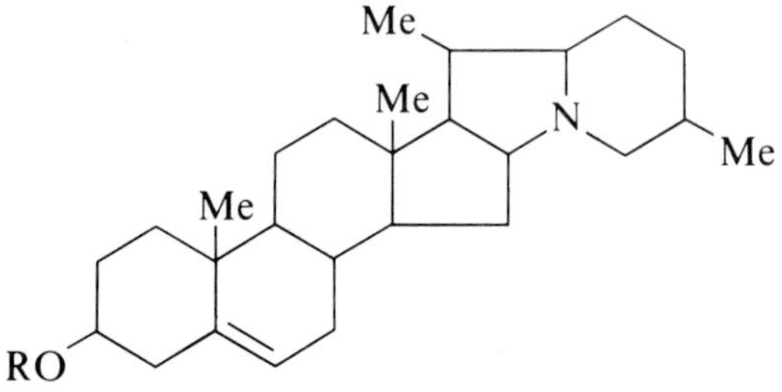

(141)

(a) Solanidine, R = H

(b) Solanine, R = —galactose⟨glucose / rhamnose⟩

(c) α-Chaconine, R = —glucose⟨rhamnose / rhamnose⟩

Metabolic data are available on two closely related glycoalkaloids, solanine (141b) and α-chaconine (141c). These structures show that both compounds consist of a common aglycone (solanidine) (141a) and differ only in one of the three sugar moieties. Nishie *et al.* (1971) administered [^{3}H]-solanine orally (5 mg/kg) and by i.p. injection (5–25 mg/kg) to rats. In the first case, nearly 80% of the radioacticity was excreted in the urine and faeces within 24 h and, of this, over 90% was lost by the latter route. About 65% of the faecal radioactivity was shown to be the aglycone solanidine. This metabolite is probably not well absorbed from the intestine as less than 1% of the dose was excreted in this form in the urine. The urinary radioactivity (about 10% of the dose) consisted mainly of two basic metabolites with polarity characteristics intermediate between those of solanine and solanidine. These were believed to be intermediate products of hydrolysis containing one or two hexose units. After i.p. injection, roughly 15–20% of the radioactivity was found in the urine and approximately an equal amount in the faeces in 24 h. Slightly less than half of the faecal radioactivity was due to solanidine (141a) and in this case, the two metabolites with polarity intermediate between unchanged compound and aglycone were also present. The urine from these animals also contained the two tentatively identified basic metabolites in addition to some unchanged compound but no solanidine. Similar experiments to those described above with solanine have been carried out with tritiated α-chaconine (141c) (Norred *et al.*, 1976). The major urinary and faecal metabolite following both oral and i.p. administration (5 mg/kg) was the aglycone solanidine (141a). Some unchanged compound appeared to be excreted in the faeces following oral dosage of the glycoalkaloid. Two minor metabolites, representing 1–5% of the administered radioactivity, were detected in the urine and faeces. These were less polar than α-chaconine but more polar than its aglycone.

References

Abrams, L. S. and Elliott, H. W. (1974). *J. Pharmac. exp. Ther.* **189**, 285–292.
Adamson, R. H. and Fouts, J. R. (1959). *J. Pharmac. exp. Ther.* **127**, 87–91.
Adir, J., Miller, R. P. and Rotenberg, K. S. (1976). *Res. Commun. chem. Path. Pharmac.* **13**, 173–183.
Adler, T. K. (1952). *J. Pharmac. exp. Ther.* **106**, 371.
Adler, T. K. (1954). *J. Pharmac. exp. Ther.* **110**, 1.
Adler, T. K. (1967). *J. Pharmac. exp. Ther.* **156**, 585–590.
Adler, T. K. and Latham, M. E. (1950). *Proc. Soc. exp. Biol. Med.* **73**, 401–404.
Adler, T. K. and Shaw, F. H. (1952). *J. Pharmac. exp. Ther.* **104**, 1–10.
Adler, T. K., Fujimoto, J. M., Way, E. L. and Baker, E. M. (1955). *J. Pharmac. exp. Ther.* **114**, 251–262.
Albanus, L., Hammarström, L., Sundwall, A., Ullberg, S. and Vangbo, B. (1968a). *Acta physiol. scand.* **73**, 447–456.
Albanus, L., Sundwall, A., Vangbo, B. and Windbladh, B. (1968b). *Acta pharmac. tox.* **26**, 571–582.
Ammon, R. and Savelsberg, W. (1949). *Hoppe-Seyler's Z. physiol. Chem.* **284**, 135–156.
Appelgren, L.-E., Hansson, E. and Schmitterlöw, C. G. (1962). *Acta physiol. scand.* **56**, 249–257.
Armitage, A. K., Dollery, C. T., George, C. F., Houseman, T. H., Lewis, P. J. and Turner, D. M. (1974). *Br. J. clin. Pharmac.* **1**, 180P–181P.
Armstrong, S. J. and Zuckerman, A. J. (1970). *Nature* Lond. **228**, 569–570.
Arnaud, M. J. (1976a). *Experientia* **32**, 1238–1240.
Arnaud, M. J. (1976b). *Biochem. Med.* **16**, 67–76.
Axelrod, J. (1955). *J. Pharmac. exp. Ther.* **115**, 259–267.
Axelrod, J. (1956a). *Biochem. J.* **63**, 634–639.
Axelrod, J. (1956b). *J. Pharmac. exp. Ther.* **117**, 322–330.
Axelrod, J. (1962a). *Life Sci.* **1**, 29–30.
Axelrod, J. (1962b). *J. Pharmac. exp. Ther.* **138**, 28–33.
Axelrod, J. and Cochin, J. (1957). *J. Pharmac. exp. Ther.* **121**, 107–112.
Axelrod, J. and Inscoe, J. K. (1960). *Proc. Soc. exp. Biol. Med.* **103**, 675–676.

Axelrod, J. and Reichenthal, J. (1953). *J. Pharmac. exp. Ther.* **107**, 519–523.
Axelrod, J., Shofer, R., Inscoe, J. K., King, W. M. and Sjoerdsma, A. (1958). *J. Pharmac. exp. Ther.* **124**, 9–15.
Beckett, A. H. and Morton, D. M. (1966a). *Biochem. Pharmac.* **15**, 937–946.
Beckett, A. H. and Morton, D. M. (1966b). *J. Pharm. Pharmac.* **18**, 88S–91S.
Beckett, A. H. and Sheikh, A. H. (1973). *J. Pharm. Pharmac.* **25**, 171P.
Beckett, A. H., Gorrod, J. W. and Jenner, P. (1970). *J. Pharm. Pharmac.* **22**, 722–723.
Beckett, A. H., Gorrod, J. W. and Jenner, P. (1971). *J. Pharm. Pharmac.* **23**, 55S–61S.
Beckett, A. H., Jenner, P. and Gorrod, J. W. (1973). *Xenobiotica* **3**, 557–562.
Beer, C. T. and Richards, J. F. (1964). *Lloydia* **27**, 352–360.
Beer, C. T., Wilson, M. L. and Bell, J. (1964a). *Can. J. Physiol. Pharmac.* **42**, 1–11.
Beer, C. T., Wilson, M. L. and Bell, J. (1964b). *Can. J. Physiol. Pharmac.* **42**, 368–373.
Beermann, B., Leander, K. and Lindström, B. (1976). *Acta chem. scand.* **B30**, 465.
Belpaire, F. M. and Bogaert, M. G. (1972). *Archs int. Pharmacodyn. Thér.* **199**, 191–192.
Belpaire, F. M. and Bogaert, M. G. (1973). *Biochem. Pharmac.* **22**, 59–66.
Belpaire, F. M. and Bogaert, M. G. (1974). *Archs int. Pharmacodyn. Thér.* **208**, 362.
Belpaire, F. M. and Bogaert, M. G. (1975a). *Xenobiotica* **5**, 421–429.
Belpaire, F. M. and Bogaert, M. G. (1975b). *Xenobiotica* **5**, 431–438.
Belpaire, F. M., Bogaert, M. G., Rosseel, M. T. and Anteunis, M. (1975). *Xenobiotica* **5**, 413–420.
Bergmann, F. and Dikstein, S. (1956). *J. biol. Chem.* **223**, 765–780.
Bernheim, F. and Bernheim, M. L. C. (1938). *J. Pharmac. exp. Ther.* **64**, 209–216.
Blaschko, H., Chou, T. C. and Wajda, I. (1947). *Br. J. Pharmac.* **2**, 108–115.
Blaschko, H., Himms, J. M. and Strömblad, B. C. R. (1955). *Br. J. Pharmac.* **10**, 442–444.
Boerner, U. and Roe, R. L. (1975). *J. Pharm. Pharmac.* **27**, 215–216.
Boerner, U., Roe, R. L. and Becker, C. E. (1974). *J. Pharm. Pharmac.* **26**, 393–398.
Boerner, U., Abbott, S. and Roe, R. L. (1975). *Drug. Metab. Rev.* **4**, 39–73.
Booth, J. and Boyland, E. (1970). *Biochem. Pharmac.* **19**, 733–742.
Booth, J. and Boyland, E. (1971). *Biochem. Pharmac.* **20**, 407–415.
Börner, U. and Abbot, S. (1973). *Experientia* **29**, 180–181.
Boulanger, P. and Osteux, R. (1960). *Hoppe-Seyler's Z. physiol. Chem.* **321**, 79–86.
Bowman, E. R. and McKennis, H. (1962). *J. Pharmac. exp. Ther.* **135**, 306–311.
Bowman, E. R., Turnbull, L. B. and McKennis, H. (1959). *J. Pharmac. exp. Ther.* **127**, 92–95.
Bowman, E. R., Hansson, E., Turnbull, L. B., McKennis, H. and Schmitterlöw, C. G. (1964). *J. Pharmac. exp. Ther.* **143**, 301–308.
Boyland, E. and Nery, R. (1969). *Biochem. J.* **113**, 123–130.
Brodie, B. B., Baer, J. E. and Craig, L. C. (1951). *J. biol. Chem.* **188**, 567–581.
Brodie, B. B., Axelrod, J. and Reichenthal, J. (1952). *J. biol. Chem.* **194**, 215–222.
Brunk, S. F. and Delle, M. (1974). *Clin. Pharmac. Ther.* **16**, 51–57.
Buchanan, O. H., Christman, A. A. and Block, W. D. (1945). *J. biol. Chem.* **157**, 189–201.

Bull, L. B., Culvenor, C. C. J. and Dick, A. T. (1968). "The Pyrrolizidine Alkaloids". North-Holland Publishing Co., Amsterdam.

Burke, M. D. and Upshall, D. G. (1976). *Xenobiotica* 6, 321–328.

Burg, A. W. (1975). *Drug. Metab. Rev.* 4, 199–228.

Burg, A. W. and Stein, M. E. (1972). *Biochem. Pharmac.* 21, 909–922.

Burg, A. W. and Werner, E. (1972). *Biochem. Pharmac.* 21, 923–936.

Burg, A. W., Burrows, R. and Kensler, C. J. (1974). *Toxic. appl. Pharmac.* 28, 162–166.

Capel, I. D., Millburn, P. and Williams, R. T. (1974). *Xenobiotica* 4, 601–615.

Carroll, F. I., Smith, D., Wall, M. E. and Moreland, C. G. (1974). *J. med. Chem.* 17, 985–987.

Cartoni, G. P. and Giarusso, A. (1972). *J. Chromat.* 71, 154–158.

Castle, M. C., Margileth, D. A. and Oliverio, V. T. (1976). *Cancer Res.* 36, 3684–3689.

Chadwick, M., Platz, B. B. and Liss, R. H. (1971). *Proc. Am. Ass. Cancer Res.* 12, 34.

Chagas, C. (1962). *In* "Curare and Curare-like Agents". (A. V. S. De Reuck, Ed.), Ciba Foundation Study Group No. 12, pp. 2–10. J. and A. Churchill, London.

Chattopadhyay, D., Ghosh, N. C., Chattopadhyay, H. and Banerjee, S. (1953). *J. biol. Chem.* 201, 529–534.

Chernov, H. I. and Woods, L. A. (1965). *J. Pharmac. exp. Ther.* 149, 146–155.

Chesney, C. F. and Allen, J. R. (1973). *Toxic. appl. Pharmac.* 26, 385–392.

Chesney, C. F., Hsu, I. C. and Allen, J. R. (1974). *Res. Commun. chem. Path. Pharmac.* 8, 567–570.

Chin, K.-C., Wang, Y-N., Pao, T-Y. and Hsu, P. (1965). *Sheng Li Hsueh Pao* 28, 72–81. (*Chem. Abstr.* (1965) 63, 1215e.)

De Clercq, M. and Truhaut, R. (1962). *Bull. Soc. Chim. biol.* 44, 227–234.

Clouet, D. H. (1962). *Life Sci.* 1, 31–34.

Clouet, D. H. (1963). *Biochem. Pharmac.* 12, 967–972.

Clouet, D. H., Ratner, M. and Kurzman, M. (1963). *Biochem. Pharmac.* 12, 957–966.

Cohen, E. N., Corbascio, A. and Fleischli, G. (1965). *J. Pharmac. exp. Ther.* 147, 120–129.

Cohen, E. N., Brewer, H. W. and Smith, D. (1967). *Anesthesiology* 28, 309–317.

Cohen, E. N., Hood, N. and Golling, R. (1968). *Anesthesiology* 29, 987–993.

Cooper, N. and Hatcher, R. A. (1934). *J. Pharmac. exp. Ther.* 51, 411–420.

Cornish, H. H. and Christman, A. A. (1957). *J. biol. Chem.* 228, 315–323.

Crankshaw, D. P. and Cohen, E. N. (1975). *In* "Monographs in Anesthesiology, Muscle Relaxants" (R. L. Katz, Ed.), Vol. 3, pp. 125–141. Excerpta Medica, Amsterdam, London, New York.

Creasey, W. A. and Marsh, J. C. (1973). *Proc. Am. Ass. Cancer Res.* 14, 57.

Culvenor, C. C. J., Downing, D. T., Edgar, J. A. and Jago, M. V. (1969). *Ann. N. Y. Acad. Sci.* 163, 837–847.

Culvenor, C. C. J., Edgar, J. A., Smith, L. W., Jago, M. W. and Peterson, J. E. (1971). *Nature New Biol.* 229, 255–256.

Cunningham, H. M. (1970). *Can. J. Anim. Sci.* 50, 49–54.

Czok, G., Schmidt, B. and Lang, K. (1969). *Z. ErnahrWiss.* 9, 109–117.

Dagne, E. and Castagnoli, N. (1972a). *J. med. Chem.* 15, 356–360.

Dagne, E. and Castagnoli, N. (1972b). *J. med. Chem.* 15, 840–841.

Dagne, E., Gruenke, L. and Castagnoli, N. (1974). *J. med. Chem.* 17, 1330–1333.

Daly, J. and Witkop, B. (1963). *Angew. Chem., Int. Ed. Engl.* **2**, 421–440.
Daly, J., Inscoe, J. K. and Axelrod, J. (1965). *J. med. Chem.* **8**, 153–157.
Dajani, R. M., Gorrod, J. W. and Beckett, A. H. (1972). *Biochem. J.* **130**, 88P.
Dajani, R. M., Gorrod, J. W. and Beckett, A. H. (1975a). *Biochem. Pharmac.* **24**, 109–117.
Dajani, R. M., Gorrod, J. W. and Beckett, A. H. (1975b). *Biochem. Pharmac.* **24**, 648–650.
Decker, K. and Sammeck, R. (1964). *Biochem. Z.* **340**, 326–336.
Dengler, H. J., Eichelbaum, M. and Spiteller, G. (1977). Personal communication.
Dhar, M. M., Kohli, J. D. and Srivastava, S. K. (1955). *J. scient. ind. Res.* **14C**, 179–181.
Dhar, M. M., Kohli, J. D. and Srivastava, S. K. (1956). *Indian J. Pharm.* **18**, 293–296.
Dick, A. T., Dann, A. T., Bull, L. B. and Culvenor, C. C. J. (1963). *Nature Lond.* **197**, 207–208.
Earle, D. P. (1946). *Fedn Proc. Fedn. Am. Socs exp. Biol.* **5**, 175–176.
Earle, D. P., Welch, W. J. and Shannon, J. A. (1948). *J. clin. Invest.* **27**, 87–92.
Ebbighausen, W. O. R., Mowat, J. and Vestergaard, P. (1973). *J. pharm. Sci.* **62**, 146–148.
Eichelbaum, M., Spannbrucker, N. and Dengler, H. J. (1975). *Naunyn-Schmiedeberg's Arch. Pharmac.* **287**, R94.
Elek, S. R. and Bergman, H. C. (1953). *J. appl. Physiol.* **6**, 168–172.
Eling, T. E. (1968). *Diss. Abstr. B* **29**, 1778–1779.
Elison, C. and Elliott, H. W. (1963). *Biochem. Pharmac.* **12**, 1363–1366.
Elison, C. and Elliott, H. W. (1964). *J. Pharmac. exp. Ther.* **144**, 265–275.
Elliott, H. W., Tolbert, B. M., Adler, T. K. and Anderson, H. H. (1954). *Proc. Soc. exp. Biol. Med.* **85**, 77–81.
Ellis, P. P., Littlejohn, K. and Deitrich, R. A. (1972). *Invest. Ophthal.* **11**, 747–751.
Ellis, S. (1948). *J. Pharmac. exp. Ther.* **94**, 130–135.
Estevez, V. C., Ho, B. T. and Englert, L. F. (1977). *Res. Commun. chem. Path. Pharmac.* **17**, 179–182.
von Euler, U. S. (1965). "Tobacco Alkaloids and Related Compounds". Pergamon Press, Oxford, London, Edinburgh, New York, Paris and Frankfurt.
Evertsbusch, V. and Geiling, E. M. K. (1956). *Archs. int. Pharmacodyn. Thér.* **105**, 175–192.
Ezer. E. and Szporny, L. (1967). *Kiserl. Orvostud.* **19**, 67–70. (*Chem. Abstr.* (1967) **67**, 52404n.)
Finnegan, J. K., Larson, P. S. and Haag, H. B. (1947). *J. Pharmac. exp. Ther.* **91**, 357–361.
Fish, F. and Wilson, W. D. C. (1969). *J. Pharm. Pharmac.* **21**, 135S–138S.
Fisher, R. S., Algeri, E. J. and Walker, J. T. (1949). *J. biol. Chem.* **179**, 71–79.
Fishman, J., Hahn, E. F. and Norton, B. I. (1976). *Nature Lond.* **261**, 64–65.
Flury, F. (1911). *Arch. exp. Path. Pharmak.* **64**, 105–125.
Foldes, F. F. (1957). *Acta anaesth. scand.* **1**, 63–79.
Franklin, M. (1965). *Can. J. Biochem.* **43**, 1053–1062.
Führ, J. and Kaczmarczyk, J. (1955). *Arzneimittel-Forsch.* **5**, 705–709.
Fumimoto, J. M. and Haarstad, V. B. (1969). *J. Pharmac. exp. Ther.* **165**, 45–51.
Fujimoto, J. M. and Way, E. L. (1954). *Fedn Proc. Fedn. Am. Socs exp. Biol.* **13**, 356.
Fujimoto, J. M. and Way, E. L. (1957). *J. Pharm. exp. Ther.* **121**, 340–346.

Furuya, T. (1956). *Bull. Osaka med. Sch.* **2**, 18–24. (*Chem. Abstr.* (1958) **52**, 2274e.)
Gabourel, J. D. and Gosselin, R. E. (1958). *Archs int. Pharmacodyn. Thér.* **115**, 416–432.
Gaudette, L. E. and Brodie, B. B. (1959). *Biochem. Pharmac.* **2**, 89–96.
Gimble, A. I., Davison, C. and Smith, P. K. (1952). *J. Pharmac. exp. Ther.* **94**, 431–438.
Glazko, A. J., Dill, W. A., Wolf, L. M. and Kazenko, A. (1956). *J. Pharmac. exp. Ther.* **118**, 377–387.
Glick, D. and Glaubach, S. (1941). *J. gen. Physiol.* **25**, 197–205.
Glick, D., Glaubach, S. and Moore, D. H. (1942). *J. biol. Chem.* **144**, 525–528.
Godeaux, J. and Tønnesen, M. (1949). *Acta pharmac. tox.* **5**, 95–109.
Gorrod, J. W. and Jenner, P. (1975). *In* "Essays in Toxicology" (W. J. Hayes, Ed.), Vol. 6, pp. 35–78. Academic Press, New York, San Francisco and London.
Gorrod, J. W., Jenner, P., Keysell, G. and Beckett, A. H. (1971). *Chem.-Biol. Interact.* **3**, 269–270.
Gosselin, R. E. Gabourel, J. D., Kalser, S. C. and Wills, J. H. (1955). *J. Pharmac. exp. Ther.* **115**, 217–229.
Gosselin, R. E., Gabourel, J. D. and Wills, J. H. (1960). *Clin. Pharmac. Ther.* **1**, 597–603.
Goth, R. and Cleaver, J. E. (1976). *Mutat. Res.* **36**, 105–114.
Greenius, H. F., McIntyre, R. W. and Beer, C. T. (1968). *J. med. Chem.* **11**, 254–257.
Gross, E. G. and Thompson, V. (1940). *J. Pharmac. exp. Ther.* **68**, 413–418.
Gruhn, E. (1925). *Naunyn-Schmiedebergs Arch. exp. Path. Pharmak.* **106**, 115–125.
Gutierrez, I. F. and Flaine, L. A. (1958). *Anales fac. quim. farm., Univ. Chile* **10**, 191–197. (*Chem. Abstr.* (1960) **54**, 1445i.)
Guttman, D. E., Kostenbauder, H. B., Wilkinson, G. R. and Dubé, P. H. (1974). *J. pharm. Sci.* **63**, 1625–1626.
Gvishiani, G. S. (1960). *Soobscheniya Akad. Nauk Gruzinskoi SSR* **24**, 225–230. (*Chem. Abstr.* (1961) **55**, 1912c.)
Haag, H. B. and Larson, P. S. (1942). *J. Pharmac. exp. Ther.* **76**, 235–239.
Haag, H. B., Larson, P. S. and Schwartz, J. J. (1943). *J. Pharmac. exp. Ther.* **79**, 136–139.
Hakim, S. A. E., Mijović, V. and Walker, J. (1961). *Nature Lond.* **189**, 201–204.
Hansson, E. and Schmitterlöw, C. G. (1962). *J. Pharmac. exp. Ther.* **137**, 91–102.
Hansson, E., Hoffman, P. C. and Schmitterlöw, C. G. (1964). *Acta physiol. scand.* **61**, 380–392.
Hardesty, C. T., Chaney, N. A. and Mead, J. A. R. (1972). *Cancer Res.* **32**, 1884–1889.
Harke, H-P. and Frahm, B. (1976). *Toxicology* **6**, 125–128.
Harke, H-P., Frahm, B., Schultz, C. and Dontenwill, W. (1970). *Biochem. Pharmac.* **19**, 495–498.
Harke, H-P., Chevalier, H-J. and Frahm, B. (1974a). *Experientia* **30**, 883–884.
Harke, H-P., Schüller, D., Frahm, B. and Mauch, A. (1974b). *Res. Commun. Chem. Path. Pharmac.* **9**, 595–599.
Hatcher, R. A. and Eggleston, C. (1917). *J. Pharmac. exp. Ther.* **10**, 281–319.
Hawks, R. L., Kopin, I. J., Colburn, R. W. and Thoa, N. B. (1974). *Life Sci.* **15**, 2189–2195.

Hawksworth, G. and Scheline, R. R. (1975). *Xenobiotica* **5**, 389–399.
Hayashi, Y. (1966). *Fedn Proc. Fedn. Am. Socs exp. Biol.* **25**, 688.
Heim, F. and Haas, A. (1950). *Naunyn-Schmiedebergs Arch. exp. Path. Pharmak.* **211**, 458–461.
Heimans, R. L. H., Fennessy, M. R. and Gaff, G. A. (1971). *J. Pharm. Pharmac.* **23**, 831–836.
Ho, A. K. S., Hoffman, D. B., Gershon, S. and Loh, H. H. (1971a). *Archs int. Pharmacodyn. Thér.* **194**, 304–315.
Ho, B. T., Estevez, V., Fritchie, G. E., Tansey, L. W., Idäpään-Heikkilä, J. and McIsaac, W. M. (1971b). *Biochem. Pharmac.* **20**, 1313–1319.
Hucker, H. B. and Larson, P. S. (1958). *J. Pharmac. exp. Ther.* **123**, 259–262.
Hucker, H. B., Gilette, J. R. and Brodie, B. B. (1959). *Nature* Lond. **183**, 47.
Hucker, H. B., Gillette, J. R. and Brodie, B. B. (1960). *J. Pharmac. exp. Ther.* **129**, 94–100.
Îda, S., Oguri, K. and Yoshimura, H. (1975a). *J. pharm. Soc.* Japan **95**, 564–569.
Îda, S., Oguri, K. and Yoshimura, H. (1975b). *J. pharm. Soc.* Japan **95**, 570–573.
Idänpään-Heikkilä, J. E. (1968). *Annls Med. exp. Biol. Fenn.* **46**, 201–216.
Iven, H. (1977). *Naunyn-Schmiedeberg's Arch. Pharmac.* **298**, 43–50.
Iwamoto, K. and Klaassen, C. D. (1976). *J. Pharmac. exp. Ther.* **200**, 236–244.
Iwatsubo, K. (1965). *Jap. J. Pharmac.* **15**, 244–256.
Jago, M. V., Lanigan, G. W., Bingley, J. B., Piercy, D. W. T., Whittem, J. H. and Titchen, D. A. (1969). *J. Path.* **98**, 115–128.
Jago, M. V., Edgar, J. A., Smith, L. V. and Culvenor, C. C. J. (1970). *Molec. Pharmac.* **6**, 402–406.
Jenne, J. W., Nagasawa, H. T. and Thompson, R. D. (1976). *Clin. Pharmac. Ther.* **19**, 375–381.
Jenner, P. and Gorrod, J. W. (1973). *Res. Commun. Chem. Path. Pharmac.* **6**, 829–843.
Jenner, P., Gorrod, J. W. and Beckett, A. H. (1973a). *Xenobiotica* **3**, 341–349.
Jenner, P., Gorrod, J. W. and Beckett, A. H. (1973b). *Xenobiotica* **3**, 563–572.
Jenner, P., Gorrod, J. W. and Beckett, A. H. (1973c). *Xenobiotica* **3**, 573–580.
Johns, S. R. and Wright, S. E. (1964). *J. med. Chem.* **7**, 158–161.
Johnson, R. K. and Jondorf, W. R. (1973). *Xenobiotica* **3**, 85–95.
Johnson, R. K., Wynn, W. T. and Jondorf, W. R. (1971). *Biochem. J.* **125**, 26P–27P.
Jovanović, J., Remberg, G., Ende, M. and Spiteller, G. (1976). *Archs Toxicol.* **35**, 137–139.
Kalow, W. (1953). *J. Pharmac. exp. Ther.* **109**, 74–82.
Kalow, W. (1959). *Anesthesiology* **20**, 505–518.
Kalser, S. C. and McLain, P. L. (1970). *Clin. Pharmac. Ther.* **11**, 214–227.
Kalser, S. C., Wills, J. H., Gabourel, J. D., Gosselin, R. E. and Epes, C. F. (1957). *J. Pharmac. exp. Ther.* **121**, 449–456.
Kalser, S. C., Kelvington, E. J., Randolph, M. M. and Santomenna, D. M. (1965a). *J. Pharmac. exp. Ther.* **147**, 252–259.
Kalser, S. C., Kelvington, E. J., Randolph, M. M. and Santomenna, D. M. (1965b). *J. Pharmac. exp. Ther.* **147**, 260–269.
Kamei, K., Matsuda, M. and Momose, A. (1975). *Chem. pharm. Bull.* Tokyo **23**, 683–685.
Kametani, T., Ohta, T., Takemura, M., Ihara, M. and Fukumoto, K. (1977). *Heterocycles* **6**, 415–421.

Kasé, Y., Kataoka, M. and Miyata, T. (1967). *Life Sci.* **6**, 2427–2431.
Kato, R., Chiesara, E. and Vassanelli, P. (1962). *Biochem. Pharmac.* **11**, 913–922.
Keeser, E., Oelkers, H. A. and Raetz, W. (1933). *Naunyn-Schmiedebergs Arch. exp. Path. Pharmak.* **173**, 622–632.
Kelsey, F. E. and Oldham, F. K. (1943). *J. Pharmac. exp. Ther.* **79**, 77–80.
Kelsey, F. E., Geiling, E. M. K., Oldham, F. K. and Dearborn, E. H. (1944). *J. Pharmac. exp. Ther.* **80**, 391–392.
Khanna, K. L., Rao, G. S. and Cornish, H. H. (1972). *Tox. appl. Pharmac.* **23**, 720–730.
King, L. J., Parke, D. V. and Williams, R. T. (1966). *Biochem. J.* **98**, 266–277.
Klutch, A. (1974). *Drug Metab. Disposit.* **2**, 23–30.
Koelle, G. B. (1975). *In* "The Pharmacological Basis of Therapeutics" (L. S. Goodman and A. Gilman, Eds), pp. 467–476. Macmillan Publishing Co., New York.
Kohlrausch, A. (1912). *Z. Biol.* **57**, 273–308.
Knox, W. E. (1946). *J. biol. Chem.* **163**, 699–711.
Krenitsky, T. A., Neil, S. M., Elion, G. B. and Hitchings, G. H. (1972). *Archs Biochem. Biophys.* **150**, 585–599.
Kruger, M. (1899). *Ber. dt. chem. Ges.* **32**, 2818–2824.
Kruger, M. and Schmidt, P. (1899). *Ber. dt. chem. Ges.* **32**, 2677–2682.
Kruger, M. and Schmidt, P. (1901). *Arch. exp. Path. Pharmak.* **45**, 259–261.
Kuhn, H. F. and Friebel, H. (1962). *Med. Exptl.* **7**, 255–261. (*Chem. Abstr.* (1963) **59**, 2072f.)
Lang, K. and Keuer, H. (1957). *Biochem. Z.* **329**, 277–282.
Langecker, H. and Lewit, K. (1938). *Naunyn-Schmiedebergs Arch. exp. Path. Pharmak.* **190**, 492–499.
Lanigan, G. W. (1970). *Aust. J. agric. Res.* **21**, 633–639.
Lanigan, G. W. and Smith, L. W. (1970). *Aust. J. agric. Res.* **21**, 433–500.
Larson, P. S. and Haag, H. B. (1942). *J. Pharmac. exp. Ther.* **76**, 240–244.
Larson, P. S. and Silvette, H. (1968). "Tobacco: Experimental and Clinical Studies. Supplement I". Williams and Wilkins, Baltimore.
Larson, P. S. and Silvette, H. (1971). "Tobacco: Experimental and Clinical Studies. Supplement II". Williams and Wilkins, Baltimore.
Larson, P. S. and Silvette, H. (1975). "Tobacco: Experimental and Clinical Studies. Supplement III". Williams and Wilkins, Baltimore.
Larson, P. S., Haag, H. B. and Silvette, K. (1961). "Tobacco: Experimental and Clinical Studies". Williams and Wilkins, Baltimore.
Latham, M. E. and Elliott, H. W. (1951). *J. Pharmac. exp. Ther.* **101**, 259–267.
Lavallee, W. F. and Rosenkrantz, H. (1966). *Biochem. Pharmac.* **15**, 206–210.
Lazare, R. and Horlington, M. (1975). *Exptl Eye Res.* **21**, 281–287.
Leighty, E. G. and Fentiman, A. F. (1974). *Res. Commun. Chem. Path. Pharmac.* **8**, 65–74.
Lemberger, L. and Rubin, A. (1976). "Physiological Disposition of Drugs of Abuse". Spectrum Publications, New York.
Lesca, P., Lecointe, P., Paoletti, C. and Mansuy, D. (1976). *C.r. hebd. Séanc. Acad. Sci. Paris* **282D**, 1457–1460.
Lévy, J. (1945). *Bull. Soc. Chim. biol.* **27**, 578–584.
Li, C-H., Ho, S-H. and Wang, S-H. (1963). *Yao Hsueh Hsueh Pao* **10**, 581–586. (*Chem. Abstr.* (1964) **60**, 7309f.)
Lohmann, S. M. and Miech, R. P. (1976). *J. Pharmac. exp. Ther.* **196**, 213–225.
Löhr, J. P. and Bartsch, G. G. (1975). *Arzneimittel-Forsch.* **25**, 870–873.

Maass, A. R., Jenkins, B., Shen, Y. and Tannenbaum, P. (1969). *Clin. Pharmac. Ther.* **10**, 366–371.

McLean, E. K. (1970). *Pharmac. Rev.* **22**, 429–483.

McKennis, H. (1965). *In* "Tobacco Alkaloids and Related Compounds" (U.S. von Euler, Ed.), pp. 55–74. Pergamon Press, Oxford, London, Edinburgh, New York, Paris and Frankfurt.

McKennis, H., Turnbull, L. B. and Bowman, E. R. (1957). *J. Am. chem. Soc.* **79**, 6342–6343.

McKennis, H., Turnbull, L. B. and Bowman, E. R. (1958). *J. Am. chem. Soc.* **80**, 6597–6600.

McKennis, H., Turnbull, L. B., Bowman, E. R. and Wada, E. (1959). *J. Am. chem. Soc.* **81**, 3951–3954.

McKennis, H., Bowman, E. R. and Turnbull, L. B. (1960). *J. Am. chem. Soc.* **82**, 3974–3976.

McKennis, H., Bowman, E. R. and Turnbull, L. B. (1961). *Proc. Soc. exp. Biol. Med.* **107**, 145–148.

McKennis, H., Turnbull, L. B., Bowman, E. R. and Schwartz, S. L. (1962a). *J. Am. chem. Soc.* **84**, 4598–4599.

McKennis, H., Turnbull, L. B., Schwartz, S. L., Takami, E. and Bowman, E. R. (1962b). *J. biol. Chem.* **237**, 541–546.

McKennis, H., Turnbull, L. B. and Bowman, E. R. (1963). *J. biol. Chem.* **238**, 719–723.

McKennis, H., Schwartz, S. L. and Bowman, E. R. (1964). *J. biol Chem.* **239**, 3990–3996.

Maga, J. A. and Sizer, C. E. (1973). *J. agric. Fd Chem.* **21**, 22–30.

Maggiolo, C. and Haley, T. J. (1964). *Proc. Soc. exp. Biol. Med.* **115**, 149–151.

Maggiolo, C. and Huidobro, F. (1967). *Arch. Biol. Med. exp.* (Chile) **4**, 99–104.

Mahfouz, M. (1949). *Brit. J. Pharmac.* **4**, 295–303.

Mannering, G. J., Dixon, A. C., Baker, E. M. and Asami, T. (1954). *J. Pharmac. exp. Ther.* **111**, 142–146.

March, C. H. and Elliott, H. W. (1952). *Fedn Proc. Fedn. Am. Socs exp. Biol.* **11**, 373.

March, C. H. and Elliott, H. W. (1954). *Proc. Soc. exp. Biol. Med.* **86**, 494–497.

Maronde, R. F., Haywood, L. J., Feinstein, D. and Sobel, C. (1963). *J. Am. med. Ass.* **184**, 7–10.

Marsh, D. F. (1952). *J. Pharmac. exp. Ther.* **105**, 299–316.

Mattocks, A. R. (1968). *Nature* Lond. **217**, 723–728.

Mattocks, A. R. (1970). *Nature* Lond. **228**, 174–175.

Mattocks, A. R. (1971). *Xenobiotica* **1**, 563–565.

Mattocks, A. R. (1972a). *In* "Phytochemical Ecology" (J. B. Harborne, Ed.), pp. 179–200. Academic Press, London and New York.

Mattocks, A. R. (1972b). *Chem.-Biol. Interactions* **5**, 227–242.

Mattocks, A. R. and White, I. N. H. (1971a). *Nature New Biol.* **231**, 114–115.

Mattocks, A. R. and White, I. N. H. (1971b). *Chem-Biol. Interactions* **3**, 383–396.

Mattocks, A. R. and White, I. N. H. (1973). *Chem.-Biol. Interactions* **6**, 297–306.

Mazel, P. and Henderson, J. F. (1965). *Biochem. Pharmac.* **14**, 92–94.

Meacham, R. H., Bowman, E. R. and McKennis, H. (1972). *J. biol. Chem.* **247**, 902–908.

Meacham, R. H., Sprouse, C. T., Bowman, E. R. and McKennis, H. (1973). *Fedn. Proc. Fedn. Am. Socs exp. Biol.* **32**, 511.

Mead, J. and Koepfli, J. B. (1944). *J. biol. Chem.* **154**, 507–515.

Meijer, D. K. F., Weitering, J. G. and Vonk, R. J. (1976). *J. Pharmac. exp. Ther.* **198**, 229–239.
Mellett, L. B. and El Dareer, S. M. (1971). *Pharmacoligist* **13**, 294.
Mellett, L. B. and Woods, L. A. (1956). *J. Pharmac. exp. Ther.* **116**, 77–83.
Mellett, L. B. and Woods, L. A. (1961). *Proc. Soc. exp. Biol. Med.* **106**, 221–223.
Mikami, K. (1951). *Folia jap. Pharmac.* **47**, 51. (*Chem. Abstr.* (1952) **46**, 7656f.)
Milthers, K. (1962a). *Acta pharmac. tox.* **19**, 149–155.
Milthers, K. (1962b). *Acta pharmac. tox.* **19**, 235–240.
Misra, A. L. (1972). *In* "Chemical and Biological Aspects of Drug Dependence" (S. J. Mulé and H. Brill, Eds), pp. 219–276. CRC Press, Cleveland, Ohio.
Misra, A. L. and Mitchell, C. L. (1971). *Biochem. Med.* **5**, 379–383.
Misra, A. L. and Mulé, S. J. (1972). *Biochem. Pharmac.* **21**, 103–107.
Misra, A. L. and Woods, L. A. (1970). *Nature Lond.* **228**, 1226–1227.
Misra, A. L., Jacoby, H. I. and Woods, L. A. (1961a). *J. Pharmac. exp. Ther.* **132**, 311–316.
Misra, A. L., Mulé, S. J. and Woods, L. A. (1961b). *J. Pharmac. exp. Ther.* **132**, 317–322.
Misra, A. L., Yeh, S. Y. and Woods, L. A. (1970). *Biochem. Pharmac.* **19**, 1536–1539.
Misra, A. L., Vadlamani, N. L., Pontani, R. B. and Mulé, S. J. (1973). *Biochem. Pharmac.* **22**, 2129–2139.
Misra, A. L., Pontani, R. B. and Mulé, S. J. (1974a). *Xenobiotica* **4**, 17–32.
Misra, A. L., Vadlamani, N. L., Bloch, R., Nayak, P. K. and Mulé, S. J. (1974b). *Res. Commun. Chem. Path. Pharmac.* **8**, 55–63.
Misra, A. L., Nayak, P. K., Patel, M. N., Vadlalmani, N. L. and Mulé, S. J. (1974c). *Experientia* **30**, 1312–1314.
Misra, A. L., Nayak, P. K., Bloch, R. and Mulé, S. J. (1975). *J. Pharm. Pharmac.* **27**, 784–786.
Misra, A. L., Giri, V. V., Patel, M. N., Alluri, V. R., Pontani, R. B. and Mulé, S. J. (1976a). *Res. Commun. chem. Path. Pharmac.* **13**, 579–584.
Misra, A. L., Patel, M. N., Alluri, V. R., Mulé, S. J. and Nayak, P. K. (1976b). *Xenobiotica* **6**, 537–552.
Misra, A. L., Pontani, R. B. and Mulé, S. J. (1976c). *Experientia* **32**, 895–897.
Mori, M-A., Oguri, K., Yoshimura, H., Shimomura, K. Kamata, O. and Ueki, S. (1972). *Life Sci.* **11**, 525–533.
Morselli, P. L., Ong, H. H., Bowman, E. R. and McKennis, H. (1967). *J. med. Chem.* **10**, 1033–1036.
Moss, M. S. (1977). *In* "Drug Metabolism—from Microbe to Man" (D. V. Parke and R. L. Smith, Eds), pp. 263–280. Taylor and Francis, London.
Mulé, S. J. (1971). *In* "Narcotic Drugs. Biochemical Pharmacology" (D. H. Clouet, Ed.), pp. 99–121. Plenum Press, New York and London.
Mulé, S. J., Cassella, G. A. and Misra, A. L. (1976). *Life Sci.* **19**, 1585–1596.
Mulder, G. J. and Bleeker, B. (1975). *Biochem. Pharmac.* **24**, 1481–1484.
Mulder, G. J. and Hagedoorn, A. H. (1974). *Biochem. Pharmac.* **23**, 2101–2109.
Mulder, G. J. and Pilon, A. H. E. (1975). *Biochem. Pharmac.* **24**, 517–521.
Mulder, G. J., Hayen-Keulemans, K. and Sluiter, N. E. (1975). *Biochem. Pharmac.* **24**, 103–107.
Murphy, P. J. (1973). *J. biol. Chem.* **248**, 2796–2800.
Myers, V. C. and Hanzal, R. F. (1946). *J. biol. Chem.* **162**, 309–323.
Myers, V. C. and Wardell, E. L. (1928). *J. biol. Chem.* **77**, 697–722.

Nariyuki, H. and Asaki, K. (1959). *Kagaku to Sosa* **12**, 175–179. (*Chem. Abstr.* (1961) **55**, 1921i.)

Nayak, K. P., Brochmann-Hanssen, E. and Way, E. L. (1965). *J. pharm. Sci.* **54**, 191–194.

Nayak, P. K. (1966). *Diss. Abstr.* **26**, 5735.

Nayak, P. K., Misra, A. L. and Mulé, S. J. (1974). *Fedn Proc. Fedn. Am. Socs exp. Biol.* **33**, 527.

Nayak, P. K., Misra, A. L. and Mulé, S. J. (1976). *J. Pharmac. exp. Ther.* **196**, 556–569.

Nery, R. (1971). *Biochem. J.* **122**, 503–508.

Newsome, D. A. and Stern, R. (1974). *Am. J. Ophthal.* **77**, 918–922.

Nguyen, T-L., Gruenke, L. D. and Castagnoli, N. (1976). *J. med. Chem.* **19**, 1168–1169.

Nieschulz, O. and Schmersahl, P. (1968). *Arzneimittel-Forsch.* **18**, 222–225.

Nishie, K., Gumbmann, M. R. and Keyl, A. C. (1971). *Toxic. appl. Pharmac.* **19**, 81–92.

Norred, W. P., Nishie, K. and Osman, S. F. (1976). *Res. Commun. chem. Path. Pharmac.* **13**, 161–171.

Numerof, P., Gordon, M. and Kelly, J. M. (1955). *J. Pharmac. exp. Ther.* **115**, 427–431.

Numerof, P., Virgona, A. J., Cranswick, E. H., Cunningham, R. N. and Kline, N. S. (1958). *Psychiat. Res. Rep.* **9**, 139–142.

Oberst, F. W. (1940). *J. Pharmac. exp. Ther.* **69**, 240–251.

Oberst, F. W. (1941). *J. Pharmac. exp. Ther.* **73**, 401–404.

Oelkers, H. A. and Vincke, E. (1935). Naunyn-Schmiedebergs *Arch. exp. Path. Pharmak.* **179**, 341–348.

Oesch, F. (1977). Personal communication.

Oguri, K., Îda, S., Yoshimura, H. and Tsukamoto, H. (1970). *Chem. pharm. Bull.* Tokyo **18**, 2414–2419.

Oldham, F. K. and Kelsey, F. E. (1943). *J. Pharmac. exp. Ther.* **79**, 81–84.

Opdyke, D. L. J. (1974). *Fd Cosmet. Toxicol.* **12**, 807–1016.

Ortiz, R. V. (1966). *Anales fac. quim. farm.*, (Univ. Chile) **18**, 15–19. (*Chem. Abstr.* (1968) **68**, 48055c.)

Ortiz, V. C. (1952). *Bull. Narcot.* **4**, 26–33.

Otomo, T. (1959). *Nichidae Igaku Zasshi* **18**, 77–86. (*Chem. Abstr.* (1964) **61**, 9914g.)

Otorii, T. (1969). *Acta Med. Biol.* (Niigata) **17**, 173–175. (*Chem. Abstr.* (1970) **73**, 43499q.)

Owen, F. B. and Larson, P. S. (1958). *Archs int. Pharmacodyn. Thér.* **115**, 402–407.

Pærregaard, P. (1958). *Acta pharmac. tox.* **14**, 394–399.

Palmer, K. H., Martin, B., Baggett, B. and Wall, M. E. (1969). *Biochem. Pharmac.* **18**, 1845–1860.

Papadopoulos, N. M. (1964a). *Can. J. Biochem.* **42**, 435–442.

Papadopoulos, N. M. (1964b). *Archs Biochem. Biophys.* **106**, 182–185.

Papadopoulos, N. M. and Kintzios, J. A. (1963). *J. Pharmac. exp. Ther.* **140**, 269–277.

Phillipson, J. D., Handa, S. S. and Gorrod, J. W. (1976). *J. Pharm. Pharmac.* **28**, 687–691.

Plowman, J., Cysyk, R. L. and Adamson, R. H. (1976). *Xenobiotica* **6**, 281–294.

Poole, A. and Urwin, C. (1976). *Biochem. Pharmac.* **25**, 281–283.

Posner, H. S., Mitoma, C. and Udenfriend, S. (1961). *Archs Biochem. Biophys.* **94**, 269–279.

Ramos-Aliaga, R. and Chiriboga, J. (1970). *Archos lat.-am. Nutr.* **20**, 415–428. (*Chem. Abstr.* (1971) **74**, 108861y.)

Rao, G. S., Khanna, K. L. and Cornish, H. H. (1973). *Experientia* **29**, 953–955.

Redmond, N. and Parker, J. M. (1963). *Can. J. Biochem. Physiol.* **41**, 243–245.

Reinhold, V. N. and Bruni, R. J. (1976). *Biomed. Mass Spectrom.* **3**, 335–339.

Reinhold, V., Bittman, L., Bruni, R., Thrun, K. and Silveira, D. (1975). *Proc. Am. Ass. Cancer Res.* **16**, 135.

Rimington, C. (1946). *Biochem. J.* **40**, 669–677.

Robelet, A., Bizard-Gregoire, N. and Bizard, J. (1964). *C. r. Séanc. Soc. Biol.* **158**, 1100–1103.

Roerig, S., Fujimoto, J. M. and Wang, R. I. H. (1973). *Proc. Soc. exp. Biol. Med.* **143**, 230–233.

Rothstein, M. and Greenberg, D. M. (1959). *J. Am. Chem. Soc.* **81**, 4756–4757.

Rothstein, M. and Miller, L. L. (1954). *J. biol. Chem.* **211**, 851–858.

Rosazza, J. P., Kammer, M., Youel, L., Smith, R. V., Erhardt, P. W., Truong, D. H. and Leslie, S. W. (1977). *Xenobiotica* **7**, 133–143.

Russell, G. R. and Smith, R. M. (1968). *Aust. J. biol. Sci.* **21**, 1277–1290.

Sánchez, C. A. (1957). *Anales fac. farm bioquim.* (Univ. nacl. mayor San Marcos) **8**, 82–86. (*Chem. Abstr.* (1959), **53**, 2246h.)

Sanchez, E. and Tephly, T. R. (1973). *Fedn Proc. Fedn. Am. Socs exp. Biol.* **32**, 763.

Sano, M. and Hakusui, H. (1974). *Chem. pharm. Bull.* Tokyo **22**, 696–706.

Sax, S. M. and Lynch, H. J. (1964). *J. Pharmac. exp. Ther.* **145**, 113–121.

Schaumlöffel, E. (1974). *Med. Welt* **25**, 2008–2014.

Schein, F. T. and Hanna, C. (1960). *Archs int. Pharmacodyn. Thér.* **124**, 317–325.

Schoental, R. (1970). *Nature* Lond. **227**, 401–402.

Schonberg, S. S. and Ellis, P. P. (1969). *Archs Ophthal.* N.Y. **82**, 351–355.

Schmidt, G. and Schoyerer, R. (1966). *Dt. Z. gerichtl. Med.* **57**, 402–409.

Schutz, H. and Hempel, J. (1974). *Archs Toxicol.* **32**, 143–148.

Schwartz, D. E. and Herrero, J. (1965). *Am. J. trop. Med.* **14**, 78–83.

Schwartz, S. L. and McKennis, H. (1963). *J. biol. Chem.* **239**, 1807–1812.

Schwartz, S. L. and McKennis, H. (1964). *Nature* Lond **202**, 594–595.

Scrafani, J. T. and Clouet, D. H. (1971). *In* "Narcotic Drugs. Biochemical Pharmacology" (D. H. Clouet, Ed.), pp. 137–158. Plenum Press, New York and London.

Seibert, R. A., Williams, C. E. and Huggins, R. A. (1954). *Science* N.Y. **120**, 222–223.

Seiler, N., Kameniková, L. and Werner, G. (1968a). *Hoppe-Seyler's Z. physiol. Chem.* **349**. 692–698.

Seiler, N., Kameniková, L. and Werner, G. (1968b). *Colln Czech. chem. Commun.* **34**, 719–723.

Severi, A., Longhi, A. and Pomarelli, P. (1967). *Atti Accad. med. lomb.* **22**, 485–488. (*Chem. Abstr.* (1969) **71**, 29032p.)

Shen, W-C. and Van Vunakis, H. (1974a). *Res. Commun. Chem. Path. Pharmac.* **9**, 405–412.

Shen, W-C. and Van Vunakis, H. (1974b). *Biochemistry* N.Y. **13**, 5362–5367.

Sheppard, H. and Tsien, W. H. (1955). *Proc. Soc. exp. Biol Med.* **90**, 437–440.

Sheppard, H., Lucas, R. C. and Tsien, W. H. (1955). *Archs. int. Pharmacodyn. Thér.* **103**, 256–268.

Sheppard, H., Tsien, W. H., Sigg, E. B., Lucas, R. A. and Plummer, A. J. (1957). *Archs int. Pharmacodyn. Thér.* **113**, 160–168.

Shull, L. R., Buckmaster, G. W. and Cheeke, P. R. (1976). *J. Anim. Sci.* **43**, 1247–1253.

Slotkin, T. and DiStefano, V. (1970a). *Biochem. Pharmac.* **19**, 125–131.

Slotkin, T. A. and DiStefano, V. (1970b). *J. Pharmac. exp. Ther.* **174**, 456–462.

Slotkin, T. A., DiStefano, V. and Au, W. Y. W. (1970). *J. Pharmac. exp. Ther.* **173**, 26–30.

Smith, D. S., Peterson, R. E. and Fujimoto, J. M. (1973). *Biochem. Pharmac.* **22**, 485–492.

Smith, R. L. (1973). "The Excretory Function of Bile. The Elimination of Drugs and Toxic Substances in Bile". Chapman and Hall, London.

Stålhandske, T. (1970). *Acta physiol. scand.* **78**, 236–248.

Stawarz, R. J. and Stitzel, R. E. (1974). *Pharmacology* **11**, 178–191.

Stitzel, R. E., Wagner, L. A. and Stawarz, R. J. (1972). *J. Pharmac. exp. Ther.* **182**, 500–506.

Strominger, J. L., Kalckar, H. M., Axelrod, J. and Maxwell, E. S. (1954). *J. Am. chem. Soc.* **76**, 6411–6412.

Sullivan, H. R., Billings, R. E., Jansen, C. J., Occolowitz, J. L., Boaz, H. E., Marshall, F. J. and McMahon, R. E. (1969). *Pharmacologist* **11**, 241.

Sved, S., Hossie, R. D. and McGilverary, I. J. (1976). *Res. Commun. chem. Path. Pharmac.* **13**, 185–192.

Taggart, J. V., Earle, D. P., Berliner, R. W., Zubrod, C. G., Welch, W. J., Wise, N. B., Schroeder, E. F., London, I. M. and Shannon, J. A. (1948). *J. clin. Invest.* **27**, 80–86.

Takemori, A. E. and Mannering, G. J. (1958). *J. Pharmac. exp. Ther.* **123**, 171–179.

Tampier, L. and Penna-Herreros, A. (1966). *Zrch. Biol. Med. Exp* (Chile) **3**, 146–147.

Taylor, D., Estevez, V. S., Englert, L. F. and Ho, B. T. (1976). *Res. Commun. chem. Path. Pharmac.* **14**, 249–257.

Testa, B., Jenner, P., Beckett, A. H. and Gorrod, J. W. (1976). *Xenobiotica* **6**, 553–556.

Tønnesen, M. (1950). *Acta pharmac. tox.* **6**, 147–164.

Truhaut, R. and De Clercq, M. (1959). *Bull. Soc. Chim. biol.* **41**, 1693–1705.

Truhaut, R. and Yonger, J. (1967). *C. r. hebd. Séanc. Acad. Sci.* Paris, Ser. D. **264**, 2526–2528.

Tsukamoto, H., Oguri, K., Watabe, T. and Yoshimura, H. (1964a). *J. Biochem.* Tokyo **55**, 394–400.

Tsukamoto, H., Yoshimura, H., Watabe, T. and Oguri, R. (1964b). *Biochem. Pharmac.* **13**, 1577–1586.

Tsunoda, N., Yoshimura, H. and Kozuka, H. (1976). *Eisei Kagaku* **22**, 280–285. (*Chem. Abstr.* (1977) **86**, 133338p.)

Turner, D. M. (1969). *Biochem. J.* **115**, 889–896.

Turner, D. M. (1971). *Br. J. Pharmac.* **41**, 521–529.

Turner, D. M., Armitage, A. K., Briant, R. H. and Dollery, C. T. (1975). *Xenobiotica* **5**, 539–551.

Uehleke, H. (1963). *Naunyn-Schmiedebergs Arch. exp. Path. Pharmak.* **246**, 34.

Usanova, M. I. and Snol, S. E. (1955). *Byull. exp. Biol. Med.* **40**, 41–44. (*Chem. Abstr.* (1956) **50**, 488c.)

Valanju, N. N., Baden, M. M., Valanju, S. N., Mulligan, D. and Verma, S. K. (1973). *J. Chromat.* **81**, 170–173.
Vedsö, S. (1961). *Acta pharmac. tox.* **18**, 157–164.
Villeneuve, A. and Sourkes, T. L. (1966). *Revue can. Biol.* **25**, 231–239.
Wada, E., Bowman, E. R., Turnbull, L. B. and MacKennis, H. (1961). *J. med. pharm. Chem.* **4**, 21–30.
Walsh, C. T. and Levine, R. R. (1975). *J. Pharmac. exp. Ther.* **195**, 303–310.
Waser, P. G. and Lüthi, U. (1968). *J. nucl. biol. Med.* **12**, 4–11.
Waser, P., Schmid, H. and Schmid, K. (1954). *Naunyn-Schmiedebergs Arch. exp. Path. Pharmak.* **96**, 368–405.
Watabe, T. and Kiyonaga, K. (1972). *J. Pharm. Pharmac.* **24**, 625–630.
Watabe, T., Yoshimura, H. and Tsukamoto, H. (1964). *Chem. pharm. Bull.* Tokyo **12**, 1151–1158.
Watson, W. C. (1964). *Biochem. J.* **90**, 3P.
Way, E. L. and Adler, T. K. (1961). *Bull. WHO* **25**, 227–262.
Way, E. L. and Adler, T. K. (1962). *Bull. WHO* **26**, 51–66.
Weinfeld, H. (1952). *Diss. Abstr.* **12**, 243–244.
Weinfeld, H. and Christman, A. A. (1953). *J. biol. Chem.* **200**, 345–355.
Weissmann, B., Bromberg, P. A. and Gutman, A. B. (1954). *Proc. Soc. exp. Biol. Med.* **87**, 257–260.
Werle, E. and Meyer, A. (1950). *Biochem. Z.* **321**, 221–235.
Werle, E., Koebke, K. and Meyer, A. (1950). *Biochem. Z.* **320**, 189–198.
Werner, G. (1961). *Planta med.* **9**, 293–316.
Werner, G. and Brehmer, G. (1963). *Abh. dt. Akad. Wiss. Berl.* **4**, 217–233. (*Chem. Abstr.* (1964) **61**, 6213b.)
Werner, G. and Brehmer, G. (1967). *Hoppe-Seyler's Z. physiol. Chem.* **348**, 1640–1652.
Werner, G. and Schmidt, H-L. (1968). *Hoppe-Seyler's Z. physiol. Chem.* **349**, 677–691.
White, I. N. H. (1977). *Chem-Biol. Interactions* **16**, 169–180.
Wiechowski, W. (1901). *Arch. exp. Path. Pharmak.* **45**, 155–162.
Williams, R. T. (1959). "Detoxication Mechanisms". Chapman and Hall, London.
Winbladh, B. (1973). *Acta pharmac. tox.* **32**, 46–64.
Wong, K. P. and Sourkes, T. L. (1967). *Analyt. Biochem.* **21**, 444–453.
Wong, K. P. and Sourkes, T. L. (1968). *Biochem. J.* **110**, 99–104.
Woo, J. T. C., Gaff, G. A. and Fennessy, M. R. (1968). *J. Pharm. Pharmac.* **20**, 763–767.
Woods, L. A. (1954). *J. Pharmac. exp. Ther.* **112**, 158–175.
Woods, L. A. and Chernov, H. I. (1966). *Pharmacologist* **8**, 206.
Woods, L. A., McMahon, F. G. and Seevers, M. H. (1951). *J. Pharmac. exp. Ther.* **101**, 200–204.
Woods, L. A., Muehlenbeck, H. E. and Mellett, L. B. (1956). *J. Pharmac. exp. Ther.* **117**, 117–125.
Yao, P-P. and Sung, C-Y. (1975). *Clin. Med. J.* **1**, 205–215.
Yeh, S. Y. (1973). *Fedn Proc. Fedn. Am. Socs exp. Biol.* **32**, 763.
Yeh, S. Y. (1974). *Experientia* **30**, 264–266.
Yeh, S. Y. (1975a). *J. Pharmac. exp. Ther.* **192**, 201–210.
Yeh, S. Y. (1975b). *J. Pharm. Pharmac.* **27**, 214–215.
Yeh, S. Y. (1976). *Pharmacologist* **18**, 178.
Yeh, S. Y. and Woods, L. A. (1969). *J. Pharmac. exp. Ther.* **166**, 86–95.

Yeh, S. Y. and Woods, L. A. (1970a). *J. Pharmac. exp. Ther.* **173**, 21–25.
Yeh, S. Y. and Woods, L. A. (1970b). *J. Pharmac. exp. Ther.* **175**, 69–74.
Yeh, S. Y. and Woods, L. A. (1971). *Archs int. Pharmacodyn. Thér.* **191**, 231–242.
Yeh, S. Y., Chernov, H. I. and Woods, L. A. (1971). *J. Pharm. Sci.* **60**, 469–471.
Yeh, S. Y., McQuinn, R. L. and Gorodetzky, C. W. (1977). *Drug Metab. Disposit.* **5**, 335–342.
Yi, J. M., Sprouse, C. T., Bowman, E. R. and McKennis, H. (1977). *Drug. Metab. Disposit.* **5**, 355–362.
Yoshimura, H., Oguri, K. and Tsukamoto, H. (1969). *Biochem. Pharmac.* **18**, 279–286.
Yoshimura, H., Mori, M-A., Oguri, K. and Tsukamoto, H. (1970). *Biochem. Pharmac.* **19**, 2353–2360.
Yosikawa, N. (1940). *Jap. J. med. Sci.* **12**, 74–75 (*Chem. Abstr.* (1940) **34**, 74294.)
Zahn, K. (1915). *Biochem. Z.* **68**, 444–476.
Zauder, H. L. (1952). *J. Pharmac. exp. Ther.* **104**, 11–19.
Zetler, G., Back, G. and Iven, H. (1974). *Naunyn-Schmiedeberg's Arch. Pharmac.* **285**, 273–292.
Ziegler, D. M., Mitchell, C. H. and Jollow, D. (1969). *In* "Microsomes and Drug Oxidation" (J. R. Gillette, A. H. Conney, G. J. Cosmides, R. W. Estabrook, J. R. Fouts and G. J. Mannering, Eds), pp. 173–187. Academic Press, New York and London.

10

METABOLISM OF SULPHUR COMPOUNDS

Sulphur is ubiquitous in plants as it is a component of vitamins, coenzymes and proteins which are essential in intermediary metabolism. In addition, some plants contain secondary organic sulphur compounds, usually in small amounts, which often are of interest because of their powerful and characteristic odours. Many of these have been shown to be simple aliphatic sulphur compounds including thiols (mercaptans) (see Maga, 1976), sulphides, disulphides and trisulphides. These types are especially common in species of *Brassica* (e.g. cabbage, kale, asparagus) and of *Allium* (e.g. onion and garlic). Other aliphatic sulphur compounds from plants include sulphonium compounds, sulphoxides and sulphones. The basic structures of these compounds are illustrated in Fig. 10.1. Derivatives in which sulphur is found in a higher oxidation state include sulphonic acids and sulphates, however, these types appear to be found mainly in lower plants. A noteworthy class consists of thiophene derivatives containing substituent groups in the 2- and 5-positions (Fig. 10.1) which include highly unsaturated moieties (e.g. 1-propynyl, $R = -C \equiv C - CH_3$). One of the most important types of sulphur compound is that of glucosinolates or mustard oil glycosides. Their general structure is shown in Fig. 10.1 and R may include simple alkyl groups, various aromatic substituents or more complex groups. These anionic compounds are normally found as potassium salts, however salts of organic bases also occur as with the basic compound sinapine found in sinalbin. Recently, many thiazole (Fig. 10.1) derivatives have been identified as volatile aromatic constituents of foods (see Maga, 1975). These are generally fairly simple derivatives containing one, two or three alkyl groups attached to the ring carbons. Benzothiazole, the 2-phenyl derivative, also occurs naturally. Various other sulphur compounds have also been reported, including thioketones and sulphur-containing alkaloids.

The above brief comments indicate that organic sulphur compounds from plants constitute a diverse group of substances. It seems reasonable to assume that studies on their metabolism should provide a fertile field which could be expected to furnish many interesting results. Nonetheless, the most appropriate characteristic which can be applied to the subject at the

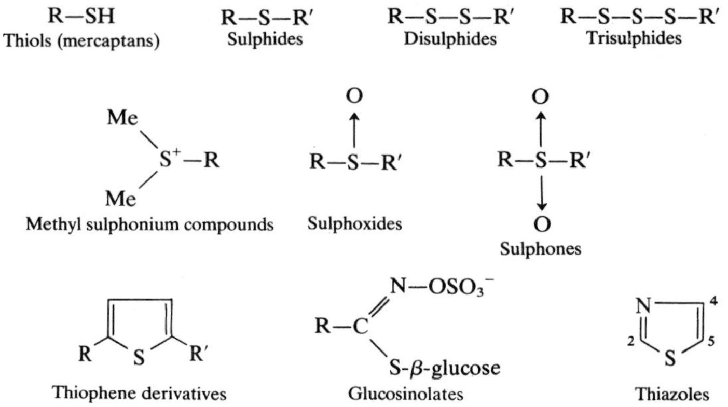

R—SH
Thiols (mercaptans)

R—S—R'
Sulphides

R—S—S—R'
Disulphides

R—S—S—S—R'
Trisulphides

$$\begin{array}{c} Me \\ \diagdown \\ \diagup S^{+}{-}R \\ Me \end{array}$$

Methyl sulphonium compounds

$$\begin{array}{c} O \\ \uparrow \\ R{-}S{-}R' \end{array}$$

Sulphoxides

$$\begin{array}{c} O \\ \uparrow \\ R{-}S{-}R' \\ \downarrow \\ O \end{array}$$

Sulphones

Thiophene derivatives

$$\begin{array}{c} N{-}OSO_3^{-} \\ \diagup\!\!/ \\ R{-}C \\ \diagdown \\ S{-}\beta{-}glucose \end{array}$$

Glucosinolates

Thiazoles

FIG. 10.1. General structures of some sulphur-containing plant compounds.

present time is that of paucity of information. At best, metabolic information is available on only a few members of each group while with other groups essentially nothing is known.

I. Thiols

The simple thiols are volatile compounds having offensive odours. The metabolism of the simplest member of the series, **methanethiol,** was studied *in vivo* in rats by Canellakis and Tarver (1953b) using ^{35}S- or ^{14}C-labelled compounds. The latter compound was given intraperitoneally at a dose level of about 1·7 mg/kg and the respiratory air and urine monitored for 6 h. Only about 2% of the radioactivity appeared in the urine during this period whereas about 40% was lost as respiratory CO_2, mainly during the first hour (29%). About 6% of the dose was lost as volatile sulphur, presumably unchanged compound, during the first hour only. The ^{35}S-compound was administered similarly in four divided doses totaling 10 mg/kg. The urine collected over an 8 h period ending 2 h after the last dose contained 32% of the dose as total sulphur. The values for inorganic sulphate and total sulphate were 29% and 31% of the dose, respectively. These results show that both the carbon and the sulphur of methanethiol are rapidly oxidized to CO_2 and sulphate, respectively. Considerable radioactivity from the ^{14}C-labelled compound was retained in the tissues and it was found that the methyl carbon was converted to the β-carbon of serine and the methyl groups of methionine, choline and creatinine. On the other hand, the experiments using [^{35}S]-methanethiol indicated that the sulphur did not appear to any significant extent in liver methionine or cysteine. This study indicates that the fate of the methyl group is similar to that of methyl alcohol, entering into the l-carbon metabolic pool. A metabolic possibility not studied in this investigation is the S-methylation of the thiol followed by oxidation to dimethyl sulphone, a reaction which occurs with the closely related ethanethiol which is formed from diethyl disulphide (see Section III). It seems likely that the method employed for the determination of total sulphur, based on oxidation to sulphate would not detect the presence of the highly stable sulphone. The non-plant aromatic thiol, thiophenol, undergoes S-methylation and oxidation in rats with the result that methylphenylsulphone is excreted in the urine (McBain and Menn, 1969).

Information on the possible mechanism of methanethiol metabolism noted above was obtained by Mazel *et al.* (1964) in a study of the S-demethylation of several S-methyl compounds by a microsomal system from rat liver. This system, requiring NADPH and O_2, converted the thiol to formaldehyde and H_2S. However, the S-demethylating system differs from the well-known O- and N-demethylating systems in some respects (e.g. inhibitor effects and inducibility). It is likely that more than one S-demethylating enzyme exists.

The metabolism of ethanethiol and l-propanethiol has not been studied, however relevant information on the fate of the former compound may be inferred from the investigation by Snow (1957) which dealt with the metabolism of diethyldisulphide and is summarized in Section III. A dithiol, **2, 2′-dithiolisobutyric acid** (1) was reported by Jansen (1948) to occur in asparagus. In view of the well-known excretion of methanethiol in the urine following ingestion of asparagus, the possibility that this dithiol might be the precursor was studied. However, when two individuals took 10 mg each of the dithiolisobutyric acid orally no odour resulted.

$$
\begin{array}{c}
\text{HS} \\
\quad \diagdown \\
\qquad \text{CH}_2 \\
\qquad\quad \diagdown \\
\qquad\qquad \text{CH—COOH} \\
\qquad\quad \diagup \\
\qquad \text{CH}_2 \\
\quad \diagup \\
\text{HS} \qquad\qquad (1)
\end{array}
$$

2, 2′-Dithiolisobutyric acid

II. Sulphides

Examples of simple sulphides including the dimethyl, diallyl and dibutyl derivatives occur in cabbage and garlic while the methyl and ethyl esters of

3-methylthiopropionic acid are found in pineapple. Metabolic data on this group is very limited. Maw (1953) reported that **dimethyl sulphide,** when given orally or by injection to rats, did not result in an increase in the urinary sulphate excretion. Some unchanged compound was detected in the expired air 1 h following injection. Expiration of unchanged dimethyl sulphide was similarly reported in rabbits by Williams *et al.* (1966) who also found that some oxidation to the corresponding sulphoxide and sulphone occurred. In this experiment a total dose of dimethyl sulphide of about 3·4 g (1·4 g/kg) was injected subcutaneously in four daily portions and the urine collected for six days. Roughly 20% of the dose in the pooled urine was recovered as dimethyl sulphoxide and a further 10% was present as dimethyl sulphone. Several sulphides including dimethylsulphide were studied by Mazel *et al.* (1964) using the microsomal *S*-demethylating system from rat liver. As noted in Section I, this system was active in demethylating the corresponding thiol, however no demethylation was observed when dimethylsulphide was used as the substrate.

$$\overset{\displaystyle NH_2}{\underset{\displaystyle |}{Me-S-CH_2-CH-COOH}}$$

(2)

S-methylcysteine

The aforementioned liver microsomal system also demethylates **S-methylcysteine** (2), however several pathways in the metabolism of this compound are available. Binkley (1950) described its cleavage by rat liver preparations to form methanethiol and Canellakis and Tarver (1953a) reported the formation of the same metabolite by a different system from liver mitochondria. Horner and Kuchinskas (1959) administered *S*-methylcysteine labelled with ^{14}C in the methyl group to rats but were unable to detect methanethiol in the expired air. However, it was believed that this metabolite would be rapidly oxidized to CO_2 and, in fact, nearly 40% of the administered radioactivity was lost as $^{14}CO_2$ in 24 h. The degradation of *S*-methylcysteine by rumen contents is covered in Section V.

III. Disulphides

The reported disulphides, mainly found in cabbage, onion and garlic, are simple symmetrical or unsymmetrical derivatives containing small groups (e.g. methyl, ethyl, propyl, allyl and butyl). Only one of these, **diethyl disulphide,** has been studied metabolically (Snow, 1957). Using ^{35}S-labelled compound, mice were injected subcutaneously twice with a total dose of 800 mg/kg. Excretion of radioactivity occurred mainly in the urine and little was detected in the faeces. As much as 14% of the dose was found in the respiratory air and it was believed this was due to the reduction product, ethanethiol. In both mice and guinea pigs the 24 h urines contained 40–65% of the dose as relatively non-volatile radioactive compounds as well as an undetermined amount of ethanethiol. Sulphate accounted for 80–90% of the urinary radioactivity and two organic metabolites were also detected chromatographically. The major metabolite was shown to be ethyl methyl sulphone and the other remained un-identified although it was not the corresponding sulphoxide. The sulphone was also found in mouse tissues. The radioactive components separated on chromatograms of rabbit urine were similar except that an additional substance with radioactivity about equal to that in the sulphone peak and having slightly greater polarity was observed. In view of these findings it seems likely that the major metabolic pathways of diethyl disulphide are those shown in Fig. 10.2.

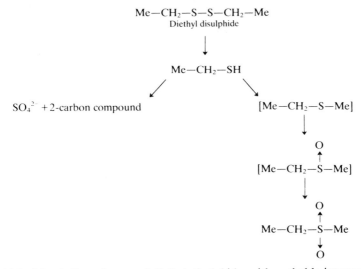

FIG. 10.2. Metabolic pathways of diethyl disulphide, with probable intermediates shown in brackets.

IV. Sulphonium Compounds

This group is represented by the thetins, the general formula of which is shown in Structure (3). Past interest in these compounds derived partly from the fact that they share some of the properties of quaternary

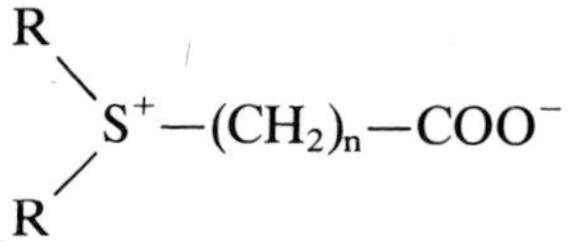

$$\begin{array}{c} R \\ \diagdown \\ S^{+}-(CH_2)_n-COO^{-} \\ \diagup \\ R \end{array}$$

(3)

ammonium compounds and also that they can enter into transmethylation reactions. Maw (1953) studied the metabolism of four simple alkyl thetins including **dimethyl-β-propiothetin** (4) which has been reported to occur in some lower plants. In thetins containing one or two methyl groups, oxidation to sulphate was observed. With dimethyl-β-propiothetin two-thirds of the sulphur was excreted in the urine as inorganic sulphate within two days of an oral dose (about 13 mg S/rat).

Maw (1953) also directed attention to the possible intermediates in sulphate formation from the thetins, noting earlier work which showed that one methyl group may undergo transmethylation and subsequent oxidation to CO_2. The products thus formed are S-alkyl derivatives of acetic or propionic acid. In the former case the metabolites (e.g. S-methyl- or S-ethylthioglycollate), when themselves administered to rats, gave rise to appreciable excretion of urinary sulphate. This result of high inorganic sulphate excretion was also obtained with the corresponding thiol, thiolglycollic acid, a finding which was subsequently confirmed by Freeman *et al.* (1956). Interestingly, the latter investigation showed that species differences in thioglycollate metabolism occur as monkeys excreted the sulphur mainly in the neutral sulphur fraction whereas rabbits excreted it in the neutral sulphur and organic sulphate fractions. In view of the results obtained in rats, the probable metabolic pathway for dimethyl-β-propiothetin is that shown in Fig. 10.3.

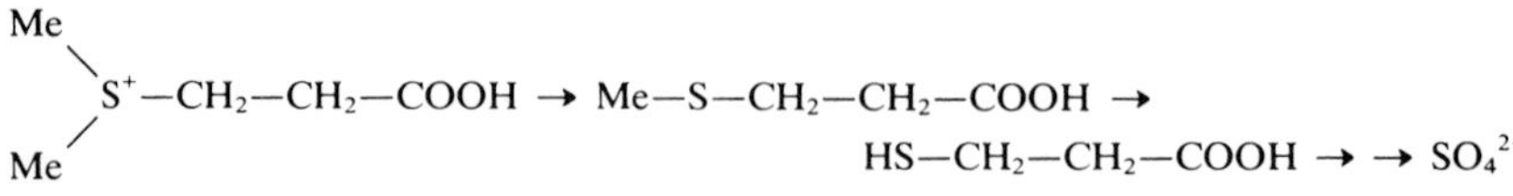

$$\begin{array}{c} Me \\ \diagdown \\ S^{+}-CH_2-CH_2-COOH \rightarrow Me-S-CH_2-CH_2-COOH \rightarrow \\ \diagup \\ Me HS-CH_2-CH_2-COOH \rightarrow \rightarrow SO_4^{2-} \end{array}$$

(4)

Dimethyl-β-propiothetin

FIG. 10.3. Probable metabolic pathway of dimethyl-β-propiothetin in rats.

The participation of the intestinal microflora in the metabolism of thetins must also be considered and Zikakis and Salsbury (1969) found that dimethyl thetin (Structure (3), $R = Me$, $n = 1$) was converted to dimethyl sulphide when incubated with bovine rumen microorganisms.

V. Sulphoxides

The best known sulphoxides from plants are derivatives of cysteine. The lachrymatory factor in onion results from the breakdown of *S*-(prop-1-enyl) cysteine sulphoxide. Its isomer, the corresponding *S*-allyl derivative (alliin), is the precursor of the active principle of garlic. The only metabolic data from animals available deals with the simpler analogue, **S-methyl-cysteine sulphoxide** (5), which occurs in onions and several crucifers including turnips and kale. This information will be summarized in this section rather than with the non-protein amino acids (Chapter 8, Section IV) since the significant metabolic changes occur with the sulphur-containing portion of the molecule.

$$Me-\overset{\overset{\textstyle O}{\uparrow}}{S}-CH_2-\overset{\overset{\textstyle NH_2}{|}}{C}H-COOH$$

(5)

S-Methylcysteine sulphoxide

In an investigation designed to elucidate the factors responsible for the toxic effects of kale in ruminants, Smith *et al.* (1974) studied the ability of goat rumen contents to metabolize *S*-methylcysteine sulphoxide (5) *in vitro*. Analysis of the head-space gases from these incubations showed the presence of dimethyl disulphide, methanethiol and dimethyl sulphide. The latter compound, a minor component, showed a proportionate increase in later samples, as did methanethiol. The disulphide was the dominant component during the earlier stages. Similar studies using the corresponding sulphide, **S-methylcysteine** (2), which also occurs in kale, showed an initial production of about equal amounts of the disulphide and the thiol, with smaller amounts of dimethylsulphide. This suggests that reduction of the sulphoxide to the sulphide is an early step in the metabolism of the former compound to volatile sulphur compounds by the rumen microflora. The formation of methanethiol from *S*-methylcysteine by bovine rumen microorganisms was reported by Zikakis and Salsbury (1969).

VI. Glucosinolates

The glucosinolates or mustard oil glycosides are important flavour constituents in many plants, especially crucifers (e.g. mustard, horseradish and water cress). Also present in the plant is an enzyme system, known as myrosinase or thioglucosidase, which is released upon crushing of the plant and effects the hydrolysis of the glycoside. As shown in Fig. 10.4 this

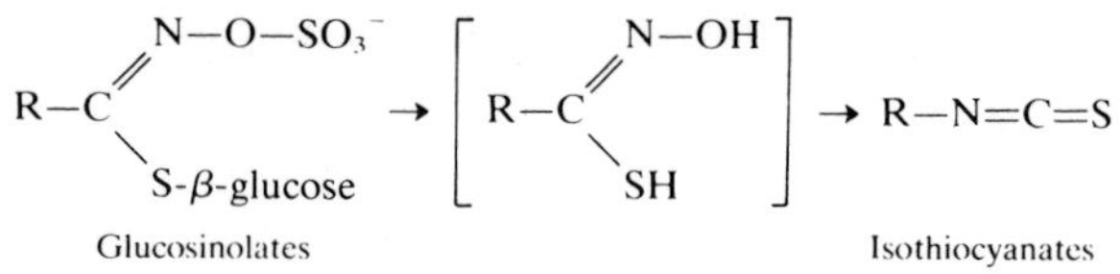

$$R-C \overset{\displaystyle N-O-SO_3^-}{\underset{\displaystyle S\text{-}\beta\text{-glucose}}{\Big\langle}} \rightarrow \left[R-C \overset{\displaystyle N-OH}{\underset{\displaystyle SH}{\Big\langle}} \right] \rightarrow R-N=C=S$$

Glucosinolates   Isothiocyanates

FIG. 10.4. Hydrolysis of glucosinolates to isothiocyanates.

reaction leads to the formation of isothiocyanates or mustard oils. These secondary products are responsible for the notable physiological properties of this group of compounds. In addition to the major route leading to isothiocyanates, breakdown of the glucosinolates can also lead to thiocyanates ($R-S=C=N$), nitriles ($R-C\equiv N$) and even heterocyclic ring compounds (e.g. goitrin from progoitrin). Information dealing with the chemical and biological properties of the thioglucosides with emphasis on their goitrogenic activity has been reviewed by Van Etten (1969) and Briggs and Briggs (1974).

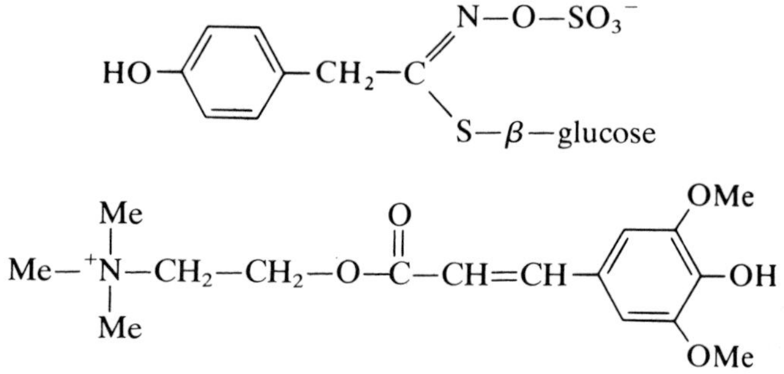

$$HO-\!\!\!\left\langle\bigcirc\right\rangle\!\!\!-CH_2-C \overset{\displaystyle N-O-SO_3^-}{\underset{\displaystyle S-\beta-glucose}{\Big\langle}}$$

$$Me-\overset{\displaystyle Me}{\underset{\displaystyle Me}{\overset{+}{N}}}-CH_2-CH_2-O-\overset{\displaystyle O}{\overset{\|}{C}}-CH=CH-\!\!\!\left\langle\bigcirc\right\rangle\!\!\!\begin{matrix}OMe\\ -OH\\ OMe\end{matrix}$$

(6)

Sinalbin

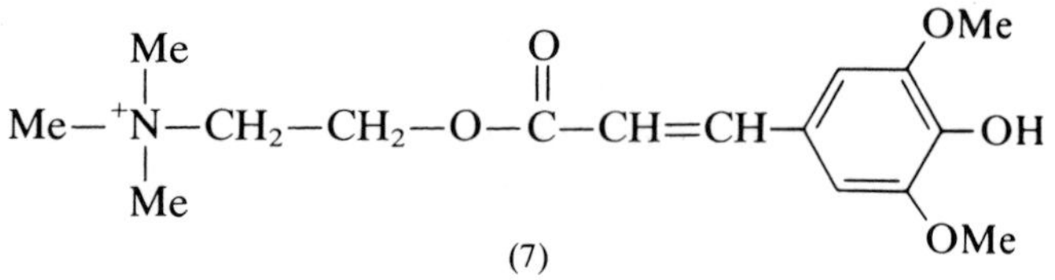

$$Me-\overset{\displaystyle Me}{\underset{\displaystyle Me}{\overset{+}{N}}}-CH_2-CH_2-O-\overset{\displaystyle O}{\overset{\|}{C}}-CH=CH-\!\!\!\left\langle\bigcirc\right\rangle\!\!\!\begin{matrix}OMe\\ -OH\\ OMe\end{matrix}$$

(7)

Sinapine

$$HOOC-CH=CH-\!\!\!\left\langle\bigcirc\right\rangle\!\!\!\begin{matrix}OMe\\ -OH\\ OMe\end{matrix}$$

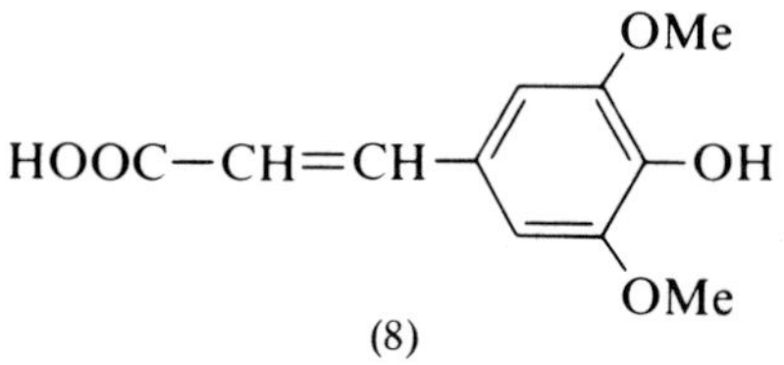

(8)

HOOC—CH₂—CH₂—(benzene ring with OMe top, OH right, OMe bottom)

HOOC—CH₂—CH₂—(benzene ring with OH top, OMe bottom)

$$HOOC-CH_2-CH_2-\underset{OMe}{\overset{OMe}{\bigcirc}}-OH \qquad HOOC-CH_2-CH_2-\underset{OMe}{\overset{OH}{\bigcirc}}$$

(9)   (10)

Relatively little is known of the mammalian metabolism of gluco-sinolates and their secondary products, perhaps due in part to the irritating properties of some of these compounds. Griffiths (1969) gave **sinalbin** (6) orally to rats at a dose level of about 400 mg/kg. Interest was shown mainly in the sinapine (7) moiety of sinalbin and it was found that it underwent ester hydrolysis with the result that sinapic (8) and dihy-drosinapic (9) acids were excreted in the 24 h urines. On the second day a further metabolite, 3-hydroxy-5-methoxyphenylpropionic acid (10) was excreted. A discussion of the metabolism of these and related methoxyl-ated C_6-C_3 acids, including the role played by the intestinal microflora, is found in Chapter 5, (Section I,C). A fourth metabolite, p-hydroxybenzoic acid, was detected in the 24 h urines and it seems evident that this compound arises from the glucosinalbate (4-hydroxy-benzylglucosinolate) moiety. Information on the mechanism involved in the metabolism of the latter is not available and it is not known if the degradation involves an isothiocyanate intermediate analogous to that noted above with the plant thioglucosidase or if another pathway, perhaps leading to the thiocyanate, predominates. The site of thioglucoside hydrolysis is probably the intestinal lumen, the reaction resulting from the activities of the intestinal bacteria. The ionized nature of the thioglucoside would be expected to appreciably reduce the chances for its absorption from the intestine, thus allowing it to reach the lower gut where it could be metabolized by bacterial enzymes. This conclusion was reached by Greer (1962) who studied the conversion of **progoitrin** (11) to the active goitrogenic compound goitrin (12) in rats and humans. Oginsky et al. (1965) isolated representatives from several genera of human faecal bacteria which were able to carry out this reaction *in vitro*. Thioglycosidase activity appears to be widely distributed among intestinal bacteria and, in the strains tested,

$$CH_2=CH-\underset{OH}{\overset{}{CH}}-CH_2-C\overset{N-OSO_3^-}{\underset{S-\beta\text{-glucose}}{}}$$

(11)

Progoitrin

$$CH_2=CH-\underset{O}{\overset{}{CH}}\underset{C}{\overset{}{}}\underset{\underset{S}{\parallel}}{\overset{}{}}CH_2\atop NH$$

(12)

Goitrin

was higher in *Paracolobactrum, Proteus vulgaris* and *Bacillus subtilis.*
Similar findings, although in fowls, were reported by Marangos and Hill
(1974). The caecal contents of the birds showed a high thioglucosidase
activity towards the thioglucosides present in rapeseed.

A clue to the possible further metabolism of isothiocyanate derivatives
was obtained by Bachelard and Trikojus (1960) in a study of the metabolism of **cheiroline** (13), a goitrogenic compound present in an Australian
grazing plant. They found that incubation of cheiroline with rumen liquor
resulted in the formation of di-cheiroline thiourea (14). It is not known if
this intestinal reaction of cheiroline is a general reaction if isothiocyanates.

$$\text{Me}-\overset{\overset{\displaystyle O}{\uparrow}}{\underset{\underset{\displaystyle O}{\downarrow}}{S}}-\text{CH}_2-\text{CH}_2-\text{CH}_2-\text{N}{=}\text{C}{=}\text{S}$$

(13)

Cheiroline

$$\text{Me}-\overset{\overset{\displaystyle O}{\uparrow}}{\underset{\underset{\displaystyle O}{\downarrow}}{S}}-(\text{CH}_2)_3-\text{NH}-\overset{\overset{\displaystyle S}{\|}}{C}-\text{NH}-(\text{CH}_2)_3-\overset{\overset{\displaystyle O}{\uparrow}}{\underset{\underset{\displaystyle O}{\downarrow}}{S}}-\text{Me}$$

(14)

Another possible metabolic pathway for isothiocyanate derivatives
liberated from glucosinolates involves conjugation with glutathione and
excretion in the urine as a mercapturic acid. Brüsewitz *et al.* (1977)
investigated this pathway using **benzyl isothiocyanate** (15). It reacted
rapidly with glutathione both spontaneously and enzymically to form the
tripeptide conjugate. The latter was degraded to the cysteine conjugate
and then acetylated to give the corresponding mercapturic acid (16). This
sequence of reactions was observed *in vitro* when rat liver or kidney
homogenates were used and metabolite (16) was excreted in the urine of
rats given benzyl isothiocyanate (10 mg/kg, p.o.). Brüsewitz *et al.* noted
unpublished observations which indicated that the mercapturic acid (16) is
also formed from the corresponding cysteine conjugate in dogs, pigs and
humans but not in guinea pigs or rabbits. This investigation also studied the

$$\text{C}_6\text{H}_5-\text{CH}_2-\text{N}{=}\text{C}{=}\text{S}$$

(15)

Benzyl isothiocyanate

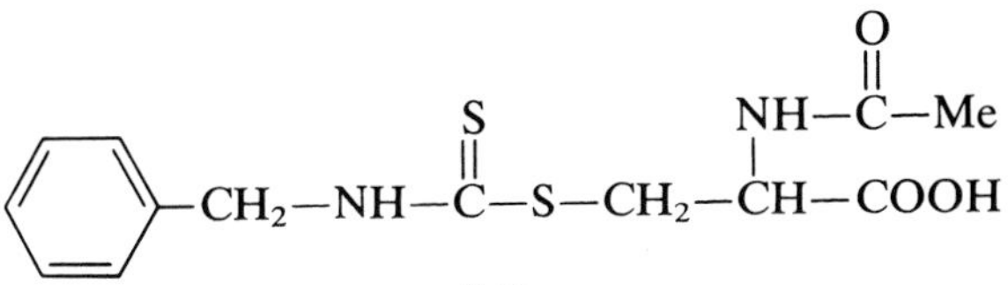

$$\text{C}_6\text{H}_5-\text{CH}_2-\text{NH}-\overset{\overset{\text{S}}{\|}}{\text{C}}-\text{S}-\text{CH}_2-\overset{\overset{\text{NH}-\overset{\overset{\text{O}}{\|}}{\text{C}}-\text{Me}}{|}}{\text{CH}}-\text{COOH}$$

(16)

metabolism of the cysteine and other amino acid conjugates of ben-zylisothiocyanate.

Brüsewitz *et al.* (1977) reported that rats also formed mercapturic acids from **allyl isothiocyanate.** As the corresponding metabolite was not pro-duced from α-naphthyl isothiocyanate, they suggested that mercapturic acid formation may only occur when the isothiocyanate moiety is attached to reactive positions such as those furnished by benzyl or allyl groups.

As noted above, glucosinolates may undergo degradation to thio-cyanates and it is possible that the mammalian metabolism of the thio-glucosides also involves these intermediates. Ohkawa and Casida (1971) and Ohkawa *et al.* (1972) studied the ability of mouse liver homogenates and subcellular fractions to liberate HCN from numerous synthetic organic thiocyanates. These included several lower alkyl thiocyanates and **benzyl thiocyanate.** The latter compound is a degradation product of thio-glucosides and the former group show similarity to the naturally occurring allyl thiocyanate. These studies showed that the formation of HCN from the thiocyanates was catalysed by glutathione *S*-transferases which bring about the attack of glutathione at the thiocyanate sulphur. This activity was found in the soluble and not the microsomal fraction and the reaction products detected were HCN, oxidized glutathione and the thiol derivative of the original thiocyanate. This reaction sequence is illustrated in Fig. 10.5. Habig *et al.* (1975) subsequently studied this reaction using homo-genous preparations of glutathione *S*-transferases A, B and C from rat liver. Employing ethyl thiocyanate, octyl thiocyanate and benzyl thio-cyanate, they found that the reaction products were HCN and the mixed disulphide formed from the substrate and glutathione. Thus, these purified preparations carry out the first step shown in Fig. 10.5.

The sequence of metabolic events occurring with organic thiocyanates *in vivo* is not known, however Ohkawa *et al.* (1972) also administered the

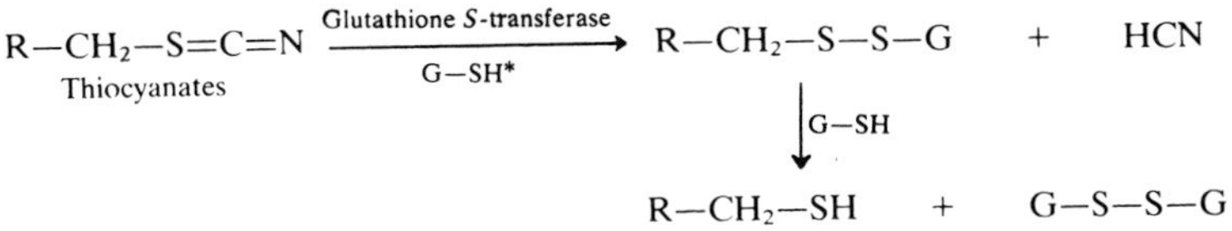

$$\underset{\text{Thiocyanates}}{\text{R}-\text{CH}_2-\text{S}=\text{C}=\text{N}} \xrightarrow[\text{G}-\text{SH}^*]{\text{Glutathione }S\text{-transferase}} \text{R}-\text{CH}_2-\text{S}-\text{S}-\text{G} \quad + \quad \text{HCN}$$

$$\downarrow \text{G}-\text{SH}$$

$$\text{R}-\text{CH}_2-\text{SH} \quad + \quad \text{G}-\text{S}-\text{S}-\text{G}$$

FIG. 10.5. Metabolism of organic thiocyanates. * G—SH = glutathione.

compounds noted above to mice. After 15 min they measured the concentrations of HCN and its metabolite thiocyanate (SCN⁻) present in the liver and brain. High levels of these compounds were found, especially in the liver, with most of the thiocyanates tested and these results indicate that HCN liberation occurs *in vivo*. The fate of the disulphides and thiols formed is not known, however it is possible that these compounds may be metabolized along the pathways shown above in Fig. 10.2 for diethyl disulphide and its thiol derivative.

VII. Thiazoles

As noted above, numerous thiazole derivatives are known to be volatile aroma constituents in foods. While no published reports on the mammalian metabolism of these compounds are available, Scheline (1973) found mass

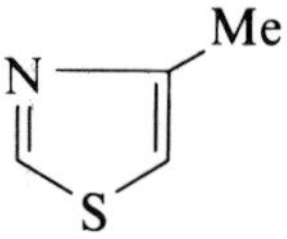

(17)

4-Methylthiazole

spectrometric evidence for the presence of thiazole-4-carboxylic acid in the urine of rats treated orally with 4-methylthiazole (17) at a dose level of 200 mg/kg.

References

Bachelard, H. S. and Trikojus, V. M. (1960). *Nature* Lond. **185**, 80–82.
Binkley, F. (1950). *J. biol. Chem.* **186**, 287–296.
Briggs, M. and Briggs, M. (1974). "The Chemistry and Metabolism of Drugs and Toxins" pp. 277–281. William Heinemann Medical Books, London.
Brüsewitz, G., Cameron, B. D., Chasseaud, L. F., Görler, K., Hawkins, D. R., Koch, H. and Mennicke, W. H. (1977). *Biochem. J.* **162**, 99–107.
Canellakis, E. S. and Tarver, H. (1953a). *Archs Biochem. Biophys.* **42**, 387–398.
Canellakis, E. S. and Tarver, H. (1953b). *Archs Biochem. Biophys.* **42**, 446–455.
Freeman, M. V., Draize, J. H. and Smith, P. K. (1956). *J. Pharmac. exp. Ther.* **118**, 304–308.
Greer, M. A. (1962). *Recent Prog. Horm. Res.* **18**, 187–219.
Griffiths, L. A. (1969). *Biochem. J.* **113**, 603–609.
Habig, W. H., Keen, J. H. and Jakoby, W. B. (1975). *Biochem. biophys. Res. Commun.* **64**, 501–506.
Horner, W. H. and Kuchinskas, E. J. (1959). *J. biol. Chem.* **234**, 2935–2937.

Jansen, E. F. (1948). *J. biol. Chem.* **176**, 657–664.
McBain, J. B and Menn, J. J. (1969). *Biochem. Pharmac.* **18**, 2282–2285.
Maga, J. A. (1975). *CRC Critical Rev. Food Sci. Nutr.* **6**, 153–176.
Maga, J. A. (1976). *CRC Critical Rev. Food Sci. Nutr.* **7**, 147–192.
Marangos, A. and Hill, R. (1974). *Proc. Nutr. Soc.* **33**, 90A.
Maw, G. A. (1953). *Biochem. J.* **55**, 42–46.
Mazel, P., Henderson, J. F. and Axelrod, J. (1964). *J. Pharmac. exp. Ther.* **143**, 1–6.
Oginsky, E. L., Stein, A. E. and Greer, M. A. (1965). *Proc. Soc. exp. Biol. Med.* **119**, 360–364.
Ohkawa, H. and Casida, J. E. (1971). *Biochem. Pharmac.* **20**, 1708–1711.
Ohkawa, H., Ohkawa, R., Yamamoto, I. and Casida, J. E. (1972). *Pestic. Biochem. Physiol.* **2**, 95–112.
Scheline, R. R. (1973). Unpublished observations.
Smith, R. H., Earl, C. R. and Matheson, N. A. (1974). *Biochem. Soc. Trans.* **2**, 101–104.
Snow, G. A. (1957) *Biochem. J.* **65**, 77–82.
Van Etten, C. H. (1969). *In* "Toxic Constituents of Plant Foodstuffs" (I. E. Liener, Ed.), pp. 103–142. Academic Press, New York and London.
Williams, K. I. H., Burstein, S. H. and Layne, D. S. (1966). *Archs Biochem. Biophys.* **117**, 84–87.
Zikakis, J. P. and Salsbury, R. L. (1969). *J. Dairy Sci.* **52**, 2014–2019.

SUBJECT INDEX